Advanced Oxidation Processes for Emerging Contaminant Removal

Advanced Oxidation Processes for Emerging Contaminant Removal

Editors

Dionysios (Dion) Demetriou Dionysiou
Yujue Wang
Huijiao Wang

MDPI • Basel • Beijing • Wuhan • Barcelona • Belgrade • Manchester • Tokyo • Cluj • Tianjin

Editors

Dionysios (Dion) Demetriou Dionysiou
Department of Chemical and Environmental Engineering (ChEE)
University of Cincinnati
Cincinnati
United States

Yujue Wang
School of Environment
Tsinghua University
Beijing
China

Huijiao Wang
School of Chemical and Environmental Engineering
China University of Mining and Technology (Beijing)
Beijing
China

Editorial Office
MDPI
St. Alban-Anlage 66
4052 Basel, Switzerland

This is a reprint of articles from the Special Issue published online in the open access journal *Water* (ISSN 2073-4441) (available at: www.mdpi.com/journal/water/special_issues/Advanced_Oxidation_Processes_Removal).

For citation purposes, cite each article independently as indicated on the article page online and as indicated below:

LastName, A.A.; LastName, B.B.; LastName, C.C. Article Title. *Journal Name* **Year**, *Volume Number*, Page Range.

ISBN 978-3-0365-6611-5 (Hbk)
ISBN 978-3-0365-6610-8 (PDF)

Contents

About the Editors

Dionysios (Dion) Demetriou Dionysiou

Dr. Dionysios (Dion) D. Dionysiou is currently a Herman Schneider Professor of Environmental Engineering and Distinguished Research Professor (STEMM) at the University of Cincinnati. He has served as a UNESCO co-Chair Professor on "Water Access and Sustainability". He teaches courses and performs research in the areas of drinking water quality and treatment, advanced unit operations for water treatment, advanced oxidation technologies and nanotechnologies, and physical-chemical processes for water quality control. He has received funding from NSF, US EPA, NASA, NOAA/CICEET, USGS, USDA, Ohio Sea Grant, USAID, and DuPont. He is currently one of the Executive Editors of Chemical Engineering Journal, Editor-in-Chief of the Journal of Environmental Engineering (ASCE), co-Editor of Current Opinion in Chemical Engineering, and member of the Editorial Boards of several other journals. Dr. Dionysiou is the author or co-author of over 630 refereed journal publications, over 87 conference proceedings, 39 book chapter publications, 38 editorials, and more than 650 presentations. He has edited/co-edited seven books on water quality, water reuse, ferrates, photocatalysis and treatment of harmful algal blooms and cyanotoxins. Dr. Dionysiou's work received over 71,000 citations with an H factor of 137 (Google Scholar). He is a Highly Cited Researcher (in Engineering, Environment/Ecology, and Chemistry based on Clarivate Analytics, Web of Science, 2018, 2019, 2020, 2021 and 2022, and in Environmental Science and Engineering and Chemical Engineering based of Shanghai Ranking's Global Ranking of Academic Subjects by Elsevier, 2016).

Yujue Wang

Dr. Yujue Wang is currently a full professor at the School of Environment, Tsinghua University. He has served as a Vice Director of Beijing Key Laboratory for Emerging Organic Contaminants Control since 2017, and is an associate editor of the journal *Water* and editorial board of *Journal of Porous Materials*. Dr. Wang's research mainly focuses on advanced oxidation processes and physicochemical processes for water and wastewater treatment. He has authored or co-authored more than 150 referred journal papers, which have been cited more than 9400 times with an H factor of 60 (Google Scholar). In addition, Dr. Wang is the author or co-author of three book chapters and more than 50 presentations. Dr. Wang is ranked as the world's top 2% scientist by Stanford University in 2019, 2020, and 2021. His research has been granted several distinguished awards, including the First Prize of Natural Science of Ministry of Education of China, the Gold Medal of International Exhibition of Inventions of Geneva, and Cutting-edge Academic Achievements of Beijing.

Huijiao Wang

Dr. Huijiao Wang is currently an associate professor at School of Chemical and Environmental Engineering, China University of Mining and Technology (Beijing). Her main research area is advanced oxidation technologies and kinetic modelling for water and wastewater treatment. Dr. Wang is the author or co-author of more than 30 referred journal papers, and her work has received more than 1400 citations with an H factor of 20 (Google Scholar). Additionally, she is the editorial board member of the journal *Water*, and has received the funding of China National Postdoctoral Program for Innovative Talents.

water

MDPI

Editorial

Advanced Oxidation Processes for Removal of Emerging Contaminants in Water

Huijiao Wang [1,*], Yujue Wang [2,*] and Dionysios D. Dionysiou [3,*]

[1] School of Chemical and Environmental Engineering, China University of Mining and Technology (Beijing), Beijing 100083, China
[2] School of Environment, Tsinghua University, Beijing 100084, China
[3] Environmental Engineering and Science Program, Department of Chemical and Environmental Engineering (ChEE), University of Cincinnati, Cincinnati, OH 45221-0012, USA
* Correspondence: wanghuijiao@cumtb.edu.cn (H.W.); wangyujue@tsinghua.edu.cn (Y.W.); dionysdd@ucmail.uc.edu (D.D.D.)

Abstract: This Special Issue includes manuscripts on mechanistic understanding, development, and implementation of advanced oxidation processes (AOPs) for the removal of contaminants of emerging concern in water and wastewater treatment. The main goal was successfully achieved under the joint effort of authors, anonymous reviewers, and editorial managers. Totally, one review and 15 research papers are included in the Special Issue. These are mainly focused on catalyst synthesis, reactor design, treatment performance, kinetic modeling, reaction mechanisms, and by-product formation during electrochemical, photocatalytic, plasma, persulfate, chlorine, ozone-based, and Fenton-related AOPs at different scales. This Special Issue received attention from researchers from different parts of the world such as Argentina, Brazil, Canada, China, Germany, India, Mexico, and the USA. The guest editors are happy to see that all papers presented are innovative and meaningful, and hope that this Special Issue can promote mechanistic understanding and engineering applications of AOPs for the removal of contaminants of emerging concern in water.

Keywords: advanced oxidation; concern; contaminant; electrochemical; emerging; hydroxyl radical; oxidation; ozonation; persulfate; photocatalysis; radical; UV; chlorine; UV/chlorine; Fenton; plasma; water

Citation: Wang, H.; Wang, Y.; Dionysiou, D.D. Advanced Oxidation Processes for Removal of Emerging Contaminants in Water. *Water* **2023**, *15*, 398. https://doi.org/10.3390/w15030398

Received: 3 January 2023
Revised: 6 January 2023
Accepted: 9 January 2023
Published: 18 January 2023

1. Introduction

Efficient and cost-effective removal of various contaminants in water matrices is a major challenge in water and wastewater treatment [1,2]. In this regard, advanced oxidation processes (AOPs) have been considered a promising option because highly reactive radicals such as hydroxyl and sulfate radicals generated in AOPs can effectively oxidize a broad range of emerging contaminants; other radicals such as reactive chlorine and nitrogen radicals can also play significant roles [1,3–5]. Nevertheless, the practical application of AOPs is challenged by the high energy demand, formation of harmful oxidation byproducts, difficulty in scaling up, etc. [3,6,7]. Therefore, both novel mechanistic understanding and improved engineering designs are needed to overcome these challenges and thus bridge academic research with practical applications.

In this Special Issue, we attempted to focus on the mechanistic understanding, development, and implementation of AOPs for the removal of emerging contaminants in water and wastewater treatment. Ozone-, UV-, H_2O_2-, chlorine-, persulfate-based AOPs, electricity-driven AOPs, and photocatalytic AOPs were among the technologies of interest in this Special Issue. Relevant topics of interest included reaction kinetics, catalyst fabrication, model simulation, theoretical calculations, by-product formation, and degradation mechanisms. Topics on reactor design, economic evaluation, and experiments at different scales (lab- and pilot-scale) were also of interest.

2. Summary of This Special Issue

Two papers were published on the topic of persulfate technology. In Liu et al. [8], the authors investigated the degradation of acyclovir and atenolol by the UV/peroxydisulfate (UV/PDS) process from the perspective of degradation kinetics, model simulations, and reaction pathway. Results show that the UV/PDS process could effectively generate sulfate radicals ($SO_4^{\bullet-}$) and hydroxyl radicals ($^{\bullet}OH$) to remove the two micropollutants, with $SO_4^{\bullet-}$ playing a more significant role in the process. In the other study, Mo et al. [9] focused on the synthesis of a copper-magnesium oxide/carbon nitride composite (CM/g-C_3N_4) to activate peroxymonosulfate (PMS). The CM/g-C_3N_4 presented superior catalytic performance and reusability for PMS activation and Rhodamine B (RhB) degradation, and $SO_4^{\bullet-}$ and singlet oxygen ($^1O_2$) were found to be important for RhB removal. Therefore, different oxidant species can be generated by various activation methods during the persulfate oxidation process, which is beneficial for the degradation of micropollutants with varying reactivities.

The electrochemical oxidation process is another effective technology for micropollutant removal. For instance, Yanagida et al. [10] used the boron-doped diamond (BDD) electrode to electrochemically oxidize PFAS in contaminated water and then scaled up the technology for the treatment of 189 L of PFOA and PFOS-contaminated water. LC/MS/MS analysis results show that micrograms per liter (ppb) PFAS could be easily degraded by BDD electrochemical oxidation. Considering the great importance of electrode material, da Silva et al. [11] evaluated the performance of three anodes (Ni/BDD, Ti/Pt, Ti/RuO_2) to treat groundwater contaminated by petroleum-derived fuel, with the Ti/RuO_2 anode achieving the highest chemical oxygen demand (COD) degradation efficiency and lowest energy consumption. Besides, a pilot-flow plant was established to further verify the viability of electrochemical treatment at a larger scale.

In addition to electrochemical oxidation, a heterogeneous electro-Fenton (HEF) process using $MnFe_2O_4$-GO catalyst was employed to remove Rhodamine B from aqueous solution [12]. This study focused on the efficiency of electrodes and catalyst, as well as their application in real textile wastewater treatment. Significant color reduction and obvious biodegradability enhancement were observed after treatment. Two other Fenton-related studies were also reported in this Special Issue. In the study of Olea-Mejia et al. [13], a Cu_2O/Al_2O_3 catalyst was synthesized to improve Bisphenol A (BPA) oxidation and mineralization during the photo-Fenton process driven by UV radiation and visible light. Besides, Lin et al. [14] applied the Fenton process to treat acrylic manufacturing wastewater. The results show that total organic carbon and nitrogen can be effectively removed to meet related discharge standards, providing a successful example of industrial wastewater treatment by the Fenton technology.

Three papers on the photocatalytic process were included in this Special Issue. In Juárez-Cortazar et al. [15], TiO_2 was doped with metal waste (door key) to improve its photocatalytic efficiency, and a synergistic effect of the dopants and TiO_2 was achieved for diclofenac mineralization. Meanwhile, Mehling et al. [16] investigated the energetic efficiency of TiO_2 photocatalysis from a different perspective, i.e., reactor design. Three reactor systems were evaluated, with catalyst arrangement and irradiation power identified as the major influencing parameters on energy consumption performance. Other than TiO_2 doping and reactor design, Manassero et al. [17] focused more on radiation modeling and kinetics in different photocatalytic reactors. In their study, a strategy was proposed to obtain intrinsic kinetic parameters independent of reactor geometry, reactor size, and irradiation conditions. The results indicate that the radiation model can be employed for photocatalytic reactor design, optimization, and scaling-up, thus bridging the gap between laboratory experiments and real applications.

Ozone-based AOP, as a promising research and development option, was investigated from two different perspectives in this Special Issue. In the study of Luo et al. [18], ozonation was combined with electro-coagulation (i.e., the electro-hybrid ozonation-coagulation process) to remove surfactant and microplastics from laundry wastewater. In addition,

Zhang et al. [19] prepared a mesoporous CeO_2 by the nano-casting method and applied the catalyst for the catalytic ozonation of atrazine. The well-ordered mesoporous structure, high surface area, and redox Ce^{3+}/Ce^{4+} cycling contributed to the superior activity of the synthetical CeO_2. Both studies present the effectiveness and important role of ozone-based AOPs in the removal of emerging contaminants in water.

Plasma technology was also reviewed and studied in this Special Issue. In the research paper of Liu et al. [20], a novel reactor was designed for simulated dye wastewater treatment by plasma in the presence of various catalysts, and the results show that the plasma/PS/Fe^{2+} system achieved the best synergy and highest removal rate. In the review paper of He et al. [21], they summarized recent research progress on non-thermal plasma technique for remediation of water and soil contaminated by emerging organic pollutants in terms of pollutant degradation mechanism, the synergy of non-thermal plasma with other techniques, bottlenecks, and suggestions to promote plasma technology toward practical applications.

Besides, one paper investigated the removal of emerging contaminants by novel material adsorption [22]. Specifically, a carbon material was derived from the nitrogen-rich bio-based metal-organic framework (MOF) and was evaluated as an absorbent for pharmaceutical elimination from the water environment. The high surface area and abundant mesoporous structure of the obtained MOF contributed greatly to hydrophobic pharmaceutical removal.

In addition to contaminant removal, the study of Li et al. [23] paid attention to disinfection by-product (DBP) formation during medium-pressure UV/chlorine AOP. Results show that DBP formation is highly dependent on the precursor activity, solution pH, and the presence of Br^-. The authors suggest that the UV/chlorine-induced change in total chlorine demand might be taken as an indicator to predict the change in DBP formation potential.

Author Contributions: Writing—original draft preparation, H.W.; writing—review and editing, D.D.D., Y.W. and H.W.; supervision, D.D.D. and Y.W. All authors have read and agreed to the published version of the manuscript.

Funding: This research received no external funding.

Acknowledgments: Thanks to all who have contributed to the Special Issue, the authors, anonymous reviewers, as well as the editorial managers. All the guest editors are very happy with the review process and management of the Special Issue and offer their special thanks.

Conflicts of Interest: The authors declare no conflict of interest.

References

1. von Sonntag, C.; von Gunten, U. *Chemistry of Ozone in Water and Wastewater Treatment: From Basic Principles to Applications*; IWA Publishing: London, UK, 2012.
2. Schwarzenbach, R.; Escher, B.; Fenner, K.; Hofstetter, T.; Johnson, C.; von Gunten, U.; Wehrli, B. The challenge of micropollutants in aquatic systems. *Science* **2006**, *313*, 1072–1077. [CrossRef] [PubMed]
3. Wang, Y.; Yu, G.; Deng, S.; Huang, J.; Wang, B. The electro-peroxone process for the abatement of emerging contaminants: Mechanisms, recent advances, and prospects. *Chemosphere* **2018**, *208*, 640–654. [CrossRef] [PubMed]
4. Oturan, M.; Aaron, J. Advanced Oxidation Processes in Water/Wastewater Treatment: Principles and Applications. A Review. *Crit. Rev. Environ. Sci. Technol.* **2014**, *44*, 2577–2641. [CrossRef]
5. Waclawek, S.; Lutze, H.; Grubel, K.; Padil, V.; Cernik, M.; Dionysiou, D. Chemistry of persulfates in water and wastewater treatment: A review. *Chem. Eng. J.* **2017**, *330*, 44–62. [CrossRef]
6. von Gunten, U. Oxidation Processes in Water Treatment: Are We on Track? *Environ. Sci. Technol.* **2018**, *52*, 5062–5075. [CrossRef] [PubMed]
7. Radjenovic, J.; Sedlak, D. Challenges and Opportunities for Electrochemical Processes as Next-Generation Technologies for the Treatment of Contaminated Water. *Environ. Sci. Technol.* **2015**, *49*, 11292–11302. [CrossRef] [PubMed]
8. Liu, Z.; Qin, W.; Sun, L.; Dong, H.; Yuan, X.; Pan, F.; Xia, D. Insights into the Kinetics, Theoretical Model and Mechanism of Free Radical Synergistic Degradation of Micropollutants in UV/Peroxydisulfate Process. *Water* **2022**, *14*, 2811. [CrossRef]
9. Mo, Y.; Xu, W.; Zhang, X.; Zhou, S. Enhanced Degradation of Rhodamine B through Peroxymonosulfate Activated by a Metal Oxide/Carbon Nitride Composite. *Water* **2022**, *14*, 2054. [CrossRef]

10. Yanagida, A.; Webb, E.; Harris, C.; Christenson, M.; Comfort, S. Using Electrochemical Oxidation to Remove PFAS in Simulated Investigation-Derived Waste (IDW): Laboratory and Pilot-Scale Experiments. *Water* **2022**, *14*, 2708. [CrossRef]
11. da Silva, J.; Solano, A.; Segundo, I.B.; Santos, E.; Martínez-Huitle, C.; da Silva, D. Achieving Sustainable Development Goal 6 Electrochemical-Based Solution for Treating Groundwater Polluted by Fuel Station. *Water* **2022**, *14*, 2911. [CrossRef]
12. Anil, G.; Scaria, J.; Nidheesh, P. Removal of Synthetic Dye from Aqueous Solution Using $MnFe_2O_4$-GO Catalyzed Heterogeneous Electro-Fenton Process. *Water* **2022**, *14*, 3350. [CrossRef]
13. Olea-Mejia, O.; Brewer, S.; Donkor, K.; Amado-Piña, D.; Natividad, R. Photo-Fenton Catalyzed by Cu_2O/Al_2O_3: Bisphenol (BPA) Mineralization Driven by UV and Visible Light. *Water* **2022**, *14*, 3626. [CrossRef]
14. Lin, Z.; Zhang, C.; Su, P.; Lu, W.; Zhang, Z.; Wang, X.; Hu, W. Fenton Process for Treating Acrylic Manufacturing Wastewater: Parameter Optimization, Performance Evaluation, Degradation Mechanism. *Water* **2022**, *14*, 2913. [CrossRef]
15. Juárez-Cortazar, D.; Torres-Torres, J.; Hernandez-Ramirez, A.; Arévalo-Pérez, J.; Cervantes-Uribe, A.; Godavarthi, S.; de los Monteros, A.; Silahua-Pavón, A.; Cordero-Garcia, A. Doping of TiO_2 Using Metal Waste (Door Key) to Improve Its Photocatalytic Efficiency in the Mineralization of an Emerging Contaminant in an Aqueous Environment. *Water* **2022**, *14*, 1389. [CrossRef]
16. Mehling, S.; Schnabel, T.; Londong, J. Investigation on Energetic Efficiency of Reactor Systems for Oxidation of Micro-Pollutants by Immobilized Active Titanium Dioxide Photocatalysis. *Water* **2022**, *14*, 2681. [CrossRef]
17. Manassero, A.; Alfano, O.; Satuf, M. Degradation of Emerging Pollutants by Photocatalysis: Radiation Modeling and Kinetics in Packed-Bed Reactors. *Water* **2022**, *14*, 3608. [CrossRef]
18. Luo, J.; Jin, X.; Wang, Y.; Jin, P. Advanced Treatment of Laundry Wastewater by Electro-Hybrid Ozonation—Coagulation Process: Surfactant and Microplastic Removal and Mechanism. *Water* **2022**, *14*, 4138.
19. Zhang, J.; Zhuang, T.; Liu, S.; Sun, S.; Wang, Y.; Liu, X.; Wang, J.; Liu, R. Catalytic Ozonation of Atrazine Enhanced by Mesoporous CeO2: Morphology, Performance and Intermediates. *Water* **2022**, *14*, 3431. [CrossRef]
20. Liu, Y.; Song, J.-W.; Bao, J.; Shen, X.-J.; Li, C.-L.; Wang, X.; Shao, L.-X. Optimized Removal of Azo Dyes from Simulated Wastewater through Advanced Plasma Technique with Novel Reactor. *Water* **2022**, *14*, 3152. [CrossRef]
21. He, Y.; Sang, W.; Lu, W.; Zhang, W.; Zhan, C.; Jia, D. Recent Advances of Emerging Organic Pollutants Degradation in Environment by Non-Thermal Plasma Technology: A Review. *Water* **2022**, *14*, 1351. [CrossRef]
22. Meng, Y.; Li, X.; Wang, B. Efficient Removal of Micropollutants by Novel Carbon Materials Using Nitrogen-Rich Bio-Based Metal-Organic Framework (MOFs) as Precursors. *Water* **2022**, *14*, 3413. [CrossRef]
23. Li, W.; Shu, S.; Zhu, Y.; Wu, L.; Wang, Q.; Gao, N. Effect of Medium Pressure Ultraviolet/Chlorine Advanced Oxidation on the Production of Disinfection by-Products from Seven Model Benzene Precursors. *Water* **2022**, *14*, 3775. [CrossRef]

water

MDPI

Article

Advanced Treatment of Laundry Wastewater by Electro-Hybrid Ozonation–Coagulation Process: Surfactant and Microplastic Removal and Mechanism

Jiahao Luo [1], Xin Jin [2], Yadong Wang [1] and Pengkang Jin [1,2,*]

1 School of Environmental and Municipal Engineering, Xi'an University of Architecture and Technology, Xi'an 710055, China
2 School of Human Settlements and Civil Engineering, Xi'an Jiaotong University, Xi'an 710049, China
* Correspondence: pkjin@xjtu.edu.cn

Abstract: Laundry wastewater is supposed to be one of the most important sources of surfactants and microplastics in the wastewater treatment plant. Consequently, the aim of the study was evaluating the performance and mechanism of the electro-hybrid ozonation–coagulation (E-HOC) process for the removal of surfactants and microplastics. In this study, the efficiency of the E-HOC process for surfactant and microplastic removal was examined at different current densities and ozone dosages. Under the optimal reaction conditions (current density 15 $mA \cdot cm^{-2}$, ozone dosage 66.2 $mg \cdot L^{-1}$), both the removal efficiency of surfactant and microplastic can reach higher than 90%. Furthermore, the mechanism of surfactant and microplastic removal was investigated by electron paramagnetic resonance (EPR) and Fourier transform infrared spectroscopy (FT-IR). The results showed that the E-HOC (carbon fiber cathode) system can produce more reactive oxygen species (ROS), which can significantly improve the removal of the contaminants. In addition, the shape, size and abundance of the microplastics were analyzed. It was found that the shape of the microplastics in laundry wastewater is mainly fiber. Microplastics less than 50 μm account for 46.9%, while only 12.4% are larger than 500 μm. The abundance of microplastics in laundry wastewater ranges between 440,000 and 1,080,000 items per 100 L. The analysis of microplastics by FT-IR showed that most of the microplastics in laundry wastewater were polyethylene, nylon and polyester. These results indicated that the E-HOC process can effectively remove surfactants and microplastics from laundry wastewater.

Keywords: electro-hybrid ozonation–coagulation (E-HOC); laundry wastewater; hydroxyl radical; microplastic; surfactant

Citation: Luo, J.; Jin, X.; Wang, Y.; Jin, P. Advanced Treatment of Laundry Wastewater by Electro-Hybrid Ozonation–Coagulation Process: Surfactant and Microplastic Removal and Mechanism. *Water* **2022**, *14*, 4138. https://doi.org/10.3390/w14244138

Academic Editors: Dionysios (Dion) Demetriou Dionysiou, Yujue Wang and Huijiao Wang

Received: 14 November 2022
Accepted: 8 December 2022
Published: 19 December 2022

Publisher's Note: MDPI stays neutral with regard to jurisdictional claims in published maps and institutional affiliations.

1. Introduction

Domestic laundry wastewater is a kind of greywater [1]. In China, the average annual water consumption of each washing machine is about 10 m^3 [2]. According to statistics, the total amount of laundry wastewater discharged annually in China is about 10.71 billion tons [3]. If such a large amount of laundry wastewater is directly discharged into the environment without treatment, it will cause great burden and harm to the environment. Direct discharge of laundry wastewater may not only cause eutrophication, but also affect wastewater treatment plant performance [4]. Most of the laundry wastewater enters the wastewater treatment plant directly through the municipal wastewater plumbing, which can save water resources if it can be reused after treatment.

Domestic laundry wastewater often contains oils and fats, suspended solids, surfactants as well as microplastics, etc. [1,5]. Due to the addition of detergents, there is an increase in alkalinity and pH in laundry wastewater, which is rich in sodium, nitrogen, phosphorus and surfactants but has a low BOD [6]. Laundry wastewater has a COD, BOD and turbidity level of 375~4155 mg/L, 48~1200 mg/L and 14~400 NTU, respectively [3]. In addition, surfactants and microplastics are two typical pollutants in laundry wastewater. Surfactants will cause potential harm to the environment after entering the environment. Studies have pointed out that surfactants not only cause damage to the morphology and physiology of plants, but also cause a significant decrease in sludge dewaterability [7,8]. Microplastic is another typical pollutant. The microplastics dropped from the clothes in the laundry process is also an important pollutant. Microplastics are defined as plastic particles and fibers less than 5 mm [9,10]. The study estimated that 9.1×10^{10} microplastic items are discharged from domestic wastewater to wastewater treatment plants every day in China [11]. Studies have shown that when activated sludge is chronically exposed to microplastics, both millimeter sized microplastics and nano-sized microplastics can cause a decline in sludge dehydration capacity [12]. In addition, the presence of surfactants also enhances the migration behavior of microplastics in the environment. [13]. This may contaminate ground water, rivers and lakes, etc. There have been many studies on the removal of surfactants and microplastics, mainly divided into physicochemical (membrane filtration, adsorption and electrocoagulation) and biological (aerobic, anaerobic, biofilter and wetland). They are briefly mentioned below.

For physicochemical treatments, both ultrafiltration and microfiltration membrane processes were highly effective, removing more than 90% of NPs/MPs fragment particles. Although the membrane filtration had a high efficiency, it needed to be cleaned frequently [14]. Using micron iron oxide powder (magnetite and hematite) as an adsorbent can remove 90% of cationic surfactants within 10 min, but a large amount of adsorbents are required [15]. Furthermore, studies have shown that about 90% of the surfactant can be removed by using electrocoagulation/electroflotation [16]. Another study showed that the removal efficiencies of surfactants and microplastics are 94% and 98%, respectively, by the electrocoagulation/electroflotation process [17].

For biological treatments, the aerobic moving bed bioreactor (MBBR) can stably remove 85~96% of the surfactant after 4 months of operation [18]. Moreover, an artificial wetland to treat greywaters was developed. Their research demonstrated 77% removal of LAS after 15 days of hydraulic retention time (HRT) [19]. Meanwhile, the removal of microplastics currently mainly relies on wastewater treatment plants, and studies have shown that oxidation ditch (OD) and MBR can remove 53.6% and 82.1% of microplastics, respectively. This may be due to the improved removal of microplastics by membrane filtration in the MBR [20].

The treatment process of laundry wastewater can generally be divided into three types: physicochemical process, biological process and the combination of physicochemical and biological process [2]. Specific processes include moving bed bioreactor (MBBR) [18], membrane filtration [21–25], solar photo-Fenton [26], electrocoagulation, ozonation [27], membrane bioreactor (MBR) [28], UV/H_2O_2 [29], biological treatment combined with UV/O_3 [30], etc. In these processes, the electrocoagulation process has been widely used because of its simple process, no need for manual dosing, low cost and high treatment efficiency [31]. There have been many studies on the treatment of laundry wastewater by electrocoagulation, which can effectively remove turbidity and chromaticity in water, and has a good removal efficiency on COD [16,17,32–34]. In addition, electrocoagulation process has a very good removal efficiency on microplastics. However, the electrocoagulation process has limited removal efficiency although it is able to remove part of the surfactant. It has been pointed out that the presence of surfactants can reduce the removal efficiency of microplastic fibers during coagulation [35]. Obviously, the single electrocoagulation process could not satisfy the treatment of all the wastewater. If it was combined with other advanced oxidation processes, the desired efficiency could be obtained. Advanced oxidation processes (AOPs) are increasingly used to remove surfactants from wastewater [36–40]. The combination of electrocoagulation process and electrooxidation process realizes the efficient removal of COD and TOC [41,42]. The coupling of the electrocoagulation process and ozone oxidation or electrocoagulation process in a series with ozone oxidation can achieve efficient removal of textile wastewater [43].

The research group has studied the removal mechanism of the relevant electro-hybrid ozonation–coagulation (E-HOC) process in the early stage. The research shows that the synergistic effects between ozone and coagulants (SOC) is helpful to improve the generation of hydroxyl radicals (•OH) in the reaction system [44]. In addition, the efficiency of the E-HOC process for pollutant removal was significantly improved in bipolar connection and the dual-coagulant generation mode [45]. Moreover, the E-HOC (composite anode, CA^+) process achieves high efficiency and simultaneous removal of organics and ammonia in leachate. The reactive chlorine species (RCS) generated in the E-HOC system promoted the generation of hydroxyl radicals [46]. However, the efficiency and mechanism of the E-HOC process for laundry wastewater treatment are still unclear.

In this study, the removal efficiency of C-E-HOC with a carbon fiber electrode as cathode and an aluminium electrode as anode on COD_{cr}, turbidity, LAS and microplastics in laundry wastewater was studied. The microplastics were analyzed by fluorescence microscopy for characterization, shape, size and abundance. The role of the carbon fiber electrode in the E-HOC process for production of reactive oxygen species analysis (ROS) was assessed through electron spin resonance (EPR), Fourier transform infrared spectroscopy (FT-IR) and H_2O_2 quantify to elucidate the mechanism of the simultaneous surfactant and microplastic removal in the C-E-HOC process.

2. Materials and Methods

2.1. Raw Water

Two kinds of raw waters are used in this study. One is synthetic water in the laboratory, and the other is taken from the drainage during the machine washing of household clothes (including laundry wastewater, primary rinsing water and secondary rinsing water). The raw water quality parameters were listed in Table 1 (Sodium linear alkylbenzene sulfonate (LAS) is a widely used anionic surfactant in synthetic detergents [47]). Synthetic water in the laboratory was used for the removal efficiency experiment of COD_{cr}, turbidity and LAS, and another raw water is used for the relevant research of microplastics.

Table 1. Raw water quality parameters.

Raw Water	COD_{cr} (mg/L)	Turbidity (NTU)	LAS (mg/L)
synthetic wastewater	800 ± 50	90 ± 20	100 ± 20
laundry wastewater	800 ± 100	97.5 ± 17.5	800 ± 50
primary rinsing water	140 ± 40	22.5 ± 7.5	15.5 ± 4.5
secondary rinsing water	80 ± 20	10.5 ± 4.5	8 ± 2

2.2. Experimental Agents

Dimethyl pyridine N-oxide (DMPO) used in EPR experiment is superior pure and purchased from Sigma Aldrich (St. Louis, MO, USA). The rest were purchased from Kermel, Tianjin, China. The potassium bromide (KBr) used in the experiment is spectral grade. Other reagents used in this study are analytical grade, such as potassium dichromate ($K_2Cr_2O_7$), mercury sulfate ($HgSO_4$), silver sulfate (Ag_2SO_4), concentrated sulfuric acid, potassium titanium oxalate ($C_4K_2O_9Ti \cdot 2H_2O$), chloroform ($CHCl_3$), sodium hydroxide ($NaOH$), sodium dihydrogen phosphate monohydrate ($NaH_2PO_4 \cdot H_2O$) and phenolphthalein ($C_{20}H_{14}O_4$). Methylene blue is indicator level. All solutions were prepared with ultrapure water.

2.3. The E-HOC Experiment

The experiment was carried out in a cubic glass container with a reaction volume of 500 mL at room temperature. In this system, carbon fiber or stainless steel was used as the cathode and aluminium was used as the anode. The length, width and thickness of the electrodes are 7 cm, 6 cm and 0.1 cm, respectively. The effective reaction area of the electrode was 84 cm^2. The reaction time was 60 min, and slow stirring was always performed at 100 rpm during the reaction. The ozone generator converts the oxygen source to ozone, which is fed through an aerated head to the reaction container. The ozone generator was purchased from Sankang, Jinan, China. The electrode was connected with a direct current (DC) power supply, which was purchased from ANNAISI, Wuxi, China. A small amount of Na_2SO_4 was added to the water sample before the reaction to adjust the electrical conductivity. The electrocoagulation reaction was started simultaneously with the ozonation reaction. The flocs were allowed to settle sufficiently by standing for 30 min at the end of the reaction before the assay was performed. The removal efficiencies of the E-HOC processes for COD_{cr}, turbidity and LAS were determined at different current density or different ozone dosage.

2.4. Analytical Methods

2.4.1. Determination of Physicochemical Indexes

The COD was analyzed by fast digestion-spectrophotometry [48]. Before the determination, the interference of suspended solids in the sample should be minimized. The appropriate scale was selected according to the water quality of the water samples to be tested. The turbidity was measured by a portable turbidimeter, purchased from HANNA, Padovana, Italy. The LAS was determined by methylene blue spectrophotometry [49]. Water samples need to be pre-filtered by medium-speed qualitative paper filter to remove suspended solids in order to remove insoluble anionic surfactants in the water samples before the determination. H_2O_2 was determined by potassium titanate oxalate spectrophotometry. A sample was taken at 10 min (adjusted pH to 3), and 20 mL of the water sample to be measured was placed in a 25 mL colourimetric tube. Next, 1 mL of 0.05 mol/L potassium oxalate color developer was added to the water sample and subsequently diluted to the scale, shaken well and left for 8 min. The sample was poured into a 10 mm cuvette and measured with an ultraviolet spectrophotometer (UV-4802) at 385 nm wavelength.

2.4.2. Fourier Transform Infrared Spectroscopy (FT-IR) Analysis

The flocs after completion of the reaction were first frozen in the refrigerator and subsequently cold dried in a vacuum freeze dryer until the moisture completely disappeared. The sample was obtained by mixing 10 mg of the cold dried floc and 990 mg of KBr and grinding in an agate mortar. The samples were analyzed using a FT-IR spectrometer (Model Nicolet 6700, Thermo Fisher Scientific, Waltham, MA, USA) at the range from 4000 to 500 cm^{-1}.

2.4.3. EPR Analysis

EPR experiments trap hydroxyl radicals in the reaction system at room temperature by trapping agent DMPO. Experiments were performed on a Bruker EMXmicro spectrometer (Bruker Corporation, Bremen, Germany). Water samples were filtered through a 0.45 μm filter before the determination. Next, 30 mM of the radical trapping agent DMPO were added to each 500 μL of the water sample. The instrument determined the following parameters: resonance frequency of 9.77 GHz, microwave power of 25.18 mW, modulation frequency of 100 kHz, modulation amplitude of 1.0 G, sweep width of 700 G, time constant of 81.92 ms, sweep time of 35 s, and receiver gain of 30 dB.

2.4.4. Microplastics Analysis

In all, 500 mL water samples were passed through a glass fiber filter membrane (0.7 μm. 47 mm Ø, Xingya, Shanghai, China) and the samples were filtered. The filtered filter membrane was placed in a Petri dish and wrapped in aluminium foil. Several fine holes were tied on the surface of aluminium foil with needles, and the dish was placed in an oven (60 °C) to dry for 1 h and then observed under a fluorescence microscope. The microplastics observed under the microscope were classified and counted according to shape and size.

3. Results and Discussion

3.1. Removal Performance

The E-HOC process at the different current densities or different ozone dosages for the removal efficiency of COD_{cr} (a_1, b_1), turbidity (a_2, b_2), as well as LAS (a_3, b_3) in laundry wastewater is shown in Figure 1. According to Figure 1, the E-HOC process resulted in a superior COD_{cr} removal efficiency compared with the electrocoagulation (EC) and ozonation processes. The use of the EC process or ozonation process as the separate treatment process faces important practical limitations. Neither the electrocoagulation process nor the ozonation process can effectively remove COD_{cr}, turbidity and LAS. Obviously, a single treatment process cannot meet the treatment standards [50]. By contrast, the E-HOC process can remove COD_{cr}, turbidity and LAS efficiently. Our previous study found that Al-based coagulants can not only enhance·OH generation during ozonation, but also the addition of ozone promotes the conversion of the Al salt coagulant hydrolysis form to the higher polymerized Al species. Consequently, there are synergistic effects between ozone and coagulants (SOC) in the E-HOC process. [44]. In addition, contrasting Figure 1a$_2$ with Figure 1b$_2$, it can be seen that the removal of turbidity was mainly dependent on the EC process. This is similar to the previous research results. EC has an outstanding efficiency on removing turbidity in the water samples [51–54].

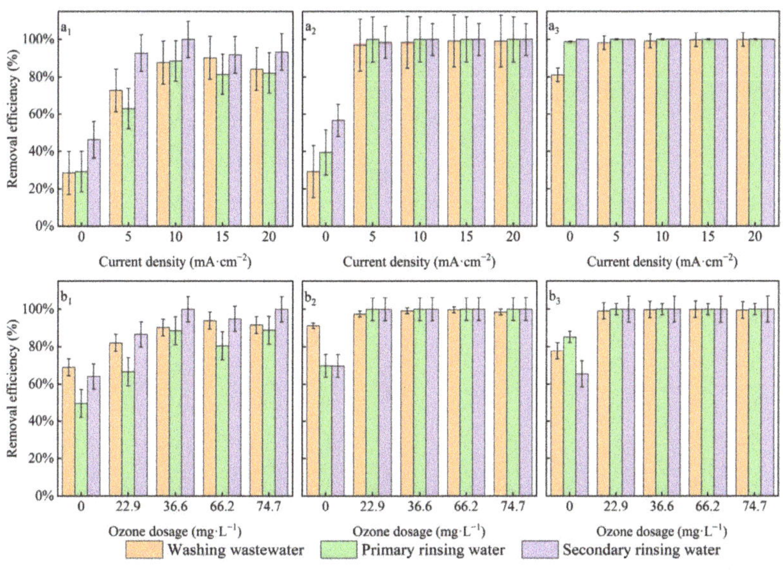

Figure 1. The E-HOC process at different current densities (**a$_1$,a$_2$,a$_3$**) or different ozone dosages (**b$_1$,b$_2$,b$_3$**) for the removal efficiency of COD_{cr} (**a$_1$,b$_1$**), turbidity (**a$_2$,b$_2$**), LAS (**a$_3$,b$_3$**).

However, the EC process is less effective for the removal of LAS than the ozonation process. Contrasting Figure 1a$_3$ with Figure 1b$_3$, it can be seen that the ozonation process can remove LAS from laundry wastewater with high efficiency. There are two ways for ozone to oxidize organic compounds in laundry wastewater: (i) direct reaction between ozone and organic molecules or (ii) indirect reaction by attack of free radical species produced by ozone decomposition. Studies have shown that ozone has the highest contribution rate in the process of removing surfactant. The radical reaction plays a dominant role in the oxidation of surfactant [55]. In addition, studies have shown that the presence of low concentrations of surfactants can contribute to the ozone dissolution and indirect generation of hydroxyl radicals in the ozone process [56]. Therefore, the removal of LAS mainly depends on the ozonation process.

Ultimately, the optimal reaction conditions of the E-HOC process in three laundry wastewaters were obtained ((washing wastewater, 15 mA·cm^{-2}, 66.2 mg·L^{-1}) (primary rinsing water, 10 mA·cm^{-2}, 36.6 mg·L^{-1}) (secondary rinsing water, 10 mA·cm^{-2}, 36.6 mg·L^{-1})). Under the optimal reaction conditions, the removal rates of COD_{cr}, turbidity and LAS in washing wastewater are 93.9%, 99.7% and 99.9%, respectively. The pollutants in laundry wastewater are mainly long-chain alkanes, aromatic compounds, esters and ethers. Long-chain alkanes are firstly decomposed into short-chain alkanes. These substances subsequently further decomposed into small molecule alkanes as well as small molecule organic acids. Finally, long-chain alkanes, alkenes, ethers and esters have been completely degraded [57,58].

3.2. *Microplastic Analysis*

3.2.1. Microplastics Removal Performance

The removal efficiency of the EC and E-HOC processes for microplastics from three laundry wastewaters are shown in Figure 2. According to Figure 2, both E-HOC and EC had a significant efficiency in removing the microplastics. In addition, a slightly lower microplastics removal efficiency for the E-HOC process can be seen compared to the EC process at the same reaction conditions. This is probably due to the addition of ozone. Ozone causes the large flocs formed in the EC process to be broken into small flocs, resulting in part in the microplastics falling off the flocs and returning back to the water sample. Compared with the EC process, the flocs produced by E-HOC process are looser and smaller in particle size.

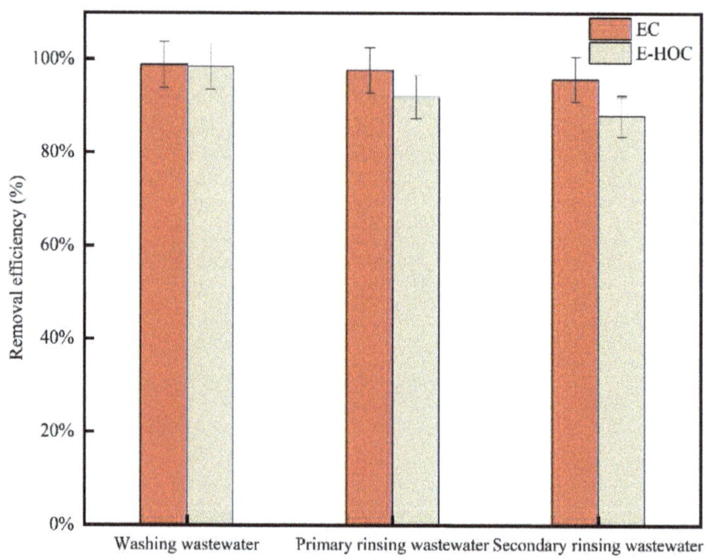

Figure 2. The EC and E-HOC processes on the removal efficiency of microplastics from three laundry wastewaters.

The process of EC can produce aluminium salt coagulant in situ at the Al electrode, which synergistically removes microplastics from water bodies through a variety of mechanisms including charge neutralization, adsorptive bridging and sweep flocculation [59]. There have been many previous studies that have shown that EC has excellent removal efficiency for microplastics. The removal efficiency can reach 96.5% for microplastics under the reaction conditions with a current density of 2.88 mA·cm^{-2} and an initial pH of 4 for the real wastewater samples [60]. In addition, the aluminium anode is superior to the iron anode in removing microplastics. When the pH is within the range from 3 to 10, the removal efficiency of EC for microplastics can reach more than 82%, and when the pH is 7.2, the maximum removal efficiency for microplastics can reach 98.4% [61]. A considerable number of studies have proved that the removal efficiency of microplastics by the EC process can reach more than 90% [17,62,63].

3.2.2. Shape and Size Analysis

Microplastics are divided into fibers, particles, fragments and films based on the shape [64–67]. Before collecting laundry wastewater, wash the washing machine twice without adding clothes to reduce interference. The shape of the microplastics observed under the fluorescence microscope is shown in Figure 3. As shown in Figure 3, fiber-shaped microplastics account for the vast majority of them. Clothing releases a large number of microplastic fibers during its washing, and laundry wastewater is an important source of microplastic fibers in wastewater treatment plants and the environment [68].

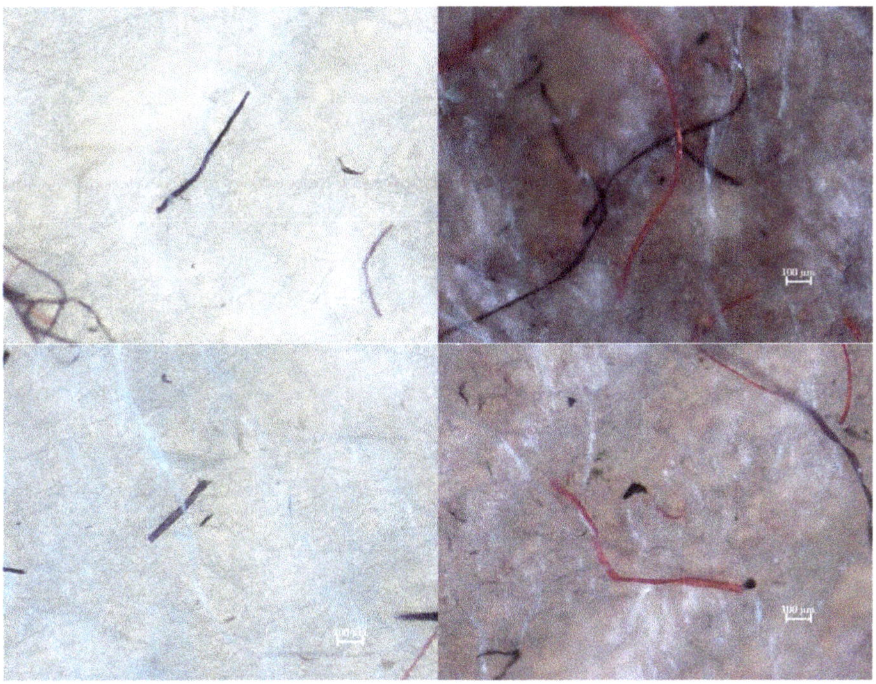

Figure 3. Microplastics observed by fluorescence microscopy.

To further study the characteristics of microplastics in laundry wastewater, the shape and size distribution of the microplastics is shown in Figure 4. As shown in Figure 4, 73.8% of the microplastics detected in laundry wastewater were fibers. According to the microplastics size, it was divided into four ranges: less than 50 μm, more than 50 μm and less than 200 μm, more than 200 μm and less than 500 μm, more than 500 μm. According to Figure 4, 46.9% of the microplastics were less than 50 μm, Only 12.4% of the microplastics were larger than 500 μm. The microplastics were available in sizes ranging from 6.1 μm to 4000 μm. A study found that the average length of the microplastics found in household laundry wastewater was 2258.59 μm [69]. In addition, only 7% of the microplastics fibers were longer than 500 μm, 40% were between 100 and 500 μm and 53% were between 50 and 100 μm [70]. In household laundry wastewater, the study found that 48.64% of the released microfibers were shorter than 5 μm in length, and only 11.49% of the microfibers were longer than 500 μm [71]. This is similar to our findings.

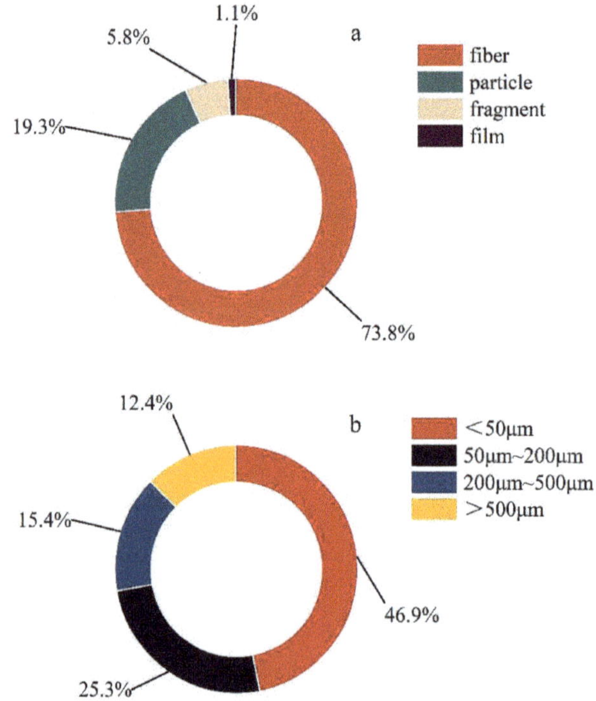

Figure 4. Shape (**a**) and size (**b**) distribution of microplastics.

3.2.3. Quantitative Analysis

Due to the consideration of the factor that the amount of water in laundry wastewater is large, even low concentration of microplastics have many microplastics discharged to the environment. As shown in Figure 5, overall, the amount of microplastics in the washing wastewater was the largest, followed by the primary rinsing wastewater and finally the secondary rinsing wastewater. In terms of the overall quantity, sample 1 released 44 × 104 items of microplastics per 100 L of water, and sample 4 released 108 × 104 items of microplastics per 100 L of water. This can be caused by many factors including the fabric, the textile process as well as the thickness of the clothing. Studies have confirmed that the interlacing coefficient of textiles affects the number of particles in wastewater. The results show that the higher the yarn's density/cm in the fabric, the lower the fiber content in the wastewater [72]. One study has shown that more than 150,000 fibers are released per 1 kg of laundry wash [11]. Cleaning PET knitted fibers in the washing machine releases 4489.93 fibers/L or 368,094.07 fibers per kilogram [73].

It can be seen that the quantity of microplastics is easily affected by many factors, but considering that the total amount of microplastics discharged into the environment is still large, its potential harm to the environment cannot be ignored. We can think of a front-end design to remove microplastics before they enter the wastewater treatment plant or the environment.

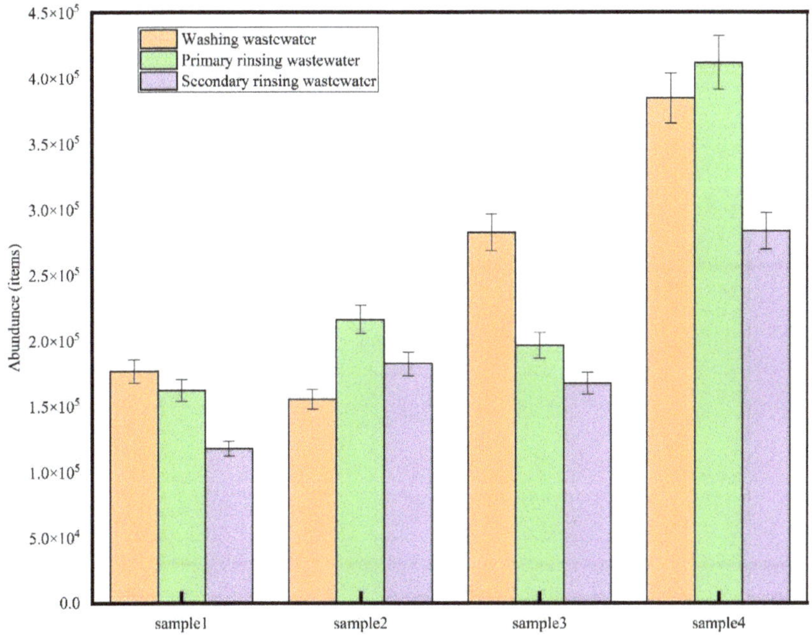

Figure 5. Total drainage microplastics emissions in different stages.

3.2.4. FT-IR Analysis of Microplastics

The FT-IR spectra of the microplastics detected from the different stages of laundry wastewater are shown in Figure 6. According to Figure 6, the kinds of microplastics in laundry wastewater at the different stages did not change obviously. Comparing with FT-IR spectra of standards [74], it is found that the microplastics in laundry wastewater are mainly polyethylene (PE), nylon (Polyamide, PA) and polyester (PET). The peaks of PE, PA and PET are 2914 and 2850 cm^{-1}, 1650 and 1540 cm^{-1}, 1700 and 1060 cm^{-1}, respectively. These are the plastic polymers commonly used in textiles [69]. PE may also come from personal care products such as toothpaste and facial cleanser [75]. This is similar to the composition of microplastic fibers in wastewater treatment plants [76].

3.3. Mechanism Analysis

3.3.1. Reactive Oxygen Species Analysis

The production of •OH in the reaction system was compared by EPR when the cathode was a carbon fiber electrode and a stainless steel electrode, respectively (Figure 7). According to Figure 7, there were four peaks centered at 3500 G in the EPR spectrum with a peak height ratio close to 1:2:2:1, which was a typical DMPO-•OH profile [77,78]. •OH was detected in both laundry wastewater and ultrapure water, but it was not obviously detected in ultrapure water by the carbon fiber cathode system. Compared with Figure 7a,b, the EPR peak of •OH generated by the carbon fiber cathode system is higher than that of the stainless steel cathode system, which is because the carbon fiber cathode will generate H_2O_2 in the system (Equation (1)) [79], and H_2O_2 will further react and convert into •OH. This indicates that more •OH is produced in the carbon fiber cathode system than in the stainless steel cathode system. Previous studies of our group have shown that $•O_2^-$ can also be produced in the reaction system [44,45], which can be further converted to •OH in the system (Equations (2) and (3)) [80,81]. The change of pH in the system before and after the reaction was measured. The carbon fiber system changed from 8.3 to 8.9, and the stainless steel system changed from 8.26 to 8.78.

$$O_2 + 2H^+ + 2e^- \rightarrow H_2O_2, \quad (1)$$

$$\bullet O_2{}^- + H^+ \rightarrow H_2O_2 + O_2, \quad (2)$$

$$\bullet O_2{}^- + H_2O_2 + H^+ \rightarrow O_2 + \bullet OH + H_2O, \quad (3)$$

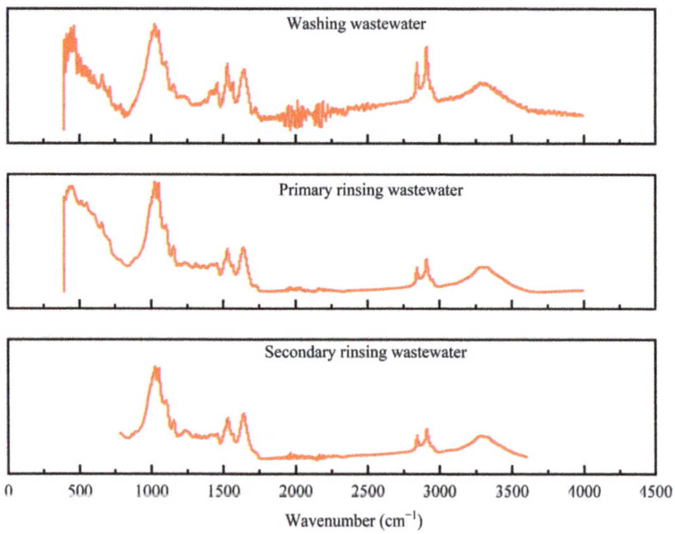

Figure 6. FT-IR spectra of microplastics from laundry wastewater.

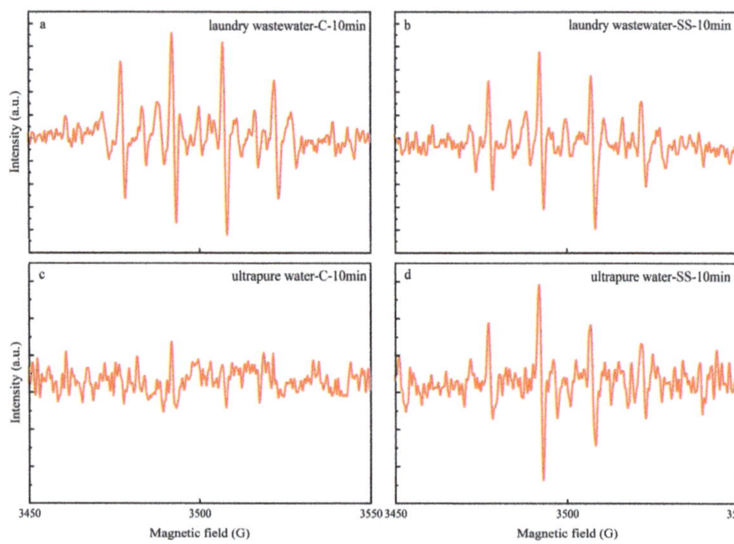

Figure 7. EPR spectra detected in the E-HOC processes. Reaction conditions: current density 15 $mA \cdot cm^{-2}$, ozone dosage 66.2 $mg \cdot L^{-1}$. Samples for EPR test were taken at the 10th min during the reactions. Experimental condition: raw water (laundry wastewater (**a**,**b**), ultrapure water (**c**,**d**)), cathode (carbon fiber (**a**,**c**), stainless steel (**b**,**d**)).

To further verify the ability of the carbon fiber cathode system to produce H_2O_2, the concentration of H_2O_2 in the reaction system at different current densities or under different reaction conditions was determined in the washing wastewater. There are two groups of reactions. One is to keep the ozone dosage constant, only changing the current density, while the other is to keep the current density constant, only changing the ozone dosage.

Figure 8 shows the ability of the carbon fiber cathode and the stainless steel cathode to produce H_2O_2 under different current densities or different ozone dosages by using titanium salt photometry. According to Figure 8, under different reaction conditions, the carbon fiber cathode had a stronger ability to produce H_2O_2 in the reaction system than the stainless steel cathode. The best reaction conditions were further demonstrated by the fact that the carbon fiber electrode has the highest capacity to produce H_2O_2 in the reaction system with a current density of 15 $mA \cdot cm^{-2}$ or ozone dosage of 66.2 $mg \cdot L^{-1}$. Under the optimal reaction conditions (current density of 15 $mA \cdot cm^{-2}$, ozone dosage of 66.2 $mg \cdot L^{-1}$), the carbon fiber system produced H_2O_2 at a concentration 32.2% higher than that of the stainless steel system. This also verifies the result that the peak height of the carbon fiber cathode system is higher than the stainless steel cathode system in EPR experiments.

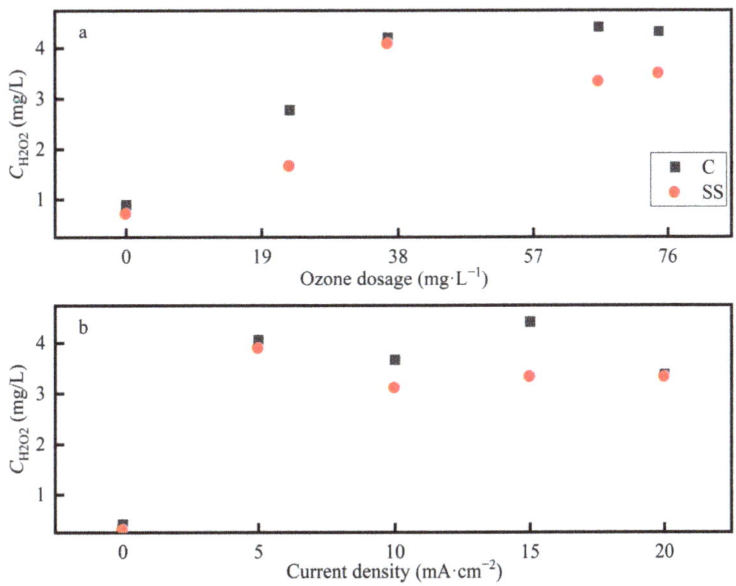

Figure 8. Concentration of H_2O_2 under different cathode electrodes (**a**) at different ozone dosages (0, 22.9, 36.6, 66.2 and 74.7 $mg \cdot L^{-1}$) (**b**) at different current densities (0, 5, 10, 15 and 20 $mA \cdot cm^{-2}$).

Based on the previous research, •OH, H_2O_2 and $\bullet O_2^-$ were generated in the E-HOC process. The •OH quenching experiences indicated that •OH accounted for approximately 90% contribution for the E-HOC process, which indicated •OH played a dominant role in pollutant removal [45].

3.3.2. FT-IR Analysis of Flocs

In order to further define the characteristics of the E-HOC process with the carbon fiber electrode as the cathode, the functional group changes of the flocs after the reaction of the contrasting carbon fiber system with the stainless steel system were determined. As shown in Figure 9, the peak around 1630 cm^{-1} is mainly caused by the stretching vibration of the -C=C- or O-H bending of the absorbed water molecules [82–85]. The peak around 3400 cm^{-1} results from the stretching of the O-H bond of the hydroxyl group [86]. The peak around 1400 cm^{-1} may be due to the formation of the H bond with the aluminium hydroxide precipitates [87].

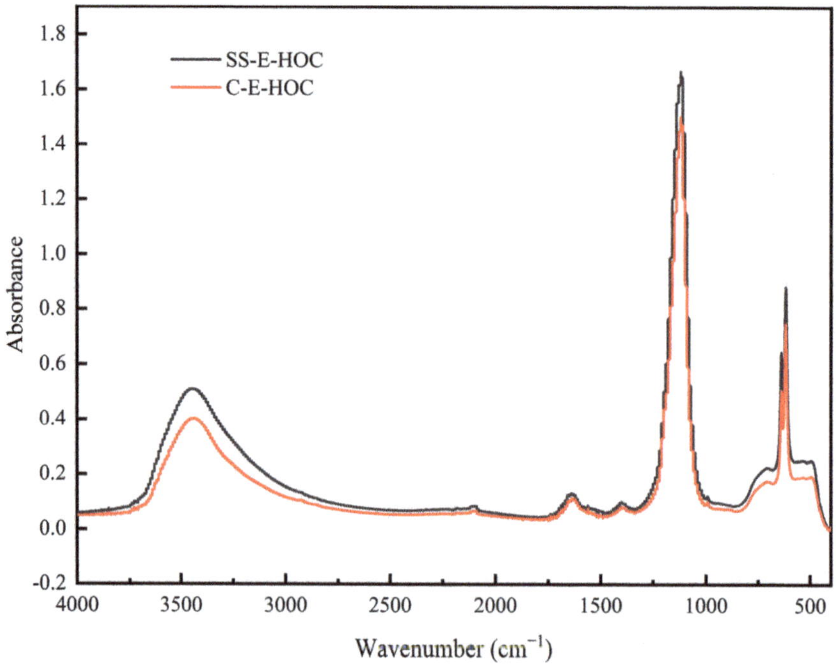

Figure 9. FT-IR spectra of flocs in the reaction system of washing wastewater treatment by E-HOC process under different cathode electrode conditions.

Comparing the FT-IR spectra of the flocs between the carbon fiber cathode system and the stainless steel cathode system, the peaks of both basically coincided, but the peaks in the carbon fiber cathode system was a little lower than that in the stainless steel cathode system. In a previous study, the O_3 addition during the EC process decreased the peak intensity of hydroxyl groups on the surface of the aluminium salt coagulants generated in situ at the anode, indicating that there is an interaction between the surface hydroxyl groups and O_3, and that there is a complexation between the coagulant and the surface hydroxyl groups of organics in the system [88]. Since the carbon fiber cathode system showed higher removal efficiency for the contaminants than the stainless steel cathode system with a large consumption of reactive oxygen species, the intensity of the peaks would be lower than that of the stainless steel cathode system.

4. Discussion

The removal efficiency of COD_{cr}, turbidity and LAS in the C-E-HOC process is higher than that in the traditional electrocoagulation process and the ozonation process. The optimal reaction conditions of C-E-HOC for washing wastewater are a current density of 15 $mA \cdot cm^{-2}$ and an ozone dosage of 66.2 $mg \cdot L^{-1}$. The removal efficiency of microplastics by the C-E-HOC process is slightly lower than that by the EC process because ozone aeration causes the large flocs to break down, causing a portion of the microplastics to fall off from the flocs. In addition, 73.8% of the microplastics were shaped as fibers, and nearly half of the microplastics were less than 50 μm in size. In the process of household laundry, 440,000 to 1,080,000 microplastics will be released to the environment for per 100 L of water. According to FT-IR analysis, the microplastics in laundry wastewater are mainly polyethylene, nylon and polyester. By comparing the E-HOC processes of different types of cathodes, the C-E-HOC process generated more •OH during the reaction compared with SS-E-HOC. In addition, more H_2O_2 could be generated from the carbon fiber cathode in the system, and H_2O_2 would be further converted to •OH.

Author Contributions: Methodology, J.L. and X.J.; formal analysis, J.L.; investigation, J.L. and Y.W.; data curation, J.L.; writing—original draft preparation, J.L.; writing—review and editing, X.J. and J.L.; project administration, P.J.; funding acquisition, P.J. All authors have read and agreed to the published version of the manuscript.

Funding: This work was supported by the National Natural Science Foundation of China (No. 52230001, 52070151, 52170052), the Key Research and Development Project of Shaanxi Province (No. 2021ZDLSF05-06), Technology Innovation Leading Program of Shaanxi (No. 2020CGXNG-021).

Institutional Review Board Statement: Not applicable.

Informed Consent Statement: Not applicable.

Data Availability Statement: Data will be made available on request.

Conflicts of Interest: The authors declare no conflict of interest.

References

1. Deshayes, S.; Eudes, V.; Bigourie, M.; Droguet, C.; Moilleron, R. Alkylphenol and phthalate contamination of all sources of greywater from French households. *Sci. Total Environ.* **2017**, *599*, 883–890. [CrossRef] [PubMed]
2. Ho, K.C.; Teow, Y.H.; Sum, J.Y.; Ng, Z.J.; Mohammad, A.W. Water pathways through the ages: Integrated laundry wastewater treatment for pollution prevention. *Sci. Total Environ.* **2021**, 760. [CrossRef] [PubMed]
3. Manouchehri, M.; Kargari, A. Water recovery from laundry wastewater by the cross flow microfiltration process: A strategy for water recycling in residential buildings. *J. Clean. Prod.* **2017**, *168*, 227–238. [CrossRef]
4. Santiago, D.E.; Hernández Rodriguez, M.J.; Pulido-Melian, E. Laundry Wastewater Treatment: Review and Life Cycle Assessment. *J. Environ. Eng.* **2021**, *147*, 143966. [CrossRef]
5. Cesa, F.S.; Turra, A.; Baruque-Ramos, J. Synthetic fibers as microplastics in the marine environment: A review from textile perspective with a focus on domestic washings. *Sci. Total Environ.* **2017**, *598*, 1116–1129. [CrossRef]
6. Shi, K.-W.; Wang, C.-W.; Jiang, S.C. Quantitative microbial risk assessment of Greywater on-site reuse. *Sci. Total Environ.* **2018**, *635*, 1507–1519. [CrossRef]
7. Nicolotti, G.; Rettori, A.; Paoletti, E.; Gullino, M.L. Morphological and physiological damage by surfactant-polluted seaspray on *Pinus pinea* and *Pinus halepensis*. *Environ. Monit. Assess.* **2005**, *105*, 175–191. [CrossRef]
8. Wang, L.-F.; Wang, L.-L.; Li, W.-W.; He, D.-Q.; Jiang, H.; Ye, X.-D.; Yuan, H.-P.; Zhu, N.-W.; Yu, H.-Q. Surfactant-mediated settleability and dewaterability of activated sludge. *Chem. Eng. Sci.* **2014**, *116*, 228–234. [CrossRef]
9. Sun, J.; Zhu, Z.R.; Li, W.H.; Yan, X.; Wang, L.K.; Zhang, L.; Jin, J.; Dai, X.; Ni, B.J. Revisiting Microplastics in Landfill Leachate: Unnoticed Tiny Microplastics and Their Fate in Treatment Works. *Water Res.* **2021**, *190*, 116784. [CrossRef]
10. Li, J.Y.; Liu, H.H.; Chen, J.P. Microplastics in freshwater systems: A review on occurrence, environmental effects, and methods for microplastics detection. *Water Res.* **2018**, *137*, 362–374. [CrossRef]
11. Tang, N.; Liu, X.N.; Xing, W. Microplastics in wastewater treatment plants of Wuhan, Central China: Abundance, removal, and potential source in household wastewater. *Sci. Total Environ.* **2020**, *745*, 141026. [CrossRef] [PubMed]
12. Xu, J.; Wang, X.; Zhang, Z.; Yan, Z.; Zhang, Y. Effects of chronic exposure to different sizes and polymers of microplastics on the characteristics of activated sludge. *Sci. Total Environ.* **2021**, *783*, 146954. [CrossRef] [PubMed]
13. Jiang, Y.J.; Zhou, S.; Fei, J.; Qin, Z.; Yin, X.; Sun, H.; Sun, Y. Transport of different microplastics in porous media: Effect of the adhesion of surfactants on microplastics. *Water Res.* **2022**, *215*, 118262. [CrossRef] [PubMed]
14. Pramanik, B.K.; Pramanik, S.K.; Monira, S. Understanding the fragmentation of microplastics into nano-plastics and removal of nano/microplastics from wastewater using membrane, air flotation and nano-ferrofluid processes. *Chemosphere* **2021**, 282. [CrossRef]
15. Borghi, C.C.; Fabbri, M.; Fiorini, M.; Mancini, M.; Ribani, P.L. Magnetic removal of surfactants from wastewater using micrometric iron oxide powders. *Sep. Purif. Technol.* **2011**, *83*, 180–188. [CrossRef]
16. Dimoglo, A.; Sevim-Elibol, P.; Dinc, O.; Gökmen, K.; Erdoğan, H. Electrocoagulation/electroflotation as a combined process for the laundry wastewater purification and reuse. *J. Water Process Eng.* **2019**, *31*, 100877. [CrossRef]
17. Akarsu, C.; Deniz, F. Electrocoagulation/Electroflotation Process for Removal of Organics and Microplastics in Laundry Wastewater. *Clean-Soil Air Water* **2021**, *49*, 2000146. [CrossRef]
18. Bering, S.; Mazur, J.; Tarnowski, K.; Janus, M.; Mozia, S.; Morawski, A.W. The application of moving bed bio-reactor (MBBR) in commercial laundry waste water treatment. *Sci. Total Environ.* **2018**, *627*, 1638–1643. [CrossRef]
19. Perez-Lopez, M.E.; Arreola-Ortiz, A.E.; Malagon Zamora, P. Evaluation of detergent removal in artificial wetlands (biofilters). *Ecol. Eng.* **2018**, *122*, 135–142. [CrossRef]

20. Lv, X.; Dong, Q.; Zuo, Z.; Liu, Y.; Huang, X.; Wu, W.M. Microplastics in a municipal wastewater treatment plant: Fate, dynamic distribution, removal efficiencies, and control strategies. *J. Clean. Prod.* **2019**, *225*, 579–586. [CrossRef]
21. Bilad, M.R.; Nawi, N.I.M.; Subramaniam, D.D.; Shamsuddin, N.; Khan, A.L.; Jaafar, J.; Nandiyanto, A.B.D. Low-pressure submerged membrane filtration for potential reuse of detergent and water from laundry wastewater. *J. Water Process Eng.* **2020**, *36*, 101264. [CrossRef]
22. Kogut, I.; Szwast, M.; Hussy, S.; Polakb, D.; Gerhardtsa, A.; Piątkiewiczd, W. Evaluation of wastewater reuse in commercial laundries: A pilot field study. *Desalination Water Treat.* **2021**, *214*, 39–48. [CrossRef]
23. Park, C.; Kim, S. Fouling behavior and cleaning strategies of ceramic ultrafiltration membranes for the treatment and reuse of laundry wastewater. *J. Water Process Eng.* **2022**, *48*, 102840.
24. Sumisha, A.; Arthanareeswaran, G.; Thuyavan, Y.L.; Ismail, A.F.; Chakraborty, S. Treatment of laundry wastewater using polyethersulfone/polyvinylpyrollidone ultrafiltration membranes. *Ecotoxicol. Environ. Saf.* **2015**, *121*, 174–179. [CrossRef] [PubMed]
25. Szwast, M.; Polak, D. New membranes for industrial laundry wastewater treatment. *Przem. Chem.* **2018**, *97*, 439–441.
26. Esteban Garcia, A.B.; Szymanski, K.; Mozia, S.; Sánchez Pérez, J.A. Treatment of laundry wastewater by solar photo-Fenton process at pilot plant scale. *Environ. Sci. Pollut. Res.* **2021**, *28*, 8576–8584. [CrossRef]
27. Joseph, C.G.; Farm, Y.Y.; Taufiq-Yap, Y.H.; Pang, C.K.; Nga, J.L.; Puma, G.L. Ozonation treatment processes for the remediation of detergent wastewater: A comprehensive review. *J. Environ. Chem. Eng.* **2021**, *9*, 106099. [CrossRef]
28. Bering, S.; Mazur, J.; Tarnowski, K.; Dąbkowska, N.; Janus, M.; Mozia, S.; Morawski, A.W. Removal of organic pollutants and surfactants from laundry wastewater in membrane bioreactor (MBR). *Desalination Water Treat.* **2018**, *134*, 281–288. [CrossRef]
29. Zotesso, J.P.; Cossich, E.S.; Janeiro, V.; Tavares, C.R.G. Treatment of hospital laundry wastewater by UV/H_2O_2 process. *Environ. Sci. Pollut. Res.* **2017**, *24*, 6278–6287. [CrossRef]
30. Benis, K.Z.; Behnami, A.; Aghayani, E.; Farabi, S.; Pourakbar, M. Water recovery and on-site reuse of laundry wastewater by a facile and cost-effective system: Combined biological and advanced oxidation process. *Sci. Total Environ.* **2021**, *789*, 148068. [CrossRef]
31. Hakizimana, J.N.; Gourich, B.; Chafi, M.; Stiriba, Y.; Vial, C.; Drogui, P.; Naja, J. Electrocoagulation process in water treatment: A review of electrocoagulation modeling approaches. *Desalination* **2017**, *404*, 1–21. [CrossRef]
32. Ramcharan, T.; Bissessur, A. Treatment of laundry wastewater by biological and electrocoagulation methods. *Water Sci. Technol.* **2017**, *75*, 84–93. [CrossRef] [PubMed]
33. Pratiwi, N.I.; Mukimin, A.; Zen, N.; Septarina, I. Integration of electrocoagulation, adsorption and wetland technology for jewelry industry wastewater treatment. *Sep. Purif. Technol.* **2021**, *279*, 119690. [CrossRef]
34. Medrano-Hurtado, Z.Y.; Medina-Aguirre, J.C.; Marcelo-Medrano, H.; Castillón-Barraza, A.; Zamora-Alarcón, R.; Casillas-Lamadrid, M.E.; Jumilla-Corral, A.A.; Mayorga-Ortiz, P. Domestic wastewater treatment by electrocoagulation system using photovoltaic solar energy. *Rev. Mex. Ing. Quim.* **2022**, *21*, IA2809. [CrossRef]
35. Li, J.; Dagnew, M.; Ray, M.B. Effect of coagulation on microfibers in laundry wastewater. *Environ. Res.* **2022**, *212*, 113401. [CrossRef]
36. Da Silva, S.W.; Klauck, C.R.; Siqueira, M.A.; Bernardes, A.M. Degradation of the commercial surfactant nonylphenol ethoxylate by advanced oxidation processes. *J. Hazard. Mater.* **2015**, *282*, 241–248. [CrossRef]
37. Mondal, B.; Adak, A.; Datta, P. Anionic surfactant degradation by UV-H_2O_2 advanced oxidation process and optimization of process parameters. *J. Indian Chem. Soc.* **2020**, *97*, 1328–1335.
38. Rios, F.; Olak-Kucharczyk, M.; Gmurek, M.; Ledakowicz, S. Removal efficiency of anionic surfactants from water during UVC photolysis and advanced oxidation process in H_2O_2/UVC system. *Arch. Environ. Prot.* **2017**, *43*, 20–26. [CrossRef]
39. Bai, X.; Wang, Y.; Zheng, X.; Zhu, K.; Long, A.; Wu, X.; Zhang, H. Remediation of phenanthrene contaminated soil by coupling soil washing with Tween 80, oxidation using the UV/S2O82- process and recycling of the surfactant. *Chem. Eng. J.* **2019**, *369*, 1014–1023. [CrossRef]
40. Wang, X.-J.; Song, Y.; Mai, J.-S. Combined Fenton oxidation and aerobic biological processes for treating a surfactant wastewater containing abundant sulfate. *J. Hazard. Mater.* **2008**, *160*, 344–348. [CrossRef]
41. Linares Hernandez, I.; Barrera Diaz, C.; Valdes Cerecero, M.; Almazan Sanchez, P.T.; Castaneda Juarez, M.; Lugo Lugo, V. Soft drink wastewater treatment by electrocoagulation-electrooxidation processes. *Environ. Technol.* **2017**, *38*, 433–442. [CrossRef] [PubMed]
42. Nayir, T.Y.; Kara, S. Container washing wastewater treatment by combined electrocoagulation-electrooxidation. *Sep. Sci. Technol.* **2018**, *53*, 1592–1603. [CrossRef]
43. Bilinska, L.; Blus, K.; Gmurek, M.; Ledakowicz, S. Coupling of electrocoagulation and ozone treatment for textile wastewater reuse. *Chem. Eng. J.* **2019**, *358*, 992–1001. [CrossRef]
44. Jin, X.; Xie, X.; Liu, Y.; Wang, Y.; Wang, R.; Jin, P.; Yang, C.; Shi, X.; Wang, X.C.; Xu, H. The role of synergistic effects between ozone and coagulants (SOC) in the electro-hybrid ozonation-coagulation process. *Water Res.* **2020**, *177*, 115800. [CrossRef] [PubMed]
45. Jin, X.; Xie, X.; Zhang, S.; Yang, C.; Xu, L.; Shi, X.; Jin, P.; Wang, X.C. Insights into the electro-hybrid ozonation-coagulation process-Significance of connection configurations and electrode types. *Water Res.* **2021**, *204*, 117600. [CrossRef] [PubMed]

46. Yang, C.; Jin, X.; Guo, K.; Diao, Y.; Jin, P. Simultaneous removal of organics and ammonia using a novel composite magnetic anode in the electro-hybrid ozonation-coagulation (E-HOC) process toward leachate treatment. *J. Hazard. Mater.* **2022**, *439*, 129664. [CrossRef] [PubMed]
47. Ramcharan, T.; Bissessur, A. Analysis of Linear Alkylbenzene Sulfonate in Laundry Wastewater by HPLC-UV and UV-Vis Spectrophotometry. *J. Surfactants Deterg.* **2016**, *19*, 209–218. [CrossRef]
48. Zhou, Z.Q.; Yu, T.; Dong, H.L.; Huang, L.; Chu, R.K.; Tolic, N.; Wang, X.; Zeng, Q. Chemical oxygen demand (COD) removal from bio-treated coking wastewater by hydroxyl radicals produced from a reduced clay mineral. *Appl. Clay Sci.* **2019**, *180*, 105199. [CrossRef]
49. Hu, N.; Li, Y.F.; Yang, C.Y.; Wu, Z.; Liu, W. In-Situ activated nanoparticle as an efficient and recyclable foam stabilizer for enhancing foam separation of LAS. *J. Hazard. Mater.* **2019**, *379*, 120843. [CrossRef]
50. Kumar, S.; Mostafazadeh, A.K.; Kumar, L.R.; Tyagi, R.D.; Drogui, P.; Brien, E. Advancements in laundry wastewater treatment for reuse: A review. *J. Environ. Sci. Health Part A-Toxic/Hazard. Subst. Environ. Eng.* **2022**, *57*, 927–946. [CrossRef]
51. Rookesh, T.; Samaei, M.R.; Yousefinejad, S.; Hashemi, H.; Derakhshan, Z.; Abbasi, F.; Jalili, M.; Giannakis, S.; Bilal, M. Investigating the Electrocoagulation Treatment of Landfill Leachate by Iron/Graphite Electrodes: Process Parameters and Efficacy Assessment. *Water* **2022**, *14*, 205. [CrossRef]
52. Rodriguez, A.Z.; Wang, H.; Hu, L.; Zhang, Y.; Xu, P. Treatment of Produced Water in the Permian Basin for Hydraulic Fracturing: Comparison of Different Coagulation Processes and Innovative Filter Media. *Water* **2020**, *12*, 770. [CrossRef]
53. Llanos, J.; Cotillas, S.; Canizares, P.; Rodrigo, M.A. Effect of bipolar electrode material on the reclamation of urban wastewater by an integrated electrodisinfection/electrocoagulation process. *Water Res.* **2014**, *53*, 329–338. [CrossRef] [PubMed]
54. Khor, C.M.; Wang, J.; Li, M.; Oettel, B.A.; Kaner, R.B.; Jassby, D.; Hoek, E.M. Performance, Energy and Cost of Produced Water Treatment by Chemical and Electrochemical Coagulation. *Water* **2020**, *12*, 3426. [CrossRef]
55. Rivera-Utrilla, J.; Bautista-Toledo, M.I.; Sanchez-Polo, M.; Méndez-Díaz, J.D. Removal of surfactant dodecylbenzenesulfonate by consecutive use of ozonation and biodegradation. *Eng. Life Sci.* **2012**, *12*, 113–116. [CrossRef]
56. Chu, W.; Chan, K.H.; Graham, N.J.D. Enhancement of ozone oxidation and its associated processes in the presence of surfactant: Degradation of atrazine. *Chemosphere* **2006**, *64*, 931–936. [CrossRef]
57. Ma, Y. Study on Treatment and Reuse of Laundry Wastewater by Electrocoagulation-Ozone. Master's Thesis, Jiangsu University, Zhenjiang, China, 2020. (In Chinese).
58. Motteran, F.; Nascimento, R.F.; Nadai, B.M.; Titato, G.M.; dos Santos Neto, Á.J.; Silva, E.L.; Varesche, M.B.A. Identification of Anionic and Nonionic Surfactant and Recalcitrants Compounds in Commercial Laundry Wastewater by GC-MS Analysis after Anaerobic Fluidized Bed Reactor Treatment. *Water Air Soil Pollut.* **2019**, *230*, 301. [CrossRef]
59. Tang, W.H.; Li, H.; Fei, L.Y.; Wei, B.; Zhou, T.; Zhang, H. The removal of microplastics from water by coagulation: A comprehensive review. *Sci. Total Environ.* **2022**, *851*, 158224. [CrossRef]
60. Elkhatib, D.; Oyanedel-Craver, V.; Carissimi, E. Electrocoagulation applied for the removal of microplastics from wastewater treatment facilities. *Sep. Purif. Technol.* **2021**, *276*, 118877. [CrossRef]
61. Shen, M.C.; Zhang, Y.X.; Almatrafi, E.; Hu, T.; Zhou, C.; Song, B.; Zeng, Z.; Zeng, G. Efficient removal of microplastics from wastewater by an electrocoagulation process. *Chem. Eng. J.* **2022**, *428*, 131161. [CrossRef]
62. Xu, R.Y.; Yang, Z.N.; Niu, Y.X.; Xu, D.; Wang, J.; Han, J.; Wang, H. Removal of microplastics and attached heavy metals from secondary effluent of wastewater treatment plant using interpenetrating bipolar plate electrocoagulation. *Sep. Purif. Technol.* **2022**, *290*, 120905. [CrossRef]
63. Perren, W.; Wojtasik, A.; Cai, Q. Removal of Microbeads from Wastewater Using Electrocoagulation. *Acs Omega* **2018**, *3*, 3357–3364. [CrossRef] [PubMed]
64. Qiao, R.X.; Deng, Y.F.; Zhang, S.H.; Wolosker, M.B.; Zhu, Q.; Ren, H.; Zhang, Y. Accumulation of different shapes of microplastics initiates intestinal injury and gut microbiota dysbiosis in the gut of zebrafish. *Chemosphere* **2019**, *236*, 124334. [CrossRef] [PubMed]
65. Choi, D.; Hwang, J.; Bang, J.; Han, S.; Kim, T.; Oh, Y.; Hwang, Y.; Choi, J.; Hong, J. In Vitro toxicity from a physical perspective of polyethylene microplastics based on statistical curvature change analysis. *Sci. Total Environ.* **2021**, *752*, 142242. [CrossRef]
66. Chia, R.W.; Lee, J.Y.; Kim, H.; Jang, J. Microplastic pollution in soil and groundwater: A review. *Environ. Chem. Lett.* **2021**, *19*, 4211–4224. [CrossRef]
67. Ding, J.F.; Li, J.X.; Sun, C.J.; Jiang, F.; Ju, P.; Qu, L.; Zheng, Y.; He, C. Detection of microplastics in local marine organisms using a multi-technology system. *Anal. Methods* **2019**, *11*, 78–87. [CrossRef]
68. Tian, Y.; Chen, Z.; Zhang, J.; Wang, Z.; Zhu, Y.; Wang, P.; Zhang, T.; Pu, J.; Sun, H.; Wang, L. An innovative evaluation method based on polymer mass detection to evaluate the contribution of microfibers from laundry process to municipal wastewater. *J. Hazard. Mater.* **2021**, *407*, 124861. [CrossRef]
69. Praveena, S.M.; Asmawi, M.S.; Chyi, J.L.Y. Microplastic emissions from household washing machines: Preliminary findings from Greater Kuala Lumpur (Malaysia). *Environ. Sci. Pollut. Res.* **2021**, *28*, 18518–18522. [CrossRef]
70. Galvao, A.; Aleixo, M.; De Pablo, H.; Lopes, C.; Raimundo, J. Microplastics in wastewater: Microfiber emissions from common household laundry. *Environ. Sci. Pollut. Res.* **2020**, *27*, 26643–26649. [CrossRef]
71. Sudheshna, A.A.; Srivastava, M.; Prakash, C. Characterization of microfibers emission from textile washing from a domestic environment. *Sci. Total Environ.* **2022**, *852*, 158511. [CrossRef]

72. Berruezo, M.; Bonet-Aracil, M.; Montava, I.; Bou-Belda, E.; Díaz-García, P.; Gisbert-Payá, J. Preliminary study of weave pattern influence on microplastics from fabric laundering. *Text. Res. J.* **2021**, *91*, 1037–1045. [CrossRef]
73. Ozkan, I.; Gundogdu, S. Investigation on the microfiber release under controlled washings from the knitted fabrics produced by recycled and virgin polyester yarns. *J. Text. Inst.* **2021**, *112*, 264–272. [CrossRef]
74. Prajapati, S.; Beal, M.; Maley, J.; Brinkmann, M. Qualitative and quantitative analysis of microplastics and microfiber contamination in effluents of the City of Saskatoon wastewater treatment plant. *Environ. Sci. Pollut. Res.* **2021**, *28*, 32545–32553. [CrossRef] [PubMed]
75. Ziajahromi, S.; Neale, P.A.; Rintoul, L.; Leusch, F.D. Wastewater treatment plants as a pathway for microplastics: Development of a new approach to sample wastewater-based microplastics. *Water Res.* **2017**, *112*, 93–99. [CrossRef]
76. Hongprasith, N.; Kittimethawong, C.; Lertluksanaporn, R.; Eamchotchawalit, T.; Kittipongvises, S.; Lohwacharin, J. IR microspectroscopic identification of microplastics in municipal wastewater treatment plants. *Environ. Sci. Pollut. Res.* **2020**, *27*, 18557–18564. [CrossRef]
77. Fontmorin, J.M.; Castillo, R.C.B.; Tang, W.Z.; Sillanpää, M. Stability of 5,5-dimethyl-1-pyrroline-N-oxide as a spin-trap for quantification of hydroxyl radicals in processes based on Fenton reaction. *Water Res.* **2016**, *99*, 24–32. [CrossRef]
78. Xu, L.; Yang, L.; Bai, X.; Du, X.; Wang, Y.; Jin, P. Persulfate activation towards organic decomposition and Cr(VI) reduction achieved by a novel CQDs-TiO_{2-x}/rGO nanocomposite. *Chem. Eng. J.* **2019**, *373*, 238–250. [CrossRef]
79. Wang, H.J.; Yuan, S.; Zhan, J.H.; Wang, Y.; Yu, G.; Deng, S.; Huang, J.; Wang, B. Mechanisms of enhanced total organic carbon elimination from oxalic acid solutions by electro-peroxone process. *Water Res.* **2015**, *80*, 20–29. [CrossRef]
80. Li, M.; Xia, J.; Tian, R.; Wang, J.; Fan, J.; Du, J.; Long, S.; Song, X.; Foley, J.W.; Peng, X. Near-Infrared Light-Initiated Molecular Superoxide Radical Generator: Rejuvenating Photodynamic Therapy against Hypoxic Tumors. *J. Am. Chem. Soc.* **2018**, *140*, 14851–14859. [CrossRef]
81. Malik, R.; Tomer, V.K.; Joshi, N.; Dankwort, T.; Lin, L.; Kienle, L. Au-TiO_2-Loaded Cubic gC_3N_4 Nanohybrids for Photocatalytic and Volatile Organic Amine Sensing Applications. *ACS Appl. Mater. Interfaces* **2018**, *10*, 34087–34097. [CrossRef]
82. Zhou, R.; Liu, F.Y.; Du, X.Y.; Zhang, C.; Yang, C.; Offiong, N.A.; Bi, Y.; Zeng, W.; Ren, H. Removal of metronidazole from wastewater by electrocoagulation with chloride ions electrolyte: The role of reactive chlorine species and process optimization. *Sep. Purif. Technol.* **2022**, *290*, 120799. [CrossRef]
83. Karatas, O.; Gengec, N.A.; Gengec, E.; Khataee, A.; Kobya, M. High-performance carbon black electrode for oxygen reduction reaction and oxidation of atrazine by electro-Fenton process. *Chemosphere* **2022**, *287*, 132370. [CrossRef]
84. Qiu, R.L.; Zhang, D.D.; Diao, Z.H.; Huang, X.; He, C.; Morel, J.L.; Xiong, Y. Visible light induced photocatalytic reduction of Cr(VI) over polymer-sensitized TiO_2 and its synergism with phenol oxidation. *Water Res.* **2012**, *46*, 2299–2306. [CrossRef] [PubMed]
85. Gomes, A.J.G.; Atambo, D.O.; Das, K.K.; Cocke, D.L.; Das, K.P. Electrochemical remediation of chicken processing plant wastewater. *J. Environ. Chem. Eng.* **2018**, *6*, 6028–6036. [CrossRef]
86. Asfaha, Y.G.; Zewge, F.; Yohannes, T.; Kebede, S. Application of hybrid electrocoagulation and electrooxidation process for treatment of wastewater from the cotton textile industry. *Chemosphere* **2022**, *302*. [CrossRef]
87. Kumari, S.; Kumar, R.N. River water treatment using electrocoagulation for removal of acetaminophen and natural organic matter. *Chemosphere* **2021**, *273*, 128571. [CrossRef]
88. Jin, X.; Liu, Y.; Wang, Y.; Zhang, S.; Zhang, W.; Jin, P.; Xu, L.; Shi, X.; Wang, X.C.; Lv, S. Towards a comparison between the hybrid ozonation-coagulation (HOC) process using Al- and Fe-based coagulants: Performance and mechanism. *Chemosphere* **2020**, *253*, 126625. [CrossRef]

water

MDPI

Article

Effect of Medium Pressure Ultraviolet/Chlorine Advanced Oxidation on the Production of Disinfection by-Products from Seven Model Benzene Precursors

Wanting Li [1], Shihu Shu [1], Yanping Zhu [1,*], Linjing Wu [1], Qiongfang Wang [2] and Naiyun Gao [3]

1 College of Environmental Science and Engineering, Donghua University, Shanghai 201620, China
2 College of Chemistry and Chemical Engineering, Shanghai University of Engineering Science, Shanghai 201600, China
3 State Key Laboratory of Pollution Control Reuse, Tongji University, Shanghai 200092, China
* Correspondence: yanpingzhu@dhu.edu.cn

Highlights:

- MPUV/chlorine increased the chlorine demand and DBPFP of benzoic acid and nitrobenzene.
- MPUV/chlorine showed no further activation on active DBP precursors.
- MPUV/chlorine at pH 6 showed greater impact on precursors' DBPFP than at pH 8.
- MPUV/chlorine increased the BSF of THMs in the presence of bromine.
- Linear correlation was observed between changes in precursors' chlorine demand and DBPFP.

Abstract: UV/chlorine advanced oxidation process (AOP), as a potential alternative to UV/H_2O_2 in water treatment, may pose a potential risk of increased disinfection by-product (DBP) formation and is of great concern. In this paper, seven benzene derivatives, containing two chlorine-inert and five chlorine-active compounds, were selected as typical model DBP precursors, and the effects of medium pressure UV/chlorine (MPUV/chlorine) on their chlorine demand and DBP formation potential (DBPFP) were evaluated. The results showed that MPUV/chlorine could significantly increase the chlorine demand and DBPFP of the two inert precursors. For the four slow but active DBP precursors, MPUV/chlorine may accelerate their short-term DBP formation, whereas it showed an insignificant effect or even reduced their chlorine demand and DBPFP. For the only fast and active DBP precursor, MPUV/chlorine showed an insignificant effect on its short-term DBP formation or DBPFP. The overall effect of MPUV/chlorine was more significant at pH 6 than at pH 8. In the presence of Br^-, MPUV/chlorine significantly increased the bromine substitution factors of THMs. In addition, linear fitting results indicated that the UV/chlorine-induced change in overall chlorine demand may be considered as a potential indicator for the prediction of DBPFP alteration.

Keywords: medium pressure UV/chlorine; advanced oxidation process; disinfection by-products; benzene derivatives; bromide (Br^-)

Citation: Li, W.; Shu, S.; Zhu, Y.; Wu, L.; Wang, Q.; Gao, N. Effect of Medium Pressure Ultraviolet/Chlorine Advanced Oxidation on the Production of Disinfection by-Products from Seven Model Benzene Precursors. *Water* **2022**, *14*, 3775. https://doi.org/10.3390/w14223775

Academic Editors: Dionysios (Dion) Demetriou Dionysiou, Yujue Wang and Huijiao Wang

Received: 25 September 2022
Accepted: 18 November 2022
Published: 20 November 2022

1. Introduction

Ultraviolet (UV) irradiation and chlorination are commonly used disinfection strategies for the inactivation of water-borne pathogens. In recent years, with the rapid urbanization and industrialization of China, the problem of micro-pollution in source water has become prominent, and the removal of refractory micro-pollutants has posed a major challenge for conventional drinking water treatment processes. UV-based advanced oxidation process (AOP) can generate radicals via UV irradiation of oxidants and effectively remove refractory pollutants from water. Among the state-of-the-art UV-AOPs (e.g., H_2O_2, O_3, persulfate, chlorine, and peracetic acid) [1–6], UV/chlorine AOP can generate a diverse spectrum of reactive radicals including nonselective hydroxyl radicals ($HO^•$) and selective reactive chlorine species (RCS), such as $Cl^{\bullet}$, $ClO^{\bullet}$, and [7] $Cl_2^{\bullet}-$ [7,8], contributing to its comparable or even better performance in eliminating refractory micro-pollutants (e.g., pharmaceuticals and personal care products, endocrine disrupting chemicals, and algal toxins) compared to UV/H_2O_2 AOP [9–11]. Consequently, UV/chlorine AOP is emerging as a potential alternative to the conventional UV/H_2O_2 AOP for tertiary water treatment.

However, the relatively high chlorine dose used in UV/chlorine AOP compared to traditional disinfection (e.g., 2–10 mg/L vs. 0.2–2 mg/L) and the involvement of RCS oxidation may pose the risk of enhanced DBP formation during the process [12–14]. In addition, UV/chlorine oxidation may change the structure of organics in water while removing refractory pollutants, and thus affect the chlorine demand and disinfection by-product (DBP) formation of treated water during subsequent chlorine disinfection. Previous studies have found that the short-term (2–60 s) UV/chlorine treatment often does not directly generate a large amount of DBPs [15,16]. However, UV/chlorine oxidation may significantly increase the DBP formation potential (DBPFP) of effluent after the subsequent chlorine disinfection [17,18]. At present, relevant researches were mostly targeted to specific source water or natural organic matter (NOM) from specific sources [13,19–21]. Considering that the structure of DBP precursors in water is complex and varies greatly with different water sources, the research results have certain limitations. Therefore, this study attempts to disassemble a complex structure of aromatic precursors in the water body and select seven benzene derivatives with different substituents as the model precursors, in order to explore the relationship between the chemical structure of the precursors and the UV/chlorine-induced changing patterns of DBPs. Furthermore, in this study, medium pressure mercury lamp (MPUV) was used as UV light source, which has high power to obtain the UV dose for AOP within seconds and is widely applied in practical water plants, in order to better simulate the actual reaction conditions of UV/chlorine [13,22].

In summary, this study aimed to select seven benzene derivatives with different substituents as model DBP precursors, and evaluate the effect of MPUV/chlorine oxidation on their chlorine demand and DBP formation. Considering the pH-dependence of radical production in MPUV/chlorine system [23], both MPUV/chlorine at pH 6 and 8 were evaluated [24]. In addition, given that common co-existing bromide (Br^-) may promote the formation of highly toxic brominated disinfection by-products (Br^-DBPs), the effect of MPUV/chlorine on Br^-DBP formation in the presence of bromide was also investigated. The results of this study would provide insights into the relationship between MPUV/chlorine-induced DBP formation and precursor structure, and be helpful in predicting the DBP variation trend caused by MPUV/chlorine oxidation.

2. Materials and Methods

2.1. Chemicals and Materials

A total of seven benzene derivatives including phenol (PN), resorcinol (RSC), benzoic acid (BA), nitrobenzene (NB), o-chlorophenol (2-MCP), p-chlorophenol (4-MCP), and 2,4,6-trichlorophenol (2,4,6-TCP) were selected as model precursors, structures, and characteristics, which were shown in Table 1. Model precursors, sodium hypochlorite (NaOCl, 4.0–4.99%), and ascorbic acid were of analytical grade and obtained from Shanghai Titan Scientific Co., Ltd. (Shanghai, China). Analytical grade potassium hydrogen phosphate (K_2HPO_4), potassium dihydrogen phosphate (KH_2PO_4), sodium bromide, and anhydrous sodium sulfate (Na_2SO_4) were obtained from Sinopharm Chemical Reagent Co., Ltd. (Shanghai, China). All reaction solutions were prepared with Milli-Q® ultrapure water. DBP standards, including THMs, HAAs, HANs, HALs, and TCNM were purchased from CanSyn Chemical Corp. (Toronto, ON, Canada). Internal standards (1,2-dibromopropane and 2,3,4,5-tetrafluorobenzoic acid) were purchased from Sigma-Aldrich (Saint Louis, MO, USA). HPLC grade methyl-tert-butyl ether (MTBE) and methanol were provided by Thermo Fisher (Waltham, MA, USA).

Table 1. List of seven benzene derivatives as model precursors.

Name (Symbol)	Molecular Formula	Structure	Substituted Groups
Benzoic acid (BA)	$C_7H_6O_2$	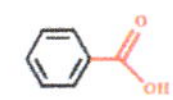	-COOH
Nitrobenzene (NB)	$C_6H_5NO_2$	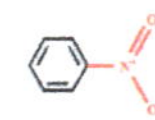	$-NO_2$
Phenol (PN)	C_6H_5OH	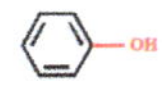	-OH
Resorcinol (RSC)	$C_6H_6O_2$	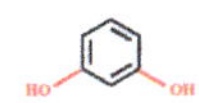	-OH, -OH
o-chlorophenol (2-MCP)	C_6H_5ClO	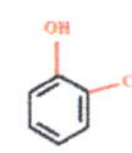	-OH, -Cl
p-chlorophenol (4-MCP)	C_6H_5ClO	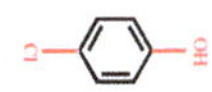	-OH, -Cl
2,4,6-trichlorophenol (2,4,6-TCP)	$C_6H_3Cl_3O$	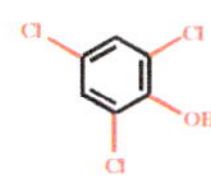	-OH, -Cl, -Cl, -Cl

2.2. Experimental Procedures

2.2.1. Short-Term Oxidation in the UV Reactor

MPUV/chlorine illumination experiment was carried out by medium pressure UV quasi-parallel beam, as shown in Figure 1. Spectra of the MP mercury lamp (1 kW, Heraeus, Hanau, Germany) were shown in Figure S1. The average irradiance (10 mW/cm^2) was obtained using the calibrated Ocean Optics Spectro radiometer (USB4000), and the UV dose (600 mJ/cm^2, a typical dose used practically) [13,25,26] was obtained by multiplying the average irradiation intensity with the irradiation time. A 100 mL working solution containing a certain precursor (3.0 mg C/L) was dosed with 10 mg/L free chlorine and simultaneously irradiated at 22 ± 1 °C with stirring for 60 s. The reaction solution was adjusted to pH 6.0 or 8.0 using 10 mM phosphate buffer. Additionally, 1 mL of the solution was taken for the analysis of total chlorine residual after 60 s reaction. The effect of Br^- on their DBP formation with regard to UV/chlorine treatment was evaluated by spiking 1 mg/L Br^- to 100 mL precursor solution as discussed. Dark chlorination tests without UV irradiation were carried out as control under parallel conditions.

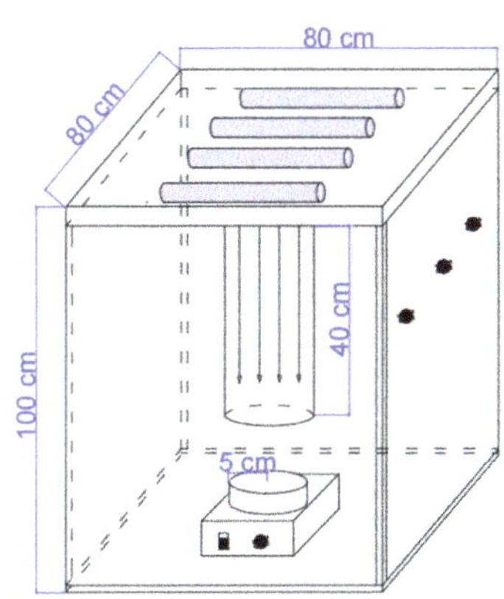

Figure 1. Schematic diagram of the ultraviolet collimated beam apparatus.

2.2.2. DBP Formation Potential Tests

After the pre-chlorination test, the samples were transferred to headspace-free amber glass bottles and subject to the recommended uniform formation condition (UFC) tests [27]. The working conditions were as follows: The pH value was adjusted to 8.0, and the initial residual chlorine value was adjusted by adding chlorine or quenching agent (ascorbic acid) to ensure that the residual chlorine was 1 ± 0.4 mg /L after standing at 22 ± 1 °C for 24 h under dark conditions. The amount of chlorine or quench agent was determined by preliminary tests. Chlorine demand (total active chlorine) and DBP yields (i.e., DBPFP) were analyzed after 24 h post-chlorination. All experiments were conducted in duplicate with the averaged values being reported.

2.3. *Analytical Methods*

Concentrations of total chlorine in working solutions were determined using N,N diethyl-p-phenylenediamine (DPD) colorimetric method with a HACH DR/2500 spectrophotometer [28]. Additionally, 1,2-dibromopropane was used as the internal standard of THMs, HALs, HANs, and TCNM. Moreover, 2,3,4,5-tetrafluorobenzoic acid was used as the internal standard of HAA samples. After the addition of internal standards, samples of THMs, HALs, HANs, and TCNM were pretreated by liquid-liquid extraction according to USEPA 551.1 [29]. HAA samples were measured by GC-ECD after liquid-liquid extraction and derivatization according to USEPA 552.1 [30]. All DBPs were measured using a gas chromatograph (GC-2010, Shimadzu, Kyoto, Japan), which was equipped with HP-5 capillary column (30 m × 0.25 mm, 0.25 mm film thickness, J&W, USA) and electron capture detector. For HAAs, the temperature program was initially set at 40 °C for 5 min, ramped to 65 °C at 2.5 °C/min, then to 85 °C at 10 °C/min and held for 3 min, and finally to 230 °C at 20 °C/min and held for 3 min. For non-HAA DBPs, the temperature program was set at 30 °C for 10 min, ramped to 80 °C at 7 °C/min and held for 3 min, then to 110 °C at 5 °C /min and held for 3 min, and finally to 200 °C at 20 °C /min and held for 2.5 min. The limits of quantification (LOQ) for THMs, HAAs, HALs, HANs, and TCNM were less than 0.5, 1.5, 1.6, 2.4, and 0.05 μg/L. The effect of Br^- on formation of brominated THMs and HAAs were analyzed by calculating the bromine substitution factors (BSFs) of trihalomethane and haloacetic acid (Equations (1) and (2)):

$$BSF_{THMs} = \frac{\sum_{n=1}^{3} n \times CHCl_{(3-n)}Br_n}{3 \times \sum_{n=0}^{n=3} CHCl_{(3-n)}Br_n} \quad (1)$$

$$BSF_{HAAs} = \frac{\sum_{n=1}^{3}\sum_{m=1}^{4-n} n \times C_2H_mO_2Cl_{(4-m-n)}Br_n}{\sum_{n=0}^{3}\sum_{m=1}^{3-n} (4-m) \times C_2H_mO_2Cl_{(4-m-n)}Br_n} \quad (2)$$

3. Results and Discussion

3.1. *Chlorine Demand*

The chlorine consumption of the seven model compounds in 60 s dark chlorination and MPUV/chlorine oxidation is shown in Table 2. During dark chlorination, RSC reacted with chlorine at an extremely fast rate among those precursors, with almost exhaustion of 10 mg/L chlorine within 60 s. However, the other six benzene derivatives reacted with chlorine relatively slowly, with the chlorine consumption ranging between 0.10 and 0.27 mg Cl/mg C. As UV was turned on, their chlorine consumption increased to different extents (except for RSC), which followed the order of NB < BA < 2,4,6-TCP < PN < 2-MCP < 4-MCP. The increase in chlorine consumption may be attributed to the photolysis of chlorine, and probably partly due to the higher reactivity of the intermediate products toward chlorine than the parent precursor in the MPUV/chlorine-treated system. The greater increase in chlorine consumption at pH 8 than at pH 6 may be related to the different dominant existing forms at different pHs (HOCl and OCl^- at pH 6 and 8, respectively) and higher photolysis rate of OCl^- than HOCl [31].

Table 2. Chlorine decay of the seven model compounds during the 60 s pre-oxidation under the following conditions: Chlorine dose = 10 mg Cl_2/L, pH = 6.0 or 8.0, [model benzene precursors]$_0$ = 3 mg-C/L, UV fluence rate = 10 mW/cm^2, [Br^-] = 1 mg/L.

Model Precursors	pH 6 (mg Cl/mg C)				pH 8 (mg Cl/mg C)			
	Ambient		Bromide-spiked		Ambient		Bromide-spiked	
	Cl_2 alone	UV/Chlorine	Cl_2 alone	UV/Chlorine	Cl_2 alone	UV/Chlorine	Cl_2 alone	UV/Chlorine
BA	0.05 ± 0.0	0.37 ± 0.0	0.01 ± 0.0	0.38 ± 0.0	0.02 ± 0.0	0.96 ± 0.1	0.01 ± 0.0	0.98 ± 0.1
NB	0.01 ± 0.0	0.33 ± 0.0	0.01 ± 0.0	0.30 ± 0.0	0.01 ± 0.0	0.63 ± 0.0	0.01 ± 0.0	0.66 ± 0.0
PN	0.13 ± 0.0	0.93 ± 0.1	0.12 ± 0.0	0.98 ± 0.1	0.2 ± 0.0	1.34 ± 0.1	0.24 ± 0.0	1.47 ± 0.2
RSC	3.34 ± 0.3	3.38 ± 0.3	3.30 ± 0.3	3.37 ± 0.3	3.30 ± 0.2	3.37 ± 0.3	3.30 ± 0.3	3.37 ± 0.4
2-MCP	0.23 ± 0.0	1.07 ± 0.0	0.53 ± 0.0	1.37 ± 0.0	0.27 ± 0.0	1.53 ± 0.1	0.55 ± 0.0	1.65 ± 0.1
4-MCP	0.10 ± 0.0	1.33 ± 0.1	0.08 ± 0.0	1.57 ± 0.1	0.13 ± 0.0	1.73 ± 0.1	0.17 ± 0.0	1.83 ± 0.1
2,4,6-TCP	0.20 ± 0.0	0.70 ± 0.0	0.23 ± 0.0	1.37 ± 0.1	0.17 ± 0.0	1.15 ± 0.1	0.21 ± 0.0	1.37 ± 0.1

The overall chlorine demands of the seven model compounds after post-chlorination were shown in Table 3. Compared with dark chlorination, MPUV/chlorine oxidation showed an activation effect on the two inert precursors (i.e., BA and NB) especially at pH 6, which increased the chlorine demands of BA and nitrobenzene from 0.02–0.05 mg Cl/mg C and 0.01 mg Cl/mg C to 1.50–2.07 mg Cl/mg C and 0.97–1.23 mg Cl/mg C, respectively. The greater activation effect at pH 6 indicated the probably higher contribution of $HO^\bullet$/$Cl^\bullet$ than $ClO^\bullet$ considering that the concentration of $HO^\bullet$/$Cl^\bullet$ in the system was higher under acidic conditions than under basic conditions [32]. On the contrary, MPUV/chlorine oxidation reduced or showed little effect on the chlorine demands of the five chlorine-active precursors. Additionally, MPUV/chlorine oxidation promoted the chlorination rate of these five precursors (Table 2), but was not likely to further increase their overall chlorine demands. Moreover, it can be seen from Tables 2 and 3 that addition of 1 mg/L Br^- may increase the chlorine demands of certain precursors (e.g., PN and 2-MCP), which may be probably due to the consumption of chlorine via the oxidation of Br^- by chlorine [33]. However, Br^- addition showed a minor effect on the overall changing trends of these precursors' short-term chlorine decay or chlorine demand.

Table 3. Chlorine demands of the seven model compounds after the 60 s short-term oxidation under the following conditions: Chlorine dose = 10 mg Cl_2/L, pH = 6.0 or 8.0, [model benzene precursors]$_0$ = 3 mg-C/L, UV fluence rate = 10 mW/cm^2, [Br^-] = 1 mg/L.

Model Precursors	pH 6 (mg Cl/mg C)				pH 8 (mg Cl/mg C)			
	Ambient		Bromide-spiked		Ambient		Bromide-spiked	
	Cl_2 alone	UV/Chlorine	Cl_2 alone	UV/Chlorine	Cl_2 alone	UV/Chlorine	Cl_2 alone	UV/Chlorine
BA	0.05 ± 0.0	2.07 ± 0.1	0.12 ± 0.0	2.07 ± 0.1	0.02 ± 0.0	1.50 ± 0.1	0.10 ± 0.0	1.50 ± 0.1
NB	0.01 ± 0.0	0.97 ± 0.1	0.01 ± 0.0	1.01 ± 0.0	0.01 ± 0.0	1.23 ± 0.1	0.01 ± 0.0	1.29 ± 0.1
PN	6.71 ± 0.3	6.47 ± 0.2	7.42 ± 0.2	6.89 ± 0.2	6.90 ± 0.2	6.66 ± 0.2	7.68 ± 0.1	6.78 ± 0.2
RSC	6.11 ± 0.2	6.32 ± 0.3	6.27 ± 0.2	6.31 ± 0.3	6.30 ± 0.1	6.45 ± 0.0	6.42 ± 0.0	6.54 ± 0.1
2-MCP	6.87 ± 0.3	6.72 ± 0.1	7.40 ± 0.2	7.22 ± 0.1	7.59 ± 0.2	6.81 ± 0.2	7.93 ± 0.2	7.36 ± 0.1
4-MCP	7.19 ± 0.2	6.65 ± 0.2	7.14 ± 0.2	7.2 ± 0.1	7.29 ± 0.3	6.76 ± 0.0	7.63 ± 0.1	7.14 ± 0.2
2,4,6-TCP	6.72 ± 0.2	6.17 ± 0.1	6.71 ± 0.1	6.22 ± 0.2	6.10 ± 0.0	6.30 ± 0.1	6.43 ± 0.1	6.25 ± 0.0

In summary, for benzene precursors with low chlorination rate but high chlorine demand, MPUV/chlorine tends to promote their chlorine consumption rate but reduce their chlorine demand, while for chlorine-inert precursors, MPUV/chlorine showed a significant activation effect, increasing both their chlorination rate and overall chlorine demands especially under acidic conditions.

3.2. *Effect of MPUV/Chlorine Oxidation on DBP Formation from Seven Benzene Precursors*

3.2.1. Effect of MPUV/Chlorine Oxidation on DBP Formation from Inert Benzene Precursors without Bromide

As shown in Figure 2, the 60 s short-term DBP formation and DBPFP of BA and NB were within 5 μg/mg C, indicating that both BA and NB are typical inert DBP precursors. Turning on UV increased their short-term DBP formation to 8.41–8.41 μg/mg C and 5.91–7.62 μg/mg C, respectively, and their DBPFP to 20.98–42.47 μg/mg C and 17.45–33.53 μg/mg C, with more significant activation due to MPUV/chlorine oxidation at pH 6 than at pH 8. The different activation extents of the two precursors may be attributed to their different reactivity with $HO^\bullet$ and $RCS^\bullet$, for example, BA has a high reactivity with both $HO^\bullet$ and $Cl^\bullet$, while NB mainly reacts with $HO^\bullet$ [7,34]. In terms of specific DBP species, after 60 s MPUV/chlorine and post-chlorination, BA mainly formed 18.71–18.72 μg THMs/mg C, 13.73–27.47 μg HAAs/mg C, and 0.53–1.28 μg HALs/mg C, while NB produced 4.00–5.22 μg THMs/mg C, 1.74–4.52 μg HALs/mg C, and 7.41 μg TCNM/mg C. Prior study found that $HO^\bullet$ can convert BA to salicylic acid, which is more chlorine-reactive and was considered as an important HAA and/or HAL precursor [35,36]. Nitrobenzene would be converted to nitrophenol during the reaction with $HO^\bullet$, and nitrophenol further generated DBPs, such as THMs, HALs, and TCNM by chlorination [35]. Among them, the significant promotion of TCNM formation due to MPUV/chlorine oxidation of NB in this study is consistent with the previously reported finding that UV/chlorine treatment may potentially lead to increased TCNM formation of nitrogenous precursors [37], which needs to be of concern.

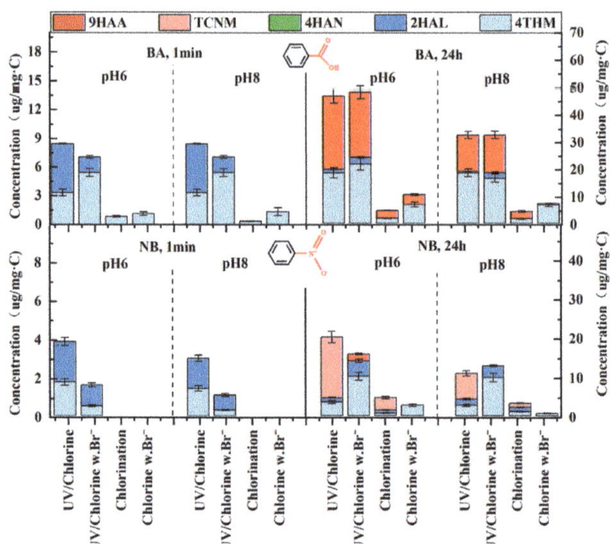

Figure 2. The formation of 4THM, 9HAA, 4HAN, 2HAL, and TCNM during the 1 min short-term oxidation and 24 h post-chlorination from two inert benzene derivatives. Short-term oxidation conditions: Chlorine dose = 10 mg chlorine/L, pH = 6.0 or 8.0, [model benzene precursors]$_0$ = 3 mg-C/L, UV fluence rate = 10 mW/cm^2, [bromide]$_0$= 1 mg/L. Post-chlorination conditions: pH = 8.0, chlorination time = 24 h, chlorine residual = 1.0 ± 0.4 mg/L, dark, 22 °C.

3.2.2. Effects of Bromide on DBP Formation from Inert Benzene Precursors

The coexisting Br^- in water would react rapidly with HOCl to form HOBr (Reaction 1) [38] and potentially further generate Br^-DBPs due to the bromination reaction between HOBr and organic matter [39]. During MPUV/chlorine oxidation, HOBr photolysis would generate reactive bromine radicals (RBS) and HO$^{\bullet}$ (Reaction 2) [40], and on the other side would consume HO$^{\bullet}$ via Reaction (3) [33]. Nevertheless, the coexistence of bromide may affect the composition of radical species during the MPUV/chlorine treatment. The effect of bromide on the DBP formation of BA and NB in the MPUV/chlorine-treated system was shown in Figure 3. For BA, the coexisting bromide slightly promoted the production of DBAA and BCAA during dark chlorination and post-chlorination. With UV on, the coexisting bromide significantly increased the production of Br^-THMs (DBCM and TBM) and Br^-HAAs (DBAA and BCAA) by 10.34–14.42 μg/mg C and 4.30–5.33 μg/mg C, respectively, although this caused a minor change in the total DBPFP. Similarly, the coexisting bromide caused rare Br^-THM formation of NB in the dark, while, with UV on, this promoted the formation of 3.40–3.58 μg DBCM/mg C and 6.83–8.71 μg TBM/mg C, respectively. This promotion effect was probably attributed to the reaction between RBS/HOBr and precursors/intermediates [33].

$$HClO + Br^- \rightarrow HOBr + Cl^- \tag{R1}$$

$$HOBr \overset{uv}{\rightarrow} HO^{\bullet} + Br^{\bullet} \tag{R2}$$

$$HOBr + HO^{\bullet} \rightarrow BrO^{\bullet} + H_2O \tag{R3}$$

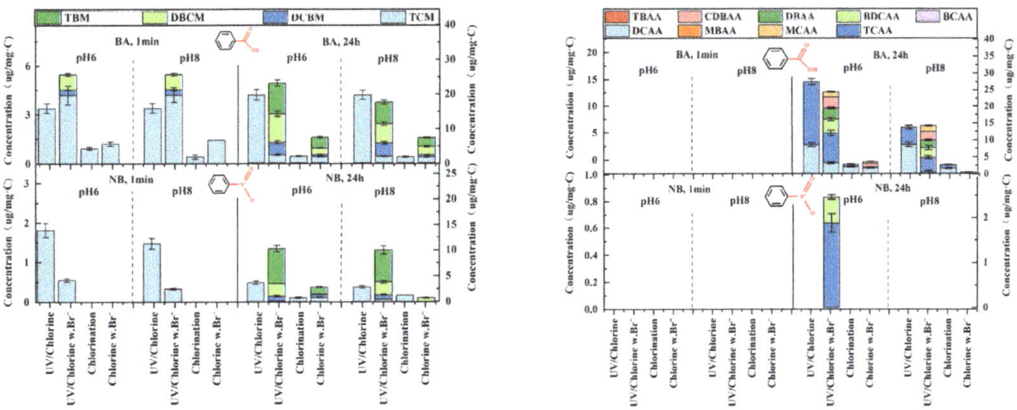

(**a**) (**b**)

Figure 3. The formation of THMs (**a**) and HAAs (**b**) during the 1 min short-term oxidation and after 24 h post-chlorination from two inert benzene derivatives. Short-term oxidation conditions: Chlorine dose = 10 mg chlorine /L, pH = 6.0 or 8.0, [model benzene precursors]$_0$ = 3 mg-C/L, UV fluence rate = 10 mW/cm^2, [bromide]$_0$= 1 mg/L. Post-chlorination conditions: pH = 8.0, chlorination time = 24 h, chlorine residual = 1.0 ± 0.4 mg/L, dark, 22 °C.

The BSFs of THMFP and HAAFP under different treatment conditions were shown in Table 4. Turning on UV increased BSF_{THMs} of BA and NB from 1.79–1.85 and 1.76–2.00 to 1.93–2.03 and 2.45–2.55, respectively, and BSF_{HAAs} from 0.73–1.00 and 0.00 to 0.81–1.04 and 0.00–0.48, respectively, compared with dark chlorination. This trend was more significant under pH 6 than pH 8. Overall, the effect of coexisting bromide on BSF_{THMs} was greater relative to BSF_{HAAs}, which is consistent with the reported bromide effect on actual water and model amino acid precursors in prior UV/chlorine studies [17,41].

Table 4. Bromine incorporation factors of THMs and HAAs from the seven benzenes under the following conditions: Chlorine dose = 10 mg Cl_2/L, pH = 6.0 or 8.0, [model benzene precursors]$_0$ = 3 mg-C/L, UV fluence rate = 10 mW/cm^2, [Br^-] = 1 mg/L.

Benzene Derivatives	Cl/Br(mg/mg-C)				UV/Cl/Br(mg/mg-C)			
	PH6		PH8		PH6		PH8	
	THM	HAA	THM	HAA	THM	HAA	THM	HAA
BA	1.85	0.73	1.78	1.00	2.03	0.81	1.93	1.04
NB	1.76	0.00	2.00	0.00	2.55	0.48	2.45	0.00
PN	1.13	0.74	1.19	0.77	1.33	0.75	1.26	0.78
RSC	0.28	1.08	0.33	1.04	0.56	1.10	0.47	1.06
2-MCP	1.25	0.38	1.33	0.34	1.47	0.44	1.51	0.47
4-MCP	1.30	0.26	1.20	0.27	1.44	0.28	1.33	0.28
2,4,6-TCP	1.39	0.29	1.21	0.28	1.52	0.35	1.23	0.35

3.3. *Effect of MPUV/Chlorine Oxidation on DBP Formation from Reactive Benzene Precursors*

3.3.1. Effect of MPUV/Chlorine Oxidation on DBP Formation from Reactive Benzene Precursors without Bromide

The DBP formation during chlorination of the contained phenolic compounds first reacted with benzene ring to form chlorine substitution products, and then disinfection by-products, such as trihalomethanes and haloacetic acids through ring opening and hydrolysis [42]. The DBP formation and DBPFP of five highly chlorine-reactive benzene derivatives were shown in Figure 4. As a significant DBP precursor with high chlorine-reactivity, RSC formed 29.81–73.14 μg THMs/mg C and 1.68–1.96 μg HAAs/mg C during 60 s dark chlorination, and 356.82–422.99 μg THMFP/mg C, 29.71–47.41 μg HAAFP/mg C, and 3.05–4.12 μg HALFP/mg C after post-chlorination. The values were comparable to the reported DBPFP of RSC [43]. Turning on UV showed no significant effect on its 60 s short-term DBP formation or its DBPFP. This is consistent with UV/chlorine-induced changing trends of its chlorine demand (Table 4), indicating that UV/chlorine would not likely accelerate or increase DBP formation of precursors with extremely high chlorine reactivity and fast chlorination rate. Similar DBP changing trends were also observed in our previous work on active amino acid-type precursors [41]. For DBP species, HAAFP increased by 62.84–91.77 μg/mg C probably at the expense of decreased THMFP (by 89.56–106.43 μg/mg C).

The four slow DBP precursors showed very limited DBP formation (0–44.88 μg/mg C) during the 60 s dark chlorination but high DBPFP (474.95–1071.96 μg/mg C) after post-chlorination. Their DBPFP increased with the number of chlorine substitutions on the benzene ring; 2,4,6-TCP had the highest DBPFP, followed by 2-MCP and 4-MCP, and PN the lowest. Turning on UV accelerated their DBP formation within 60 s, while it showed a minor effect or even slight deactivation on their DBPFP. For example, the 60 s DBP formation of 2,4,6-TCP increased from 44.88 to 37.07 μg/mg C, while its DBPFP slightly decreased from 1071.96 to 862.11 μg/mg C, with a more significant effect at pH 6.0 than at pH 8.0. For the variation of individual DBP species, short-term THM and HAA formation in the reactor increased from 0 to 12.01–15.82 μg/mg C and 24.28–28.33 μg/mg C, respectively, due to UV/chlorine oxidation, whereas the HAAFP and HALFP increased from 377.72 and 2.03 μg/mg C to 899.08 and 37.07 μg/mg C, respectively, along with THMFP decreasing from 270.01 to 41.17 μg/mg C. The enhanced HAA and/or HAL formation has been previously observed in HO$^\bullet$-based AOP systems [44,45]. Generally, the effect of UV/chlorine under acidic pH conditions was stronger than under basic conditions, which indirectly reflected the greater contribution of HO$^\bullet$ and/or Cl$^\bullet$ than ClO$^\bullet$ to the DBP changes. Previous studies on the mechanism of radical action also found that the reaction mechanism of HO$^\bullet$ and Cl$^\bullet$ which were dominant radicals under acidic conditions is mainly an addition and electron transfer, while the ClO$^\bullet$ with relatively high concentrations under alkaline conditions is mainly an electron transfer, and thus may lead to a relatively smaller effect on DBP production under alkaline conditions [14,46–48].

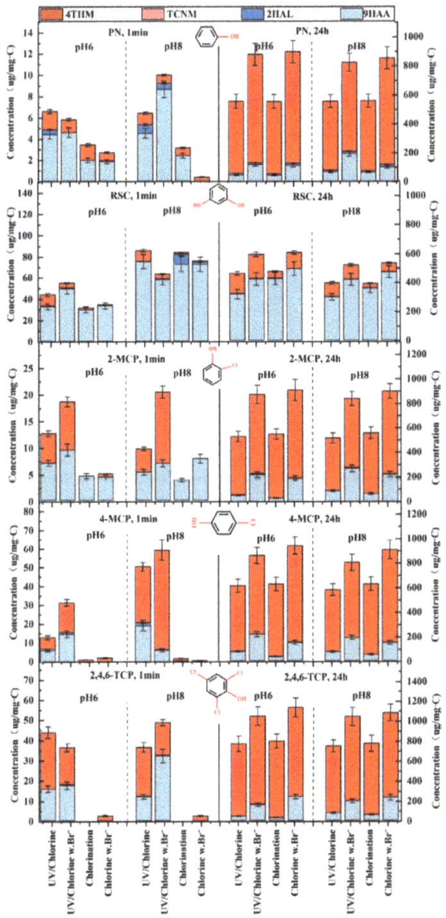

Figure 4. The formation of 4THM, 9HAA, 2HAL, and TCNM during the 1 min short-term oxidation and 24 h post-chlorination from five active benzene derivatives. Short-term oxidation conditions: Chlorine dose = 10 mg chlorine/L, pH = 6.0 or 8.0, [model benzene precursors]$_0$ = 3 mg-C/L, UV fluence rate = 10 mW/cm^2, [bromide]$_0$= 1 mg/L. Post-chlorination conditions: pH = 8.0, chlorination time = 24 h, chlorine residual = 1.0 ± 0.4 mg/L, dark, 22 °C.

Above all, MPUV/chlorine-induced DBP changes in the five active precursors were generally consistent with the corresponding changes in their chlorine consumption. The results indicated that MPUV/chlorine oxidation would not likely contribute to higher DBPFP or chlorine demand for active precursors, although it may accelerate the DBP formation rate of these slow DBP precursors.

3.3.2. Effect of Bromide on DBP Formation from Reactive Benzene Precursors

The effect of bromide on the THM and HAA formation of the five active benzenes was shown in Figures 5 and 6. Generally, the presence of bromide increased the overall DBPFP of the five precursors due to the formation of bromide-contained DBPs during dark chlorination. Turning on UV showed an insignificant impact on the overall DBP changing trend in bromide-spiked systems although this caused varied DBP changing extents relative to non-bromide systems. Specifically, for PN and RSC, in the presence of bromide, minor brominated THMs/HAAs were formed within 60 s with UV on or off, while for the three Cl-substituted phenol precursors, bromide coexistence promoted brominated THM and HAA formation especially with UV on, indicating the possible contribution of RBS to the halogenation of Cl-substituted phenol precursors. Nevertheless, UV/chlorine may accelerate the brominated DBP formation for certain chlorine-substituted benzene precursors. However, after post-chlorination, the presence of bromide consistently increased their THMFP and HAAFP under dark conditions, and turning on UV did not further promote brominated DBP formation compared with the dark control. For example, for 2,4,6-TCP, turning on UV increased the production of THMs and HAAs in 60 s in the presence of bromide from 0.00 μg/mg C and 2.30–2.30 μg/mg C to 17.55–32.35 μg/mg C and 16.26–18.53 μg/mg C, but slightly decreased its THMFP and HAAFP by 175.68–212.62 μg/mg C and 125.44–134.39 μg/mg C, respectively.

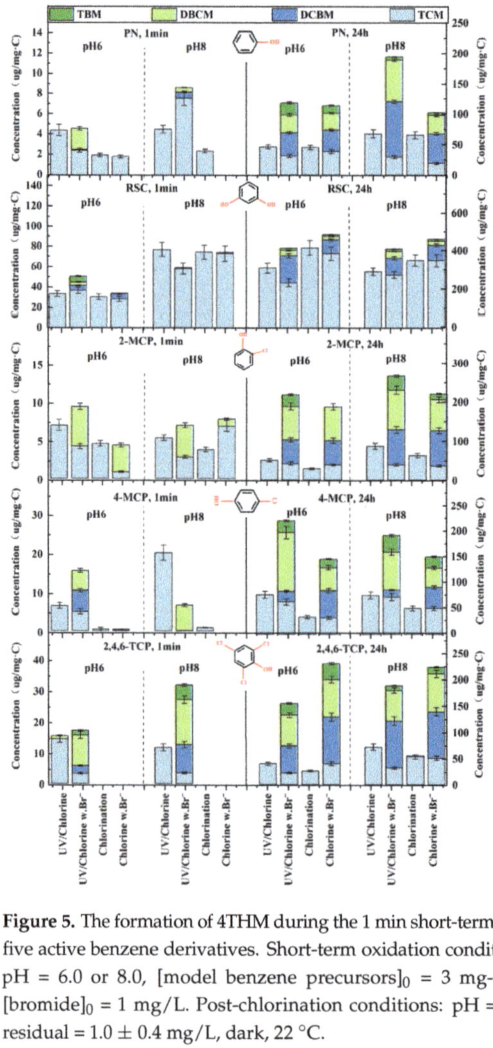

Figure 5. The formation of 4THM during the 1 min short-term oxidation and 24 h post-chlorination from five active benzene derivatives. Short-term oxidation conditions: Chlorine dose = 10 mg chlorine/L, pH = 6.0 or 8.0, [model benzene precursors]$_0$ = 3 mg-C/L, UV fluence rate = 10 mW/cm^2, [bromide]$_0$ = 1 mg/L. Post-chlorination conditions: pH = 8.0, chlorination time = 24 h, chlorine residual = 1.0 ± 0.4 mg/L, dark, 22 °C.

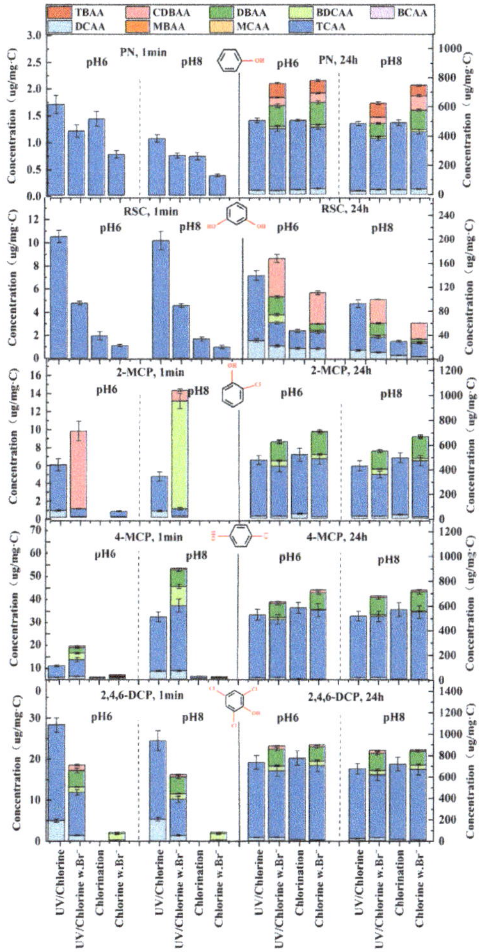

Figure 6. The formation of 9HAA during the 1 min short-term oxidation and 24 h post-chlorination from five active benzene derivatives. Short-term oxidation conditions: Chlorine dose = 10 mg chlorine/L, pH = 6.0 or 8.0, [model benzene precursors]$_0$ = 3 mg-C/L, UV fluence rate = 10 mW/cm^2, [bromide]$_0$ = 1 mg/L. Post-chlorination conditions: pH = 8.0, chlorination time = 24 h, chlorine residual = 1.0 ± 0.4 mg/L, dark, 22 °C.

The BSF_{THMs} and BSF_{HAAs} of the five precursors under different treatment conditions were shown in Table 4. UV/chlorine oxidation increased the BSF_{THMs} of PN, RSC, 2-MCP, 4-MCP, and 2,4,6-TCP from 1.13–1.19, 0.28–0.33, 1.25–1.33, 1.20–1.30, and 1.21–1.39, to 1.26–1.33, 0.47- 0.56, 1.47–1.51, 1.33–1.44, and 1.23–1.52, respectively. In contrast, BSF_{HAAs} remained almost unchanged. This was basically consistent with the BSF changing trends of the two inert benzene precursors in Section 3.2.2, further reflecting the effect of MPUV/chlorine on brominated THM formation in bromide-contained systems.

3.4. Relationship between MPUV/Chlorine-Induced Chlorine Demand Changes and Corresponding DBPFP Changing Trends of Model Precursors

To help in the prediction of MPUV/chlorine-induced changing trends in treated water's DBPFP (Δ(DBPFP)), the linear correlation between the change in chlorine demand (Δ(chlorine demand)) of the seven model precursors and their DBPFP changes (Δ(DBPFP)) was fitted and shown in Figure 7. A positive correlation was observed in the seven precursors' Δ(chlorine demand) and their Δ(DBPFP) under each treatment condition, with R^2 values distributed between 0.70 and 0.86. Therefore, considering that the measurement of chlorine demand is more convenient and cost-effective than DBPFP, the change in chlorine demand may be considered as a potential indicator to predict the effect of MPUV/chlorine AOP treatment on treated water's DBPFP and help in the determination of whether the process may pose a potential threat to DBPFP.

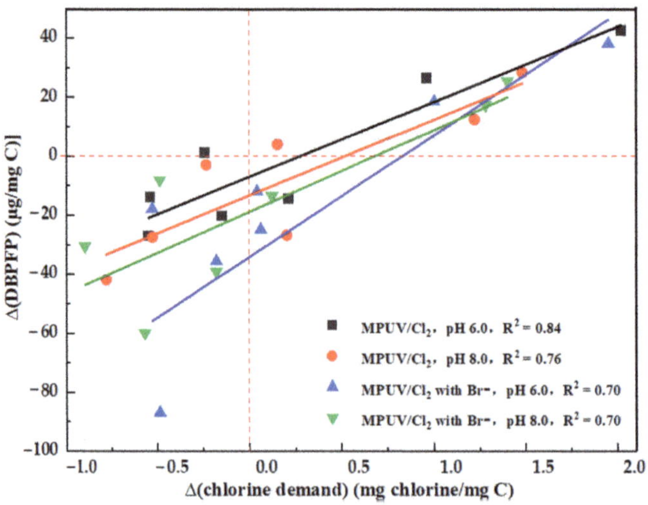

Figure 7. Linear fitting between Δ(chlorine demand) and Δ(DBPFP) caused by MPUV/chlorine AOP oxidation.

4. Conclusions

In this paper, the relationship between the DBP changing pattern and the structural characteristics of precursors was investigated by evaluating MPUV/chlorine oxidation-induced alteration of DBP formation and DBPFP from seven model benzene precursors containing different substituents. According to the results in this study, for the two inert precursors with low chlorine demand, MPUV/chlorine activated them and increased their chlorine demand and DBPFP, with the activation more significant under acidic conditions. In contrast, for active DBP precursors, MPUV/chlorine showed no significant effect or even exhibited slight deactivation on their DBPFP. For the benzene precursors with low chlorination rate but high chlorine demand, MPUV/chlorine accelerated their chlorine consumption and DBP formation rates, but did not further increase their chlorine demand or DBPFP. For bromide-spiked samples, the results showed that bromide addition consistently increased their THMFP and HAAFP under dark conditions, and turning on UV did not further promote brominated DBP formation compared with the dark control. Bromide caused significant increases in BSF_{THMs} of the seven substances, which should be of concern. MPUV/chlorine-induced changes in the seven precursors' chlorine demand was linearly correlated to their DBPFP changes, indicating that Δ(chlorine demand) may be considered as a potential indicator in the prediction of the changing trends of DBPFP for MPUV/chlorine-treated water.

Supplementary Materials: The following supporting information can be downloaded at: https://www.mdpi.com/article/10.3390/w14223775/s1, Figure S1: spectrogram of a medium-pressure mercury lamp; Table S1: Chlorine decay during 60 s pre-oxidation of the seven model compounds.

Author Contributions: W.L.: Data curation, Writing—Original draft preparation; S.S.: Visualization, Investigation, Funding acquisition; Y.Z.: Conceptualization, Writing—Original draft preparation, Supervision, Funding acquisition; L.W.: Data curation, Writing—Original draft preparation; Q.W.: Methodology, Writing—Reviewing and Editing; N.G.: Writing—Reviewing and Editing. All authors have read and agreed to the published version of the manuscript.

Funding: This research received no external funding.

Acknowledgments: This work was supported by the National Natural Science Foundation of China (No. 52000023), Shanghai Committee of Science and Technology (No. 19DZ1204400), Fundamental Research Funds for Center Universities (21D111311), and the Shanghai Sailing Program (20YF1401200).

Conflicts of Interest: The authors declare no conflict of interest.

Abbreviations

DBPs, disinfection by-products; DBPFP, DBP formation potential; THMs, trihalomethanes; TCM, trichloromethane; BDCM, bromodichloromethane; DBCM, dibromochloromethane; TBM, tribromomethane; HAAS, haloacetic acids; MCAA, monochloroacetic acid; MBAA, monobromoacetic acid; DCAA, dichloroacetic acid; DBAA, dibromoacetic acid; BCAA, bromochloroacetic acid; TCAA, trichloroacetic acid; BDCAA, bromodichloroacetic acid; CDBAA, chlorodibromoacetic acid; TBAA, tribromoacetic acid; HANs, haloacetonitriles; DCAN, dichloroacetonitrile; BCAN, bromochloroacetonitrile; DBAN, dibromoacetonitrile; TCAN, trichloroacetonitrile; HALS, haloacetaldehydes; DCAL, dichloroacetaldehyde; CH, trichloroacetaldehyde; TCNM, trichloronitromethane; RCS, reactive chlorine species; NOM, natural organic matter; PN, phenol; RSC, resorcinol; BA, benzoic acid; NB, nitrobenzene; 2-MCP, o-chlorophenol; 4-MCP, p-chlorophenol; 2,4,6-TCP, 2,4,6-trichlorophenol; MTBE, methyl-tert-butyl ether; UFC, uniform formation condition; BSFs, bromine substitution factors.

References

1. Miklos, D.B.; Wang, W.L.; Linden, K.G.; Drewes, J.E.; Hübner, U. Comparison of UV-AOPs (UV/H2O2, UV/PDS and UV/Chlorine) for TOrC removal from municipal wastewater effluent and optical surrogate model evaluation. *Chem. Eng. J.* **2019**, *362*, 537–547. [CrossRef]
2. Zhang, Y.L.; Wang, W.L.; Lee, M.Y.; Yang, Z.W.; Wu, Q.Y.; Huang, N.; Hu, H.Y. Promotive effects of vacuum-UV/UV (185/254 nm) light on elimination of recalcitrant trace organic contaminants by UV-AOPs during wastewater treatment and reclamation: A review. *Sci. Total Environ.* **2022**, *818*, 151776. [CrossRef]
3. VVaranasi, L.; Coscarelli, E.; Khaksari, M.; Mazzoleni, L.R.; Minakata, D. Transformations of dissolved organic matter induced by UV photolysis, Hydroxyl radicals, chlorine radicals, and sulfate radicals in aqueous-phase UV-Based advanced oxidation processes. *Water Res.* **2018**, *135*, 22–30. [CrossRef]
4. Wang, J.; Deng, J.; Du, E.; Guo, H. Reevaluation of radical-induced differentiation in UV-based advanced oxidation processes (UV/hydrogen peroxide, UV/peroxydisulfate, and UV/chlorine) for metronidazole removal: Kinetics, mechanism, toxicity variation, and DFT studies. *Sep. Purif. Technol.* **2022**, *3*, 121905. [CrossRef]
5. Laat, J.D.; Stefan, M.I. UV/chlorine process. In *Advanced Oxidation Processes for Water Treatment*; Stefan, M.I., Ed.; IWA Publishing: London, UK, 2018; Chapter 9.
6. Guo, K.; Wu, Z.; Chen, C.; Fang, J. UV/Chlorine process: An efficient advanced oxidation process with multiple radicals and functions in water treatment. *Acc. Chem. Res.* **2022**, *55*, 286–297. [CrossRef]
7. Fang, J.; Fu, Y.; Shang, C. The roles of reactive species in micropollutant degradation in the UV/free chlorine system. *Environ. Sci. Technol.* **2014**, *48*, 1859–1868. [CrossRef] [PubMed]
8. Lei, Y.L.; Cheng, S.; Luo, N.; Yang, X.; An, T. Rate constants and of mechanisms for the reactions of Cl and Cl2- with trace organic contaminants. *Environ. Sci. Technol.* **2019**, *55*, 11170–11182. [CrossRef]
9. Gao, Y.Q.; Zhang, J.; Li, C.; Tian, F.X.; Gao, N.Y. Comparative evaluation of metoprolol degradation by UV/chlorine and UV/H2O2 processes. *Chemosphere* **2020**, *243*, 125325. [CrossRef] [PubMed]

10. Guo, K.; Wu, Z.; Yan, S.; Yao, B.; Song, W.; Hua, Z.; Zhang, X.; Kong, X.; Li, X.; Fang, J. Comparison of the UV/chlorine and UV/H2O2 processes in the degradation of PPCPs in simulated drinking water and wastewater: Kinetics, radical mechanism and energy requirements. *Water Res.* **2018**, *147*, 184–194. [CrossRef]
11. Pai, C.-W.; Wang, G.-S. Treatment of PPCPs and disinfection by-product formation in drinking water through advanced oxidation processes: Comparison of UV, UV/Chlorine, and UV/H2O2. *Chemosphere* **2021**, *287*, 132171. [CrossRef] [PubMed]
12. Watts, M.J.; Linden, K.G. Chlorine photolysis and subsequent OH radical production during UV treatment of chlorinated water. *Water Res.* **2007**, *41*, 2871–2878. [CrossRef] [PubMed]
13. Wang, D.; Bolton, J.R.; Andrews, S.A.; Hofmann, R. Formation of disinfection by-products in the ultraviolet/chlorine advanced oxidation process. *Sci. Total Environ.* **2015**, *518–519*, 49–57. [CrossRef]
14. Bulman, D.M.; Remucal, C.K. Role of reactive halogen species in disinfection byproduct formation during chlorine photolysis. *Environmental Sci. Technol.* **2020**, *54*, 9629–9639. [CrossRef] [PubMed]
15. Wang, C.; Moore, N.; Bircher, K.; Andrews, S.; Hofmann, R. Full-scale comparison of UV/H2O2 and UV/Cl2 advanced oxidation: The degradation of micropollutant surrogates and the formation of disinfection byproducts. *Water Res.* **2019**, *161*, 448–458. [CrossRef] [PubMed]
16. Yang, X.; Sun, J.; Fu, W.; Shang, C.; Li, Y.; Chen, Y.; Gan, W.; Fang, J. PPCP degradation by UV/chlorine treatment and its impact on DBP formation potential in real waters. *Water Res.* **2016**, *98*, 309–318. [CrossRef]
17. Hua, Z.; Li, D.; Wu, Z.; Wang, D.; Cui, Y.; Huang, X.; Fang, J.; An, T. DBP formation and toxicity alteration during UV/chlorine treatment of wastewater and the effects of ammonia and bromide. *Water Res.* **2021**, *188*, 116549. [CrossRef]
18. Luo, Y.; Feng, L.; Liu, Y.; Zhang, L. Disinfection by-products formation and acute toxicity variation of hospital wastewater under different disinfection processes. *Sep. Purif. Technol.* **2020**, *238*, 116405. [CrossRef]
19. Li, T.; Jiang, Y.; An, X.; Liu, H.; Hu, C.; Qu, J. Transformation of humic acid and halogenated byproduct formation in UV/chlorine processes. *Water Res.* **2016**, *102*, 421–427. [CrossRef]
20. Shah, A.D.; Dotson, A.D.; Linden, K.G.; Mitch, W.A. Impact of UV disinfection combined with chlorination/chloramination on the formation of halonitromethanes and haloacetonitriles in drinking water. *Environ. Sci. Technol.* **2011**, *45*, 3657–3664. [CrossRef]
21. Wang, W.-L.; Zhang, X.; Wu, Q.-Y.; Du, Y.; Hu, H.-Y. Degradation of natural organic matter by UV/chlorine oxidation: Molecular decomposition, formation of oxidation byproducts and cytotoxicity. *Water Res.* **2017**, *124*, 251–258. [CrossRef]
22. Bolton, R.J.; Hofmann, J.R.; Ding, W.; Andrews, A.S. UV/chlorine control of drinking water taste and odour at pilot and full-scale. *Chemosphere: Environ. Toxicol. Risk Assess.* **2015**, *136*, 239–244.
23. Wu, Z.; Guo, K.; Fang, J.; Yang, X.; Xiao, H.; Hou, S.; Kong, X.; Shang, C.; Yang, X.; Meng, F.; et al. Factors affecting the roles of reactive species in the degradation of micropollutants by the UV/chlorine process. *Water Res.* **2017**, *126*, 351–360. [CrossRef] [PubMed]
24. Wang, C.; Ying, Z.; Ma, M.; Huo, M.; Yang, W. Degradation of micropollutants by UV chlorine treatment in reclaimed water: pH effects, formation of disinfectant byproducts, and toxicity assay. *Water* **2019**, *11*, 2639. [CrossRef]
25. Bolton, J.R.; Linden, K.G. Standardization of methods for fluence (UV dose) determination in bench-scale UV experiments. *J. Environ. Eng.* **2003**, *129*, 209–215. [CrossRef]
26. Chuang, Y.H.; Szczuka, A.; Shabani, F.; Munoz, J.; Aflaki, R.; Hammond, S.D.; Mitch, W.A. Pilot-scale comparison of microfiltration/reverse osmosis and ozone/biological activated carbon with UV/hydrogen peroxide or UV/free chlorine AOP treatment for controlling disinfection byproducts during wastewater reuse. *Water Res.* **2019**, *152*, 215–225. [CrossRef]
27. Summers, R.; Hooper, S.; Shukairy, H.; Solarik, G.; Owen, D. Assessing DBP yield: Uniform formation conditions. *J. Am. Water Work. Assoc.* **1996**, *88*, 80–93. [CrossRef]
28. APHA; AWWA; WEF. *Standard Methods for the Examination of Water and Wastewater*, 21st ed.; 4500-Cl, G. DPD Colorimmetric Method; American Public Health Association: Washington, DC, USA, 2005; pp. 4–67.
29. Munch, D.; Hautman, D. *Method 551.1: Determination of Chlorination Disinfection Byproducts, Chlorinated Solvents, and Halogenated Pesticides/Herbicides in Drinking Water by Liquid-Liquid Extraction and Gas Chromatography with Electron-Capture Detection*; National Exposure Research Laboratory Office of Research and Development U.S. Environmental Protection Agency: Cincinnati, OH, USA, 1995.
30. Hodgeson, J.; Collins, J.; Barth, R. *Method 552.2. Determination of Haloacetic Acids in Drinking Water by Liquid Liquid Extraction and Gas Chromatography with Electron Capture Detection*; Technical Support Center Office of Ground Water and Drinking Water U. S. Environmental Protection Agency: Cincinnati, OH, USA, 1995.
31. Bulman, D.M.; Mezyk, S.P.; Remucal, C.K. The impact of pH and Irradiation wavelength on the production of reactive oxidants during chlorine photolysis. *Environ. Sci. Technol.* **2019**, *53*, 4450–4459. [CrossRef]
32. An, Z.; Li, M.; Huo, Y.; Jiang, J.; Zhou, Y.; Jin, Z.; Xie, J.; Zhan, J.; He, M. The pH-dependent contributions of radical species during the removal of aromatic acids and bases in light/chlorine systems. *Chem. Eng. J.* **2022**, *433*, 133493. [CrossRef]
33. Cheng, S.; Zhang, X.; Yang, X.; Shang, C.; Song, W.; Fang, J.; Pan, Y. The multiple role of bromide ion in PPCPs degradation under UV/Chlorine treatment. *Environ. Sci. Technol.* **2018**, *52*, 1806–1816. [CrossRef]
34. Guo, K.; Wu, Z.; Shang, C.; Yao, B.; Hou, S.; Yang, X.; Song, W. Radical chemistry and structural relationships of PPCP degradation by UV/chlorine treatment in simulated drinking water. *Environ. Sci. Technol.* **2017**, *51*, 10431–10439. [CrossRef]
35. Zhao, Q.; Shang, C.; Zhang, X.; Ding, G.; Yang, X. Formation of halogenated organic byproducts during medium-pressure UV and chlorine coexposure of model compounds, NOM and bromide. *Water Res.* **2011**, *45*, 6545–6554. [CrossRef] [PubMed]

36. Liang, L.; Singer, P.C. Factors influencing the formation and relative distribution of haloacetic acids and trihalomethanes in drinking water. *Environ. Sci. Technol.* **2003**, *37*, 2920–2928. [CrossRef] [PubMed]
37. Deng, L.; Huang, C.H.; Wang, Y.L. Effects of combined UV and chlorine treatment on the formation of trichloronitromethane from amine precursors. *Environ. Sci. Technol.* **2014**, *48*, 2697–2705. [CrossRef] [PubMed]
38. Kumar, K.; Margerum, D.W. ChemInform abstract: Kinetics and mechanism of general-acid-assisted oxidation of bromide by hypochlorite and hypochlorous acid. *ChemInform* **1987**, *18*, 2706–2711. [CrossRef]
39. Langsa, M.; Heitz, A.; Joll, C.A.; von Gunten, U.; Allard, S. Mechanistic aspects of the formation of adsorbable organic bromine during chlorination of bromide-containing synthetic waters. *Environ. Sci. Technol.* **2017**, *51*, 5146–5155. [CrossRef]
40. Guo, K.; Zheng, S.; Zhang, X.; Zhao, L.; Ji, S.; Chen, C.; Wu, Z.; Wang, D.; Fang, J. Roles of bromine radicals and hydroxyl radicals in the degradation of micropollutants by the UV/bromine process. *Environ. Sci. Technol.* **2020**, *54*, 6415–6426. [CrossRef]
41. Zhu, Y.; Li, W.; Shu, S.; Wang, Q.; Gao, N. Effects of MPUV/chlorine oxidation and coexisting bromide, ammonia, and nitrate on DBP formation potential of five typical amino acids. *Sci. Total Environ.* **2022**, *821*, 153221. [CrossRef]
42. Vermeulen, A.; Welvaert, K.; Vercammen, J. Evaluation of a dedicated gas chromatography–mass spectrometry method for the analysis of phenols in water. *J. Chromatogr. A* **2005**, *1071*, 41–46. [CrossRef]
43. Bond, T.; Henriet, O.; Goslan, E.H.; Parsons, S.A.; Jefferson, B. Disinfection byproduct formation and fractionation behavior of natural organic matter surrogates. *Environ. Sci. Technol.* **2009**, *43*, 5982–5989. [CrossRef]
44. Xiang, Y.; Gonsior, M.; Schmitt-Kopplin, P.; Shang, C. Influence of the UV/H2O2 advanced oxidation process on dissolved organic matter and the connection between elemental composition and disinfection byproduct formation. *Environ. Sci. Technol.* **2020**, *54*, 14964–14973. [CrossRef]
45. Li, J.; Zhang, Z.; Xiang, Y.; Jiang, J.; Yin, R. Role of UV-based advanced oxidation processes on NOM alteration and DBP formation in drinking water treatment: A state-of-the-art review. *Chemosphere* **2022**, *311*, 136870. [CrossRef]
46. Buxton, G.V.; Greenstock, C.L.; Helman, W.P.; Ross, A.B. Critical Review of rate constants for reactions of hydrated electrons, hydrogen atoms and hydroxyl radicals (·OH/·O in Aqueous Solution. *J. Phys. Chem. Ref. Data* **1988**, *17*, 513–886. [CrossRef]
47. Minakata, D.; Kamath, D.; Maetzold, S. Mechanistic insight into the reactivity of chlorine-derived radicals in the aqueous-phase UV chlorine advanced oxidation process: Quantum mechanical calculations. *Environ. Sci. Technol.* **2017**, *51*, 6918–6926. [CrossRef] [PubMed]
48. Alfassi, Z.B.; Huie, R.E.; Mosseri, S.; Neta, P. Kinetics of one-electron oxidation by the ClO radical. *Int. J. Radiat. Appl. Instrumentation. Part C Radiat. Phys. Chem.* **1988**, *32*, 85–88. [CrossRef]

water

MDPI

Article

Photo-Fenton Catalyzed by Cu_2O/Al_2O_3: Bisphenol (BPA) Mineralization Driven by UV and Visible Light

Oscar Olea-Mejia [1,*], Sharon Brewer [2], Kingsley Donkor [2], Deysi Amado-Piña [1] and Reyna Natividad [1,*]

[1] Centro Conjunto de Investigación en Química Sustentable UAEM-UNAM, Carretera Toluca-Atlacomulco Km 14.5, San Cayetano, Toluca 50200, Mexico
[2] Department of Chemistry, Thompson Rivers University, Kamloops, BC V2C 0C8, Canada
* Correspondence: oleaoscar@yahoo.com.mx (O.O.-M.); rnatividadr@uaemex.mx (R.N.)

Abstract: This work aimed to demonstrate Cu_2O/Al_2O_3 as a catalyst of the photo-Fenton process in the UV and visible spectra. Cu_2O nanoparticles were synthesized by laser ablation in liquid and supported on Al_2O_3. The catalytic activity of the resulting solid was assessed in the mineralization of bisphenol A (BPA). The studied variables were type of Al_2O_3 (α and γ), Cu content (0.5 and 1%), and H_2O_2 concentration (1, 5, and 10 times the stoichiometric amount). The response variables were BPA concentration and total organic carbon (TOC) removal percentage. The presence of Cu_2O nanoparticles (11 nm) with an irregular sphere-like shape was confirmed by transmission electron microscopy (TEM) and their dispersion over the catalytic surface was verified by energy-dispersed spectroscopy (EDS). These particles improve ·OH radical production, and thus a 100% removal of BPA is achieved along with ca. 91% mineralization in 60 min. The BPA oxidation rate is increased one order of magnitude compared to photolysis and doubles that for H_2O_2 + UV. An increase of 40% in the initial oxidation rate of BPA was observed when switching from α-Al_2O_3 to γ-Al_2O_3. 4-hydroxybenzaldehyde, 4-hydroxybenzoic acid, acetaldehyde, and acetic acid are the BPA oxidation by-products identified using LC/MS and based on this a reaction pathway was proposed. Finally, it was also concluded that the synthesized catalyst exhibits catalytic activity not only in the UV spectrum but also in the visible one under circumneutral pH. Therefore, Cu_2O/Al_2O_3 can be recommended to conduct a solar photo-Fenton reaction that can degrade other types of molecules.

Keywords: BPA degradation; photo-Fenton; nanocomposite catalysts; bisphenol A mineralization; solar photo-Fenton

Citation: Olea-Mejia, O.; Brewer, S.; Donkor, K.; Amado-Piña, D.; Natividad, R. Photo-Fenton Catalyzed by Cu_2O/Al_2O_3: Bisphenol (BPA) Mineralization Driven by UV and Visible Light. *Water* **2022**, *14*, 3626. https://doi.org/10.3390/w14223626

Academic Editors: Huijiao Wang, Dionysios (Dion) Demetriou Dionysiou and Yujue Wang

Received: 30 September 2022
Accepted: 7 November 2022
Published: 10 November 2022

1. Introduction

Bisphenol A (BPA) is ubiquitous in the environment. Potential sources of human exposure to bisphenol A are air, water, soil, sediments, indoor dust, and human tissues. Bisphenol A is an emerging environmental contaminant. The continuous release of bisphenol A into the environment results in the continuous exposure of both plant and animal life to it. Furthermore, BPA has been determined to be an endocrine disruptor chemical, potentially associated with numerous diseases such as breast cancer, infertility, cognitive dysfunction, diabetes, cardiovascular diseases, and obesity. As a response to these concerns, several countries have banned the use of BPA in various consumer products. The abovementioned effects of BPA on human health and the environment prompted us to choose it as the research object in the present work and develop the method presented here, which can be used to remove or minimize its presence [1,2].

There are various techniques to remove pollutants from contaminated streams; these include adsorption [3,4], biodegradation, and thermal destruction, among others. However, these methods also have important disadvantages, such as only transferring the pollutant from one material to another, high energy requirements, and slow process. In this context, advanced oxidation processes (AOPs) are known to be highly efficient in removing organic contaminants from water, the Fenton process being one that has been extensively studied [5]. However, the Fenton reaction takes place in highly acidic conditions, and iron salts are dissolved to produce the desired ·OH radicals. Evidently, the disadvantages of this technique are the loss of the iron catalyst (homogeneous catalysis) as well as the need to reverse the pH to an acceptable level after the reaction. This is usually conducted by the addition of HCl, which indicates a new problem—the increase of purification cost.

Therefore, different AOPs have been developed to overcome these disadvantages, including ozonation [6–8], heterogeneous catalysis/photocatalysis [9,10], bioprocesses [11–13], and Fenton-like reactions [14–16]. Among these methods, the heterogeneous photo-Fenton-like is a promising one. It has the advantages of re-using catalysts and can be conducted at circumneutral pH. Here, the generation of the ·OH radicals is due to the presence of H_2O_2, UV light, and the metal ion that is capable of facilitating the radical production.

The degradation of bisphenols using Fenton-like processes has been studied extensively in recent years, and several methods have been proposed, including sono-photo-Fenton [17] and photo-Fenton using UV [18–24] and visible radiation [10,25,26]. Furthermore, the use of metallic nanoparticles in Fenton-like reactions has also been studied recently, with iron being the most-used metal [6,8,15,27–33]. Copper has also been used but to a much lesser extent [34,35], and its use in the degradation of bisphenols has been barely assessed [18,36]. Therefore, copper was elected in this work because of its potential as a Fenton-like reactant and due to the very little information regarding its use in the degradation of bisphenols.

Metallic and Cu oxide nanoparticles can be produced by several methods. In this work, Cu_2O nanoparticles were produced via laser ablation in liquid (LAL). LAL is a physical method that can produce nanoparticles from a pure metal target immersed in a liquid medium. In short, a high-energy pulsed laser is focused on the surface of the target, the material is abruptly heated, and plasma is formed. Subsequently, the plasma condenses in the form of nanoparticles inside the liquid [37]. The main advantage when using LAL is that the obtained nanoparticles are free from any chemical by-products [38,39]. This is relevant to catalytic applications because the surface contamination-free nanoparticles are readily available to interact with the chemical species in the reaction of interest.

This work aimed to investigate the effect of the amount of H_2O_2 and the amount of Cu incorporated in the catalysts in the degradation of BPA via a photo-Fenton-like reaction using two types of commercially available Al_2O_3. This is a well-known catalytic support that is relatively low-cost, with no environmental issues. From an engineering point of view, the support of nanoparticles is desirable to ease the catalyst separation. Other Cu composites have been used successfully to this purpose [18].

2. Materials and Methods

2.1. Cu Nanoparticle Preparation and Characterization

A Continuum Surelite II laser system was used at a wavelength of 1064 nm, pulse duration of 6 ns, and repetition rate of 10 Hz. The metal used was a 2.5 cm diameter Cu target (99.99% purity, J.K. Lesker Co., Jefferson Hills, PA, USA). The liquid medium was deionized water. The full procedure has already been published in previous works. In short, 5 mL of water was added to a Pyrex glass containing the Cu target and irradiated for the time necessary to produce the required amount of Cu nanoparticles (about 20 min). The energy output from the laser system was 90 mJ/pulse. The amount of nanoparticles was calculated from the weight difference of the target before and after the laser ablation process. A JEOL JEM-2100 transmission electron microscope (TEM) was used to study the size and shape of the nanoparticles. The same equipment was used to perform high-resolution TEM (HRTEM) to elucidate the crystal structure and metallic state of the nanoparticles.

2.2. Catalyst Preparation and Characterization

Two types of alumina (Al_2O_3) were used in this work to study the possible effects of the crystalline phase and the size and shape of the alumina powder on BPA degradation efficiency. Powders of α-Al_2O_3 and γ-Al_2O_3 (Sigma Aldrich, Burlington, MA, USA, 99% purity, 1–10 μm particle size) were used. Both types of alumina were treated prior to the addition of the nanoparticles with a solution of 1M NaOH (Sigma-Aldrich) at 60 °C for 1 h to improve the interaction with the Cu nanoparticles. After that, the alumina was dried and added to the Cu nanoparticle colloid and stirred for 1 h. Finally, the water was evaporated at 80 °C to obtain the dried Cu/Al_2O_3. The Cu concentrations used in this study were 0, 0.5, 1, and 2% weight percentages. The catalysts were studied using scanning electron microscopy (SEM), and a JEOL JSM-6510LV with an energy-dispersive X-ray spectroscopy (EDS) probe attached to it to confirm the presence of the Cu nanoparticles on the surface of the catalyst.

2.3. BPA Degradation Procedure

The BPA degradation was studied using a photo-Fenton-like reaction in an annular batch reactor with 50 mL of reaction volume. The UV irradiation was performed with a monochromatic lamp with an emission of 254 nm that was placed at the center of the reactor. All the reactions were carried out at a constant temperature of 26 °C, controlled with a water bath. The pH was not adjusted at any time and the initial pH was 6.7. The amount of catalyst was 10 mg for all experiments, and only the Cu nanoparticle concentration in the catalyst was varied, as mentioned above. The catalyst was added to the BPA solution (initial BPA concentration = 50 mg/L) and dispersed throughout the reaction via magnetic stirring. Finally, hydrogen peroxide was added to the reaction system right before the UV lamp was turned on. The amount of peroxide was varied by 1, 5, and 10 times the stoichiometric amount according to the total BPA mineralization reaction.

To establish the activity of the catalyst under visible light, the same aforementioned procedure was followed, but instead of using a 254 nm wavelength lamp, the reactor was surrounded by 4 lamps emitting light with a wavelength $\lambda > 400$ nm; 3 were placed at the perimeter of the reactor while a fourth one was placed on top of it. The lamps were TL5 Essential 14W/840, Philips Lighting.

2.4. Analytical Methods

Aliquots were taken from the reactor at regular times and centrifuged in capillary electrophoresis (CE) vials to gather the amount of catalyst in the bottom of the vial. It is important to mention that a test reaction with no UV light (only hydrogen peroxide and catalyst) was performed and no BPA degradation was registered over 1 h of reaction; therefore, it is safe to assume that even though there might be a very small amount of catalyst in the extracted aliquots there was no reaction taking place during the analytical procedures.

Capillary electrophoresis (CE) was used to study the degradation of BPA. A Beckman P/ACE System MDQ capillary electrophoresis unit was used for all CE analysis. A polyimide-coated silica capillary (50 mm long, 50 μm inner diameter) was used inside a cartridge configured for UV detection. BPA was detected via direct absorbance at 214 nm, 25 kV normal polarity, and a constant temperature of 25 °C. Before each sample injection, the capillary was rinsed with a 0.1 M NaOH solution for 5 min, with deionized water for 5 min, and with the run buffer for 8 min. The run buffer was a sodium dodecyl sulfate—tetrahydrate sodium borate (20 mM–60 mM, respectively) aqueous solution with a pH of 9.5.

Total organic carbon (TOC) measurements were performed after 20, 40, and 60 min of reaction in all cases to determine the mineralization percentage reached in each case. A TOC-VCSH analyzer (Mendel Scientific, Houston, TX, USA) was used for this purpose. The samples were sparged for 2 min with dried air to remove inorganic CO_2 from the ambient prior to the catalytic oxidation at 680 °C.

Liquid chromatography coupled with mass spectroscopy (LC/MS) was used to determine the possible oxidation intermediates at the end of the BPA degradation reactions. For this purpose, an HPLC system (Agilent Series 1200) was used, coupled with an LC/MS Q-TOF mass spectrometer (Agilent 6530) equipped with an electrospray ionization (ESI) source (gas temperature, 350 °C; drying gas, 8 L/min; nebulizer, 10 psig; sheath gas temperature, 350 °C; sheath gas flow, 8 L/min; Vcap, 3500 V). The analyses were carried out in the positive ion mode and the mass spectra were collected between 40 and 250 m/z. A sample volume of 10 μL was injected into the liquid chromatograph with a flow rate of 0.5 mL/min. Separation was achieved in a Zorbax Extend-C18 column (Agilent, Mississauga, ON, Canada) at a constant temperature of 25 °C. The mobile phase had two components: Component A was pure LC/MC-grade water with a pH of 2 adjusted with phosphoric acid, and Component B was pure methanol. A gradient elution referred to Component B (%) was used: 0% for 15 min, 60% for 1 min, 40% for 5 min, and finally 0% for 9 min.

To determine the concentration of H_2O_2 over time, a colorimetric method was applied [40]. More details about this technique can also be found in [41]. A spectrophotometer was used to determine the absorbance of samples at 408 nm, where maximum absorbance was observed.

3. Results

3.1. Cu Nanoparticles and Catalyst Characterization

Figure 1a shows the TEM image of a typical sample of Cu_2O nanoparticles produced by LAL. The particles have an average size of 11 nm and an irregular sphere-like shape. To determine the chemical state and crystalline phase of the nanoparticles, high-resolution TEM (HRTEM), as seen in Figure 1b, and electron diffraction (Figure 1c) were performed. The interplanar distance from the HRTEM image is 0.24 nm, which can be ascribed to the 111 plane in the crystalline phase of Cu_2O according to the JPCDS 01-075-1531 card. To corroborate this, Figure 1c shows the diffraction rings that correspond to the 110, 111, 200, and 211 planes of Cu_2O. It is important to note that even though the target was pure Cu, Cu_2O nanoparticles were obtained, which means that the particles oxidize when in contact with water or air after being produced by laser ablation.

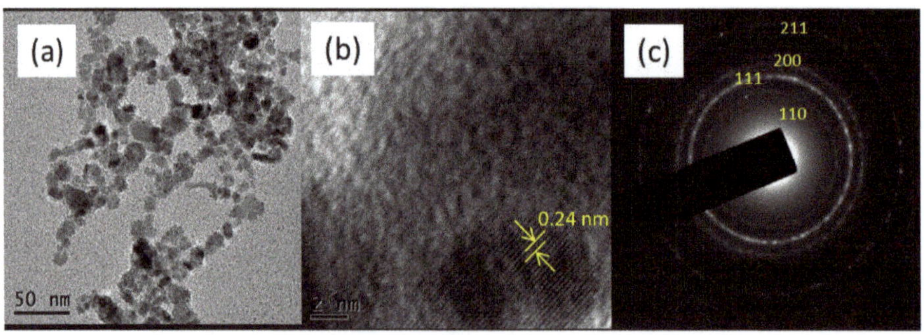

Figure 1. TEM image (**a**), HRTEM (**b**), and electron diffraction pattern (**c**) of the Cu_2O nanoparticles obtained via laser ablation in liquid.

EDS was conducted to determine the presence of Cu_2O nanoparticles on the surface of the alumina particles. Figure 2a,b show the SEM and Cu mapping for the 1% Cu/α-Al_2O_3 catalyst. As can be seen, Cu is uniformly present throughout the sample. The same observations are true for the 1% Cu/γ-Al_2O_3 catalyst (Figure 2c,d) as well as for the rest of the prepared catalyst (not shown). Having the Cu_2O nanoparticles all over the catalyst surface is very important because they can achieve their function more efficiently during the BPA degradation process, which will be further discussed in a later section.

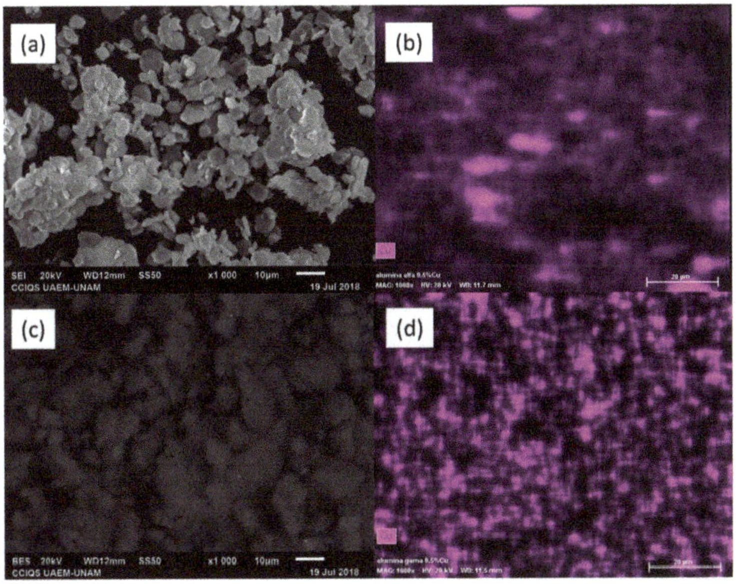

Figure 2. SEM images and Cu mapping of 1%Cu_2O/α-Al_2O_3 catalyst (**a**,**b**) and 1%Cu_2O/γ-Al_2O_3 catalyst (**c**,**d**).

3.2. BPA Degradation Results

The degradation of BPA using UV light only, UV/H_2O_2, and UV/H_2O_2/catalyst was tested to assess the possible contribution of each component in the reaction. It can be observed in Figure 3 that all the experiments led to BPA degradation, with differences in the rate and level of degradation achieved. With direct photolysis it was possible to degrade BPA to a concentration of 22.5 ppm. When H_2O_2 was added to the reaction (10 times the stoichiometric amount of the complete mineralization reaction), the BPA concentration was as low as ~1 ppm for the same reaction time (15 min), which means that almost complete degradation was reached. This considerable effect was achieved because it is known that UV light is able to form ·OH radicals from H_2O_2 (reaction 1 in Scheme 1). However, when the catalysts were added to the reaction, the complete BPA degradation occurred at shorter times. By comparing the initial BPA reaction rates ($-r_{BPA,o}$) reported in Table 1, it can be concluded that the two types of Cu_2O/Al_2O_3 catalyst show a similar behavior, γ-Al_2O_3 being better than α-Al_2O_3 by about 40%. This initial reaction rate corresponds to the first minute of reaction, where the hydroxyl radicals are being mainly consumed by BPA; after this reaction time, it can be observed in Figure 3 that there was a change in the slope of the BPA concentration profiles of the heterogeneous processes. This change can be ascribed to the competition of the oxidation by-products with the BPA for hydroxyl radicals. It can also be observed in Table 1 that the initial BPA oxidation rate aided by the catalyst was one order of magnitude higher than that with UV and hydrogen peroxide and with UV only. Table 1 also presents the pseudo-first order kinetic constants for each process; it can be observed that the specific rate constant for α-Al_2O_3 was about 70% higher than that obtained with UV/H_2O_2 and was more than twice that when the support was γ-Al_2O_3. Although the determination coefficient, R^2, is still reasonably good for the heterogeneous process, it can be observed that is not as good as for the homogeneous case. This could be ascribed to not incorporating in an explicit way the competition of the organic molecule with the catalyst for the photons in the applied kinetic power law model [42].

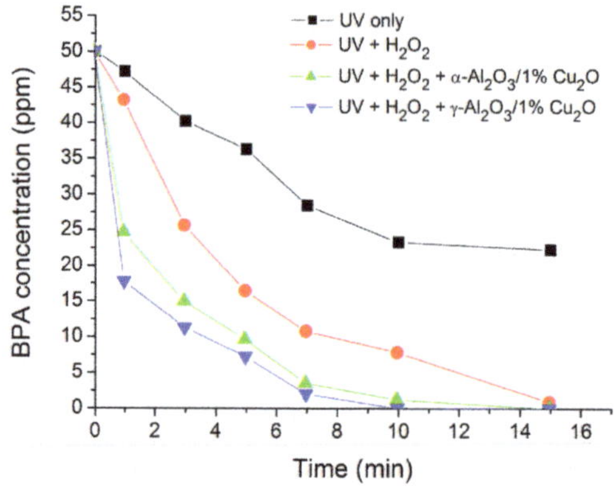

Figure 3. BPA degradation under different experimental reaction conditions.

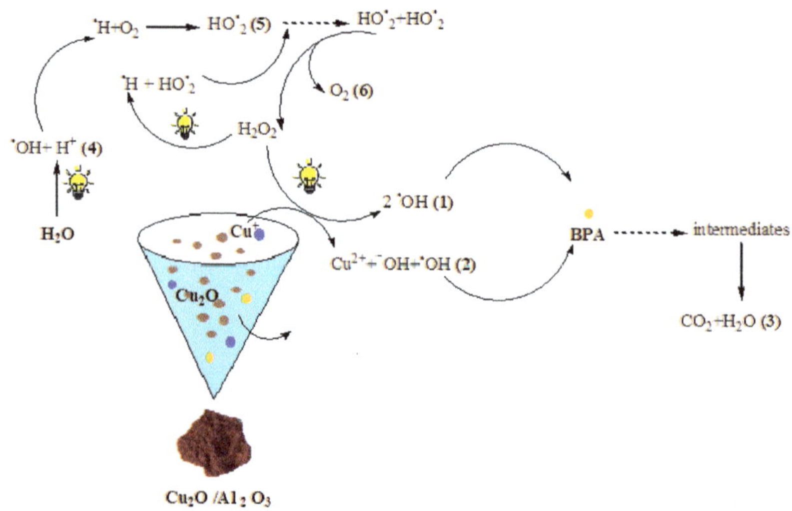

Scheme 1. Reactions occurring at the catalyst surface and in solution in the studied photo-Fenton process catalyzed by Cu_2O/Al_2O_3.

Table 1. Percent of BPA mineralization, initial BPA oxidation rate ($-r_{BPA,o}$), pseudo-first order kinetic constant (k), and determination coefficient (R^2) under different experimental conditions.

	BPA Mineralization Percentage (%)			
Time (min)	**Photolysis (UV Only)**	**UV + H_2O_2**	**UV + H_2O_2 + 1% Cu_2O/α-Al_2O_3**	**UV + H_2O_2 + 1% Cu_2O/γ-Al_2O_3**
20	2.4	17.5	19.6	24.4
40	4.7	50.5	70.1	79.6
60	7.1	80.9	86.5	90.7
	Kinetic Parameters and Determination Coefficient			
$-r_{BPA,o}$ (mol_{BPA}/L·min)	1.09×10^{-5}	1.23×10^{-5}	1.09×10^{-4}	1.4×10^{-4}
k (min^{-1})	0.074	0.1982	0.3432	0.4511
R^2	0.991	0.9969	0.9874	0.9685

In these cases, the Cu^+ ion plays a crucial role in the reaction, because the OH radicals can also be formed from the Cu ions in a similar way that Fe ions can form these radicals in a conventional Fenton reaction. Cu^+ can react with H_2O_2 to form OH and ^-OH radicals (reaction 2 in Scheme 1 [43–45]), and it is well-known that the OH radicals are the main responsible for complete mineralization of BPA through reaction 3 in Scheme 1 [46–48].

Figure 3 shows BPA degradation based solely on the BPA peak intensity measured in the CE electropherograms; however, it is not enough to determine the extent of complete mineralization of the molecule. For this purpose, Table 1 shows the TOC results for the same set of experiments. As expected, all the experiments show different mineralization values, following the tendency from Figure 3. The lowest mineralization achieved after 1 h of reaction was for the experiment when only UV light was used. This is in agreement with other reports where the activation of the BPA molecule by UV light has been suggested [48]. On the other hand, the highest mineralization was reached when the catalyst with Cu_2O nanoparticles was used. Therefore, the synthesized catalyst enhances reaction 2 (Scheme 1) in a way that not only hastens BPA degradation, but it also aids the degradation of intermediates until complete mineralization.

Figure 4 shows the consumption of hydrogen peroxide during three processes: when only adding BPA (BPA + H_2O_2); when adding BPA, UV light, and H_2O_2; and when the catalyst was added to this system. In the first case, there was no consumption of H_2O_2, and this in agreement with the BPA concentration profiles under these conditions. When the BPA + H_2O_2 was irradiated with UV light, an important consumption of H_2O_2 was observed (red dots); this was from its decomposition, as shown in Scheme 1. However, when the catalyst was added, 1500 mg/L of H_2O_2 were consumed within the system, which was significantly higher than the H_2O_2 consumed when the catalyst was absent. This is evidence of BPA oxidation occurring via the hydroxyl radicals produced in reactions 1 and 2 (see Scheme 1) as reported by [48]. Nevertheless, because of the narrow Cu_2O bandgap, a photocatalytic effect of the cupreous oxide cannot be ruled out [18]. It is worth noticing that the radicals produced are not only consumed in the degradation of BPA but also in the oxidation of the by-products.

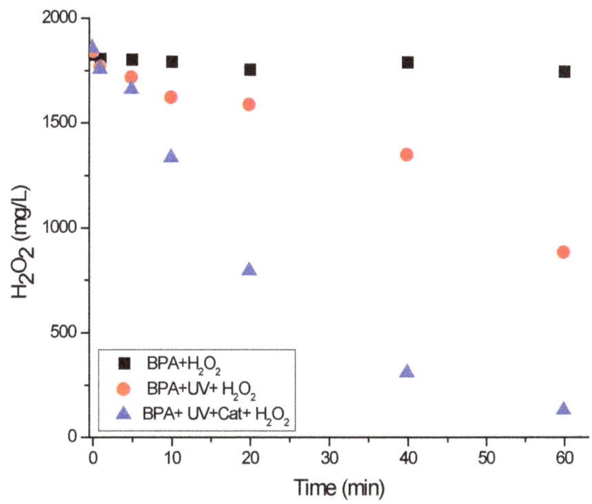

Figure 4. Hydrogen peroxide consumption profiles of BPA alone (BPA + H_2O_2), BPA and UV light (BPA + UV + H_2O_2), and by BPA, UV light, and Cu/Al_2O_3 (BPA + UV + Cat + H_2O_2).

Scheme 2 depicts a plausible BPA degradation pathway via hydroxyl radicals. The identification of acetic acid, acetaldehyde, 4-hydroxybenzaldehyde, and 4-hydroxybenzoic acid was possible by means of LC/MS.

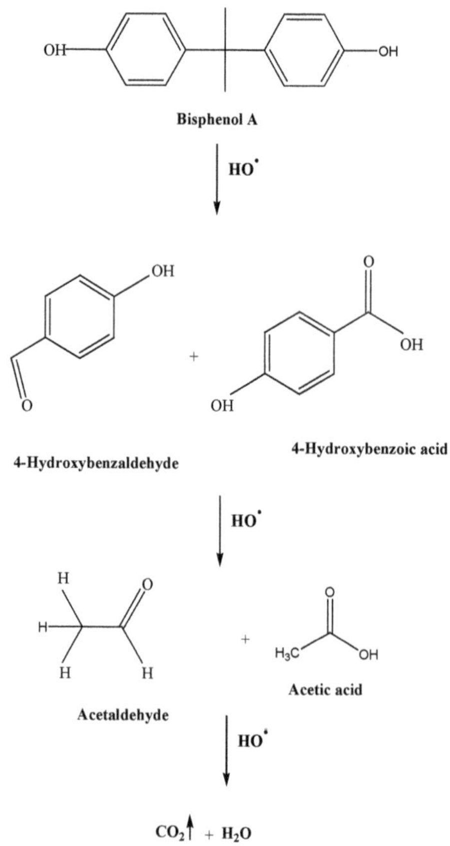

Scheme 2. Degradation of bisphenol A by a photo-Fenton reaction catalyzed by Cu_2O/Al_2O_3: proposed reaction pathway.

Because the proposed degradation pathway is via hydroxyl radicals, it is expected to observe some similarities with other reported proposals. The first step expected is a hydroxyl radical attack of the electron-rich carbon positions. According to calculations made by [46], such positions correspond to C3 and C8. The calculated frontier density for these positions was 0.206 and 0.207, respectively. The hydroxyl radical attack of C3 may actually lead to 4-hydroxybenzaldehyde, which was identified in this work by LC/MS. It is worth pointing out that, in the context of BPA oxidation by hydroxyl radicals, this intermediate has only been identified in the photocatalytic process conducted by [46]. The hydroxyl radical attack of the electron-rich positions also leads to the appearance of the phenyl group [47] that undergoes hydroxylation to form hydroquinone. According to the literature, benzoquinone is then produced via hydroquinone dehydrogenation. Then the ring opens from further attack by hydroxyl radicals [41,47], and acetaldehyde and acetic acid appear. It is worth noting that, as in other reports, other intermediates could be present at the end of the reaction; however, they were not detected, most likely because the concentration was lower than the detection limit.

Scheme 3 depicts the three plausible structures of BPA depending on pH. This directly impacts the observed intermediates and the efficiency of the process. In this work, the initial pH was 6.7 (BPA is a weak organic acid) and it only went down to around 6.3 at the end of the experiments. Therefore, structure (A) in Scheme 3 is the one reacting in the studied system.

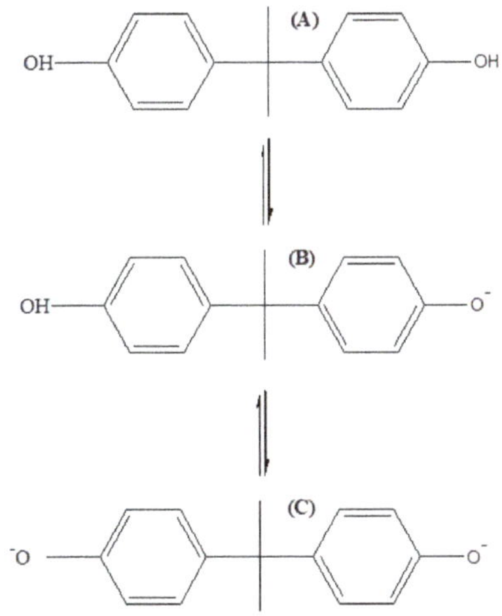

Scheme 3. BPA structures and pKa: (**A**) Neutral, (**B**) $pK_{a,1} = 9.59$, (**C**) $pK_{a,2} = 10.2$ [48].

3.3. Effect of H_2O_2 Concentration

From the previous results it is clear that H_2O_2 is important for the degradation of BPA, and three different concentrations of this component were tested with only UV light and no catalyst. The assessed concentrations were 1 time (1X), 5 times (5X), and 10 times (10X) the amount of H_2O_2 theoretically needed according to the stoichiometry of the complete mineralization reaction of BPA. It can be seen in Figure 5 that the lowest degradation occurred when using 1X H_2O_2. On the other hand, there seems not to be a considerable difference between the 5X and 10X cases, and it could be safely assumed that it is practically the same to use 5X or 10X of H_2O_2. However, the TOC results in Table 2 indicate that the mineralization achieved with 10X of H_2O_2 was considerably higher. This means that the rate of BPA degradation was very similar, but in the case of 10X the extra H_2O_2 was reacting to degrade the intermediates faster.

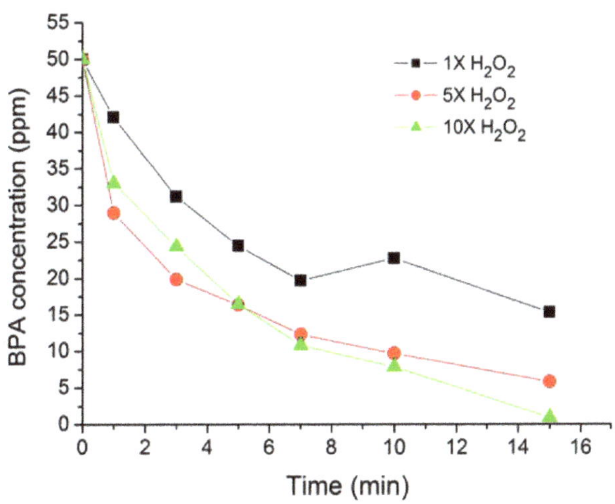

Figure 5. BPA degradation with different concentrations of H_2O_2.

Table 2. Percentage of BPA mineralization under different H_2O_2 concentrations and UV light. $C_{BPA,o}$ = 50 mg/L, pH_o = 6.7, W_{cat} = 0.0 g/L.

Time (min)	1X H_2O_2	5X H_2O_2	10X H_2O_2
20	2.3	16.8	17.5
40	15.5	25.1	50.5
60	20.6	50.0	80.9

It is worth pointing out that H_2O_2 was also being generated within the system by means of reactions 4, 5, and 6 [43,49], depicted in Scheme 1. This is plausible because of the highly energetic lamp that was being used. Figure 6 shows the H_2O_2 generation under different operating conditions. It can be observed that the accumulated amount of H_2O_2 with the lamp after 60 min was about 1.75 mg/L. It is worth clarifying that this was only part of the generated amount since there was a fraction being dissociated by UV light.

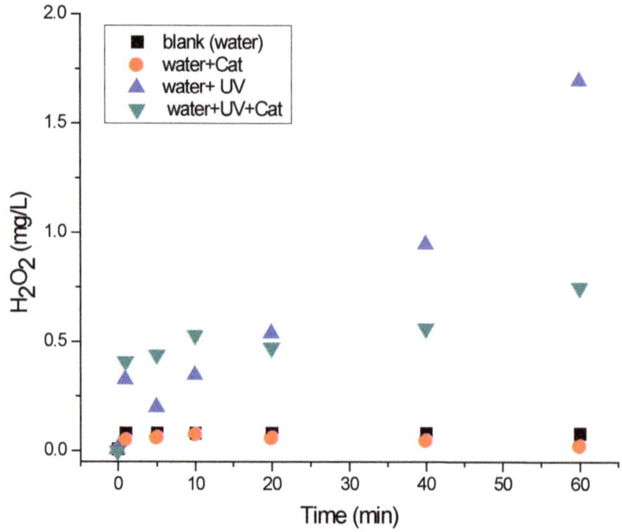

Figure 6. Production of hydrogen peroxide under different reaction conditions.

3.4. *Effect of Cu_2O Concentration on the Catalyst*

Based on the results from the previous section, γ-Al_2O_3 was elected as the support material to study the effect of Cu_2O concentration on the degradation of BPA. Figure 7 shows the BPA degradation for γ-Al_2O_3 catalysts with different quantities of Cu_2O added. As can be seen, when pure γ-Al_2O_3 was used without Cu_2O nanoparticles, the degradation curve was very similar to the case of H_2O_2 + UV light (Figure 5), which means that the γ-Al_2O_3 acted only as a support for Cu_2O nanoparticles and did not contribute to any catalytic function, nor hinder the reaction. On the other hand, when Cu_2O nanoparticles were added to the surface of the catalyst, a faster degradation of BPA was observed. This reinforces the hypothesis that the Cu_2O nanoparticles help degrade the BPA by creating ·OH groups via mechanisms such as the one proposed in reaction 2 in Scheme 1. Furthermore, there was no notable difference when 0.5% or 1.0% of Cu_2O was added to the catalyst, since the degradation profiles in both cases are very similar. In addition, the TOC results for these two cases show that the final mineralization after 1 h of reaction was similar (Table 3). This could be explained in terms of nanoparticle agglomeration on the catalyst surface: when loading a material with nanometric nanoparticles, there is a certain concentration where the particles start to agglomerate instead of being dispersed evenly on the surface of the material. If this is true, then the total surface area available on the nanoparticles does not grow proportionally with the number of nanoparticles loaded to the material.

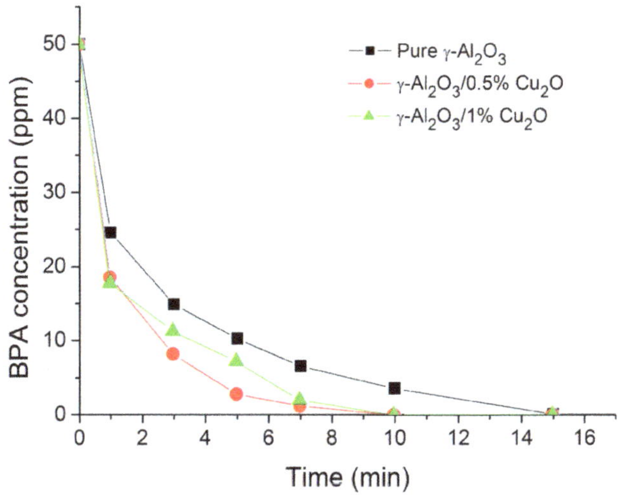

Figure 7. BPA degradation with catalysts of different Cu_2O nanoparticle concentrations.

Table 3. Percent of mineralization (TOC) of BPA by γ-Al_2O_3 with different Cu_2O content.

Time (min)	Pure γ-Al_2O_3	0.5% Cu_2O/γ-Al_2O_3	1% Cu_2O/γ-Al_2O_3
20	19.8	24.0	24.5
40	46.6	60.2	79.6
60	79.6	88.3	90.7

3.5. Effect of Radiation Source Wavelength

This variable was studied by using lamps emitting light in the visible spectrum (Vis light). Figure 8 shows the effect of different reaction variables on the TOC removal percentage after 60 min of reaction time. According to the results depicted in Figure 8, the removal of BPA by adsorption was only 5% and the removal by interaction with hydrogen peroxide was only 2.5% (gray bar). However, the TOC removal increased when Vis light was added to the system, which enhanced the TOC removal up to 18%. By comparing the results with Vis and the ones with Vis + H_2O_2, it can be concluded that dissociation of H_2O_2 does not occur readily or does not proceed at all under visible light, and therefore the addition of hydrogen peroxide alone under visible light does not improve TOC removal. On the other hand, when the catalyst and hydrogen peroxide were added to the system, the TOC removal significantly increased up to 40 and 45% for the 1% and 2 w% Cu catalyst, respectively. This is evidence of the activity of the catalyst under visible light, which is a desirable characteristic in a catalyst for the sake of sustainability. The activity of Cu_2O under visible light was expected due to its narrow bandgap (2.2 eV) [18].

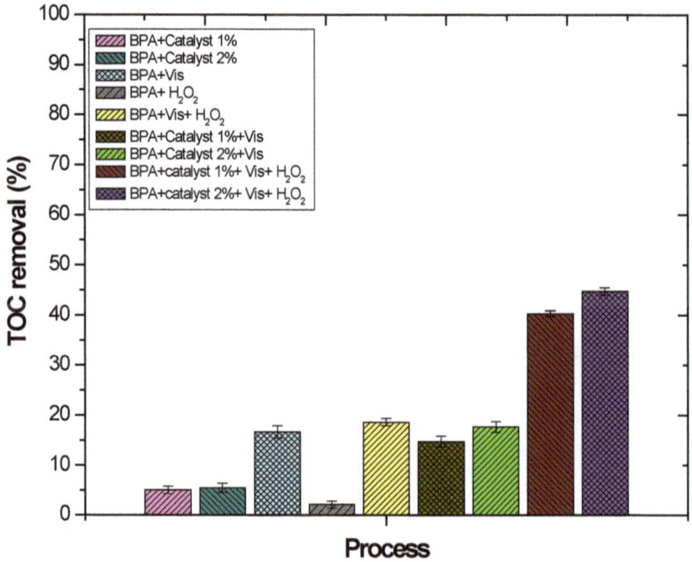

Figure 8. Effect of visible light on TOC removal percentage.

4. Conclusions

Cu_2O nanoparticles were obtained via laser ablation in liquid (LAL) and supported on Al_2O_3. This material catalyzes the photo-Fenton process to remove bisphenol A (BPA) under UV light as well as under visible light at circumneutral pH, and the need for acidification is eliminated. The BPA oxidation rate is one order of magnitude higher when using the catalyst than when applying only UV light and is more than twice that achieved when H_2O_2 is only dissociated by light. The BPA oxidation rate is affected by the crystalline structure of the catalyst support, Al_2O_3, and the mineralization extent is directly related to the initial H_2O_2 concentration. The highest mineralization percentage (ca. 91%) was found using 1%Cu_2O/γ-Al_2O_3, 10 times the stoichiometric amount of H_2O_2 at an initial pH of 6.7, and catalyst loading of 0.2 g/L.

After 1 h of treatment, the remaining compounds in solution were acetic acid, acetaldehyde, and in smaller quantities 4-hydroxybenzaldehyde and 4-hydroxybenzoic acid.

Author Contributions: Conceptualization, O.O.-M., R.N. and K.D.; methodology, O.O.-M., R.N., K.D., S.B. and D.A.-P.; validation, O.O.-M., R.N., K.D. and S.B.; investigation, O.O.-M., R.N., K.D., S.B. and D.A.-P.; writing—original draft preparation, O.O.-M. and R.N.; writing—review and editing, O.O.-M., R.N., K.D. and S.B.; supervision, O.O.-M. and R.N.; project administration, O.O.-M. and R.N.; funding acquisition, O.O.-M. and R.N. All authors have read and agreed to the published version of the manuscript.

Funding: This research received no external funding.

Data Availability Statement: Not applicable.

Conflicts of Interest: The authors declare no conflict of interest.

Abbreviations List and Nomenclature

Abbreviations List	
AOP	Advanced oxidation process
BPA	Bisphenol A
BPS	Bisphenol S
CE	Capillary electrophoresis
EDS	Energy dispersive X-ray spectroscopy
HRTEM	High resolution transmission electron microscopy
LAL	Laser ablation in liquids
LC/MS	Liquid chromatography coupled with mass spectroscopy
TEM	Transmission electron microscopy
TOC	Total organic carbon
SEM	Scanning electron microscopy
Nomenclature	
$-r_{BPA,o}$	Initial BPA oxidation rate
k	Pseudo-first order kinetic constant
R2	Determination coefficient
CBPA,o	Initial BPA concentration
pHo	Initial pH
Wcat	Catalyst concentration

References

1. Chen, M.-Y.; Ike, M.; Fujita, M. Acute toxicity, mutagenicity, and estrogenicity of bisphenol-A and other bisphenols. *Environ. Toxicol.* **2002**, *17*, 80–86. [CrossRef] [PubMed]
2. Grignard, E.; Lapenna, S.; Bremer, S. Weak estrogenic transcriptional activities of Bisphenol A and Bisphenol S. *Toxicol. Vitr.* **2012**, *26*, 727–731. [CrossRef]
3. Zhu, S.; Khan, M.A.; Kameda, T.; Xu, H.; Wang, F.; Xia, M.; Yoshioka, T. New insights into the capture performance and mechanism of hazardous metals Cr^{3+} and Cd^{2+} onto an effective layered double hydroxide based material. *J. Hazard. Mater.* **2022**, *426*, 128062. [CrossRef] [PubMed]
4. Zhu, S.; Xia, M.; Chu, Y.; Khan, M.A.; Lei, W.; Wang, F.; Muhmood, T.; Wang, A. Adsorption and Desorption of Pb(II) on l-Lysine Modified Montmorillonite and the simulation of Interlayer Structure. *Appl. Clay Sci.* **2018**, *169*, 40–47. [CrossRef]
5. Munoz, M.; de Pedro, Z.M.; Casas, J.A.; Rodriguez, J.J. Preparation of magnetite-based catalysts and their application in heterogeneous Fenton oxidation—A review. *Appl. Catal. B Environ.* **2015**, *176–177*, 249–265. [CrossRef]
6. Chand, R.; Ince, N.H.; Gogate, P.R.; Bremner, D.H. Phenol degradation using 20, 300 and 520 kHz ultrasonic reactors with hydrogen peroxide, ozone and zero valent metals. *Sep. Purif. Technol.* **2009**, *67*, 103–109. [CrossRef]
7. Torres-Blancas, T.; Roa-Morales, G.; Barrera-Díaz, C.; Ureña-Nuñez, F.; Cruz-Olivares, J.; Balderas-Hernandez, P.; Natividad, R. Ozonation of Indigo Carmine Enhanced by Fe/Pimenta dioica L. Merrill Particles. *Int. J. Photoenergy* **2015**, *2015*, 608412. [CrossRef]
8. Ziylan, A.; Ince, N.H. Catalytic ozonation of ibuprofen with ultrasound and Fe-based catalysts. *Catal. Today* **2015**, *240*, 2–8. [CrossRef]
9. Umar, K.; Haque, M.M.; Mir, N.A.; Muneer, M.; Farooqi, I.H. Titanium Dioxide-mediated Photocatalysed Mineralization of Two Selected Organic Pollutants in Aqueous Suspensions. *J. Adv. Oxid. Technol.* **2013**, *16*, 252–260. [CrossRef]
10. Haque, M.M.; Khan, A.; Umar, K.; Mir, N.A.; Muneer, M.; Harada, T.; Matsumura, M. Synthesis, Characterization and Photocatalytic Activity of Visible Light Induced Ni-Doped TiO_2. *Energy Environ. Focus* **2013**, *2*, 73–78. [CrossRef]
11. Frankowski, R.; Płatkiewicz, J.; Stanisz, E.; Grześkowiak, T.; Zgoła-Grześkowiak, A. Biodegradation and photo-Fenton degradation of bisphenol A, bisphenol S and fluconazole in water. *Environ. Pollut.* **2021**, *289*, 117947. [CrossRef] [PubMed]
12. Huang, W.-C.; Jia, X.; Li, J.; Li, M. Dynamics of microbial community in the bioreactor for bisphenol S removal. *Sci. Total Environ.* **2019**, *662*, 15–21. [CrossRef] [PubMed]
13. Yang, T.; Wang, L.; Liu, Y.; Huang, Z.; He, H.; Wang, X.; Jiang, J.; Gao, D.; Ma, J. Comparative study on ferrate oxidation of BPS and BPAF: Kinetics, reaction mechanism, and the improvement on their biodegradability. *Water Res.* **2018**, *148*, 115–125. [CrossRef]
14. Chai, F.; Li, K.; Song, C.; Guo, X. Synthesis of magnetic porous $Fe_3O_4/C/Cu_2O$ composite as an excellent photo-Fenton catalyst under neutral condition. *J. Colloid Interface Sci.* **2016**, *475*, 119–125. [CrossRef]
15. Davarnejad, R.; Azizi, J. Alcoholic wastewater treatment using electro-Fenton technique modified by Fe_2O_3 nanoparticles. *J. Environ. Chem. Eng.* **2016**, *4*, 2342–2349. [CrossRef]
16. Mahy, J.G.; Tasseroul, L.; Zubiaur, A.; Geens, J.; Brisbois, M.; Herlitschke, M.; Hermann, R.; Heinrichs, B.; Lambert, S.D. Highly dispersed iron xerogel catalysts for p-nitrophenol degradation by photo-Fenton effects. *Microporous Mesoporous Mater.* **2014**, *197*, 164–173. [CrossRef]

17. Dükkancı, M. Sono-photo-Fenton oxidation of bisphenol-A over a $LaFeO_3$ perovskite catalyst. *Ultrason. Sonochemistry* **2018**, *40*, 110–116. [CrossRef] [PubMed]
18. Noor, S.H.M.; Othman, M.H.D.; Khongnakorn, W.; Sinsamphanh, O.; Abdullah, H.; Puteh, M.H.; Kurniawan, T.A.; Zakria, H.S.; El-Badawy, T.; Ismail, A.F.; et al. Bisphenol A Removal Using Visible Light Driven Cu_2O/PVDF Photocatalytic Dual Layer Hollow Fiber Membrane. *Membranes* **2022**, *12*, 208. [CrossRef]
19. Chen, D.D.; Yi, X.H.; Ling, L.; Wang, C.C.; Wang, P. Photocatalytic Cr(VI) sequestration and photo-Fenton bisphenol A decomposition over white light responsive PANI/MIL-88A(Fe). *Appl. Organomet. Chem.* **2020**, *34*, e5795. [CrossRef]
20. Fu, H.; Song, X.X.; Wu, L.; Zhao, C.; Wang, P.; Wang, C.C. Room-temperature preparation of MIL-88A as a heterogeneous photo-Fenton catalyst for degradation of rhodamine B and bisphenol A under visible light. *Mater. Res. Bull.* **2020**, *125*, 110806. [CrossRef]
21. Huang, W.; Luo, M.; Wei, C.; Wang, Y.; Hanna, K.; Mailhot, G. Enhanced heterogeneous photo-Fenton process modified by magnetite and EDDS: BPA degradation. *Environ. Sci. Pollut. Res.* **2017**, *24*, 10421–10429. [CrossRef] [PubMed]
22. Pérez-Moya, M.; Kaisto, T.; Navarro, M.; del Valle, L.J. Study of the degradation performance (TOC, BOD, and toxicity) of bisphenol A by the photo-Fenton process. *Environ. Sci. Pollut. Res.* **2016**, *24*, 6241–6251. [CrossRef]
23. Silva, L.G.R.; Costa, E.P.; Starling, M.C.V.M.; Azevedo, T.d.S.; Bottrel, S.E.C.; Pereira, R.O.; Sanson, A.L.; Afonso, R.J.C.F.; Amorim, C.C. LED irradiated photo-Fenton for the removal of estrogenic activity and endocrine disruptors from wastewater treatment plant effluent. *Environ. Sci. Pollut. Res.* **2021**, *28*, 24067–24078. [CrossRef] [PubMed]
24. Xiao, S.; Zhou, C.; Ye, X.; Lian, Z.; Zhang, N.; Yang, J.; Chen, W.; Li, H. Solid-Phase Microwave Reduction of WO_3 by GO for Enhanced Synergistic Photo-Fenton Catalytic Degradation of Bisphenol A. *ACS Appl. Mater Inter.* **2020**, *12*, 32604–32614. [CrossRef] [PubMed]
25. Khandarkhaeva, M.; Batoeva, A.; Sizykh, M.; Aseev, D.; Garkusheva, N. Photo-Fenton-like degradation of bisphenol A by persulfate and solar irradiation. *J. Environ. Manag.* **2019**, *249*, 109348. [CrossRef]
26. Liu, Y.; Mao, Y.; Tang, X.; Xu, Y.; Li, C.; Li, F. Synthesis of Ag/AgCl/Fe-S plasmonic catalyst for bisphenol A degradation in heterogeneous photo-Fenton system under visible light irradiation. *Chin. J. Catal.* **2017**, *38*, 1726–1735. [CrossRef]
27. Bremner, D.H.; Burgess, A.E.; Houllemare, D.; Namkung, K.-C. Phenol degradation using hydroxyl radicals generated from zero-valent iron and hydrogen peroxide. *Appl. Catal. B: Environ.* **2006**, *63*, 15–19. [CrossRef]
28. Cheng, R.; Cheng, C.; Liu, G.-H.; Zheng, X.; Li, G.; Li, J. Removing pentachlorophenol from water using a nanoscale zero-valent iron/H_2O_2 system. *Chemosphere* **2015**, *141*, 138–143. [CrossRef]
29. Machado, S.; Stawiński, W.; Slonina, P.; Pinto, A.R.; Grosso, J.P.; Nouws, H.P.A.; Albergariaa, J.T.; Delerue-Matosa, C. Application of green zero-valent iron nanoparticles to the remediation of soils contaminated with ibuprofen. *Sci. Total Environ.* **2013**, *461–462*, 323–329. [CrossRef]
30. de la Plata, G.B.O.; Alfano, O.M.; Cassano, A.E. 2-Chlorophenol degradation via photo Fenton reaction employing zero valent iron nanoparticles. *J. Photochem. Photobiol. A: Chem.* **2012**, *233*, 53–59. [CrossRef]
31. Pastrana-Martínez, L.M.; Pereira, N.; Lima, R.; Faria, J.L.; Gomes, H.T.; Silva, A.M.T. Degradation of diphenhydramine by photo-Fenton using magnetically recoverable iron oxide nanoparticles as catalyst. *Chem. Eng. J.* **2015**, *261*, 45–52. [CrossRef]
32. Shahwan, T.; Abu Sirriah, S.; Nairat, M.; Boyacı, E.; Eroğlu, A.; Scott, T.; Hallam, K. Green synthesis of iron nanoparticles and their application as a Fenton-like catalyst for the degradation of aqueous cationic and anionic dyes. *Chem. Eng. J.* **2011**, *172*, 258–266. [CrossRef]
33. Shin, S.; Yoon, H.; Jang, J. Polymer-encapsulated iron oxide nanoparticles as highly efficient Fenton catalysts. *Catal. Commun.* **2008**, *10*, 178–182. [CrossRef]
34. Morales-Leal, F.J.; De la Rosa, J.R.; Lucio-Ortiz, C.J.; Martínez, D.B.; Rio, D.A.D.H.D.; Garza-Navarro, M.A.; Martínez-Vargas, D.X.; Garcia, C.D. Comparison between the catalytic and photocatalytic activities of Cu/Al_2O_3 and TiO_2 in the liquid–phase oxidation of methanol–ethanol mixtures: Development of a kinetic model for the preparation of catalyst. *Appl. Catal. A Gen.* **2018**, *562*, 184–197. [CrossRef]
35. Pradhan, A.C.; Parida, K.M.; Nanda, B. Enhanced photocatalytic and adsorptive degradation of organic dyes by mesoporous Cu/Al_2O_3–MCM-41: Intra-particle mesoporosity, electron transfer and OH radical generation under visible light. *Dalton. T.* **2011**, *40*, 7348–7356. [CrossRef]
36. Javaid, R.; Qazi, U.Y. Catalytic Oxidation Process for the Degradation of Synthetic Dyes: An Overview. *Int. J. Environ. Res. Public Health* **2019**, *16*, 2066. [CrossRef]
37. Kanitz, A.; Kalus, M.-R.; Gurevich, E.L.; Ostendorf, A.; Barcikowski, S.; Amans, D. Review on experimental and theoretical investigations of the early stage, femtoseconds to microseconds processes during laser ablation in liquid-phase for the synthesis of colloidal nanoparticles. *Plasma Sources Sci. Technol.* **2019**, *28*, 103001. [CrossRef]
38. Rehbock, C.; Merk, V.; Gamrad, L.; Streubel, R.; Barcikowski, S. Size control of laser-fabricated surfactant-free gold nanoparticles with highly diluted electrolytes and their subsequent bioconjugation. *Phys. Chem. Chem. Phys.* **2012**, *15*, 3057–3067. [CrossRef]
39. Olea-Mejía, O.; Fernández-Mondragón, M.; Rodríguez-de la Concha, G.; Camacho-López, M. SERS-active Ag, Au and Ag–Au alloy nanoparticles obtained by laser ablation in liquids for sensing methylene blue. *Appl. Surf. Sci.* **2015**, *348*, 66–70. [CrossRef]
40. Eisenberg, G. Colorimetric Determination of Hydrogen Peroxide. *Ind. Eng. Chem. Anal. Ed.* **1943**, *15*, 327–328. [CrossRef]

41. Amado-Piña, D.; Roa-Morales, G.; Molina-Mendieta, M.; Balderas-Hernández, P.; Romero, R.; Díaz, C.E.B.; Natividad, R. E-peroxone process of a chlorinated compound: Oxidant species, degradation pathway and phytotoxicity. *J. Environ. Chem. Eng.* **2022**, *10*, 108148. [CrossRef]
42. Alvarado-Rolon, O.; Natividad, R.; Ramírez-García, J.; Orozco-Velazco, J.; Hernandez-Servin, J.; Ramírez-Serrano, A. Kinetic modelling of paracetamol degradation by photocatalysis: Incorporating the competition for photons by the organic molecule and the photocatalyst. *J. Photochem. Photobiol. A Chem.* **2021**, *412*, 113252. [CrossRef]
43. Brillas, E.; Sirés, I.; Oturan, M.A. Electro-Fenton Process and Related Electrochemical Technologies Based on Fenton's Reaction Chemistry. *Chem. Rev.* **2009**, *109*, 6570–6631. [CrossRef] [PubMed]
44. Hurtado, L.; Romero, R.; Mendoza, A.; Brewer, S.; Donkor, K.; Gómez-Espinosa, R.M.; Natividad, R. Paracetamol mineralization by Photo Fenton process catalyzed by a Cu/Fe-PILC under circumneutral pH conditions. *J. Photochem. Photobiol. A Chem.* **2019**, *373*, 162–170. [CrossRef]
45. Hurtado, L.; Avilés, O.; Brewer, S.; Donkor, K.K.; Romero, R.; Gómez-Espinosa, R.M.; Alvarado, O.; Natividad, R. Al/Cu-PILC as a Photo-Fenton Catalyst: Paracetamol Mineralization. *ACS Omega* **2022**, *7*, 23821–23832. [CrossRef] [PubMed]
46. Watanabe, N.; Horikoshi, S.; Kawabe, H.; Sugie, Y.; Zhao, J.; Hidaka, H. Photodegradation mechanism for bisphenol A at the TiO_2/H_2O interfaces. *Chemosphere* **2003**, *52*, 851–859. [CrossRef]
47. Bechambi, O.; Jlaiel, L.; Najjar, W.; Sayadi, S. Photocatalytic degradation of bisphenol A in the presence of Ce–ZnO: Evolution of kinetics, toxicity and photodegradation mechanism. *Mat. Chem. Phys.* **2016**, *173*, 95–105. [CrossRef]
48. Young, T.; Geng, M.; Thagard, S.M.; Lin, L. Oxidative Degradation of Bisphenol A: A Comparison Between Fenton Reagent, UV, UV/H_2O_2 and Ultrasound. *J. Adv. Oxid. Technol.* **2013**, *16*, 89–101. [CrossRef]
49. Mendoza, A.; Romero, R.; Gutiérrez-Cedillo, G.P.; López-Tellez, G.; Lorenzo-González, O.; Gómez-Espinosa, R.M.; Natividad, R. Selective production of dihydroxyacetone and glyceraldehyde by photo-assisted oxidation of glycerol. *Catal. Today* **2019**, *358*, 149–154. [CrossRef]

water

MDPI

Article

Degradation of Emerging Pollutants by Photocatalysis: Radiation Modeling and Kinetics in Packed-Bed Reactors

Agustina Manassero, Orlando Mario Alfano and María Lucila Satuf *

Instituto de Desarrollo Tecnológico para la Industria Química (Universidad Nacional del Litoral and Consejo Nacional de Investigaciones Científicas y Técnicas), Ruta Nacional 168, Predio CCT CONICET-Santa Fe, Santa Fe 3000, Argentina
* Correspondence: mlsatuf@santafe-conicet.gov.ar

Abstract: Radiation modeling and kinetics in two different packed-bed reactors filled with TiO_2-coated glass rings are presented. The first reactor was cylindrical, irradiated from one end by a 150 W mercury lamp. It was employed to obtain the values of the intrinsic kinetic parameters of the degradation of the emerging contaminant clofibric acid (*CA*). The expression to represent the kinetics of the pollutant was derived from a proposed reaction scheme, and it includes explicitly the effect of photon absorption rate on the reaction rate. The second reactor was annular, irradiated internally and externally by 40 UV-LED lamps. The kinetic parameters calculated in the first reactor were directly employed to simulate the performance of the second one, without using any adjustable parameter. The Monte Carlo method was applied to solve the radiation models in both reactors. Good agreement was obtained between simulation results and experimental data under different operating conditions, with a percentage root-mean-square error of 4.6%. The kinetic parameters proved to be independent of the irradiation source, reactor geometry, and catalyst film thickness, and can be readily applied to design real scale devices.

Keywords: photocatalysis; emerging contaminant; intrinsic kinetics; packed bed; UV-LEDs; Monte Carlo

Citation: Manassero, A.; Alfano, O.M.; Satuf, M.L. Degradation of Emerging Pollutants by Photocatalysis: Radiation Modeling and Kinetics in Packed-Bed Reactors. *Water* **2022**, *14*, 3608. https://doi.org/10.3390/w14223608

Academic Editor: Sergi Garcia-Segura

Received: 1 October 2022
Accepted: 6 November 2022
Published: 9 November 2022

1. Introduction

Contaminants of emerging concern (CECs) are increasingly being detected at low levels in surface water, and there is widespread concern that these compounds may have a negative impact on aquatic life and human health. Generally, CECs are resistant to conventional biological treatment and current research is focused on the development of efficient methods to remove them from the environment. Heterogeneous photocatalysis has proven to be a useful technology to degrade this type of pollutant, including pharmaceuticals and personal care products [1–3]. The photocatalytic process starts when a semiconductor, usually titanium dioxide (TiO_2), is irradiated with photons with adequate energy to excite electrons to the conduction band, creating the electron/hole pairs. These charge carriers can initiate oxidation/reduction reactions that degrade adsorbed contaminants and often convert them into harmless products such as carbon dioxide, water and inorganic acids.

Despite the large number of papers available in the literature about the fundamentals and phenomenological aspects of photocatalysis, real applications are still scarce. A possible explanation is the lack of systematic approach for reactor design and scaling-up. Most contributions in this area are based on empirical methodologies that involve the gradual increase in the photoreactor size, it being a time-consuming and expensive practice [4]. A more rational approach consists of the use of rigorous mathematical modeling, including the derivation of kinetic expressions, evaluation of the rate of photon absorption in the reactors, assessment of the effect of radiation absorption on the reaction rate, and resolution of mass balances of the chemical species involved. By following this approach, intrinsic kinetic parameters, i.e., independent of the experimental conditions, can be obtained [5–7]. This modeling methodology has been reported and verified for scaling-up slurry reactors [8,9], and reactors with the catalyst immobilized over the windows [10]. The use of suspended TiO_2 is not recommended for large-scale reactors because it requires a catalyst recovery stage at the end of the treatment. Thus, the use of the catalyst immobilized onto a support represents a good choice in order to avoid the post-filtration step. Recently, Grcic et al. [11] determined intrinsic kinetic parameters for the photocatalytic oxidation of oxalic and formic acid in a slurry reactor. Then, these parameters were used to model the degradation of the carboxylic acids in a reactor with TiO_2-chitosan films immobilized on glass plates. The six-flux absorption scattering model was used to model the radiation field in both reactors. Results showed that the kinetic parameters can be used to represent the photocatalytic degradation of both compounds using either suspended or immobilized photocatalyst. Although local values of photon absorption were calculated in the reactors, the reaction rate expressions include the average value (across the reactor volume in the slurry reactor, and over the photocatalyst surface in the immobilized reactor). Additionally, to simulate the pollutants degradation in the immobilized reactor, an additional parameter, the apparent film absorption coefficient, was adjusted.

It is worth mentioning that the selection and location of the radiation source are important features to design large-scale photocatalytic reactors [12]. UV light emitting diodes (LEDs) have received great attention in recent years, especially due to its eco-friendliness, long life time, energy efficiency, low cost and flexibility for integration within a photocatalytic reactor [13].

In this work, the modeling of the photocatalytic oxidation of a CEC in two packed-bed reactors filled with TiO_2-coated glass rings is presented. Clofibric acid (*CA*), the active metabolite of many pharmaceuticals employed as blood lipid regulators, was the emerging pollutant chosen to validate the proposed methodology. The reactors have different geometry, size, configuration, and irradiation sources. A kinetic model that incorporates the effect of radiation absorption by the photocatalyst on the reaction rate was established. Kinetic parameters of the degradation of *CA* were first estimated in a simple, cylindrical reactor irradiated from one end by a mercury lamp. A 1-D radiation model was employed to calculate the photon absorption distribution. Then, the intrinsic nature of these parameters was verified by simulating the performance of an annular reactor with a more complex configuration and illuminated by UV LEDs. In this case, a 3-D radiation model was developed. The Monte Carlo method was applied to solve the radiation models in both reactors.

2. Employed Strategy

The aim of this work was to validate the use of intrinsic kinetic parameters and radiation models to predict the performance of packed-bed photocatalytic reactors. The employed strategy involved two stages, as summarized in Figure 1.

1- Intrinsic kinetic parameters were first obtained in a simple, cylindrical reactor (CR) irradiated from one end by a mercury lamp. To accomplish this, a kinetic model to represent the photocatalytic degradation of the pollutant *CA* was developed. The expression of the *CA* degradation rate includes the value of the local surface rate of photon absorption (LSRPA or $e^{a,s}$). Therefore, a 1-D radiation model was solved to obtain the LSRPA distribution inside the reactor. Then, the mass balance of *CA* was set in order to predict the evolution of the pollutant in the system. Finally, the values of the intrinsic kinetic parameters were obtained by applying an optimization algorithm to adjust model simulations to experimental data.
2- The second stage involved the use of the kinetic parameters found in the first reactor to predict the performance of the second one: an annular reactor (AR) illuminated by UV-LEDs. This task involved the resolution of the mass balance for *CA*, using the kinetic expression obtained in the CR, and the calculation of the LSRPA. Due to its particular configuration and illumination arrangement, a 3-D radiation model was

developed to calculate the LSRPA in the AR. Finally, simulation results were compared with experimental data in order to validate the modeling procedure.

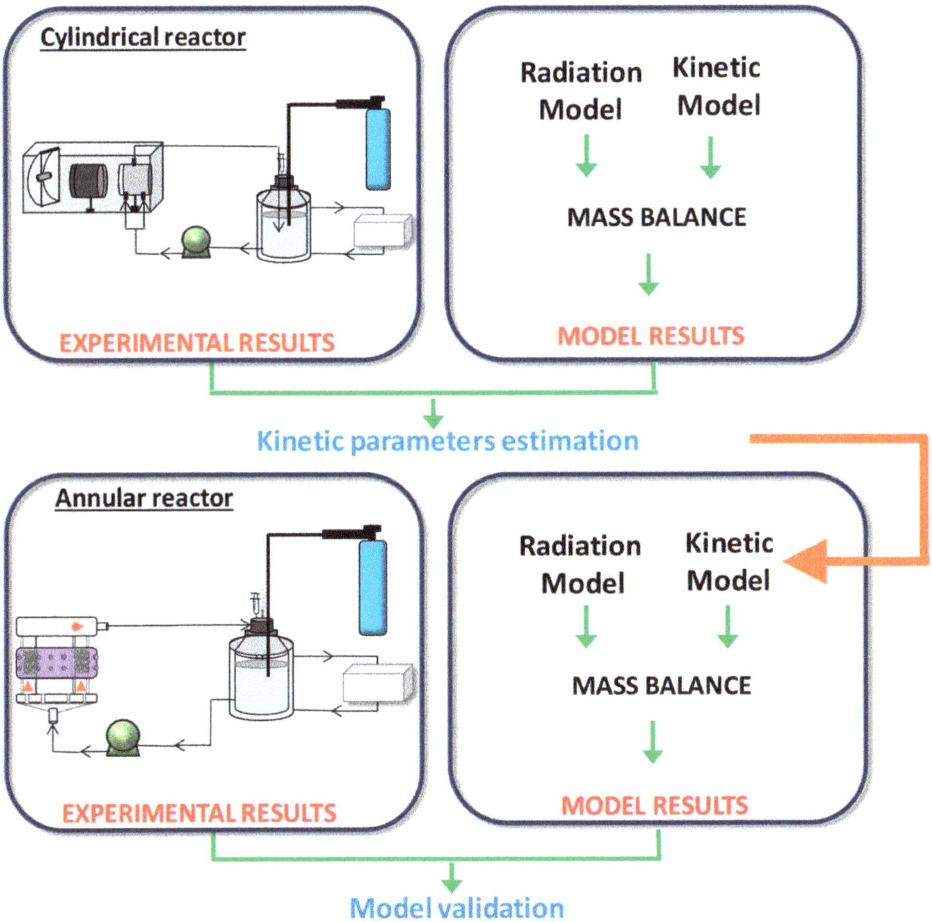

Figure 1. Schematic representation of employed methodology.

3. Experimental Devices

The packed-bed reactors employed in this study, which operate in batch recycling mode, were very different in shape, configuration and irradiation conditions (source type and distribution of the lamps). A description of the experimental devices is detailed below.

3.1. Reactors and Irradiation Sources

Cylindrical reactor: It consists of a glass tube, held horizontally, with two circular flat windows. It was irradiated from one side by one halogenated mercury lamp (Powerstar HQI-TS 150 W/NDL from OSRAM), placed at the focal axis of a parabolic reflector (Figure 2a). The illuminated window was made of Pyrex ground glass. The wavelength emission range of the lamp was 350–780 nm. To avoid visible radiation, a Pyrex glass container with a $CoSO_4$ solution of 20 g/L was interposed between the lamp and the reactor window. The wavelengths of the resulting radiation that arrived at the reactor window were between 350 nm and 410 nm, as shown in Figure 3a. This reactor was filled with 310 Pyrex glass rings of 0.5 cm (diameter) × 0.5 cm (height), in random disposition, coated with TiO_2. The liquid entered the reactor through two bottom ports and exited from a single port situated on top. The inlet ports were placed obliquely, creating a swirl-flow of the liquid inside the reactor that ensured well-mixing conditions.

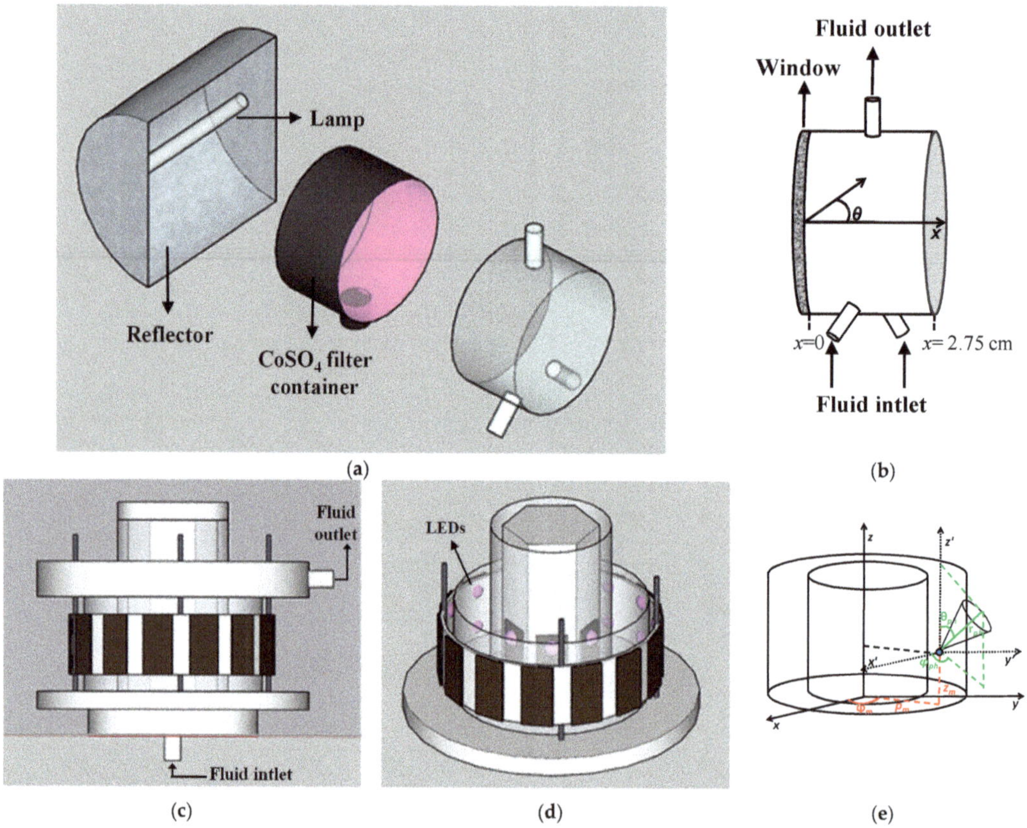

Figure 2. Schemes of the photoreactors and illumination sources: (**a**) Cylindrical reactor; (**b**) Coordinate system in the CR; (**c**) Lateral view of the annular reactor; (**d**) Top view of the annular reactor without the lid; (**e**) Coordinate systems in the AR.

Annular reactor: This reactor was made in Pyrex glass with two concentric tubes of different diameters, held vertically. It was irradiated with 40 UV-LEDs (Roithner XSL-375-TF-R2) located as two parallel strips around the external and the internal walls of the reactor. Each inner strip had 6 LEDs, and each external strip had 14 LEDs, as presented in Figure 2c,d. The maximum emission of the LED lamps was 365 nm (Figure 3b). The AR was filled with 900 Pyrex glass rings coated with TiO_2, with the same dimensions and disposition as reported for the CR. The reactor inlet contains a perforated plate that acts as a distributor to ensure uniform distribution of the fluid in the packed bed.

3.2. Recycling Setup

Each reactor was part of a recycling system consisting of a peristaltic pump and a storage tank. Each tank was fitted with a sampling port and a gas inlet through which oxygen was constantly bubbled. Isothermal conditions (25 °C) were achieved by the incorporation of a water-circulating jacket to the storage tank. A recirculation flow rate of 1.5 L/min was used in both systems.

More details of the experimental setups and reactors can be found in references [14,15].

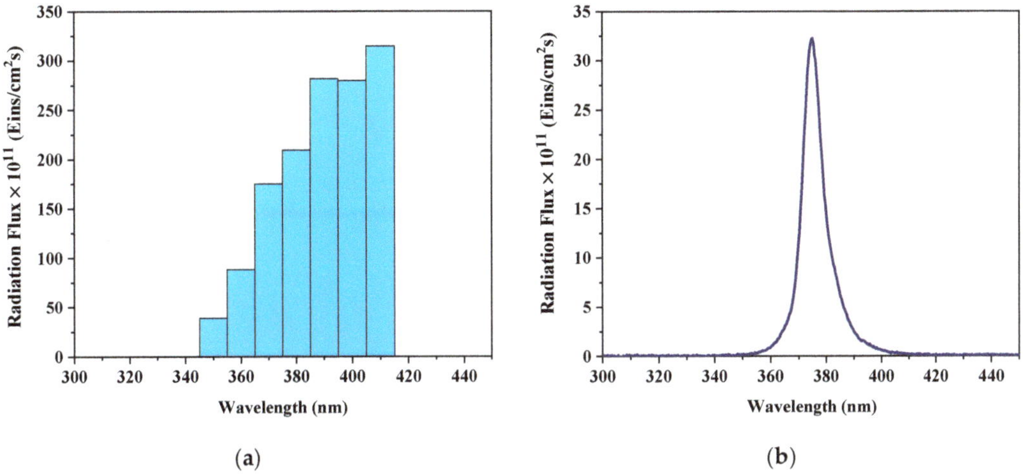

Figure 3. Radiation flux at the reactor window: (**a**) Hg lamp; (**b**) UV LED.

The main characteristics of both reactors are summarized in Table 1.

Table 1. Main characteristics of the reactors.

	Photoreactor	
Characteristics	**Cylindrical**	**Annular**
Main dimensions	Length = 2.75 cm Diameter = 5.0 cm	Length = 4.0 cm Inner diameter = 6.5 cm Outer diameter = 10.5 cm
Reactor volume	54 mL	214 mL
Total system volume	1000 mL	600 mL
Irradiation source	One mercury lamp	40 UV-LED lamps
Lamp emission range	350–410 nm	360–390 nm
Number of rings	310	900
Catalytic area	487 cm^2	1414 cm^2
Irradiation	From one side	Internal and external
Distance between lamps and irradiated windows	33 cm	Internal LEDs = 0.3 cm External LEDs = 0.6 cm
Irradiated window area	19.6 cm^2	Inner = 81.6 cm^2 Outer = 131.8 cm^2

3.3. Catalyst Immobilization

Bare Pyrex glass rings were dipped into a suspension of 150 g/L of TiO_2 Aeroxide P25 and extracted with a withdrawal velocity of 3 cm/min [16]. Then, they were dried at 110 °C for 24 h and calcined at 500 °C for 2 h with a heating rate of 5 °C/min. This procedure was repeated to obtain rings with different thicknesses of the TiO_2 film. The thickness of the resulting films (t_{TiO_2}) was estimated from SEM images with a scanning electron microscope (JEOL, JSM-35C). The phase composition of the photocatalyst after calcination was evaluated by X-ray diffraction with a Shimadzu XD-D1 diffractometer with Cu-Kα radiation. The results showed that the catalyst contains 82% anatase and 18% rutile.

4. Experimental Assays

Experiments of *CA* degradation in the photocatalytic reactors were carried out at different values of initial pollutant concentration, TiO_2 film thickness, and irradiation level, as reported in Table 2.

Table 2. Experimental conditions.

	Photoreactor	
Variable	**Cylindrical**	**Annular**
Initial CA concentration, $C_{CA,0}$ [mol/cm^3]	0.93×10^{-7}	$(0.93, 1.87, 2.34) \times 10^{-7}$
Incident radiation fluxes, q_w[Eins/(s cm^2)]	$(15.2, 9.39, 4.58) \times 10^{-9}$	$(2.29, 2.94, 3.34) \times 10^{-9}$
Thickness of TiO_2 film, t_{TiO_2} [μm]	0.27, 0.44	0.27, 3.24

In the CR, different levels of incident radiation were obtained by locating optical neutral filters at the external side of the reactor window. These filters attenuated the incident radiation without altering the spectral distribution of the lamp (100%: no filter; 62%: low optical density filter; 30%: high optical density filter). In the AR, different irradiation levels were obtained by switching on the external lamps only, the internal lamps only, and both (total illumination condition). The incident photon flux at the reactor windows (q_w) under each condition, in both reactors, was determined by ferrioxalate actinometry [17].

All the experiments were carried out following the same procedure. *CA* solutions were prepared by dissolving the required mass of *CA* in ultrapure water. The initial pH of the solutions was 5. Prior to each run, the reactor was filled up with the coated rings, and the *CA* solution was placed in the storage tank and circulated in the system for 30 min under dark conditions. During this time, the solution was saturated with pure oxygen. In the case of the CR, the Hg lamp was warmed up 30 min before the beginning of the experiment. Samples were collected from the tank at time intervals of 1 h during 6 h. Then, they were filtered through syringe filters (Anotop 25) and *CA* was quantified by HPLC with UV detection [18]. The analyses were carried out in a Waters chromatograph with a RP C-18 column (XTerra®). The mobile phase was a binary mixture of acidified water (with 0.1 % v/v phosphoric acid) and acetonitrile (50:50). The flow rate was 1.0 mL/min and the injection volume was 20 μL. Absorbance detection was made at 227 nm.

5. Reactor Modeling

5.1. Mass Balance

The theoretical evolution of the concentration of the pollutant *CA* in a batch system with recycle can be obtained by solving the mass balance of the referred chemical compound. Assuming that the conversion per pass in the reactor is differential; the system is well-mixed; mass transfer limitations are negligible; direct photolysis is insignificant; and chemical reactions occur only at the solid–liquid interface among adsorbed molecules, the mass balance expression for *CA* and the initial condition in the reacting system can be written as

$$\frac{dC_{CA}(t)}{dt} = -\frac{A_{cat}}{V_T}\langle r_{CA}(\boldsymbol{x},t)\rangle_{A_{cat}} \qquad C_{CA}(t=0) = C_{CA,0} \tag{1}$$

where C_{CA} is the molar concentration of *CA* (mol/cm^3), t refers to the reaction time (s), V_T is the total volume of the solution in the system (cm^3), $\boldsymbol{x}$ represents the position vector, and $\langle r_{CA}(\boldsymbol{x},t)\rangle_{A_{cat}}$ represents the surface degradation rate of *CA* averaged over the catalytic area, A_{cat} (cm^2).

The assumptions mentioned above and the resulting mass balance expression are valid for both systems. Calculations supporting the assumption of negligible mass transfer limitations are presented as Supplementary Material.

5.2. Kinetic Model

The expression of the reaction rate was mechanistically derived from a kinetic scheme summarized in Table 3. It involves the activation of TiO_2 upon UV irradiation with the formation of electron/hole pairs, electron and hole trapping, and the attack of *CA* molecules by hydroxyl radicals [19,20]. This mechanism leads to the formation of organic intermediates species X_i which, in turn, can be mineralized by further oxidation.

Table 3. Reaction scheme for the photocatalytic degradation of CA.

Step	Reaction	Reaction Rate
Activation	$TiO_2 + h\nu \rightarrow e^- + h^+$	r_{gs}
Recombination	$e^- + h^+ \rightarrow$ heat	$k_2[e^-][h^+]$
Electron trapping	$e^- + O_{2,ads} \rightarrow \cdot O_2^-$	$k_3[e^-][O_{2,ads}]$
Hole trapping	$h^+ + H_2O_{ads} \rightarrow \cdot OH + H^+$ $h^+ + OH^-_{ads} \rightarrow \cdot OH$	$k_4[h^+][H_2O_{ads}]$
Hydroxyl radical attack	$CA_{ads} + \cdot OH \rightarrow X_i \rightarrow\rightarrow Cl^- + CO_2 + H_2O$	$k_5[CA_{ads}][\cdot OH]$

The kinetic model also assumes that photocatalytic reactions occur at the surface of the catalyst particles among adsorbed molecules [21]; dynamic equilibrium is achieved between the bulk and the adsorbed concentrations of H_2O, O_2, inorganic species and organic compounds [22,23]; molecular oxygen and organic compounds are adsorbed on different sites of the catalyst [24,25]; there is a competitive adsorption mechanism between *CA* and its main reaction intermediates; O_2 concentration is constant and in excess with respect to the stoichiometric demand; and the concentration of water molecules, hydroxyl ions, and total adsorption sites for *CA* on the catalytic surface remains constant [26]. The resulting local reaction rate expression for *CA* degradation is

$$r_{CA}(x,t) = \alpha_2 C_{CA}(t)\left[-1 + \sqrt{1 + \alpha_1 e^{a,s}(x)}\right] \tag{2}$$

where α_1 and α_2 are intrinsic kinetic parameters, and $e^{a,s}(x)$ is the local surface rate of photon absorption (LSRPA), that is, the amount of photons absorbed per unit time and per unit area of TiO_2-coated surface. As irradiation is constant throughout the experiments, and we assume that the optical properties of the TiO_2 films do not vary during the reaction time, $e^{a,s}$ is considered to be independent of time. A detailed derivation of Equation (2) can be found in [27].

Finally, by introducing the reaction rate expression into Equation (1), the mass balance for *CA* can be written as follows:

$$\frac{dC_{CA}(t)}{dt} = -\frac{A_{cat}}{V_T}\alpha_2 C_{CA}(t)\left\langle -1 + \sqrt{-1 + \alpha_1 e^{a,s}(x)}\right\rangle_{A_{cat}} \quad C_{CA}(t=0) = C_{CA,0} \tag{3}$$

From the resolution of Equation (3), the evolution of *CA* concentration under different experimental conditions can be predicted.

It is important to highlight that the LSRPA is explicitly included in the reaction rate expression. The dependence of the reaction rate on the photon absorption rate can take two limiting forms: linear, at high levels of irradiation, and square root, at low levels of irradiation. In many cases, the radiation field in photocatalytic reactors present high non-uniformities. These situations originate different dependence orders to co-exist in the same reactor, making it inadequate to employ an averaged value of LSRPA with a single exponent in the mass balance. In such cases, the complete radiation expression $\langle -1+\sqrt{-1+\alpha_1 e^{a,s}(x)}\rangle_{A_{cat}}$ should be employed [28].

5.3. Local Surface Rate of Photon Absorption (LSRPA)

The evaluation of the *CA* reaction rate requires the estimation of the absorbed photons distribution inside the reactors. In the present study, the Monte Carlo method was applied to obtain the LSRPA. This method consists of tracking a significant amount of photons emitted by the light source, which interact with the elements of the packed-bed and the reactor walls or windows, until they are absorbed or scattered out. The reactor bed volume is discretized into spatial cells, and when photons are absorbed, the location of this event is recorded. Then, it is possible to establish the distribution of the LSRPA inside the reactor. The direction, length of the trajectory, and fate of the photons are determined by random numbers between 0 and 1. Briefly, the proposed model assumes that emitted photons travel along rectilinear trajectory until they reach the reactor walls or a TiO_2-coated ring. The photons that reach a ring can be absorbed, transmitted or reflected. If they are absorbed in the TiO_2 film, its position is stored in the corresponding spatial cell, the trajectory ends, and the tracking of a new photon starts. If the photons are reflected, the new direction is determined by considering specular reflection. If the photons are transmitted, they may interact with other rings until they are absorbed or they reach the reactor walls. Finally, the LSRPA in each cell of the reactor was calculated according to:

$$e^{a,s}(x)=\sum\nolimits_{\lambda}\frac{q_{w\lambda}A_w n_{\lambda,abs}(x)V_R}{n_{\lambda,T}V_C A_{cat}} \quad (4)$$

where $q_{w\lambda}$ is the radiation flux of wavelength λ incident at the reactor window, A_w represents the irradiated area of the reactor windows, $n_{\lambda,abs}(x)$ represents the number of photons of wavelength λ absorbed in a cell of position x, $n_{\lambda,T}$ is the total number of the photons considered in the simulation for each wavelength interval (10^7 photons), and V_C is the volume of the cell. This procedure, schematized in Figure 4, was followed in both reactors to obtain the LSRPA. However, due to the particular configuration and radiation source of each reactor, different coordinate systems and dimensions were considered to develop each model, as summarized in Table 4.

In the CR, due to the configuration of the reactor and lamp (reflector and lamp arrangement, reactor ground window, and reactor dimensions), the radiation extinction inside the photocatalytic bed occurs mainly along the longitudinal axis of the reactor, and the incoming radiation can be considered diffuse with azimuthal symmetry. Therefore, a one-dimensional radiation model with the Cartesian coordinate x was employed to assess the location of photons inside the reactor, and the polar coordinate θ was used to determine the direction of the photons (Figure 2b). In the AR, due to the particular arrangement of the LED lamps and the reactor shape, a three-dimensional radiation model was required to evaluate the LSRPA. A cylindrical coordinate system (z_m, φ_m and ρ_m) was used to define the position of the emitting LED and the location of photons with respect to the reactor, and a spherical coordinate system (r_{ph}, θ_{ph} and φ_{ph}) was employed to indicate the direction of propagation of a given photon (Figure 2e).

A detailed description of the procedure applied to solve the radiation models by the Monte Carlo method can be found in references [14,15].

Table 4. Main differences of the radiation models in each reactor.

Characteristics	Cylindrical Reactor	Annular Reactor
Spatial dimension	1D	3D
Number of spatial cells	10^3	10^6
Coordinate system for photon location	*Cartesian: x*	*Cylindrical:* z_m, φ_m, ρ_m
Coordinate system for photon direction	*Polar:* θ	*Spherical:* $r_{ph}, \theta_{ph}, \varphi_{ph}$

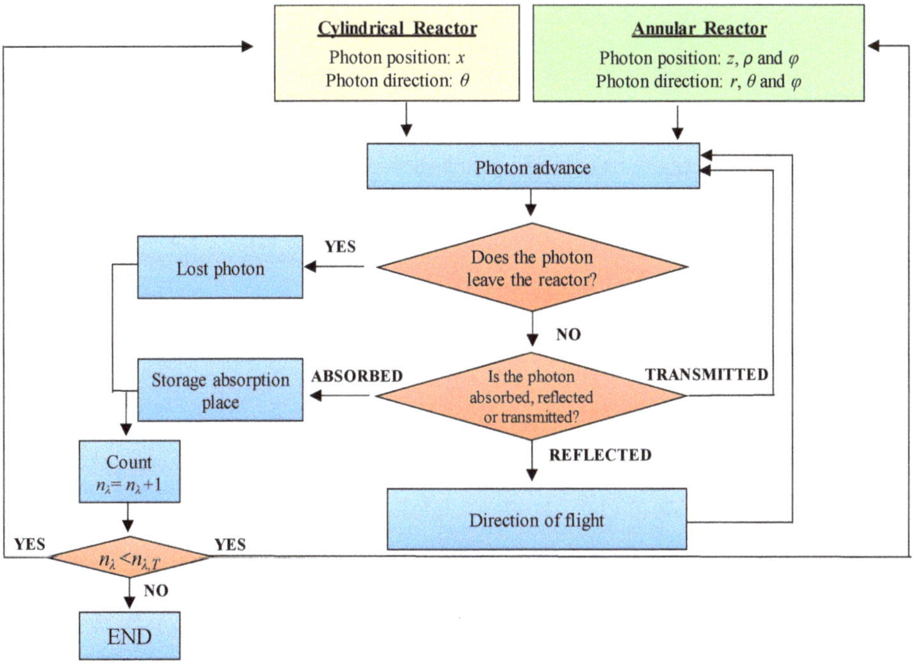

Figure 4. Flowchart of MC algorithm inside the reactors.

6. Results and Discussion

6.1. LSRPA Distribution

The most critical aspect in the proposed methodology is the evaluation of the absorbed radiation inside the reactors. Figure 5 shows the distribution of the LSRPA inside the packed beds of the CR and AR.

Figure 5a depicts the LSRPA in the CR under 100% irradiation (no filter) and t_{TiO_2} = 0.27 μm. Maximum values are achieved near the irradiated window, and radiation absorption decreases along the longitudinal axis of the reactor. Nevertheless, the decrease is smooth and radiation absorption at the opposite side of the illuminated window is still significant (1/3 of the maximum value). Figure 5b–d present the distribution of the absorbed radiation in the AR under total illumination and t_{TiO_2} = 0.27 μm. Figure 5b shows the LSRPA profile at the reactor vertical cross section corresponding to the azimuthal angle φ_m = 0°. LED strips were located at the reactor height z = 1 cm and z = 3 cm, as verified by the high values of LSRPA near these regions. The simulation also reveals that there are dark zones near the external window, at z = 2 cm, and at the bottom and top of the reactor (z = 0 cm and z = 4 cm). They are "blind" regions that are not illuminated by the external LEDs, and radiation from internal LEDs has already been extinguished. This is a usual problem when using LED lamps as an irradiation source because their viewing angle is generally narrow and it can cause high non-uniformities in the radiation distribution inside photoreactors. This effect is less evident at the internal window because external LEDs are more numerous (14 per external strip vs. 6 per internal strip) and the area of the internal window is smaller. Therefore, radiation from external LEDs sum up and reach the opposite side. Figure 5c depicts the LSRPA profile at the horizontal cross section of the reactor corresponding to z = 1 cm, where maximum values of LSRPA are reached near the irradiated windows. The positions of the LEDs are clearly individualized, mainly in the case of internal LEDs because they are located at 0.3 cm from the reactor internal window whereas external LEDs are located farther away from the external window, at 0.6 cm. In

Figure 5d, photon absorption distribution at the horizontal cross section corresponding to z = 2 cm (reactor height between the LEDs strips) is presented. The blind zone mentioned above, close to the external window, is evidenced.

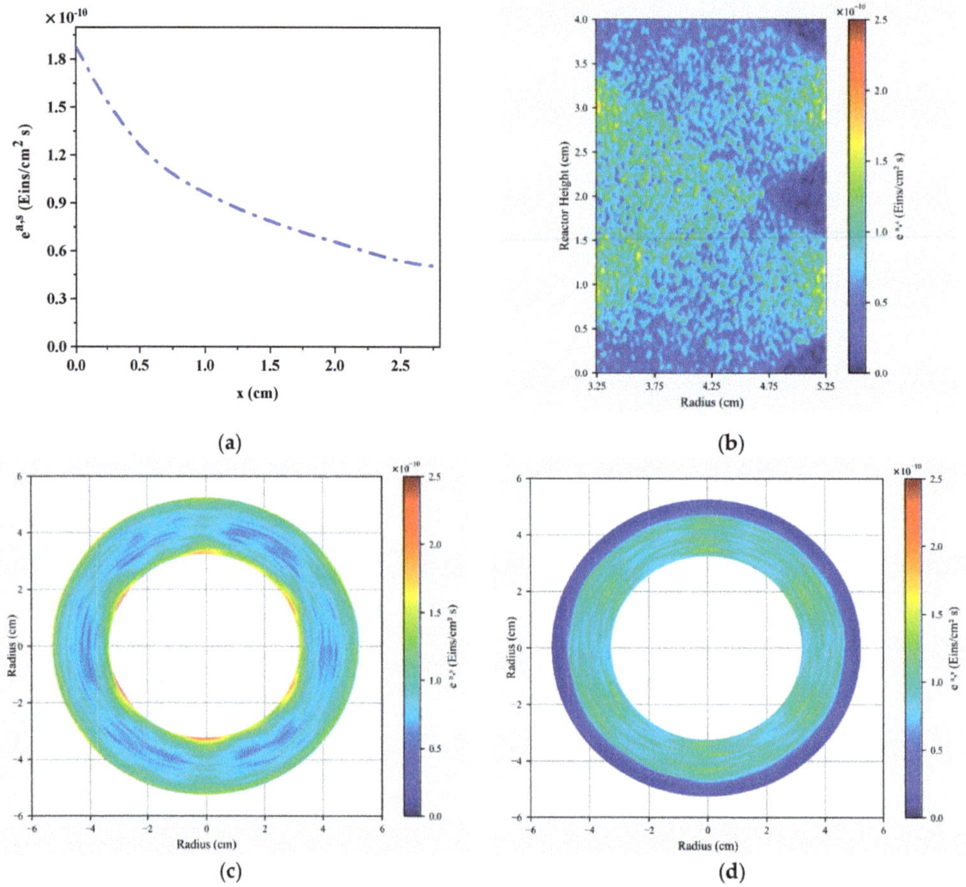

Figure 5. LSRPA distribution in the reactors under maximum irradiation and t_{TiO_2} = 0.27 μm. (**a**) 1-D plot in the CR. (**b**) 2-D plot, vertical cross section in the AR at φ_m = 0°. (**c**) 2-D plot, horizontal cross section in the AR at z = 1 cm. (**d**) 2-D plot, horizontal cross section in the AR at z = 2 cm is a figure.

6.2. Kinetic Simulations

The estimation of the kinetic parameters involved in Equation (3) was carried out employing *CA* concentration information from degradation experiments in the CR and the corresponding LSRPA values. A Levenverg–Marquardt optimization method was applied, which minimizes the differences between the estimated values of *CA* concentration and experimental data. The differential mass balance equation was solved employing a fourth-order Runge–Kutta method. The values of the kinetic parameters obtained, with the corresponding 95% confidence intervals, were: α_1 = (2.95 ± 0.31) × 10^{10} cm^2 s/Einstein, and α_2 = (3.24 ± 0.11) × 10^{-5} cm/s. As shown in Figure 6, these parameters were able to simulate with good accuracy the evolution of *CA* concentration under different irradiation conditions and film thicknesses in the CR.

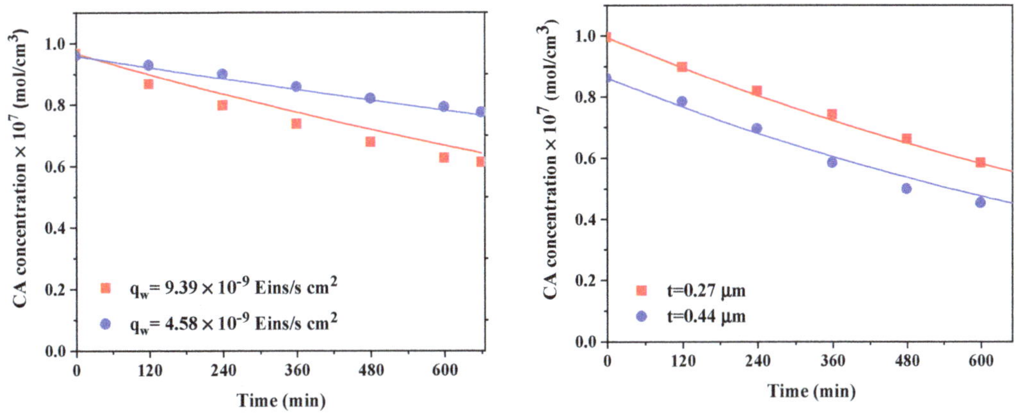

(a) (b)

Figure 6. Experimental results and model predictions for *CA* in the CR: (**a**) Different irradiation levels (t_{TiO_2} = 0.27 μm); (**b**) Different t_{TiO_2} (q_w = 15.2 × 10^{-9} Eins/s cm^2). Symbols: experimental data. Solid lines: model results.

Then, to validate the proposed strategy, the performance of the AR under different experimental conditions was simulated. The evolution of *CA* concentration in the LED reactor was calculated by solving Equation (3) with the two parameters previously estimated, and the corresponding values of LSRPA obtained from the 3D radiation model. It is important to remark that no adjustable parameters were employed in the simulations. Predicted and experimental *CA* concentrations in the AR under different conditions, for t_{TiO_2} = 0.27 μm, are shown in Figure 7.

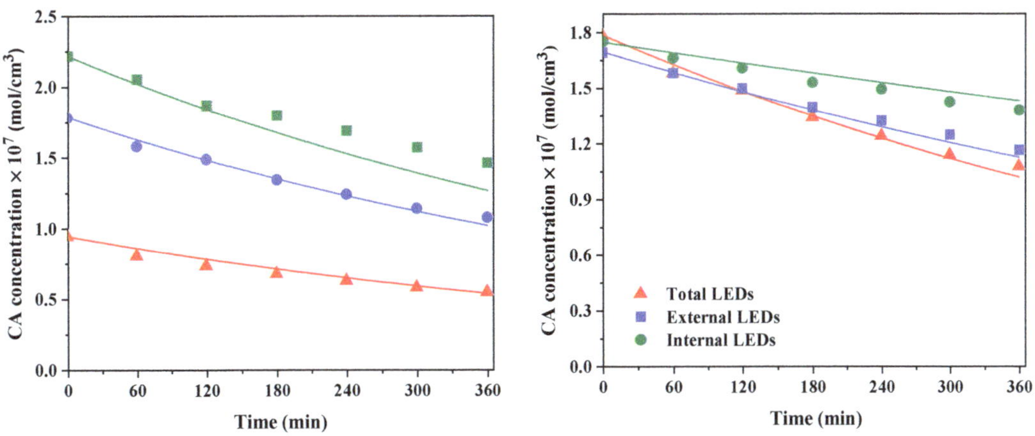

(a) (b)

Figure 7. Experimental results and model predictions for *CA* in the AR: (**a**) Different initial *CA* concentrations (total irradiation, t_{TiO_2} = 0.27 μm); (**b**) Different irradiation conditions ($C_{CA,0}$ = 1.8 × 10^{-7} mol/cm^3, t_{TiO_2} = 0.27 μm). Symbols: experimental data. Solid lines: model results.

As can be observed, very good agreement was obtained between model predictions and experimental results. The percentage root mean square error of the estimations was calculated according to the following expression:

$$RMSE\% = \sqrt{\frac{1}{N}\sum_{i=1}^{N}\left(\frac{C_{CA,\exp i} - C_{CA,i}}{C_{CA,\exp i}}\right)^2} \times 100 \quad (5)$$

where $C_{CA,\exp i}$ is the experimental concentration of CA in a defined time and experimental condition, and $C_{CA,i}$ is the corresponding predicted concentration by the model. N represents the total number of experimental points. The RMSE% for the experiments presented in Figure 7 is 4.6%.

Intrinsic kinetic parameters can be very useful to find the experimental conditions needed to achieve a defined goal. For example, if we want to obtain 80% CA conversion in 6 h under total illumination in the AR, the model indicates that a TiO_2 film thickness of 3.2 μm must be used. This prediction was experimentally verified, as presented in Figure 8. Under similar conditions but with external illumination only, 70% CA conversion was achieved after 6 h of irradiation. The corresponding experimental and simulation results under this condition are also depicted in Figure 8. The RMSE% of the estimations employing t_{TiO_2} = 3.2 μm was 12.6%. Although it was higher than the error obtained from simulations in Figure 7, the TiO_2 film thickness in the last cases was one order of magnitude higher than the t_{TiO_2} employed in the CR for the estimation of the kinetic parameters (0.27 μm and 0.44 μm). Despite this issue, reasonable predictions were obtained.

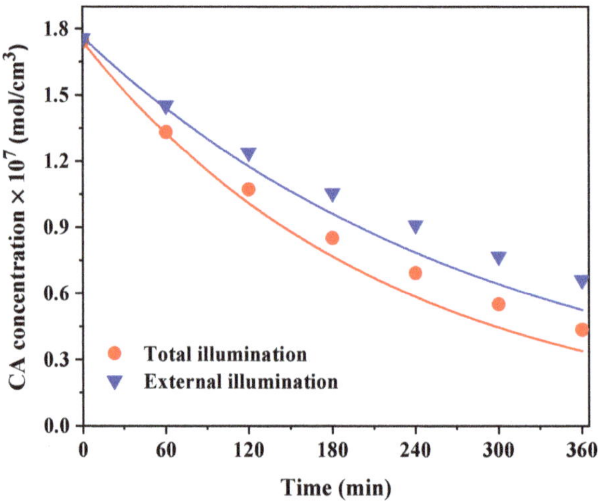

Figure 8. Experimental results and model predictions for CA in the AR under different irradiation conditions (t_{TiO_2} = 3.2 μm). Symbols: experimental data. Solid lines: Model results.

7. Conclusions

In this work, a strategy to obtain intrinsic kinetic parameters in a simple, cylindrical packed-bed photocatalytic reactor is presented. This information was used to simulate the performance of a more complex, annular packed-bed reactor, with different dimensions, types of lamp, and illumination arrangements. With only two kinetic parameters, and no adjustable factors, the proposed methodology can predict with good accuracy the behavior of the pollutant clofibric acid in the annular reactor. Because photocatalytic-reaction-rate expressions depend on the effect of radiation absorption, the key factor is to accurately model the photon absorption distribution inside the reactors. The presented results demonstrate that kinetic parameters calculated with this methodology are independent of the reactor geometry, reactor size and irradiation conditions, and that they can be employed to design, optimize and scale-up photocatalytic reactors. The rational design of photocatalytic reactors, employing intrinsic parameters and radiation models, constitutes an efficient strategy for bridging the gap between laboratory experiments and real applications.

Supplementary Materials: The following supporting information can be downloaded at: https://www.mdpi.com/article/10.3390/w14223608/s1, Evaluation of internal and external mass transfer limitations in the photocatalytic reactors. References [29–32] are cited in the supplementary materials.

Author Contributions: A.M.: Investigation, formal analysis, software, and writing—original draft preparation. O.M.A.: Overall review, editing, and fund acquisition. M.L.S.: Visualization, supervision, and writing—review and editing, and fund acquisition. All authors have read and agreed to the published version of the manuscript.

Funding: This research was funded by Universidad Nacional del Litoral, grant number CAI+D 2020 50620190100040LI from, Consejo Nacional de Investigaciones Científicas y Técnicas, grant number PIP 2021 11220200101324CO, and Agencia Nacional de Promoción de la Investigación, el Desarrollo Tecnológico y la Innovación, grant number PICT 2018-0926.

Data Availability Statement: The data presented in this study are available on request from the corresponding author.

Acknowledgments: The authors gratefully acknowledge Ing. Javier Borda Bossana for his technical assistance in the representation of simulation results.

Conflicts of Interest: The authors declare no conflict of interest.

References

1. Lee, B.C.Y.; Lim, F.Y.; Loh, W.H.; Ong, S.L.; Hu, J. Emerging Contaminants: An Overview of Recent Trends for Their Treatment and Management Using Light-Driven Processes. *Water* **2021**, *13*, 2340. [CrossRef]
2. Saravanan, A.; Senthil Kumar, P.; Jeevanantham, S.; Anubha, M.; Jayashree, S. Degradation of toxic agrochemicals and pharmaceutical pollutants: Effective and alternative approaches toward photocatalysis. *Environ. Pollut.* **2022**, *1*, 118844. [CrossRef]
3. Ahmed, S.; Saleem, F.; Khan, F.S.A.; Mubarak, N.M.; Khalid, M.; Tan, Y.; Mazari, S.A.; Karri, R.R.; Abdullah, E.C. Emerging pollutants and their removal using visible-light responsive photocatalysis—A comprehensive review. *J. Environ. Chem. Eng.* **2021**, *6*, 106643. [CrossRef]
4. Alfano, O.M.; Cassano, A.E. Scaling-Up of Photoreactors: Applications to Advanced Oxidation Processes. In *Advances in Chemical Engineering*, 1st ed.; De Lasa, H.I., Rosales, B.S., Eds.; Academic Press: Burlington, MA, USA, 2009; pp. 229–287.
5. Toepfer, B.; Gora, A.; Li Puma, G. Photocatalytic oxidation of multicomponent solutions of herbicides: Reaction kinetics analysis with explicit photon absorption effects. *App. Catal. B* **2006**, *68*, 171–180. [CrossRef]
6. Moreira, R.P.M.; Li Puma, G. Multiphysics Computational Fluid-Dynamics (CFD) Modeling of Annular Photocatalytic Reactors by the Discrete Ordinates Method (DOM) and the Six-Flux Model (SFM) and Evaluation of the Contaminant Intrinsic Kinetics Constants. *Catal. Today* **2021**, *361*, 77–84. [CrossRef]
7. Bayarri, B.; Giménez, J.; Curcó, D.; Esplugas, S. Absorbed radiation and kinetic model in photocatalysis by TiO_2. *Int. J. Chem. React. Eng.* **2022**, *20*, 3–15. [CrossRef]
8. Satuf, M.L.; Brandi, R.J.; Cassano, A.E.; Alfano, O.M. Scaling-up of slurry reactors for the photocataytic degradation of 4-chlorophenol. *Catal. Today* **2007**, *129*, 110–117. [CrossRef]
9. Casado, C.; Marugán, J.; Timmers, R.; Muñoz, M.; van Grieken, R. Comprehensive multiphysics modeling of photocatalytic processes by computational fluid dynamics based on intrinsic kinetic parameters determined in a differential photoreactor. *Chem. Eng. J.* **2017**, *310*, 368–380. [CrossRef]
10. Imoberdorf, G.E.; Irazoqui, H.A.; Alfano, O.M.; Cassano, A.E. Scaling-up from first principles of a photocatalytic reactor for air pollution remediation. *Chem. Eng. Sci.* **2007**, *62*, 793–804. [CrossRef]
11. Grcic, I.; Koprivanac, N.; Li Puma, G. Modeling the photocatalytic oxidation of carboxylic acids on aqueous TiO_2 suspensions and on immobilized TiO_2-chitosan thin films in different reactor geometries irradiated by UVA or UVC light sources. *Chem. Eng. J.* **2021**, *422*, 130104. [CrossRef]
12. Martín-Sómer, M.; Pablos, C.; van Grieken, R.; Marugán, J. Influence of light distribution on the performance of photocatalytic reactors: LED vs mercury lamps. *Appl. Catal. B* **2017**, *215*, 1–7. [CrossRef]
13. Wang, Q.; Xiao, M.; Peng, Z.; Zhang, C.; Du, X.; Wang, Z.; Wang, W. Visible LED photocatalysis combined with ultrafiltration driven by metal-free oxygen-doped graphitic carbon nitride for sulfamethazine degradation. *J. Hazard. Mat.* **2022**, *439*, 129632. [CrossRef] [PubMed]
14. Manassero, A.; Satuf, M.L.; Alfano, O.M. Photocatalytic degradation of an emerging pollutant by TiO_2-coated glass rings: A kinetic study. *Environ. Sci. Pollut. Res.* **2017**, *24*, 6031–6039. [CrossRef]

15. Manassero, A.; Alfano, O.M.; Satuf, M.L. Radiation modeling and performance evaluation of a UV-LED photocatalytic reactor for water treatment. *J. Photochem. Photobiol. A* **2022**, *436*, 114367. [CrossRef]
16. van Grieken, R.; Marugán, J.; Sordo, C.; Pablos, C. Comparison of the photocatalytic disinfection of E. coli suspensions in slurry, wall and fixed-bed reactors. *Catal. Today* **2009**, *144*, 48–54. [CrossRef]
17. Murov, S.L.; Carmichael, I.; Hug, G.L. *Handbook of Photochemistry*, 2nd ed.; Marcel Dekker: New York, NY, USA, 1993.
18. Dordio, A.; Estêvão Candeias, A.; Pinto, A.; Teixeira da Costa, C.; Palace Carvalho, A. Preliminary media screening for application in the removal of clofibric acid, carbamazepine and ibuprofen by SSF-constructed wetlands. *Ecol. Eng.* **2009**, *35*, 290–302. [CrossRef]
19. Mills, A.; Davies, R. Photomineralisation of 4-chlorophenol sensitised by titanium dioxide: A study of the intermediates. *J. Photochem. Photobiol. A* **1993**, *70*, 183–191. [CrossRef]
20. Theurich, J.; Lindner, M.; Bahnemann, D.W. Photocatalytic degradation of 4-chlorophenol in aerated aqueous titanium dioxide suspensions: A kinetic and mechanistic study. *Langmuir* **1996**, *12*, 6368–6376. [CrossRef]
21. Pelizzetti, E.; Minero, C. Mechanism of the photo-oxidative degradation of organic pollutants over TiO_2 particles. *Electrochim. Acta* **1993**, *38*, 47–55. [CrossRef]
22. Almquist, C.B.; Biswas, P. A mechanistic approach to modeling the effect of dissolved oxygen in photo-oxidation reactions on titanium dioxide in aqueous systems. *Chem. Eng. Sci.* **2001**, *56*, 3421–3430. [CrossRef]
23. Dijkstra, M.F.J.; Panneman, H.J.; Winkelman, J.G.M. Modeling the photocatalytic degradation of formic acid in a reactor with immobilized catalyst. *Chem. Eng. Sci.* **2022**, *57*, 4895–4907. [CrossRef]
24. Turchi, C.S.; Ollis, D.F. Photocatalytic degradation of organic water contaminants: Mechanisms involving hydroxyl radical attack. *J. Catal.* **1990**, *122*, 178–190. [CrossRef]
25. Terzian, R.; Serpone, N.; Minero, C.; Pelizzetti, E.; Hidaka, H. Kinetic studies in heterogeneous photocatalysis 4. The ph tomineralization of a hydroquinone and a catechol. *Photochem. Photobiol. A* **1990**, *55*, 243–249. [CrossRef]
26. Satuf, M.L.; Brandi, R.J.; Cassano, A.E.; Alfano, O.M. Quantum efficiencies of 4-chlorophenol photocatalytic degradation and mineralization in a well-mixed slurry reactor. *Ind. Eng. Chem. Res.* **2007**, *46*, 43–51. [CrossRef]
27. Manassero, A.; Satuf, M.L.; Alfano, O.M. Kinetic modeling of the photocatalytic degradation of clofibric acid in a slurry reactor. *Environ. Sci. Pollut. Res.* **2015**, *22*, 926–937. [CrossRef] [PubMed]
28. Alfano, O.M.; Cabrera, M.I.; Cassano, A.E. Photocatalytic reactions involving hydroxyl radical attack. *J. Catal.* **1997**, *172*, 370–379. [CrossRef]
29. Chen, D.; Li, F.; Ray, A.K. External and internal mass transfer effect on photocatalytic degradation. *Catal. Today* **2001**, *66*, 475–485. [CrossRef]
30. Ould-Mame, S.M.; Zahraa, O.; Bouchy, M. Photocatalytic degradation of salicylic acid on fixed TiO2 -Kinetic studies. *Int. J. Photoenergy* **2000**, *2*, 59–66. [CrossRef]
31. Cloteaux, A.; Gérardin, F.; Thomas, D.; Midoux, N., André, J.C. Fixed bed photocatalytic reactor for formaldehyde degradation: Experimental and modeling study. *Chem. Eng. J.* **2014**, *249*, 121–129. [CrossRef]
32. Onda, K.; Sada, E.; Murase, Y. Liquid-side mass transfer coefficients in packed towers. *AIChE J.* **1959**, *5*, 235–239. [CrossRef]

water

MDPI

Article

Catalytic Ozonation of Atrazine Enhanced by Mesoporous CeO_2: Morphology, Performance and Intermediates

Jianlin Zhang [1,2], Tao Zhuang [2], Shanjun Liu [2], Shan Sun [2], Yongxin Wang [2], Xinyu Liu [3], Jin Wang [4] and Rutao Liu [1,*]

1 School of Environmental Science and Engineering, China-America CRC for Environment & Health, Shandong University, Qingdao 266000, China
2 Jinan Environmental Research Academy, Jinan 250000, China
3 Shandong CRRC Huateng Environment Co. Ltd., Jinan 250000, China
4 School of Civil Engineering, Beijing Jiaotong University, Beijing 100000, China
* Correspondence: rutaoliu@sdu.edu.cn

Abstract: Heterogeneous catalytic ozonation is an alternative approach for the removal of refractory pollutants from water, and the fabrication of mesoporous materials with high dispersibility would enhance the catalytic efficiency. A mesoporous CeO_2 was prepared by the nanocasting method with SBA-15 as a hard template, and was investigated in the catalytic ozonation of atrazine. The synthetical CeO_2 nanorods have a specific surface area of 95.08 m^2/g, a diameter of 10.16 nm, and a spacing of 2.18 nm. The removal rate of atrazine was 85.5%, 64.8%, and 46.4% in the order of catalytic ozonation by synthetical CeO_2 > single ozonation > catalytic ozonation by commercial CeO_2, respectively. The superior activity of the synthetical CeO_2 could be attributed to the well-ordered mesoporous structure, the high surface area, and the redox Ce^{3+}/Ce^{4+} cycling. Moreover, eight organic intermediates were identified after one minute of catalytic ozonation of atrazine, and the cleavage of the ethylamino group was proposed as the main pathway of atrazine degradation.

Keywords: catalytic ozonation; atrazine; CeO_2; mesoporous

Citation: Zhang, J.; Zhuang, T.; Liu, S.; Sun, S.; Wang, Y.; Liu, X.; Wang, J.; Liu, R. Catalytic Ozonation of Atrazine Enhanced by Mesoporous CeO_2: Morphology, Performance and Intermediates. *Water* **2022**, *14*, 3431. https://doi.org/10.3390/w14213431

Academic Editors: Huijiao Wang, Dionysios (Dion) Demetriou Dionysiou and Yujue Wang

Received: 29 September 2022
Accepted: 25 October 2022
Published: 28 October 2022

1. Introduction

Atrazine as a chlorinated triazine herbicide has been widely used in agriculture, especially for the control of crops and grassy weeds [1]. It has a water solubility of 33 mg L^{-1} at 25 °C, an organic carbon partition coefficient value of 100 L kg^{-1}, and a half-life ranging from 21 days to two years, which makes it possible to leach into the aquatic ecosystem and persist for a long time [2]. Atrazine, even at low concentrations in water bodies, could cause serious negative effects on human health through the accumulation of it in the food chain [3], and has been listed as a priority substance by the European Commission. Notably, the detected concentration of atrazine in surface water and groundwater in Beijing ranged from 0.12 to 5.16 μg L^{-1}, which exceeded the allowable level of 3 μg L^{-1} for drinking water [4]. However, the traditional water purification methods are obviously not enough to remove atrazine from water efficiently and completely [5], and the by-products such as halogens might cause a more serious ecological impact [6]. Therefore, it is an urgent requirement to develop an effective removal method for reducing its environmental exposure.

Besides persulfate oxidation [7], photocatalytic oxidation [8], and electrochemical oxidation [9], heterogeneous catalytic ozonation technology is a promising technology for the removal of refractory pollutants [10], which could generate highly reactive hydroxyl radicals with a redox potential of 2.80 eV and rapidly and non-selectively degrade or even mineralize antibiotics with 10^6–10^{10} L mol^{-1} s^{-1} reaction rate [11]. Li et al. [12] evaluated the degradation of atrazine by the catalytic ozonation of iron scraps, and the results showed that the mineralization efficiency was as high as 86.6%, and the toxicity was efficiently removed at 30 min. Ye et al. [13] identified that the prepared catalyst could effectively improve the degradation and mineralization rate of atrazine, and the mineralization strongly depended on the solution's pH. Zeghioud et al. [14] concluded that catalytic ozonation coupled with discharge plasma could improve the removal efficiency of refractory organic pollutants. He et al. [15] combined membrane filtration and catalytic ozonation for the removal of atrazine, 99.99% of atrazine could be removed within 40 min and hydroxyl radicals were inferred to be the main reactive oxygen species in the catalytic ozonation process. Three pollutant degradation pathways are normally proposed in the process of heterogeneous catalytic ozonation [16]. Firstly, ozone is adsorbed on the catalyst surface and catalyzed to generate radical oxygen species, which further degrades the organic pollutant adsorbed on the catalyst surface or in solution. Secondly, ozone directly attacks and degrades the organic pollutants adsorbed on the surface of the catalyst or the complex formed by organic pollutants and the catalyst. Thirdly, both ozone and organic pollutants are simultaneously adsorbed and reacted on the surface of the catalyst. Obviously, heterogeneous catalytic processes are really complicated, involving gas phase (ozone), liquid phase (water) and solid phase (catalyst), where catalysts would perform various roles, such as providing active sites for adsorption and catalytic ozonation. Therefore, the textural properties of catalysts, such as surface area, pore volume and porosity, are the keys to catalytic efficiency.

Numerous studies have been devoted to designing and optimizing the properties of catalysts to improve the efficiency of catalytic ozonation. Bing et al. [17] prepared Fe_2O_3/Al_2O_3@SBA-15 by wet impregnation and calcination, and characterized that Al_2O_3 and Fe_2O_3 were highly dispersed on SBA-15, and amounts of surface Lewis acid sites existed due to the substitution of Al^{3+} for the hydrogen of Si-OH. Yuan et al. [18] synthesized microporous iron silicate with poor crystallinity and found that Fe-Si binary oxides inhibited the futile decomposition of ozone and increased the probability of hydroxyl radicals generated through the reaction between ozone and α-Fe_2O_3. Our previous work prepared a mesoporous MnO_2 by the nanocasting method and found that the ordered nanorods with a high surface area facilitated mass transport and exposure of active sites [19]. Besides MnO_2, CeO_2 is another widely studied heterogeneous catalyst for catalytic ozonation. The cerium redox pair (Ce(III) and Ce(IV)) performs a synergetic effect on catalytic ozonation, where Ce(III) could catalyze ozonolysis to generate hydroxyl radicals and would be regenerated through the reduction of Ce(IV) [20]. Notably, the surface basicity, defect density/oxygen vacancies and unsaturated coordination sites of CeO_2 crystal planes exposed are important factors in the catalytic performance, such as the ability of different crystal planes in the order of (100 + 110) > (100) > (111) to enhance the mineralization of nitrophenol [21].

In this study, an ordered mesoporous CeO_2 with a high surface area was prepared with SBA-15 as the hard template by the nanocasting method and used as an efficient catalyst for catalytic ozonation of atrazine. The textural properties of the synthesized samples were characterized by X-ray diffraction(XRD), surface area analysis techniques, scanning electron microscope (SEM), transmission electron microscope (TEM), and X-ray photoelectron spectroscopy (XPS). The performance of CeO_2 was investigated by evaluating the degradation of atrazine under various conditions compared with control samples. Moreover, the degradation pathways were assessed through the identification of intermediate products. This study may provide an approach to effectively remove refractory pollutants in the process of wastewater treatment.

2. Materials and Methods

2.1. Materials

SBA-15 was purchased from XFNANO Co., Ltd. (Nanjing, China). The commercial CeO_2 and $Ce(NO_3)_3 \bullet 6H_2O$ were supplied by Sinoparm Chemical Reagent Co., Ltd. (Shanghai, China). Atrazine was provided by the laboratory of the government chemist (Germany). NaOH was purchased from Rhawn Reagent (Shanghai, China). H_2SO_4 was obtained from Yantai Far East Fine Chemical Co., Ltd. (Yantai, China). All chemicals used were of analytical grade without further purification. Deionized water (18 MΩcm) produced by the Milli-Q purification system (Millipore, USA) was used to prepare solutions. Ozone was supplied by an ozone generator (COM-AD-01, ANSEROS), and the ozone stock solution was obtained by bubbling gaseous ozone into a 1 L cylindrical reactor filled with pure water for 30 min.

2.2. Preparation of CeO_2

The synthesis of mesoporous CeO_2 was based on a previously reported method [19]. Concisely, 4.51 g of SBA-15 was dispersed in an ethanol solution (15 mL of ultrapure water and 90 mL of ethanol) under stirring with a magnetic bar at room temperature. A total of 13 g of $Ce(NO_3)_3{\bullet}6H_2O$ was added into the suspended solution and kept stirring for 2 h. The solid was filtered from the suspension, dried for 5 h at 105 °C, and subsequently calcined at 300 °C for 3 h. The CeO_2/SBA-15 complex obtained was rinsed with 2M NaOH, then washed to neutral with ultrapure water and dried for 3 h at 105 °C.

2.3. Experimental Procedure

The experimental procedure was conducted in batch mode in a 0.6 L cylindrical reactor with a diameter of 80 mm and a height of 120 mm at room temperature. Basically, appropriate amounts of catalyst and atrazine stock solution were added to the reactor under continuous magnetic stirring. Then, appropriate ozone stock solutions were added to the reactor to initiate catalytic ozonation (the initial mixing solution contained 50 μg L^{-1} atrazine, 20 mg L^{-1} catalyst and 1 mg L^{-1} O_3). The solution of an appropriate volume was sampled from the reactor at certain intervals, filtered with a 0.22 μm prefilter and quenched with a trace of $Na_2S_2O_3$ for further analysis. Two samples were repeated for each experiment.

2.4. Characterization

The XRD patterns were recorded in the 2θ range of 5–60° using a Rigaku Ultima V diffractometer with Cu Kα as the radiation source. The surface area was measured on an automatic adsorption analyzer (ASAP2460, Micromeritics Instrument Corp., Norcross, GA, USA) and calculated by the Brunauer–Emmett–Teller (BET) method. The structure and morphology of the as-prepared CeO_2 were characterized by transmission electron microscopy (FEI Tecnai G2 F20, NL) and scanning electron microscopy (Zeiss Sigma 300, GER) equipped with an energy-dispersive spectrometer. The elementary compositions of the prepared samples were analyzed by X-ray photo-electron spectroscopy (Thermo Fisher Scientific, Waltham, MA, USA).

The atrazine concentration was detected by high-performance liquid chromatography (LC-20AT, Shimadzu, JPN) equipped with a diode array detector. A reversed-phase C18 column was used under isocratic elution with methanol and phosphoric acid solution (v/v, 55:45) at 1.0 mL/min at 35 °C. The intermediates were identified by an ultra-performance liquid chromatography tandem quadrupole time of flight mass spectrometer (LCMS-9030, Shimadzu). Gaseous O_3 concentrations were monitored using an ozone analyzer (GM-6000-OEM, ANSEROS). Aqueous O_3 concentrations were measured by iodometry and were calculated using the following equation.

$$C_{(O3)} = 24000C_{(Na2S2O3)}V_{(Na2S2O3)}/V_{(O3)} \tag{1}$$

where $C_{(O3)}$ is the aqueous O_3 concentration, mg/L; $C_{(Na2S2O3)}$ is the concentration of $Na_2S_2O_3$ standard solution, mol/L; $V_{(Na2S2O3)}$ is the consumed volume of $Na_2S_2O_3$ standard solution, mL; $V_{(O3)}$ is the titration volume of the aqueous O_3, mL.

3. Results and Discussion

3.1. Characterization of CeO_2

The XRD patterns of the CeO_2 samples are shown in Figure 1. The synthetical and commercial samples both exhibited five distinguished diffraction peaks at 28°, 33°, 47°, 56° and 59°, which could be indexed to the (111), (200), (220), (311) and (222) reflection planes of CeO_2 (JCPDS No. 78-0694) [22]. The peaks of synthetical CeO_2 were wider and weaker than those of commercial CeO_2, which implied that the crystal size of synthetical CeO_2 was smaller than that of commercial CeO_2. The average crystal size of the synthetical and commercial CeO_2 was 9.8 nm and 37.4 nm, respectively, through the calculation of the Scherrer formula as below [23].

$$D = K\lambda/\beta\cos\theta \tag{2}$$

where D is the average crystal size; K is a constant of 0.89; λ is the X-ray wavelength of 0.154 nm; β is the full width at half maximum of the diffraction peak; θ is the diffraction angle.

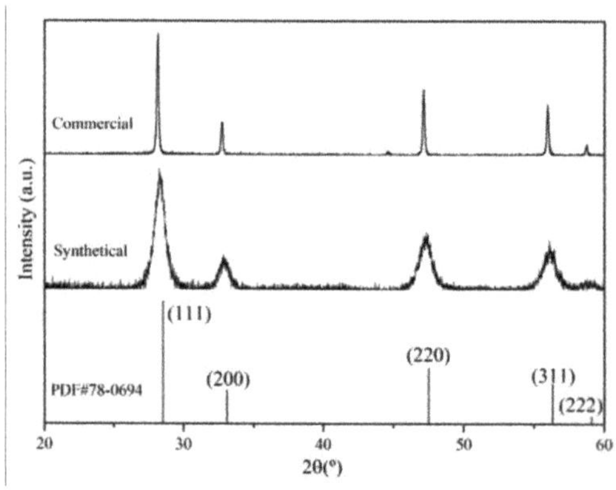

Figure 1. XRD pattern of CeO_2 samples.

The BET adsorption-desorption isotherms were measured to analyze the specific surface area and pore information of CeO_2. In Figure 2a, the hysteresis loop of the synthetical CeO_2 corresponded to type IV, indicating the existence of mesopores. The commercial CeO_2 might be nonporous or microporous since its hysteresis loop belonged to type I. Moreover, the specific surface area of the synthetical and commercial CeO_2 shown in Table S1 was 95.08 m^2/g and 1.44 m^2/g, respectively, and the specific surface area of the synthetical CeO_2 was significantly higher than that of the commercial CeO_2, which implied that the synthetical CeO_2 would provide more catalytic sites and facilitate catalytic ozonation as it could offer more active site exposure possibilities compared to the commercial CeO_2 of the same mass. Figure 2b showed that the synthetical CeO_2 had a relatively concentrated pore distribution while the commercial CeO_2 was almost non-porous, which echoed the analysis of BET adsorption-desorption isotherms and implied that the synthetical CeO_2 might be of ordered mesopores.

The morphologies of CeO_2 were characterized by SEM, and the images are shown in Figure 3. Apparently, the CeO_2 samples presented different degrees of aggregated morphology, and the synthetical CeO_2 seen in Figure 3a,c exhibited granular while the commercial CeO_2 seen in Figure 3b,d exhibited layered, which implied that the surface area of synthetical CeO_2 would be much higher than that of commercial CeO_2. Figure 4 displays the TEM images of CeO_2 within different nanometer dimensions. Through the comparison of samples, the synthetical CeO_2 exhibited higher dispersibility than the commercial CeO_2, which was consistent with the SEM results. Moreover, the synthetical CeO_2 was a nanorod with well-ordered channels, while the commercial CeO_2 was a nanocube with high agglomeration. Further, the nanorod was 10.16 nm in diameter, and the distance between each channel was 2.18 nm, implying the synthetical CeO_2 had an ordered mesoporous structure and would have large active sites exposed. The synthetical CeO_2 dominantly exposed (111) facets with an interplanar spacing of 0.31 nm, while the commercial CeO_2 mainly exposed (200) facets with an interplanar spacing of 0.19 nm. The lattice fringes revealed the high crystallinity of CeO_2, which further proved the results of XRD.

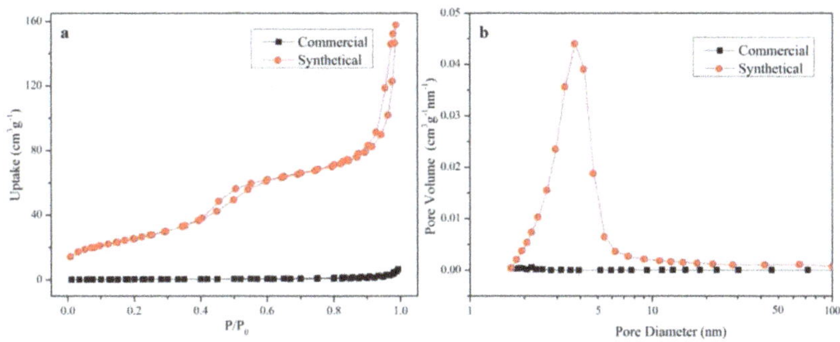

Figure 2. Nitrogen adsorption-desorption isotherms (**a**) and pore size distribution (**b**) of CeO_2.

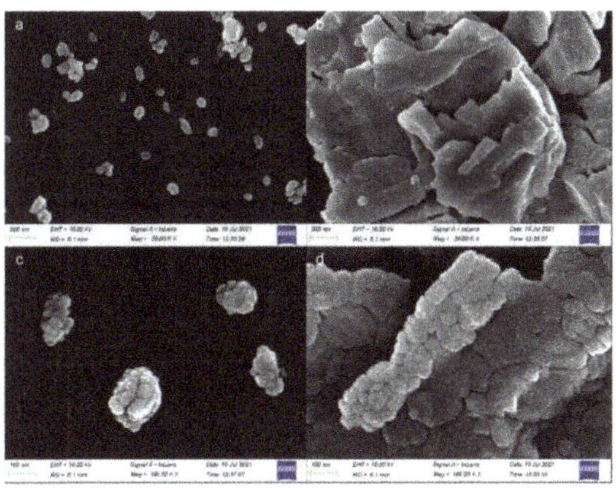

Figure 3. SEM images of synthetical (**a**,**c**) and commercial (**b**,**d**) CeO_2..

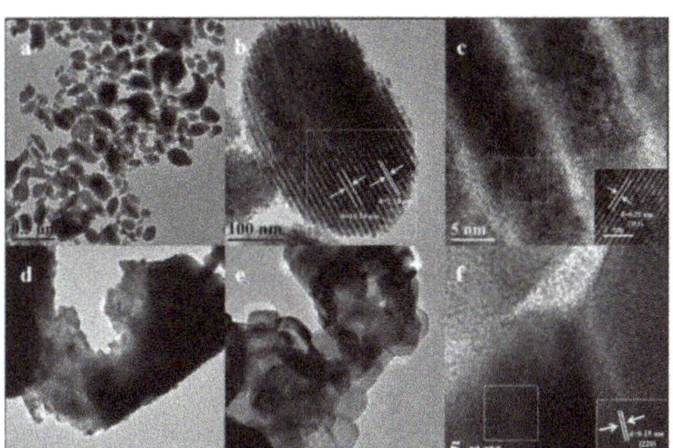

Figure 4. TEM images of synthetical (**a**–**c**) and commercial (**d**–**f**) CeO_2 within different nanometer dimensions.

The redox couple between Ce^{3+}/Ce^{4+} or O^{2-}/O_3 plays a crucial role in enhancing the catalytic activity as the electrons from Ce^{3+} would be transferred to O_3, which decomposes O_3 into reactive oxygen species, and the electrons from O^{2-} would reduce the Ce^{4+} to Ce^{3+} [21]. The chemical states of Ce/O and their concentrations on the sample surface were analyzed from the XPS spectra. As illustrated in Figure 5a, the Ce 3d XPS spectra consisted of two series of spin-orbit lines, where v represented the Ce 3d3/2 spin-orbit splitting peak while u represented the Ce 3d5/2 spin-orbit splitting peak. The spin-orbit was well-fitted with ten peaks through deconvolution [24], and the energy positions of these ten spin-orbit components were summarized in Table S2. Taking the synthetical sample as an example, three pairs of peaks, including v (882.46 eV) and u (900.98 eV), v″ (888.87 eV) and u″ (907.61 eV), and v‴(898.18 eV) and u‴ (916.83 eV), were related to the characteristic peaks of Ce^{4+}. The remaining two pairs of peaks marked as v_0 (880.38 eV) and u_0 (898.38 eV) and v′ (884.84 eV) and u′ (903.05 eV) belonged to the characteristic peaks of Ce^{3+} [24]. The results indicated that cerium existed in CeO_2 in two valence states [25], Ce^{3+} and Ce^{4+}, and the atomic ratio of Ce^{3+} to Ce^{4+} was 10.14% and 12.59% for the synthetic and commercial CeO_2, respectively. The emergence of multivalent cerium would facilitate the charge transfer and the production of reactive oxygen species during the catalytic ozone process. The presence of Ce^{3+} would introduce the occurrence of oxygen vacancies on CeO_2. As illustrated in Figure 5b, the O 1s XPS spectra exhibited a strong peak with a shoulder, and might be related to four kinds of surface oxygen states. The fitting peak at 529.47–529.51 eV could be attributed to the lattice oxygenis [26], the one at 530.51 eV could be assigned as the surface oxygen from the hydroxyl group [27], and the adjacent peaks at 531.46–531.47 eV and 531.81–531.97 eV should be characteristic of subsurface $O^-/O_2{}^{2-}$ species, which belong to oxygen substances chemically adsorbed at oxygen vacancies and would improve the catalytic activity [28].

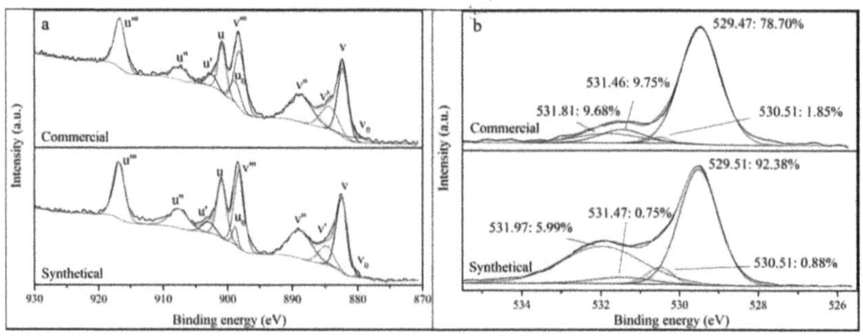

Figure 5. XPS spectra of Ce 3d (**a**) and O 1s (**b**) of CeO_2.

3.2. Catalytic Ozonation of Atrazine

To evaluate the catalytic superiority of the synthetical CeO_2 for ozonation, the atrazine degradation experiments were carried out under different conditions. The adsorption was performed for 10 min before atrazine degradation. As shown in Figure 6a, the adsorption amount of synthetical CeO_2 was as high as 9.9%, while that of the commercial one was only 4.0%. The good adsorption capacity is attributed to the high surface area of synthetical CeO_2, which would be favorable for the reaction of atrazine with hydroxyl radical generated on the catalyst surface. The atrazine removal rate with synthetical CeO_2 could reach up to 85.5% after 10 min, while the removal rates with commercial CeO_2 and ozone alone were 64.8% and 46.4%, respectively. The performance of synthetical CeO_2 on catalytic ozonation of atrazine was statistically ($p < 0.05$) different from and better than that of commercial CeO_2 or ozone alone, indicating that the synthetical CeO_2 effectively enhanced the catalytic ozonation and its catalytic activity was obviously superior to commercial CeO_2. The nano-ZnO [29] and mesoporous Fe_3O_4 [30] were also prepared as ozonation catalysts to enhance the degradation of atrazine, and in contrast, the synthetical CeO_2 could obtain the high removal efficiency of atrazine at a dose of less than an order of magnitude. To further quantitatively analyze the catalytic activity of CeO_2, the rate constants were calculated according to the pseudo-first-order kinetic equation [13]:

$$-\ln(C_t/C_0) = k_{obs}t \quad (3)$$

where C_t is the atrazine concentration at the reaction time t, C_0 is the initial atrazine concentration, and k_{obs} is the pseudo-first-order constant. As shown in Figure 6b, the apparent rate constant for ozone alone was 0.074 min^{-1}, while the apparent rate constant increased to 0.175 min^{-1} and 0.097 min^{-1} after the addition of synthetical and commercial CeO_2, which demonstrated that the applied CeO_2 could promote the removal of atrazine. Notably, the apparent rate constant of synthetical CeO_2 was significantly higher than that of commercial CeO_2, which was associated with the superior aforementioned properties such as mesoporous structure, high surface area, as well as good dispersibility.

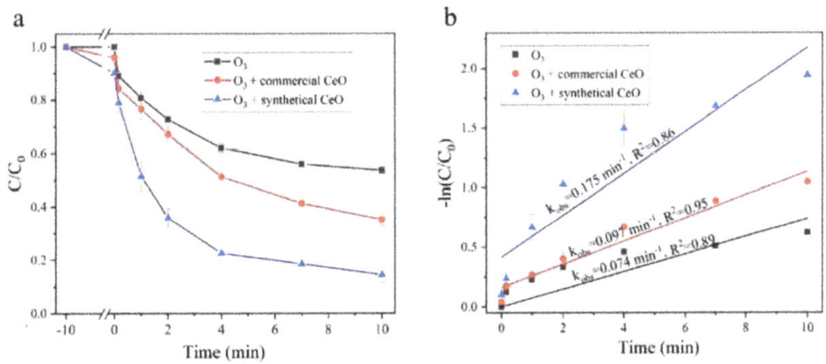

Figure 6. The curves (**a**) and pseudo-first-order plots (**b**) of atrazine degradation in different processes (atrazine concentration = 50 μg/L; catalyst dose = 20 mg/L; ozone concentration = 1 mg/L; initial pH = 7).

Combined with the experimental results and previous research [31], the atrazine degradation was a relatively complex process during the catalytic ozonation, which included adsorption, direct oxidation by molecular ozone and indirect oxidation by hydroxyl radical. The catalytic mechanism of synthetical CeO_2 on the ozonation of atrazine could be summarized as follows:

$$CeO_2 + Atrazine_{(aqu)} \rightarrow CeO_2{\equiv}Atrazine_{(ads)} \quad (4)$$

$$Ce^{3+} + O_3 + H^+ \rightarrow Ce^{4+} + HO_3\bullet \quad (5)$$

$$HO_3\bullet \rightarrow \bullet OH + O_2 \quad (6)$$

$$\bullet OH + O_3 \rightarrow O_2^- + HO_2\bullet \quad (7)$$

$$O_2^- + O_3 \rightarrow O_3^- + O_2 \quad (8)$$

$$O_3^- + H^+ \rightarrow HO_3\bullet \quad (9)$$

$$HO_3\bullet \rightarrow \bullet OH + O_2 \quad (10)$$

$$Ce^{4+} + e^- \rightarrow Ce^{3+} \quad (11)$$

$$\bullet OH + CeO_2{\equiv}Atrazine_{(ads)} \rightarrow byproducts \quad (12)$$

The pH value has an important influence on the chemical characteristics of the catalyst surface, the charge of ionic organic molecules, and the decomposition of O_3 into reactive oxygen species in the catalytic process [32]. The influence of the initial pH value on the catalytic efficiency would be taken into account in the subsequent evaluation of the environmental influence on the catalytic performance.

3.3. Transformation Products and Proposed Pathways

The molecular structure of atrazine mainly consists of the s-triazine ring, the isopropylamino group, the ethylamino group and the chlorine group, and the nitrogen atom in the amino group and the chlorine atom adjacent to the s-triazine ring are vulnerable to radical oxygen species [15]. It is widely recognized that the degradation of atrazine in catalytic ozonation involves the reaction pathways: dealkylation, dechlorination, deamination and olefination [33]. However, the priority of the pathways was basically not involved in previous studies on reaction pathways since the sampling time for the determination of intermediates was not considered.

The previous results shown in Figure 5a indicated that atrazine could be mostly degraded within 10 min, and half could be degraded in one minute. The sample after one minute of catalytic ozonation could better reflect the early reaction process and was used for the determination of intermediates. Eight organic intermediates with m/z values of 232, 174, 214, 230, 172, 212, 188 and 146 were successfully identified (mass error within 0.5 ppm), and were shown in Table S3 and Figure S1. Based on the intermediates and literature reports, the degradation pathway for the catalytic ozonation of atrazine was proposed in Figure 7. Apparently, the structural changes of atrazine first occurred in the isopropylamino and ethylamino groups. Six intermediate products were detected during the change of the ethylamino group, and their mass spectral peak responses were relatively higher, which could be inferred that the change of the ethylamino group was the main pathway of atrazine degradation. The ethylamino group was attacked by hydroxyl radicals to form the hydroxylated intermediate P1(m/z = 232), and the formed hydroxyl group could be easily attracted to form carbon-centered radicals, which could further add oxygen and rapidly eliminate hydroxyl peroxide radical to generate ketones P4(m/z = 230) [34], and further attacked by hydroxyl radicals, leading to the loss of chlorine and the formation of hydroxylated intermediate P6(m/z = 212). The intermediate P3(m/z = 232) was formed from the olefination of P1, followed by further dealkylation to form the intermediate P5(m/z = 172) [35]. It is noted that the dealkylation reactions were ubiquitous, and the intermediates P2(m/z = 174), P5, P7 and P8(m/z = 146) were the products of dealkylation reactions since the nitrogen atoms on the isopropylamino group and ethylamino group were prone to being attacked, resulting in the heterolytic cleavage of the C–N bond [15]. These intermediates were also observed in other advanced oxidation processes, such as photocatalytic oxidation, persulfate oxidation and Fenton oxidation. All the above intermediates would be further oxidized to low molecular organics and even mineralized to CO_2 and H_2O by prolonging the reaction time, which was widely investigated in previous studies [29]. The results indicated that the initiating reaction of catalytic ozonation of atrazine mainly occurred on the ethylamino group and proceeded to the dealkylation reactions.

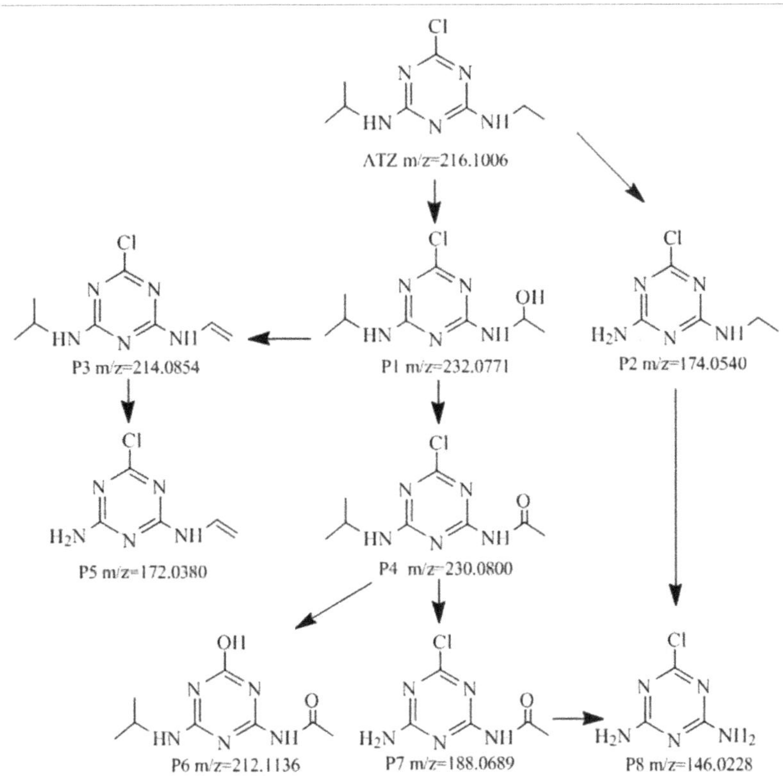

Figure 7. Proposed degradation pathway of atrazine.

4. Conclusions

The mesoporous CeO_2 was successfully obtained by the nanocasting method and investigated for catalytic ozonation of atrazine. Compared with commercial CeO_2 and single ozone, synthetical CeO_2 significantly enhanced the removal of atrazine due to its excellent properties such as high surface area, well-ordered mesoporous structure and redox Ce^{3+}/Ce^{4+} couple. Eight organic intermediates were identified, and the initiating reaction of catalytic ozonation of atrazine mainly occurred on the ethylamino group, followed by the dealkylation reactions. This study might facilitate the application of mesoporous metal oxide and the understanding of the degradation pathways of organic pollutants in catalytic ozonation.

Supplementary Materials: The following supporting information can be downloaded at: https://www.mdpi.com/article/10.3390/w14213431/s1, Figure S1. HPLC-MS/MS chromatogram of atrazine and intermediates; Table S1. Physical properties of the catalysts; Table S2. Energy positions of Ce spin orbit components in different samples (unit: eV); Table S3. UPLC-MS/MS analysis of atrazine and intermediates. Reference [36] is cited in the supplementary material.

Author Contributions: J.Z.: Methodology, Software, Investigation, Writing-original draft; T.Z.: Writing—review & editing, Supervision, Data curation; S.L.: Investigation, Writing—review & editing; S.S.: Software, Writing—review & editing; Y.W.: Writing—review & editing; X.L.: Writing—review & editing; J.W.: Writing—review & editing; R.L.: Resources, Writing—review & editing. All authors have read and agreed to the published version of the manuscript.

Funding: This research received no external funding.

Data Availability Statement: Not applicable.

Acknowledgments: The authors would like to acknowledge the support for this work by the Beijing Key Laboratory of Aqueous Typical Pollutant Control and Water Quality Safegurad.

Conflicts of Interest: The authors declare no conflict of interest.

References

1. Salazar-Ledesma, M.; Prado, B.; Zamora, O.; Siebe, C. Mobility of atrazine in soils of a wastewater irrigated maize field. *Agric. Ecosyst. Environ.* **2018**, *255*, 73–83. [CrossRef]
2. Urseler, N.; Bachetti, R.; Biolé, F.; Morgante, V.; Morgante, C. Atrazine pollution in groundwater and raw bovine milk: Water quality, bioaccumulation and human risk assessment. *Sci. Total Environ.* **2022**, *852*, 158498. [CrossRef] [PubMed]
3. Wang, F.; Liu, S.-S.; Feng, Z.; Fu, H.; Wang, M.; Wang, P.; Liu, W.; Wang, C.-C. High-efficient peroxymonosulfate activation for rapid atrazine degradation by FeS_x@MoS_2 derived from MIL-88A(Fe). *J. Hazard. Mater.* **2022**, *440*, 129723. [CrossRef] [PubMed]
4. Ge, J.; Cong, J.; Sun, Y.; Li, G.; Zhou, Z.; Qian, C.; Liu, F. Determination of Endocrine Disrupting Chemicals in Surface Water and Industrial Wastewater from Beijing, China. *Bull. Environ. Contam. Toxicol.* **2010**, *84*, 401–405. [CrossRef] [PubMed]
5. Navarra, W.; Sacco, O.; Daniel, C.; Venditto, V.; Vaiano, V.; Vignati, D.A.L.; Bojic, C.; Libralato, G.; Lofrano, G.; Carotenuto, M. Photocatalytic degradation of atrazine by an N-doped TiO_2/polymer composite: Catalytic efficiency and toxicity evaluation. *J. Environ. Chem. Eng.* **2022**, *10*, 108167. [CrossRef]
6. Yeom, Y.; Han, J.; Zhang, X.; Shang, C.; Zhang, T.; Li, X.; Duan, X.; Dionysiou, D.D. A review on the degradation efficiency, DBP formation, and toxicity variation in the UV/chlorine treatment of micropollutants. *Chem. Eng. J.* **2021**, *424*, 130053. [CrossRef]
7. Yang, L.; Jiao, Y.; Xu, X.; Pan, Y.; Su, C.; Duan, X.; Sun, H.; Liu, S.; Wang, S.; Shao, Z. Superstructures with Atomic-Level Arranged Perovskite and Oxide Layers for Advanced Oxidation with an Enhanced Non-Free Radical Pathway. *ACS Sustain. Chem. Eng.* **2022**, *10*, 1899–1909. [CrossRef]
8. Ahmed, S.; Rasul, M.G.; Sattar, M.A.; Jahirul, M.I. Phenol degradation of waste and stormwater on a flat plate photocatalytic reactor with TiO_2 on glass slide: An experimental and modelling investigation. *J. Water Process Eng.* **2022**, *47*, 102769. [CrossRef]
9. Xu, X.; Zhong, Y.; Shao, Z. Double Perovskites in Catalysis, Electrocatalysis, and Photo(electro)catalysis. *Trends Chem.* **2019**, *1*, 410–424. [CrossRef]
10. Rostami, S.; Jafari, S.; Moeini, Z.; Jaskulak, M.; Keshtgar, L.; Badeenezhad, A.; Azhdarpoor, A.; Rostami, M.; Zorena, K.; Dehghani, M. Current methods and technologies for degradation of atrazine in contaminated soil and water: A review. *Environ. Technol. Innov.* **2021**, *24*, 102019. [CrossRef]
11. Wang, J.; Chen, H. Catalytic ozonation for water and wastewater treatment: Recent advances and perspective. *Sci. Total Environ.* **2020**, *704*, 135249. [CrossRef]
12. Li, H.; Zhou, B. Degradation of atrazine by catalytic ozonation in the presence of iron scraps: Performance, transformation pathway, and acute toxicity. *J. Environ. Sci. Health B* **2019**, *54*, 432–440. [CrossRef]
13. Ye, G.; Luo, P.; Zhao, Y.; Qiu, G.; Hu, Y.; Preis, S.; Wei, C. Three-dimensional Co/Ni bimetallic organic frameworks for high-efficient catalytic ozonation of atrazine: Mechanism, effect parameters, and degradation pathways analysis. *Chemosphere* **2020**, *253*, 126767. [CrossRef]
14. Zeghioud, H.; Nguyen-Tri, P.; Khezami, L.; Amrane, A.; Assadi, A.A. Review on discharge Plasma for water treatment: Mechanism, reactor geometries, active species and combined processes. *J. Water Process Eng.* **2020**, *38*, 101664. [CrossRef]
15. He, Y.; Wang, L.; Chen, Z.; Huang, X.; Wang, X.; Zhang, X.; Wen, X. Novel catalytic ceramic membranes anchored with MnMe oxide and their catalytic ozonation performance towards atrazine degradation. *J. Membr. Sci.* **2022**, *648*, 120362. [CrossRef]
16. Von Gunten, U. Ozonation of drinking water: Part I. Oxidation kinetics and product formation. *Water Res.* **2003**, *37*, 1443–1467. [CrossRef]
17. Bing, J.; Hu, C.; Nie, Y.; Yang, M.; Qu, J. Mechanism of Catalytic Ozonation in Fe_2O_3/Al_2O_3@SBA-15 Aqueous Suspension for Destruction of Ibuprofen. *Environ. Sci. Technol.* **2015**, *49*, 1690–1697. [CrossRef]
18. Yuan, L.; Shen, J.; Yan, P.; Chen, Z. Interface Mechanisms of Catalytic Ozonation with Amorphous Iron Silicate for Removal of 4-Chloronitrobenzene in Aqueous Solution. *Environ. Sci. Technol.* **2018**, *52*, 1429–1434. [CrossRef]
19. Zhang, J.; Zhuang, T.; Liu, S.; Zhang, G.C.; Huo, K. Catalytic ozonation of phenol enhanced by mesoporous MnO_2 prepared through nanocasting method with SBA-15 as template. *J. Environ. Chem. Eng.* **2020**, *8*, 103967. [CrossRef]
20. Matheswaran, M.; Balaji, S.; Chung, S.J.; Moon, I.S. Studies on cerium oxidation in catalytic ozonation process: A novel approach for organic mineralization. *Catal. Commun.* **2007**, *8*, 1497–1501. [CrossRef]
21. Afzal, S.; Quan, X.; Lu, S. Catalytic performance and an insight into the mechanism of CeO_2 nanocrystals with different exposed facets in catalytic ozonation of p-nitrophenol. *Appl. Catal. B Environ.* **2019**, *248*, 526–537. [CrossRef]
22. Piumetti, M.; Bensaid, S.; Russo, N.; Fino, D. Nanostructured ceria-based catalysts for soot combustion: Investigations on the surface sensitivity. *Appl. Catal. B Environ.* **2015**, *165*, 742–751. [CrossRef]
23. Fan, S.; Song, J.; Xia, Y.; Dai, Q. Catalytic ozonation of thymol with a novel CoCe-MMO catalyst: Kinetics and mechanism. *Environ. Technol. Innov.* **2021**, *24*, 101881. [CrossRef]
24. Zuo, X.; Ma, S.; Wu, Q.; Xiong, J.; He, J.; Ma, C.; Chen, Z. Nanometer CeO_2 doped high silica ZSM-5 heterogeneous catalytic ozonation of sulfamethoxazole in water. *J. Hazard. Mater.* **2021**, *411*, 125072. [CrossRef] [PubMed]

25. Gao, Q.; Cui, Y.; Wang, S.; Liu, B.; Liu, C. Enhanced photocatalytic activation of peroxymonosulfate by CeO_2 incorporated ZnCo–layered double hydroxide toward organic pollutants removal. *Sep. Purif. Technol.* **2021**, *263*, 118413. [CrossRef]
26. Amri, A.; Duan, X.; Yin, C.-Y.; Jiang, Z.-T.; Rahman, M.M.; Pryor, T. Solar absorptance of copper–cobalt oxide thin film coatings with nano-size, grain-like morphology: Optimization and synchrotron radiation XPS studies. *Appl. Surf. Sci.* **2013**, *275*, 127–135. [CrossRef]
27. Marco, J.F.; Gancedo, J.R.; Gracia, M.; Gautier, J.L.; Ríos, E.; Berry, F.J. Characterization of the Nickel Cobaltite, $NiCo_2O_4$, Prepared by Several Methods: An XRD, XANES, EXAFS, and XPS Study. *J. Solid State Chem.* **2000**, *153*, 74–81. [CrossRef]
28. Gao, J.; Tang, L.; Shen, Z.; Dong, Y.; Wang, Z.; Lyu, J.; Li, J.; Yu, H.-Q. Coupling of SiC and CeO_2 nanosheets to enhance solar energy utilization and optimize catalytic ozonation. *Appl. Catal. B Environ.* **2022**, *317*, 121697. [CrossRef]
29. Yuan, X.; Yan, X.; Xu, H.; Li, D.; Sun, L.; Cao, G.; Xia, D. Enhanced ozonation degradation of atrazine in the presence of nano-ZnO: Performance, kinetics and effects. *J. Environ. Sci.* **2017**, *61*, 3–13. [CrossRef]
30. Zhu, S.; Dong, B.; Yu, Y.; Bu, L.; Deng, J.; Zhou, S. Heterogeneous catalysis of ozone using ordered mesoporous Fe_3O_4 for degradation of atrazine. *Chem. Eng. J.* **2017**, *328*, 527–535. [CrossRef]
31. Qu, Z.; Xu, X.; Ren, H.; Sun, T.; Huang, L.; Gao, Z. Effective mineralization of p-nitrophenol in water by heterogeneous catalytic ozonation using Ce-loaded sepiolite catalyst. *J. Environ. Chem. Eng.* **2022**, *10*, 108185. [CrossRef]
32. Kasprzyk-Hordern, B.; Ziółek, M.; Nawrocki, J. Catalytic ozonation and methods of enhancing molecular ozone reactions in water treatment. *Appl. Catal. B Environ.* **2003**, *46*, 639–669. [CrossRef]
33. Yang, Y.; Cao, H.; Peng, P.; Bo, H. Degradation and transformation of atrazine under catalyzed ozonation process with TiO_2 as catalyst. *J. Hazard. Mater.* **2014**, *279*, 444–451. [CrossRef]
34. Chan, K.H.; Chu, W. Model applications and mechanism study on the degradation of atrazine by Fenton's system. *J. Hazard. Mater.* **2005**, *118*, 227–237. [CrossRef]
35. Wang, D.; Xu, H.; Ma, J.; Lu, X.; Qi, J.; Song, S. Strong promoted catalytic ozonation of atrazine at low temperature using tourmaline as catalyst: Influencing factors, reaction mechanisms and pathways. *Chem. Eng. J.* **2018**, *354*, 113–125. [CrossRef]
36. Qiu, X.H.; Su, X.Y.; Li, X.J.; Li, N. Preparation of CeO_2 with Different Morphologies and Its Application to Catalytic Ozonation of Aqueous Lemon Yellow Solutions. *Adv. Mater. Res.* **2014**, *997*, 3–8. [CrossRef]

water

MDPI

Article

Efficient Removal of Micropollutants by Novel Carbon Materials Using Nitrogen-Rich Bio-Based Metal-Organic Framework (MOFs) as Precursors

Yazi Meng, Xiang Li * and Bo Wang

School of Materials, Advanced Research Institute of Multidisciplinary Science, Beijing Institute of Technology, Beijing 100081, China
* Correspondence: xiangli0369@bit.edu.cn

Abstract: Eliminating pharmaceuticals with trace concentrations in water is crucial in water purification. Developing an effective adsorbent for removing micropollutants from water has aroused great research interest. In this study, the feasibility of nitrogen-rich bio-based metal–organic framework (MOF)-derived carbon as an effective material to eliminate micropollutants from the water environment is discussed. A mixed ligand approach has been applied to synthesize IISERP-MOF27 successfully via the solvothermal method. Adenine, which is non-toxic, easily obtained, and cheap, was introduced into the structure. The novel heterogeneous porous carbon was produced by pyrolyzation with an extremely high surface area (S_{BET} = 980.5 m^2/g), which is 12.8 times higher than that of pristine MOFs. Studies show that the highest surface area and abundant mesoporous structures (V_{pore} = 0.496 cm^3/g) can be obtained when the MOFs are pyrolyzed at 900 °C. The saturated adsorption amount for sulfamethylthiazole (SMX) over MOF-derived carbon can reach 350.90 mg/g with a fast initial adsorption rate of 315.29 (mg/g·min). By adding the second linker adenine as the precursor, the adsorption performance for SMX was made extremely better than that of traditional active carbon (AC) and pyrolyzed ZIF-8(ZIF-8-C), one of the most classic Zn-MOFs. The adsorption capacity calculated by the Langmuir model (R^2 = 0.99) for SMX over bio-C-900 was 4.6 and 13.3 times more than those of AC and ZIF-8-C, respectively. The removal percentage of six representative pharmaceuticals can be well correlated to the structural parameter log K_{ow} of each pharmaceutical, indicating the hydrophobic interaction should be one of the major mechanisms for the adsorption in water. This study offers a strategy to develop novel carbon materials to remove pharmaceuticals.

Keywords: pharmaceuticals; metal–organic framework; adsorption; performance

Citation: Meng, Y.; Li, X.; Wang, B. Efficient Removal of Micropollutants by Novel Carbon Materials Using Nitrogen-Rich Bio-Based Metal-Organic Framework (MOFs) as Precursors. *Water* **2022**, *14*, 3413. https://doi.org/10.3390/w14213413

Academic Editors: Dionysios (Dion) Demetriou Dionysiou, Yujue Wang and Huijiao Wang

Received: 29 September 2022
Accepted: 21 October 2022
Published: 27 October 2022

1. Introduction

In recent years, regular testing of micropollutants in tap water, river water, and municipal wastewater has received increasing attention [1]. This kind of pollutant includes antibiotics, anti-inflammatory medications, and a variety of chemicals [2]. They have adverse environmental impacts such as mutagenicity, carcinogenicity, aquatic toxicity, and other harmful effects on both humans and the ecosystem [3,4]. For instance, the frequently used antibiotic sulfamethoxazole (SMX) is difficult entirely eliminate in wastewater treatment plants (WWTPs), leading to the transformation of its metabolites and product residues into components of surface water [5]. It has been proven that prolonged exposure to such pollutants may have negative effects on individuals, such as liver failure and genetic damage [5,6]. Therefore, it is of great significance to remove PPCPs from aquatic ecosystems greenly and efficiently.

To remove these contaminants in water, several tertiary treatments have been used, such as photocatalytic degradation [7], the Fenton reaction [8], biodegradation [9], and membrane filtration [10]. Among these, adsorption technology, which is easy to operate and low-cost, has been considered a promising advanced water treatment process for eliminating micropollutants in wastewater. To date, a large number of adsorbents have been investigated to remove micropollutants, including porous carbon [11], cyclodextrin polymer [12], and zeolites [13]. The carbon-based ones among these adsorbents are considered major materials in the water treatment process because of their high water/thermal stability and surface areas [14]. However, according to previous studies, their removal efficiency of SMX still needs to be improved compared to traditional materials (including activated carbon). Normally, about 54.34 $\pm$ 2.35% of SMX can be removed by treatment at WWTPs, but SMX content ranging from 0.3 ng L^{-1} to 783 ng L^{-1} is still detected after removal [15].

In recent years, MOFs with high specific surface areas and high porosities have been considered promising precursors to constructing adsorbents in water [16]. MOFs derived from carbon have been employed for the adsorption of water pollutants [17]. For example, results show that heavy metals, herbicides, and organic dyes can be effectively removed by carbon materials derived from Fe-MOFs [14,18–20]. Due to the especially low boiling point of the Zn atom, derived porous carbon can be created after the evaporation of metal based on Zn-based MOFs, such as ZIF-8, MOF-5, etc. [21–23]. On the other hand, bio-MOFs such as IISERP-MOF27(bio-27) and IISERP-MOF26 [24] with biological linkers should be promising in constructing MOFs because of their non-toxic, easily available, and inexpensive properties [25]. However, to the best of our knowledge, Zn-MOFs with mixed linkers-derived carbon have not been reported yet.

In this study, for the first time, bio-27-derived carbon was synthesized for the removal of micropollutants from water. The major characterization shows that this bio-MOF-derived carbon has obvious heterogeneous pores ranging from 1 nm to 6 nm. The precursors were pyrolyzed at 500~1000 °C (bio-C-500~bio-C-1000). The adsorption capacities and initial adsorption rates were evaluated by analyzing the removal performance for SMX over different bio-MOFs. The kinetic data were fitted using the Elovich model. Langmuir and Freundlich's models were applied to compare the capacities. The operational parameters were studied. To further evaluate the adsorption performance for pharmaceuticals with diverse structures, the adsorption behaviors for six pharmaceuticals—ketoprofen (KP), antipyrine (AT), ibuprofen (IBU), chloramphenicol (CAP), paracetamol (PC), and sulfamethoxazole (SMX)—were studied.

2. Experimental Section

2.1. Chemical Agents

$Zn(NO_3)_2{\cdot}6H_2O$ (CAS, 10196-18-6), dimethylformamide (DMF, 99%), and methanol (MeOH, CAS, 67-56-1) were purchased from Sinopharm (Beijing, China). Adenine (CAS, 73-24-5) and terephthalic acid (CAS, 100-21-0) were purchased from Energy Chemical (Beijing, China) (>99%). Active carbon was purchased from Energy Chemical Co. Ltd. (CAS, 7440-44-0). Sulfamethoxazole and other chemicals were all obtained from Sigma–Aldrich with a purity of >99%. Water was obtained from a Milli-Q system with 18.2 $M\Omega{\cdot}cm^{-1}$. The target compounds can be found in Table 1. K_{ow}, which is the ratio of the concentration of molecules in the octanol phase to their concentration in the aqueous phase, is widely used to describe the hydrophobicity of drugs [26]. A drug's greater hydrophobicity is indicated by a larger log K_{ow} value [27].

Table 1. Pharmaceuticals used in this study.

Full Name	Abbra.	Molecular Weight	Log K_{ow}	Chemical Structures
ketoprofen	KP	257.3	3.12	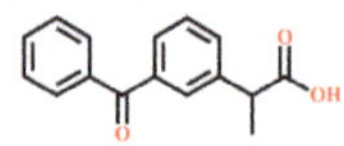
antipyrine	AT	189.1	−1.55	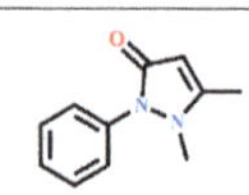

Table 1. *Cont.*

Full Name	Abbra.	Molecular Weight	Log K_{ow}	Chemical Structures
ibuprofen	IBU	205.1	0.45	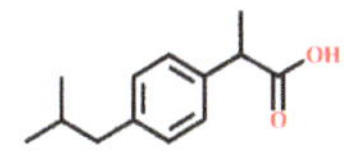
chloramphenicol	CAP	321.0	1.1	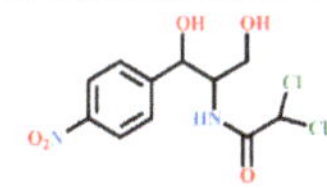
paracetamol	PC	180.1	1.58	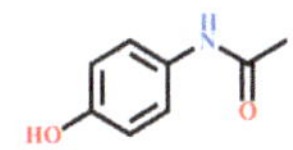
sulfamethoxazole	SMX	254.1	0.89	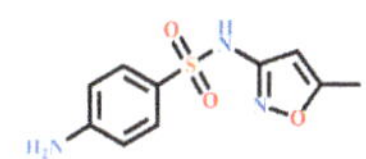

2.2. Synthesis of Bio-27 and Its Derivatives Bio-C Materials

Bio-27 and its derivatives were synthesized via a solvothermal method [24]. $Zn(NO_3)_2{\cdot}6H_2O$ (260.86 mg, 0.87 mmol), adenine (AD; 117.38 mg, 0.87 mmol), and terephthalic acid (H_2BDC; 72.60 mg, 0.435 mmol) were mixed with 10.8 mL DMF, H_2O, and MeOH (v/v/v = 1:1:0.6) and dispersed by ultrasonic processing for 30 min. Then, a white dispersion was obtained. The mixed solution was transferred to a 50 mL Teflon-lined autoclave and kept in an oven at 120 °C for 48 h. DMF and methanol were used to wash the sample before it was dried at 60 °C. A specific amount of bio-27 was evenly distributed in the center of the quartz boat and heated by 5 °C min^{-1} for 3 h in nitrogen after 6 h of vacuum activation at 120 °C. The materials obtained at the different temperatures of T (T = 500~1000 °C) are named bio-C-T. As the furnace chamber cooled to room temperature, bio-C-T compounds were generated. Some basic characterization was conducted to select a suitable material for further application in removing pharmaceuticals from water. The schematic diagram can be found in Figure 1.

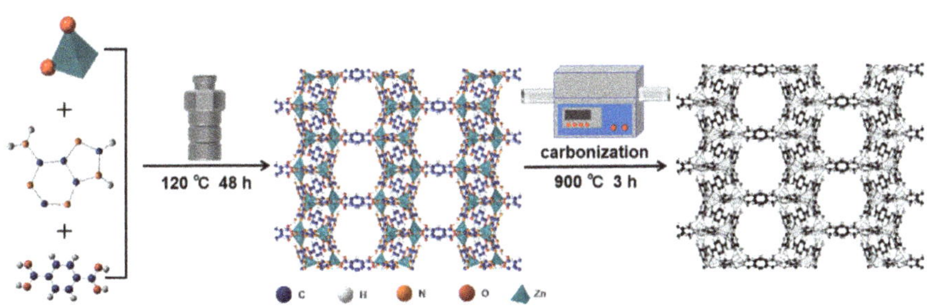

Figure 1. Schematic diagram of the synthesis of bio−27 and its derivatives, bio−C materials.

2.3. Synthesis of ZIF-8 and Its Derivatized Carbon-Based Materials (ZIF-8-C)

ZIF-8 was synthesized based on the previous report [28]. $Zn(NO_3)_2{\cdot}6H_2O$ (743.8 mg, 2.5 mmol) and dimethylimidazole (821.0 mg,10 mmol) were mixed with 50 mL MeOH and put into a 100 mL Shuniu bottle, where they were and continuously and evenly stirred for 24 h at room temperature. After standing for precipitation, the materials were separated and washed three times with fresh MeOH. Then, the materials were dried at 60 °C.

2.4. Characterization

The X-ray diffraction pattern was studied by a multipurpose, high-efficiency X-ray diffractometer (PXRD, Rigaku MiniFlEX600) (λ = 0.154 nm). The scan rate was 10°/min at room temperature. The adsorption–desorption curve was measured by an automatic specific surface and pore analyzer (Quanta chrome ASiQMVH002-5). Materials were activated in vacuum at 120 °C for 12 h. The Brunel–Emmett–Teller (BET) equation was used for measuring the specific surface area. The chemical state was obtained by X-ray photoelectron spectroscopy (XPS, PHI QUANTERA-II SXM, ULVAC-PHI, USA), and the X-ray source was Al-Ka (1486.6 eV, line width 0.68 eV). The zeta potential of bio-C was tested using a nanoparticle size and zeta potential analyzer (DLS, Malvern Zetasizer Nano ZS90, Worcestershire, UK).

2.5. Adsorption Experiment

The adsorption experiments for removing pharmaceuticals with two concentrations, 10 mg/L and 400 μg/L, were conducted separately. The solution was prepared with ultrapure water. Amounts equal to 4.5 mg of different adsorbents were added to a 50 mL glass vial containing 50 mL of a simulated solution of PPCPs at pH 7.0. All the batch adsorption experiments were performed on a multi-point stirrer at 500 rpm and 25 °C. At each time interval, the solution was sampled and then filtered with a PES filter. The adsorption experiments were conducted for 45 min. The pseudo-second-order kinetic model is shown in Equations (1) and (2), where Q_t represents the adsorption amount at time t, and μg/g is the unit. Q_e is the equilibrium adsorption amount, and μg/g is the unit. v_0 represents the initial adsorption rate, and μg/(g·min) is the unit. k is the intraparticle diffusion rate constant.

$$\frac{t}{Q_t} = \frac{1}{v_0} + \frac{t}{Qe} \tag{1}$$

$$v_0 = k \times Q_t^2 \tag{2}$$

2.6. Instrumental Analysis for Pharmaceuticals

Pharmaceuticals with concentrations of 10 mg/L were separated and detected using HPLC-UV with an Agilent 1260 system. The column was an Agilent SB-C18 column (2.7 μm, 4.6 mm × 150 mm). The mobile phase was water containing 0.1% formic acid and 1 mM ammonium acetate ($(NH_4)_2AC$) (60%), as well as MeOH (40%). The temperature in the column was 30 °C. The UV light's detecting wavelength was 270 nm. Triple quadrupole mass spectrometry (Agilent 1290, 6465 QQQ, Santa Clara, CA, USA) and ultra-performance liquid chromatography were used to measure the drugs with trace concentrations. The column was an Eclipse Plua C18 RRHD (2.1 × 50 mm 1.8 μm). The mobile phases were water with 0.1% formic acid and 2 mM ammonium acetate (A) and acetonitrile (B). The flow rate was set at 0.2 mL/min. The dilute gradient was 0–2 min (10% B), 2–10 min (10~45% B), 10~13 min (45~90% B), 13~14 min (90% B), and 14.1 min (10% B). The injection volume was 20 μL. The flow rate was 0.2 mL/min. The peak areas of each pharmaceutical were quantitively analyzed by the Quant-My-Way software, developed by Agilent.

3. Results and Discussion

3.1. Characterizations of Bio-C-T

3.1.1. Structural and Porous Properties

As shown in Figure 2a, the PXRD patterns of bio-27 were consistent with the simulation, with the main characteristic peaks at 5°, 9°, and 15° [24], indicating the metal–organic framework (MOF) called bio-27 was synthesized successfully. The PXRD pattern of bio-C-T can be found in Figure 2b, indicating that all the synthesized carbons are in an amorphous state. It is to be noted that bio-C-800, 900, and 1000 showed two broad peaks at 24° and 43°, corresponding to the diffraction peaks of the (002) crystal plane and (100) crystal plane, respectively [29]. The observed peaks move slightly to smaller angles with increased activation temperatures, indicating the distance of the interlayer could gradually increase.

According to previous studies, this shift could be caused by defects during carbonization [30]. However, bio-C-500 may not be carbonized completely based on the positions of characteristic peaks in the PXRD spectrum. On the other hand, the colors of bio-C-800, bio-C-900, and bio-C-1000 are black, consistent with the color of carbon black. This could also confirm the completeness of the carbonization of the four materials. The nitrogen adsorption isotherm in Figure 2c reveals a mixture of type I and IV isotherms. Obviously, the adsorption curve of bio-C-900 has a hysteresis loop, indicating that the material has a hierarchical porous structure and a relatively more mesoporous structure.

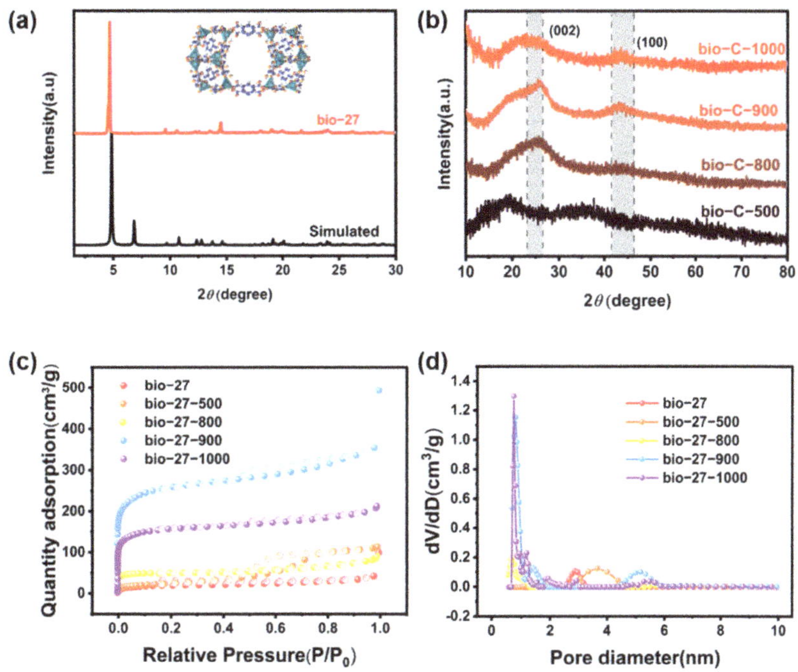

Figure 2. (**a**) PXRD patterns of IISERP−MOF−27 and its (**b**) derived carbon materials, which are pyrolyzed at temperatures ranging from 500 °C to 1000 °C. (**c**) The nitrogen adsorption and desorption isotherms of bio−27 and bio−C−T. (**d**) The pore size distribution of the as-prepared samples of bio−27 and bio−C−T (T = 500~1000 °C).

In comparison, the hysteresis loops of bio-C-800 and bio-C-1000 are not obvious, and the adsorption curves are more in line with the type I adsorption isotherm. The surface area and pore volume show an upward trend with increased carbonization temperature. Still, the specific surface area of bio-C-1000 was lower than that of bio-C-900, which may be due to the partial collapse of the mesoporous structure at high temperature due to the complete evaporation of Zn atoms (Table 2).

Table 2. Calculated surface area (S_{BET}) and porous structures.

Materials	S_{BET} (m^2/g)	V_{total} (cm^3/g)	Pore Width (nm)
Bio-27	76.5	0.054	2.890
Bio-27-500	90.2	0.160	3.698
Bio-27-800	204.2	0.108	0.783
Bio-27-900	980.4	0.496	0.852
Bio-27-1000	605.3	0.294	0.783

The pore size distribution curve calculated in the microporous region (Figure 2d) shows that all catalysts are dominated by micropores with a pore size of 0.8 nm. Although the microporous area of bio-C-500 accounts for a small proportion, the pore volume of bio-C-500 is higher than that of bio-C-800 due to the wider distribution of the mesoporous area. The surface area of the pure MOF (bio-27) is 76.4 m^2/g, as shown in Table 2. However, bio-27-C has much higher surface areas ranging from 204.2 m^2/g to 980.4 m^2/g.

On the other hand, the pore volume of bio-27-C-900 is the greatest, with a value of 0.496 cm^3/g. The pore width of the bio-C-900 is the greatest, with a value of 0.852 nm, among all the materials. The well-developed pore structure in the adsorbent may provide more adsorption sites for SMX, which should be beneficial for reducing diffusion resistance.

3.1.2. Morphology of the Samples

In Figure 3a,b, the precursor bio-27 has an irregular two-dimensional sheet structure, but the thickness is too large. With increasing pyrolysis temperatures, the edge morphology of bio-C-500 becomes irregular (Figure 3c), but it can be observed that it exists in the form of sheet-like stacking, and the scale becomes thinner. While bio-C-800, bio-C-900, and bio-C-1000 (Figure 3d–f) exhibit changes in longitudinal scale, the edges are more regular, and the ordered sheet-like stacking structure is favorable for the mass transfer process [31]. The jagged edges of bio-C-1000 may be due to the collapse of the internal structure. HRTEM was used to further characterize the morphology and microstructure of the catalyst. In the bright field image (Figure 3g), it can be observed that the surface of the catalyst with sheet-like morphology has uniformly distributed mesopores and only short-range ordered ones. N-doped carbon fringes do not exhibit any long-range ordered lattice fringes, indicating no crystalline species in the catalyst, which is mutually confirmed by the results of PXRD, as discussed above. The results show that the prepared bio-C-900 material has abundant pore structures.

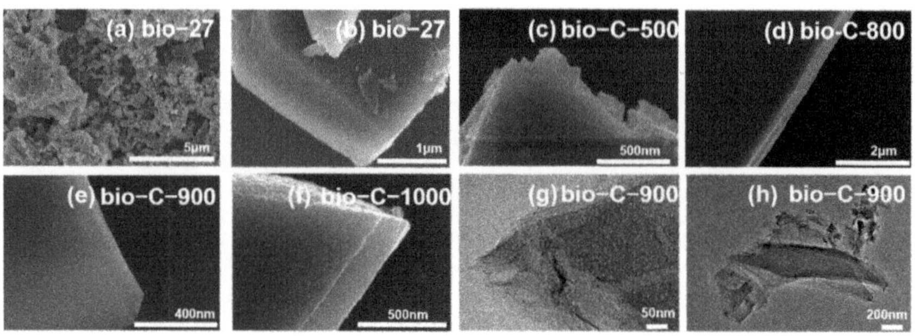

Figure 3. (**a**–**f**) SEM images of bio−27 and bio−C−T. (**g**,**h**) HRTEM of bio−C−900.

3.1.3. Fourier Transform Infrared (FTIR) Spectrum of Bio-27-Derived Carbon

Analysis via FTIR spectroscopy was carried out in order to investigate the various surface functional groups. According to Figure 4, the stretching vibrations of O-H and N-H can be found at around 3400 cm^{-1} [32]. The band at 2100 cm^{-1} is ascribed to the C=C=C stretching vibration [33]. The C=O stretching vibration is responsible for the band at 1650 cm^{-1} [32], while the N-H in-plane and out-of-plane bending vibrations are responsible for the bands at 1580 cm^{-1} and 850 cm^{-1}, respectively [32]. Furthermore, the bands at 1380 cm^{-1} and 1070 cm^{-1} are attributed to C-O stretching vibrations and C-N stretching vibrations [32,34], respectively, indicating the existence of nitrogen and oxygen functional groups on bio-C.

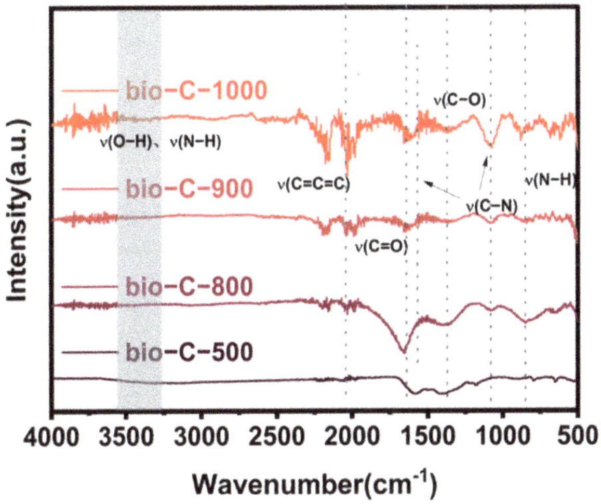

Figure 4. Fourier transform infrared (FTIR) spectrum of IISERP−MOF−27−derived carbon which is pyrolyzed at temperatures ranging from 500 °C to 1000 °C.

3.2. Adsorption Experiments

3.2.1. Pseudo-Second-Order Adsorption Kinetics of SMX

Within a contact time of 45 min, the impact of bio-C-T materials at various pyrolysis temperatures on SMX adsorption was examined. The results are displayed in Figure 5a. To fit the experimental data, a pseudo-second-order kinetic model was applied, and the relevant fit plots were produced. The model compound's adsorption of the adsorbent is indicated by the variables Q_t and Q_e in the equation at reaction time t (min)—when the reaction achieves saturation equilibrium. Q_e and v_0 represent the adsorption amount and the initial adsorption rate (mg/g·min), respectively. According to Table 3, bio-C-900 has the highest Q_e (100.85 mg/g) and the best adsorption capacity of SMX among the series of bio-C-T materials. In addition, its v_0 (315.29 mg/g·min) is the highest among all the prepared materials. The mass transfer resistance of bio-C-900 to SMX is substantially lower than that of the manufactured bio-C at other temperatures, and its adsorption sites are more readily accessible to SMX. Additionally, its v_0 is more than 30 times that of bio-C-1000. We therefore hypothesize that surface area has a close relationship with the adsorption amount of SMX in this study. A higher surface area should favor a higher adsorption amount.

Table 3. Pseudo-second-order adsorption kinetics fitting values.

Materials	k (g/mg·min)	Q_e (mg/g)	v_0 (mg/g·min)	R^2
Bio-27-500	——	1.65	——	0.99
Bio-27-800	0.0059	27.27	4.38	0.98
Bio-27-900	0.031	100.85	315.29	0.93
Bio-27-1000	0.0021	70.79	10.52	0.98

On the other hand, large pore sizes could enhance the adsorption kinetics [35]. Similar results can be found in the adsorption process by other porous material [36]. For example, Guo found that the introduction of mesopores above 2 nm can greatly increase the pore volume. The graphene oxide/carbon composite nanofibers with abundant mesopores prepared in this study achieved improved adsorption capacity for volatile organic compounds (VOCs), and the highest adsorption capacity values for benzene and butanone reached 83.2 $cm^3\ g^{-1}$ and 130.5 $cm^3\ g^{-1}$, respectively. According to Table 2, we can know that the pore volume of bio-C-T-series materials was also greatly improved after the appearance of mesopores, which is consistent with the experimental results.

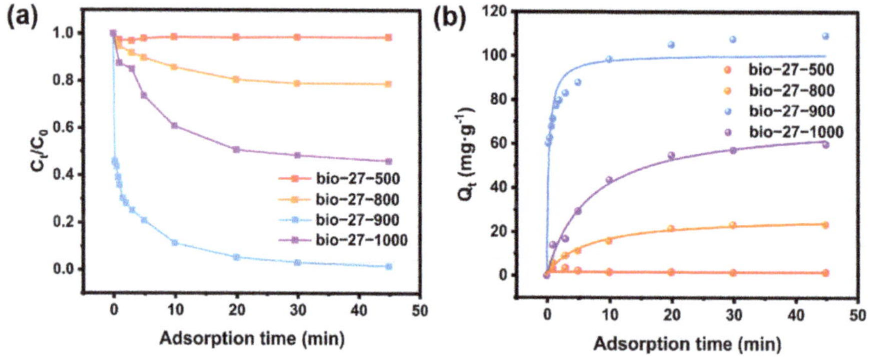

Figure 5. (**a**) Adsorption kinetics curves of bio−C−T. (**b**) The corresponding fitted pseudo-second-order adsorption kinetics. Reaction conditions: SMX = 10 mg/L, reaction volume = 50 mL, catalyst dosage = 180 mg/L, initial solution pH = 7, temperature = 25 °C, reaction time = 45 min).

3.2.2. Effect of Reaction Time

The elimination of SMX was compared over bio-C-900, AC, and ZIF-8 (Zn as the metal center). The ZIF-8-C obtained after carbonization at the same temperature was applied in an adsorption experiment for removing SMX with an initial concentration of 10 ppm for 60 min. As shown in Figure 6a, bio-C-900 showed ultra-fast adsorption behavior, with 55% removal within only 15 s. A higher removal rate of 90% could be achieved within only 10 min. After reaching adsorption saturation, the removal rate of SMX by bio-C-900 can reach 98.7%. In comparison, the removal rates for SMX over AC and ZIF-8-C were 36.5% and 9.7%, respectively. Two models, the Elovich model and the second-order kinetic model, were applied in Table 4. The Elovich kinetic fitting formula (Equation (3)) is as follows:

$$Q_t = A + b \times \ln t \quad (3)$$

where A and b are both Elovich constants. The fitting degree of the two models is very high, indicating that the adsorption process of the adsorbent includes both chemisorption with electron transfer and electron transfer during the pseudo-second-order kinetic process.

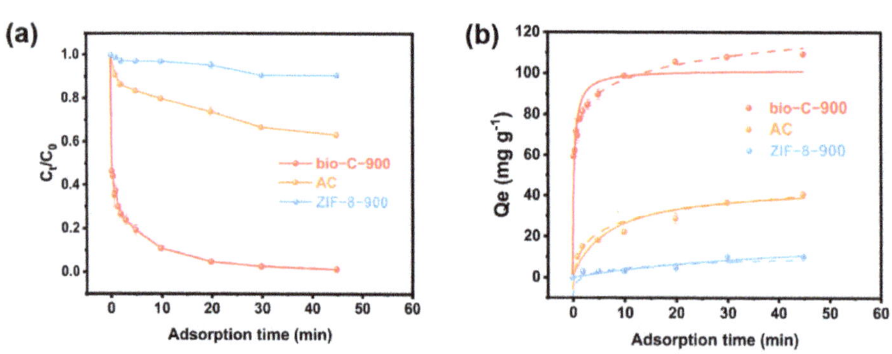

Figure 6. (**a**) Adsorption kinetics curves of all materials for SMX in this study; (**b**) The corresponding pseudo-second-order adsorption kinetic model and Elovich kinetic model were fitted. Reaction conditions: initial concentration of SMX = 10 mg/L, reaction volume = 50 mL, catalyst dosage = 4.5 mg, initial solution pH = 7, temperature = 25 °C, reaction time = 45 min.

Table 4. Langmuir and Freundlich adsorption isotherm model.

	Langmuir Model			Freundlich Model		
Sample	Q_{max} (mg/g)	K_L (L/mg)	R^2	K_F (mg g^{-1}(L $mg^{-1})^{1/n}$)	n	R^2
Bio-C-900	350.90	0.061	0.99	53.23	2.47	0.96
AC	76.96	0.025	0.98	29.79	4.03	0.82
ZIF-8-C	26.44	0.027	0.99	10.88	4.32	0.88

3.2.3. Effect of Initial Concentration

The amount of adsorption increases with increasing initial concentration and reaches saturation for bio-C-900, AC, and ZIF-8-C, as well as for adsorbents over the studied concentration (Figure 7). At initial concentration of 20 mg/L, 86.6% of SMX could be removed by the bio-C-900 sorbent. When the initial concentrations were 40, 50, 60, and 80 mg/L, SMX was partially removed with efficiencies of 58.3%, 53.1%, 44.8%, and 32.7%, respectively. Such a phenomenon can be explained by the limited adsorption sites of the same adsorbent. The adsorption capacity increases with a higher concentration of adsorbate. This demonstrates that the initial concentration should limit the adsorption amount because the adsorption sites are sufficient [37]. In addition, the data were fitted using the Langmuir and Freundlich equations.

$$Q_e = Q_{max} \times \frac{K_L C_e}{1 + K_L C_e} \tag{4}$$

$$Q_e = K_F {C_e}^{\frac{1}{n}} \tag{5}$$

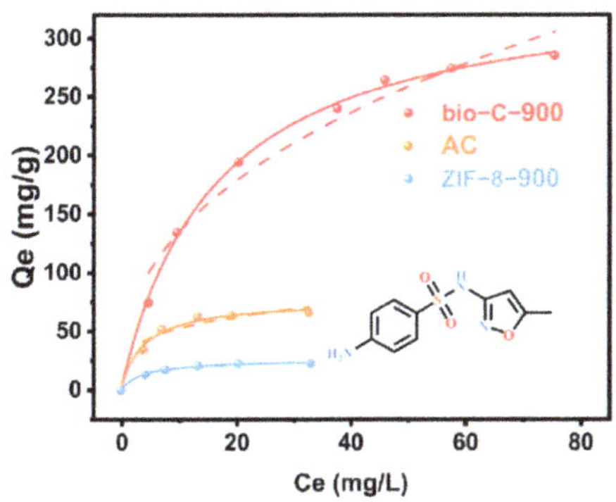

Figure 7. Adsorption isotherm for the removal of SMX over the three materials: active carbon (AC), ZIF−8−C, and bio−C−900. Reaction conditions: initial concentration of SMX = 5~80 mg/L, initial solution pH = 7, temperature = 25 °C, reaction time = 600 min.

The adsorption capacity of SMX over bio-C-900 was 4.6 and 13.3 times higher than those of AC and ZIF-8-C, respectively. The K_L value indicates that SMX is more easily removed by bio-C-900, confirming that the prepared bio-C-900 could have relatively better adsorption properties than those of conventional materials. The adsorption of SMX over bio-C-900 tends to be homogeneous monolayer adsorption [38] because the Langmuir model linear fitting result (R^2 = 0.99) is slightly better than that in the Freundlich model (R^2 = 0.96). Therefore, the material tends towards homogeneous monolayer adsorption. In addition, the maximum adsorption capacity of SMX over bio-C-900 (350.90 mg/g) was 3.6 times higher than that over AC (76.96 mg/g), which further proved that bio-C-900 is conducive to the transport of substrates on its surface and pores. The heterogeneous multi-layer adsorption effectively improves the adsorption capacity and adsorption kinetics. The specific fitting data are shown in Table 4 below.

3.3. Exploring the Adsorption Mechanism for Micropollutants

To gain more insight into the reaction mechanism, the effect of solution pH (3~9) was evaluated in Figure 8a. As seen in Figure 8, the removal percentages of SMX varied significantly with initial pH values. SMX is an amphiphilic molecule with two pK_a values (pK_{a1} = 1.8 and pK_{a2} = 5.6) [39]. When the solution pH is higher, SMX is prone to exist in its deprotonated form with a negative charge. The zeta potential can be found in Figure 8b. As a result, the adsorption amount decreased obviously, probably because of electrostatic expulsion. However, in an acid environment, SMX is usually protonated, with a positive charge, which could lead to a decrease in electrostatic interaction. However, the adsorption performance for the removal of SMX is higher at pH values ranging from 3 to 5. Thus, several major mechanisms could co-exist to contribute to the high adsorption amount. For example, the solubility of SMX is lower in an acid environment, and it is very slightly soluble in water but is soluble in alkali hydroxides, as previously reported [40,41]. The properties of hydrophilic interaction should be considered.

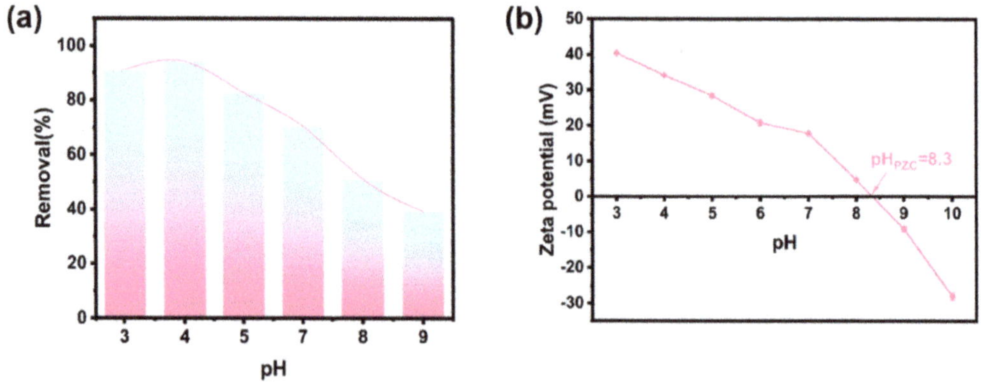

Figure 8. (**a**) Effect of pH on SMX removal. (**b**) Zeta-potential of bio−C−900.

To further evaluate the adsorption performance for pharmaceuticals with diverse structures, the adsorption behavior for six pharmaceuticals—ketoprofen (KP), antipyrine (AT), ibuprofen (IBU), chloramphenicol (CAP), paracetamol (PC), and sulfamethoxazole (SMX)—were studied. Obviously, we found that the adsorption performance (removal percentages within 45 min reaction time) correlated well (R^2 = 0.97) with the physical–chemical parameter log K_{ow} of each pharmaceutical. The log K_{ow} values of ketoprofen (KP), antipyrine (AT), ibuprofen (IBU), chloramphenicol (CAP), paracetamol (PC), and sulfamethoxazole (SMX) were 3.12, −1.55, 0.45, 1.1, 1.58, and 0.89 [42–45], respectively. Tung Xuan Bui et al. found a linear relationship between log Kow and adsorption amount at a wide range of pH values for 12 drugs, indicating that the adsorption of drugs on TMS-SBA-15 is mainly driven by hydrophobic interactions [46]. Taku Matsushita [47] et al. evaluated the adsorption capacity of nine activated carbons for geosmin and 2-methylisoborneol (MIB) and showed that hydrophobic materials enhance the adsorption. Based on the above discussion, hydrophobic interactions are dominant in this adsorption mechanism. Based on the discussion above, the proposed mechanism is shown in Figure 9b.

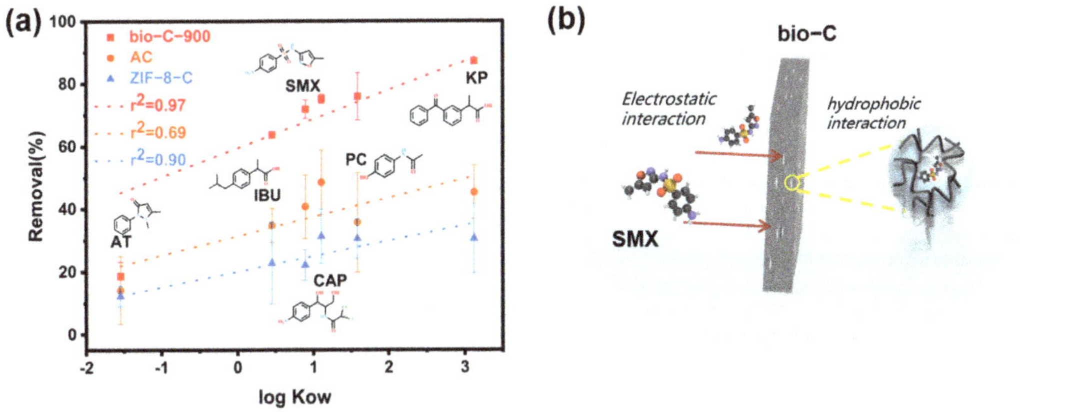

Figure 9. (**a**) Effect of K_{OW} on diverse micropollutants. (**b**) Proposed reaction mechanism.

4. Conclusions

In summary, a novel MOF-derived carbon has been designed and synthesized for the first time. A mixed ligand approach has been applied to successfully synthesize bio-27 via the solvothermal method. By pyrolyzation at different temperatures, the optimal condition at 900 °C was selected to generate a novel heterogeneous porous carbon with the highest surface area (S_{BET} = 980.5 m^2/g) and a large pore volume (0.496 cm^3/g). The maximum saturated adsorption capacity for sulfamethylthiazole (SMX) over MOF-derived carbon can reach 351.3 mg/g with a fast initial adsorption rate of 315.29 (mg/g·min). The adsorption capacity calculated by the Langmuir model (R^2 = 0.99) for SMX over bio-C-900 was 4.6 and 13.3 times more than the values for AC and ZIF-8-C, respectively. Hydrophobic interaction should be one of the major mechanisms for the material's adsorption in water. This study offers a strategy to develop novel carbon materials to remove pharmaceuticals from water.

Author Contributions: Conceptualization, X.L. and Y.M.; methodology, X.L. and Y.M.; software, Y.M.; validation, X.L. and Y.M.; formal analysis, X.L. and Y.M.; investigation, X.L. and Y.M.; resources, B.W.; data curation, X.L. and Y.M. writing—original draft preparation, Y.M.; writing—review and editing, X.L; visualization, X.L; supervision, B.W.; project administration, X.L. and B.W.; funding acquisition, X.L. and B.W.; All authors have read and agreed to the published version of the manuscript.

Funding: This work was financially supported by the National Natural Science Foundation of China (Grants 21625102, 21971017, and 21906007), China's National Key Research and Development Program (Grant 2020YFB1506300), and the Beijing Institute of Technology Research Fund Program. We gratefully acknowledge the Analysis and Testing Center of the Beijing Institute of Technology.

Conflicts of Interest: The authors declare no conflict of interest.

References

1. Yang, Y.; Ok, Y.S.; Kim, K.H.; Kwon, E.E.; Tsang, Y.F. Occurrences and removal of pharmaceuticals and personal care products (PPCPs) in drinking water and water/sewage treatment plants: A. review. *Sci. Total Environ.* **2017**, *596*, 303–320. [CrossRef]
2. Priya, A.K.; Gnanasekaran, L.; Rajendran, S.; Qin, J.Q.; Vasseghian, Y. Occurrences and removal of pharmaceutical and personal care products from aquatic systems using advanced treatment—A review. *Environ. Res.* **2022**, *204*, 15. [CrossRef]
3. Petrie, B.; Barden, R.; Kasprzyk-Hordern, B. A review on emerging contaminants in wastewaters and the environment: Current knowledge, understudied areas and recommendations for future monitoring. *Water Res.* **2015**, *72*, 3–27. [CrossRef] [PubMed]
4. Ramirez, A.J.; Brain, R.A.; Usenko, S.; Mottaleb, M.A.; O'Donnell, J.G.; Stahl, L.L.; Wathen, J.B.; Snyder, B.D.; Pitt, J.L.; Perez-Hurtado, P.; et al. Occurrence of pharmaceuticals and personal care products in fish: Results of a national pilot study in the united states. *Environ. Toxicol. Chem.* **2009**, *28*, 2587. [CrossRef] [PubMed]
5. Bonvin, F.; Omlin, J.; Rutler, R.; Schweizer, W.B.; Alaimo, P.J.; Strathmann, T.J.; McNeill, K.; Kohn, T. Direct Photolysis of Human Metabolites of the Antibiotic Sulfamethoxazole: Evidence for abiotic Back-Transformation. *Environ. Sci. Technol.* **2013**, *47*, 6746–6755. [CrossRef] [PubMed]

6. Abusin, S.; Johnson, S. Sulfamethoxazole/Trimethoprim induced liver failure: A case report. *Cases J.* **2008**, *1*, 44. [CrossRef]
7. Tong, H.; Ouyang, S.X.; Bi, Y.P.; Umezawa, N.; Oshikiri, M.; Ye, J.H. Nano-photocatalytic materials: Possibilities and challenges. *Adv. Mater.* **2012**, *24*, 229–251. [CrossRef]
8. Jain, B.; Singh, A.K.; Kim, H.; Lichtfouse, E.; Sharma, V.K. Treatment of organic pollutants by homogeneous and heterogeneous Fenton reaction processes. *Environ. Chem. Lett.* **2018**, *16*, 947–967. [CrossRef]
9. Zhang, D.Q.; Gersberg, R.M.; Ng, W.J.; Tan, S.K. Removal of pharmaceuticals and personal care products in aquatic plant-based systems: A. review. *Environ. Pollut.* **2014**, *184*, 620–639. [CrossRef]
10. Nasrollahi, N.; Vatanpour, V.; Khataee, A. Removal of antibiotics from wastewaters by membrane technology: Limitations, successes, and future improvements. *Sci. Total Environ.* **2022**, *838*, 18. [CrossRef]
11. Zhang, W.L.; Yin, J.; Wang, C.W.; Zhao, L.; Jian, W.B.; Lu, K.; Lin, H.B.; Qiu, X.Q.; Alshareef, H.N. Lignin Derived Porous Carbons: Synthesis Methods and Supercapacitor Applications. *Small Methods* **2021**, *5*, 24. [CrossRef] [PubMed]
12. Moulahcene, L.; Skiba, M.; Senhadji, O.; Milon, N.; Benamor, M.; Lahiani-Skiba, M. Inclusion and removal of pharmaceutical residues from aqueous solution using water-insoluble cyclodextrin polymers. *Chem. Eng. Res. Des.* **2015**, *97*, 145–158. [CrossRef]
13. Margeta, K.; Vojnovic, B.; Logar, N.Z. Development of Natural Zeolites for Their Use in Water-Treatment Systems. *Recent Pat. Nanotechnol.* **2011**, *5*, 89–99. [CrossRef] [PubMed]
14. Liu, X.L.; Wang, C.; Wu, Q.H.; Wang, Z. Metal-organic framework-templated synthesis of magnetic nanoporous carbon as an efficient absorbent for enrichment of phenylurea herbicides. *Anal. Chim. Acta* **2015**, *870*, 67–74. [CrossRef]
15. Rodrigues, D.A.D.; da Cunha, C.; Freitas, M.G.; de Barros, A.L.C.; Castro, P.; Pereira, A.R.; Silva, S.D.; Santiago, A.D.; Afonso, R. Biodegradation of sulfamethoxazole by microalgae-bacteria consortium in wastewater treatment plant effluents. *Sci. Total Environ.* **2020**, *749*, 9. [CrossRef]
16. Hasan, Z.; Jhung, S.H. Removal of hazardous organics from water using metal-organic frameworks (MOFs): Plausible mechanisms for selective adsorptions. *J. Hazard. Mater.* **2015**, *283*, 329–339. [CrossRef]
17. Li, S.Q.; Zhang, X.D.; Huang, Y.M. Zeolitic imidazolate framework-8 derived nanoporous carbon as an effective and recyclable adsorbent for removal of ciprofloxacin antibiotics from water. *J. Hazard. Mater.* **2017**, *321*, 711–719. [CrossRef]
18. Jiao, C.N.; Wang, Y.N.; Li, M.H.; Wu, Q.H.; Wang, C.; Wang, Z. Synthesis of magnetic nanoporous carbon from metal-organic framework for the fast removal of organic dye from aqueous solution. *J. Magn. Magn. Mater.* **2016**, *407*, 24–30. [CrossRef]
19. Wang, L.H.; Ke, F.; Zhu, J.F. Metal-organic gel templated synthesis of magnetic porous carbon for highly efficient removal of organic dyes. *Dalton Trans.* **2016**, *45*, 4541–4547. [CrossRef]
20. Chen, D.Z.; Shen, W.S.; Wu, S.L.; Chen, C.Q.; Luo, X.B.; Guo, L. Ion exchange induced removal of Pb(II) by MOF-derived magnetic inorganic sorbents. *Nanoscale* **2016**, *8*, 7172–7179. [CrossRef]
21. Salunkhe, R.R.; Young, C.; Tang, J.; Takei, T.; Ide, Y.; Kobayashi, N.; Yamauchi, Y. A high-performance supercapacitor cell based on ZIF-8-derived nanoporous carbon using an organic electrolyte. *Chem. Commun.* **2016**, *52*, 4764–4767. [CrossRef] [PubMed]
22. Yan, X.L.; Li, X.J.; Yan, Z.F.; Komarneni, S. Porous carbons prepared by direct carbonization of MOFs for supercapacitors. *Appl. Surf. Sci.* **2014**, *308*, 306–310. [CrossRef]
23. Fang, X.Z.; Jiao, L.; Yu, S.H.; Jiang, H.L. Metal-Organic Framework-Derived FeCo-N-Doped Hollow Porous Carbon Nanocubes for Electrocatalysis in Acidic and Alkaline Media. *Chem. Sus. Chem.* **2017**, *10*, 3019–3024. [CrossRef] [PubMed]
24. Maity, R.; Singh, H.D.; Yadav, A.K.; Chakraborty, D.; Vaidhyanathan, R. Water-stable Adenine-based MOFs with Polar Pores for Selective CO_2 Capture. *Chem. Asian J.* **2019**, *14*, 3736–3741. [CrossRef]
25. Bhadra, B.N.; Jhung, S.H. Adsorptive removal of wide range of pharmaceuticals and personal care products from water using bio-MOF-1 derived porous carbon. *Microporous Mesoporous Mater.* **2018**, *270*, 102–108. [CrossRef]
26. Yin, P.Q.; Yao, T.; Wu, Y.; Zheng, L.R.; Lin, Y.; Liu, W.; Ju, H.X.; Zhu, J.F.; Hong, X.; Deng, Z.X.; et al. Single Cobalt Atoms with Precise N-Coordination as Superior Oxygen Reduction Reaction Catalysts. *Angew. Chem. Int. Edit.* **2016**, *55*, 10800–10805. [CrossRef]
27. Han, S.-y.; Qiao, J.-q.; Zhang, Y.-y.; Lian, H.-z.; Ge, X. Determination of n-octanol/water partition coefficients of weak ionizable solutes by RP-HPLC with neutral model compounds. *Talanta* **2012**, *97*, 355–361. [CrossRef]
28. Kitt, J.P.; Harris, J.M. Confocal Raman Microscopy for in Situ Measurement of Octanol–Water Partitioning within the Pores of Individual C18-Functionalized Chromatographic Particles. *Anal. Chem.* **2015**, *87*, 5340–5347. [CrossRef]
29. Li, Y.H.; Shen, S.G.; Wang, C.Y.; Peng, X.; Yuan, S.J. The effect of difference in chemical composition between cellulose and lignin on carbon based solid acids applied for cellulose hydrolysis. *Cellulose* **2018**, *25*, 1851–1863. [CrossRef]
30. Shi, Y.W.; Liu, G.Z.; Wang, L.; Zhang, X.W. Efficient adsorptive removal of dibenzothiophene from model fuel over heteroatom-doped porous carbons by carbonization of an organic salt. *Chem. Eng. J.* **2015**, *259*, 771–778. [CrossRef]
31. Tsou, C.H.; An, Q.F.; Lo, S.C.; de Guzman, M.; Hung, W.S.; Hu, C.C.; Lee, K.R.; Lai, J.Y. Effect of microstructure of graphene oxide fabricated through different self-assembly techniques on 1-butanol dehydration. *J. Membr. Sci.* **2015**, *477*, 93–100. [CrossRef]
32. Yang, X.M.; Salado-Leza, D.; Porcel, E.; Gonzalez-Vargas, C.R.; Savina, F.; Dragoe, D.; Remita, H.; Lacombe, S. A Facile One-Pot Synthesis of Versatile PEGylated Platinum Nanoflowers and Their Application in Radiation Therapy. *Int. J. Mol. Sci.* **2020**, *21*, 20. [CrossRef] [PubMed]
33. Ain, Q.T.; Haq, S.H.; Alshammari, A.; Al-Mutlaq, M.A.; Anjum, M.N. The systemic effect of PEG-nGO-induced oxidative stress in vivo in a rodent model. *Beilstein J. Nanotechnol.* **2019**, *10*, 901–911. [CrossRef]

34. Wang, Z.; Ding, Y.K.; Wang, J.C. Novel Polyvinyl Alcohol (PVA)/Cellulose Nanocrystal (CNC) Supramolecular Composite Hydrogels: Preparation and Application as Soil Conditioners. *Nanomaterials* **2019**, *9*, 17. [CrossRef]
35. Idris, S.A.; Alotaibi, K.M.; Peshkur, T.A.; Anderson, P.; Morris, M.; Gibson, L.T. Adsorption kinetic study: Effect of adsorbent pore size distribution on the rate of Cr (VI) uptake. *Microporous Mesoporous Mater.* **2013**, *165*, 99–105. [CrossRef]
36. Guo, Z.Y.; Huang, J.T.; Xue, Z.H.; Wang, X.M. Electrospun graphene oxide/carbon composite nanofibers with well-developed mesoporous structure and their adsorption performance for benzene and butanone. *Chem. Eng. J.* **2016**, *306*, 99–106. [CrossRef]
37. Nielsen, L.; Biggs, M.J.; Skinner, W.; Bandosz, T.J. The effects of activated carbon surface features on the reactive adsorption of carbamazepine and sulfamethoxazole. *Carbon* **2014**, *80*, 419–432. [CrossRef]
38. Song, X.L.; Zhang, Y.; Yan, C.Y.; Jiang, W.J.; Chang, C.M. The Langmuir monolayer adsorption model of organic matter into effective pores in activated carbon. *J. Colloid Interface Sci.* **2013**, *389*, 213–219. [CrossRef]
39. Song, Y.Q.; Zeng, Y.; Liao, J.X.; Chen, J.Q.; Du, Q. Efficient removal of sulfamethoxazole by resin-supported zero-valent iron composites with tunable structure: Performance, mechanisms, and degradation pathways. *Chemosphere* **2021**, *269*, 11. [CrossRef]
40. Aronson, J.K. Sulfonamides. In *Meyler's Side Effects of Drugs*, 16th ed.; Elsevier: Oxford, UK, 2016; pp. 555–569. [CrossRef]
41. Hişmioğullari, Ş.E.; Yarsan, E. Spectrophotometric determination and stability studies of sulfamethoxazole and trimethoprim in oral suspension by classical least square calibration method. *Hacet. Univ. J. Fac. Pharm.* **2009**, *2*, 95–104.
42. Yang, L.P.; Ke, H.; Yao, H.; Jiang, W. Effective and Rapid Removal of Polar Organic Micropollutants from Water by Amide Naphthotube-Crosslinked Polymers. *Angew. Chem. Int. Ed.* **2021**, *60*, 21404–21411. [CrossRef] [PubMed]
43. Ozaki, H.; Ikejima, N.; Shimizu, Y.; Fukami, K.; Taniguchi, S.; Takanami, R.; Giri, R.R.; Matsui, S. Rejection of pharmaceuticals and personal care products (PPCPs) and endocrine disrupting chemicals (EDCs) by low pressure reverse osmosis membranes. *Water Sci. Technol.* **2008**, *58*, 73–81. [CrossRef] [PubMed]
44. Ngawhirunpat, T.; Panomsuk, S.; Opanasopit, P.; Rojanarata, T.; Hatanaka, T. Comparison of the percutaneous absorption of hydrophilic and lipophilic compounds in shed snake skin and human skin. *Pharmazie* **2006**, *61*, 331–335. [PubMed]
45. Lin, H.; Huang, L.; Gao, Z.; Lin, W.; Ren, Y. Comparative analysis of the removal and transformation of 10 typical pharmaceutical and personal care products in secondary treatment of sewage: A case study of two biological treatment processes. *J. Environ. Chem. Eng.* **2022**, *10*, 107638. [CrossRef]
46. Bui, T.X.; Pham, V.H.; Le, S.T.; Choi, H. Adsorption of pharmaceuticals onto trimethylsilylated mesoporous SBA-15. *J. Hazard. Mater.* **2013**, *254–255*, 345–353. [CrossRef] [PubMed]
47. Matsui, Y.; Nakao, S.; Sakamoto, A.; Taniguchi, T.; Pan, L.; Matsushita, T.; Shirasaki, N. Adsorption capacities of activated carbons for geosmin and 2-methylisoborneol vary with activated carbon particle size: Effects of adsorbent and adsorbate characteristics. *Water Res.* **2015**, *85*, 95–102. [CrossRef]

water

MDPI

Article

Removal of Synthetic Dye from Aqueous Solution Using $MnFe_2O_4$-GO Catalyzed Heterogeneous Electro-Fenton Process

Gayathri Anil [1,2], Jaimy Scaria [2] and Puthiya Veetil Nidheesh [2,*]

1 Sacred Heart College, Thevara, Kochi 682013, India
2 CSIR-National Environmental Engineering Research Institute, Nagpur 440020, India
* Correspondence: pv_nidheesh@neeri.res.in or nidheeshpv129@gmail.com

Abstract: In the present study, heterogeneous electro-Fenton (HEF) process using $MnFe_2O_4$-GO catalyst is employed for the successful removal of dye from aqueous solution. Pt coated over titanium and graphite felt were used as the electrodes. The study focuses on the efficiency of the electrodes and catalyst used for the successful removal of Rhodamine B (RhB) from aqueous solution and the application of the same in real textile wastewater. The effect of various operational parameters like pH, applied voltage, catalyst concentration, initial pollutant concentration and effect of ions were investigated. The optimized condition of the electrolytic system was found as pH 3, applied voltage of 3 V, and catalyst concentration of 20 mg L^{-1} for the removal of 10 ppm RhB. At the optimized condition, 97.51% $\pm$ 0.0002 RhB removal was obtained after an electrolysis time of 60 min. The role of individual systems of Fe, Mn, GO and $MnFe_2O_4$ without support were compared with that of catalyst composite. On examining the practical viability in real textile effluent, a significant colour reduction was observed (reduced by 61.24% $\pm$ 0.0261 in 60 min). Along with this, the biodegradability enhancement (BOD/COD ratio from 0.07 to 0.21) after treatment was also observed.

Keywords: heterogeneous electro-Fenton; composite catalyst; Rhodamine B dye; textile wastewater; biodegradability; advanced oxidation processes

Citation: Anil, G.; Scaria, J.; Nidheesh, P.V. Removal of Synthetic Dye from Aqueous Solution Using $MnFe_2O_4$-GO Catalyzed Heterogeneous Electro-Fenton Process. *Water* **2022**, *14*, 3350. https://doi.org/10.3390/w14203350

Academic Editors: Dionysios (Dion) Demetriou Dionysiou, Yujue Wang and Huijiao Wang

Received: 19 September 2022
Accepted: 19 October 2022
Published: 21 October 2022

Publisher's Note: MDPI stays neutral with regard to jurisdictional claims in published maps and institutional affiliations.

1. Introduction

Textile effluent, defined by a solid colour, soluble dyes, and organics, makes up half of all industrial wastewater in the world [1]. India stands second in the world in the production of textiles and fabrics [2]. Additives used in the textile industry for various softening and polishing purposes include colouring pigments, such as dyes and surfactants [3]. Synthetic dyes in the textile effluent maintain their colour and structural integrity under extreme weather situations and exhibit a strong resistance to microbial degradation [4,5]. They take a long time to degrade when exposed to the environment. These toxic organic pollutants generated from textile industries can affect living beings and can cause serious threats to humans and the environment. Synthetic dyes cause a wide range of health effects on living beings. They impart colour to the aquatic environment, even in a small concentration [6]. Synthetic dyes possess carcinogenic, mutagenic and genotoxic properties, prolonged exposure to which can cause human health effects, including skin/eye irritation, neurotoxicity, and endocrine disruption [7,8]. Based on the types of dyes and accompanying chemicals employed, the behavior of textile discharge changes [9]. Textile wastewater usually has a low BOD to COD ratio (less than 0.1) making it unfit for biodegradation [10]. The effluent has to be treated effectively before discarding into the environment [11].

Electrochemical advanced oxidation processes (EAOPs) are one of the most environmentally benign processes for wastewater treatment, because the reagent utilized is the electron, and the oxidizing agents, such as hydroxyl radical, are in-situ generated [12]. The process is considered to be very effective since it is ecologically friendly and can create good amount of hydroxyl radicals by controlling the applied current [13]. It is considered as an effective method for the treatment of dyes, as the dye compounds are readily converted into carbon dioxide, water and other inorganic ions [5].

The electro-Fenton (EF) process is the most accepted EAOP, due to the benefits such as a broad application range, excellent anti-interference ability, ease of operation, and rapid pollutant removal and mineralization efficiency [14–16]. For the past few years, EF has been the most employed EAOP method and is considered most effective due to its ability to eliminate refractory compounds [17,18]. The efficacy of the EF procedure as a tool for removing dyes from water medium has been established [19].

The EF system is supplied continuously with air or oxygen and H_2O_2 is in-situ formed by the two-electron reduction of oxygen (Equation (1)) at the cathode. The $^{\bullet}OH$ are generated in the system by the Fenton reaction (Equation (2)), by the electrogenerated H_2O_2 at the cathode and Fe^{2+} catalyst added to the system [20]. The Fe^{2+} that is used by the system is then regenerated by the cathodic reduction of Fe^{3+} (Equation (3)) [21,22]. The formation of OH takes place, preferably in acidic media [23,24].

$$O_2 + 2H^+ + 2e^- \rightarrow H_2O_2 \tag{1}$$

$$Fe^{2+} + H_2O_2 \rightarrow Fe^{3+} + {}^{\bullet}OH + OH^- \tag{2}$$

$$Fe^{3+} + e^- \rightarrow Fe^{2+} \tag{3}$$

The EF process can be classified into two types, based on the catalyst used. The process can be said to be a homogenous electro-Fenton process, if the externally added catalyst is soluble in the system. If the catalyst is not soluble or is barely soluble in the system, the process is known as heterogeneous electro-Fenton (HEF). In a homogeneous system, the catalytic process takes place in the bulk liquid phase, whereas the catalytic process occurs on the surface of the catalyst in a heterogeneous system [25,26]. The mechanism of both the process is more or less similar. In HEF, the solid iron or its minerals helps in the generation of $^{\bullet}OH$ [27].

The main advantage of HEF process over conventional EF process is that, the heterogeneous catalyst can be recovered and reused after the Fenton process. This lowers the cost of operation and reduces catalyst wastage. The HEF process could work in an extended range of pH. This way adjusting the pH of the solution in the conventional Fenton process can be avoided. All the reactions associated with the process occur at the catalyst surface. The ferric or ferrous ion which is present in the catalyst do not form hydroxyl complexes as they are mostly in the stable form. The production of iron sludge is also reduced by the HEF process [28].

HEF employing iron-based clays, zero-valent iron, iron oxide minerals and iron-containing materials, has generated a lot of interest. The material must have robust catalytic activity that is unaffected by experimental conditions, particularly pH, as well as high stability to ensure reusability across several runs [17]. The most extensively used catalysts are hematite, goethite, and magnetite because of their specific properties [29]. Iron oxides are considered as environmentally acceptable Fenton catalyst, as they are abundant in nature and are easy to synthesize, making them low cost [23]. Recently, various transition metals such as cobalt, copper, nickel, vanadium and manganese have been utilized as solid catalysts with iron/iron oxides [17]. Adding transition metals such as Cu, Mn and Ni in Fe_3O_4 is reported to be effective for the increased catalytic activity [30]. Since graphene has been discovered to considerably enhance catalyst activity, researchers have hybridized metal oxides with graphene for its immense supporting power [31]. In the present study, an approach has been made to combine the iron and transition metals with a support of graphene oxide (GO) to develop a heterogeneous catalyst and to examine its efficiency for the removal of RhB dye. For the remediation of various pollutants, such as pharmaceuticals, etc., the combination of these has shown to be successful [32,33]. When manganese was present in a magnetite structure, the rate of H_2O_2 decomposition for the breakdown of organic molecules increased noticeably [34,35].

A number of studies have been conducted on the removal of RhB from aqueous solution using HEF. Jinisha et al. [36] used iron-doped SBA-15 mesoporous silica as a heterogeneous catalyst for the effective removal of RhB from aqueous solution. The authors used graphite felt as both anode and cathode. Tian et al. [37] used sponge iron as the catalyst along with gas diffusion electrodes (GDE) for RhB removal. Zheng et al. [38] studied the removal of RhB from aqueous solution using $NiFe_2O_4/Fe_2O_3$ (a charcoal shaped catalyst) as the heterogeneous catalyst and graphite as the electrodes. The importance of Mn based ferrites for HEF is found nowhere in the literature.

This study aims to develop $MnFe_2O_4$-GO as the heterogeneous EF catalyst for the effective removal of RhB dye from aqueous solution. Pt coated over titanium was used as the anode and graphite felt was used as the cathode for the in-situ generation of H_2O_2. The effects of various operational parameters including pH, applied voltage, catalyst concentration, pollutant concentration, and the effect of ions were studied and the process was optimized. The optimized conditions were applied to the real textile wastewater and the rate of removal was investigated.

2. Materials and Methods

2.1. Chemicals and Reagents

Analytical grade chemicals including RhB ($C_{28}H_{31}ClN_2O_3$) supplied from Loba Chemie (Mumbai, India) were used for the experiments. Ferric chloride anhydrous ($FeCl_3$, ≥98%) purchased from Merck (Mumbai, India), manganese chloride ($MnCl_2\cdot 4H_2O$, ≥ 95%) purchased from Sisco Research Laboratory (Mumbai, India), sodium acetate ($CH_3\cdot COONa\cdot 3H_2O$, 99.0–101.0%) purchased from Qualigens were used for the synthesis of $MnFe_2O_4$ catalyst. Graphite powder (99.5%) for the synthesis of graphene oxide (GO), was purchased from S D Fine Chemicals (Chennai, India).

Other chemicals, including potassium dichromate ($K_2Cr_2O_7$, ≥99.0%), potassium dihydrogen phosphate (KH_2PO_4, ≥99.5%), di-potassium hydrogen phosphate (K_2HPO_4, ≥99.0%), ammonium chloride (NH_4Cl, ≥99.0%), sodium thio sulphate ($Na_2S_2O_3$, ≥99.5%), t-butanol(($(CH_3)_3COH$, ≥99%), sodium hydroxide pellets (NaOH, ≥ 97%), sodium azide (NaN_3, ≥99%), Benzoquinone (≥98%), Phenanthroline ($C_{12}H_8N_2\cdot H_2O$, ≥99.5%), copper sulphate ($HgSO_4$, ≥99.5%), silver nitrate ($AgNO_3$, ≥99.5%), silver sulphate ($AgSO_4$, ≥98%), magnesium sulphate ($MgSO_4$, ≥98%), sodium sulphate (Na_2SO_4, ≥99.0%), potassium permanganate ($KMnO_4$, ≥99.0%), and manganese sulphate ($MnSO_4\cdot H_2O$, ≥98%), were procured from Merck. Sodium chloride (NaCl, ≥99.5%), sodium bicarbonate (Na_2CO_3, ≥99.9%), di sodium hydrogen phosphate ($Na_2HPO_4\cdot 7H_2O$, ≥99.9%), magnesium sulphate ($MgSO_4\cdot 7H_2O$, ≥99.9%), calcium chloride ($CaCl_2$, ≥98%), ferric chloride ($FeCl_3\cdot 6H_2O$, ≥98%), ferrous ammonium sulphate ($(NH_4)_2(SO_4)_2\cdot 6H_2O$, ≥99%), were procured from Fisher Scientific. Ammonium acetate (CH_3COONH_4, ≥98%), hydroxyl amine (NH_2OH, ≥96%), and potassium iodide (KI, ≥99%) were purchased from Qualigens.

Solvents including ethylene glycol ($HOCH_2$- CH_2OH, ≥99.0%), glycerol and ethyl alcohol (C_2H_5OH, ≥95%) from Merck, and poly-ethylene glycol (99.9%) from Qualigens, were used for different analytical experiments. Sulphuric acid (H_2SO_4, 95.0–98%) and ferroin indicator from Merck, and phenolphthalein supplied from Rankem, were used for the experiments.

2.2. Electrodes

Pt coated over titanium plates procured from Titanium Tantalum Products Limited, Chennai, India, was used as an anode and graphite felt was used as a cathode.

2.3. Preparation of Heterogeneous Catalyst

2.3.1. Synthesis of Graphene Oxide (GO)

For the synthesis of graphene oxide (GO) from natural graphite powder, a modified version of Hummers' method combined with laboratory optimizations, as mentioned in our earlier work [39], was used. In a water bath, 1 g of the graphite powder was mixed with 68 mL of conc. H_2SO_4 and mixed thoroughly for 1 h. While keeping the ice bath conditions (to keep the temperature below 20 degrees), 4 g $KMnO_4$ was added gradually (1 g at a time). The resultant mixture was agitated at room temperature for 6 h, until the solution changed colour from green to dark brown–black. The oxidation was then stopped by drop-wise addition of 140 mL of double-distilled water (DDW), which followed by 5 mL H_2O_2 addition. Finally, the metal ions were removed by washing the mixture three times with 10% HCl.

Washing with DDW was repeated (by centrifugation at 7500 rpm) until the pH of the supernatant liquid approached neutral condition. The addition of silver nitrate confirmed the removal of chloride from the supernatant. After drying at 40 °C in a hot air oven, the resultant solid was pulverized [39].

2.3.2. Synthesis of Manganese Ferrite Supported on GO

To synthesize $MnFe_2O_4$–GO composite, the facile solvothermal reduction was used. To a 250 mL volumetric flask containing 40 mL ethylene glycol (EG), 0.0387 g of GO was added. The mixture was kept in a bath sonicator for 1 h for the effective dispersion of GO. 0.811 g ferric chloride and 0.3010 g manganese chloride was added. The mixture was kept for magnetic stirring between 650–700 rpm until the $FeCl_3$ and $MnCl_2$ added were completely dissolved. A total of 3.6 g sodium acetate ($CH_3.COONa.3H_2O$) followed by 1 mL polyethylene glycol (PEG) were added to the above mixture and agitated at 550 rpm for 1 h. The agitated mixture was then transferred to a Teflon-lined stainless steel autoclave and kept for 6 h at 200 °C. After cooling, the precipitate was then separated using a magnet. The dark brown precipitate was washed thrice with ethanol and DDW. The washed precipitate was then vacuum dried for 6 h. After cooling, it was powdered and used for experiments [39]. EG used in the preparation acted as both solvent and reducing agent [39].

To find out the role of Fe, Mn, GO and spinel structure in the catalyst, the above method was followed. Catalysts were prepared by adding Fe only ($MnCl_2$ was not added), Mn only ($FeCl_3$ was not added), GO only (both $FeCl_3$ and $MnCl_2$ was not added) and $MnFe_2O_4$ without support (GO was not added).

2.4. *Catalyst Characterization*

The structural characterization of the laboratory-synthesized catalyst was done using X-ray powder diffraction (XRD, Bruker D8 Advance). The functional groups present on the surface of the catalyst were detected by Fourier transform infrared (FTIR) spectroscopy (FTIR, Bruker, Germany). The size of the prepared catalyst was determined by transmission electron microscope (TEM, Tecnai G2 20 S-TWIN).

2.5. *Experimental Procedure*

Electrolytic experiments were carried out in a reactor of 1 L capacity. A 1000 mL solution containing of 10 ppm RhB was prepared at the beginning of each experiment. Initial pH was maintained at 3 for all the experiments, except those in which the effect of different pH was examined. pH was adjusted using 0.1 N H_2SO_4 and 0.1 N NaOH. pH was monitored using a Laqua pH meter (Horiba scientific). Two anodes and two cathodes of size 5 × 8 cm with a 1 cm inner electrode spacing were used for the experiments. Using a commercially available fish aerator, a continuous supply of air was fed into the system until the end of the experiments. After an aeration time of 5 min, catalyst was added. The electrodes were connected to a DC power supply (make: Crown). The experimental setup is shown in Figure 1. Samples were collected at 15 min intervals and stored in the refrigerator in brown vials to avoid further degradation or reaction. The residual RhB present in each sample was measured using a UV/Vis spectrophotometer (UV-1900I, Shimadzu, Japan) at 555 nm (which is the peak wavelength of RhB). All the experiments were conducted at room temperature. The effect of different operational parameters such as pH, voltage, catalyst concentration, and pollutant concentration were also studied.

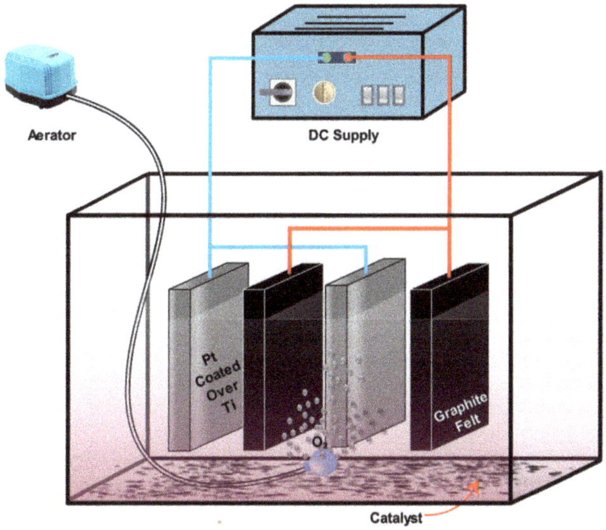

Figure 1. Schematic diagram of the experimental setup.

2.6. Real Textile Wastewater: Collection and Analysis

Real textile wastewater was collected from an industry situated in Mumbai, Maharashtra, India. A 10 L sample was collected, sealed tightly and stored in the refrigerator. Parameters including total dissolved solids (TDS), total solids (TS), chloride, sulphate, biological oxygen demand (BOD), chemical oxygen demand (COD) were analyzed as per the standard methods [40]. pH and electrical conductivity of the wastewater at room temperature were recorded using a Laqua pH meter (Horiba scientific). The absorbance of the wastewater was recorded using a UV/Vis spectrophotometer (UV 1900 I, Shimadzu, Japan). Using a total organic carbon (TOC) analyser (Shimadzu, Japan, Model: TOC Veph), the TOC content of the real textile wastewater was determined.

2.7. Energy Consumption

The specific energy consumption with different applied voltages was found using the equation mentioned in [41].

$$Q = \frac{VIt}{m} \tag{4}$$

where Q = specific energy consumption (KWh^{-1}), V = applied voltage (V), I = current (A), t = electrolysis time (min), m = change in concentration.

2.8. Leaching of Catalyst

The quantity of Fe leached was quantified using a 1.10-phenanthroline method by spectrophotometric quantification. The quantity of Mn leached from the catalyst was quantified using ICP-OES (iCAP-7400, Thermo Fisher Scientific, Waltham, MA, USA).

3. Results and Discussions

3.1. Characterization of the Catalyst

XRD, TEM and FTIR were used to characterize the $MnFe_2O_4$–GO catalyst. XRD patterns provide the crystal structure and phase information of the catalyst [42]. Figure 2a illustrates the XRD pattern of $MnFe_2O_4$–GO. The obtained diffraction peaks are similar to that of JCPDS 38-0430. The peaks observed at 2 θ values of 16.92, 29.59, 35.75, 36.32, 42.92, 52.59, 56.01, 63.96, 76.64, which can be indexed as (111), (202), (311), (222), (004), (422), (333), (440), and (533) planes, respectively. Similar patterns of $MnFe_2O_4$ were observed earlier [42,43]. The peak observed at the 2 θ value of 7.69 corresponds to the (001) characteristic plane of GO [44]. The presence of a GO peak indicates the ethylene glycol is insufficient to convert GO to rGO when more than one metal oxide is present during the solvothermal process, whereas this conversion is viable when only one metal oxide is present, as reported in our previous study [39]. This may be because the amount of solvent and time taken for the synthesis may not be enough for the reduction of GO [45].

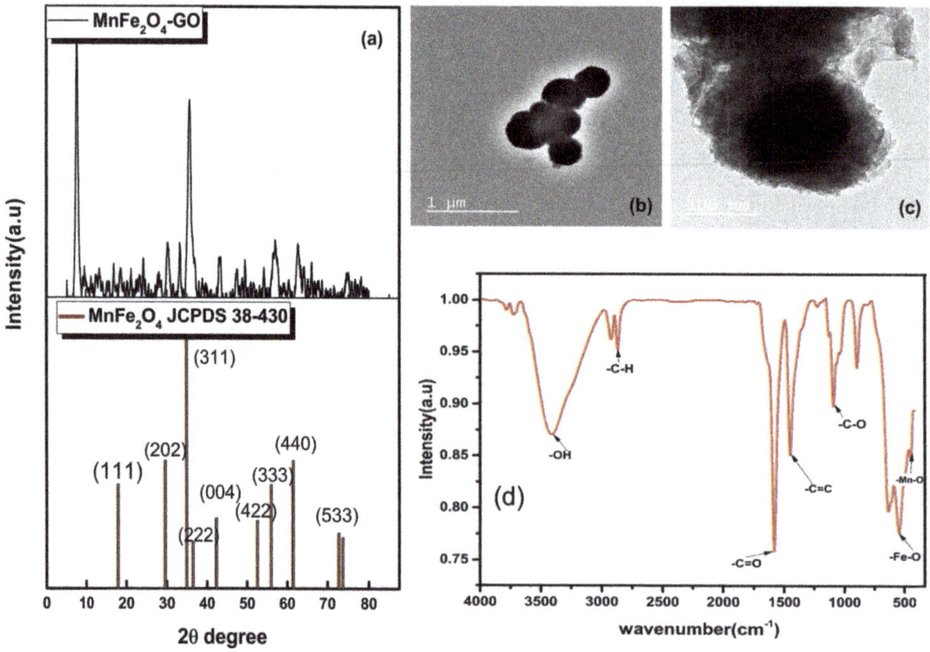

Figure 2. XRD pattern of $MnFe_2O_4$–GO (**a**) TEM images of $MnFe_2O_4$–GO (**b**), (**c**), FTIR spectrum of $MnFe_2O_4$–GO (**d**).

The shape, size and distribution of particles can be found using a TEM [46]. Figure 2b,c shows the TEM images of the catalyst composite synthesized. From Figure 2b,c, the irregular rough surface $MnFe_2O_4$ can be observed, indicating the small $MnFe_2O_4$ particles are clustered together to form a spherical aggregate, with a diameter of 132 $\pm$ 92 nm. The presence of GO sheets is evident in Figure 2c. On further examination of Figure 2b,c, the different stages of spherical structure development indicate that aggregation and particle growth are not completed and are at different phases.

The FTIR spectrum of $MnFe_2O_4$ -GO is shown in Figure 2d. As illustrated in the figure, bands of functional groups such as –OH, –C=O, C– H, C–O–C/C–O–H, –C=C, and –C–O were obtained at around 3403.9 cm^{-1}, 2917.8 cm^{-1}, 1581.6 cm^{-1}, 1125.5/1437.3 cm^{-1}, and 1081.3 cm^{-1}, respectively. The bands observed at 580 cm^{-1} and 443 cm^{-1} can be attributed to the bonds corresponding to Fe-O and Mn-O, respectively [47,48]. The presence of oxygen functionalities implies the presence of GO [45].

3.2. *Rhodamine B Degradation in $MnFe_2O_4$–GO-Based Electro-Fenton System*

3.2.1. Comparison between Different Processes

During HEF treatment of dye, a combination of various processes such as adsorption, anodic oxidation (AO), and AO+H_2O_2 may be contributed. In order to determine the contribution of these processes, degradation of RhB in different individual processes was carried out. Adsorption of dye on catalyst was analyzed in the absence of current supply and aeration. The effect of AO was investigated by applying electricity to the electrochemical system without providing catalyst and aeration. AO + H_2O_2 was carried out in the electrochemical system with electrodes, and external aeration without the addition of the catalyst. The other experimental conditions were pH 3, applied voltage of 3 V and initial pollutant concentration of 10 ppm.

Figure 3a depicts the degradation rate achieved with different processes. The different degradation percentage obtained by the individual process is as follows: 12.36% $\pm$ 0.0028 by adsorption, 64.91% $\pm$ 0.0439 by AO, 90.42% $\pm$ 0.0395 by AO + H_2O_2 and 97.52% $\pm$ 0.0002 by HEF process. On examining the kinetics, all the above individual process follows first-order kinetics. The first-order rate constant of each process are as follows: adsorption (0.0024 min^{-1}), anodic oxidation (0.019 min^{-1}), anodic oxidation combined with aeration (0.04 min^{-1}) and HEF (0.079 min^{-1}). Even though AO + H_2O_2 and HEF showed comparable dye removal within 60 min, the reaction rate of HEF is found to be double than that of AO combined with aeration. Thus, the addition of catalyst can significantly enhance the pollutant removal in EF.

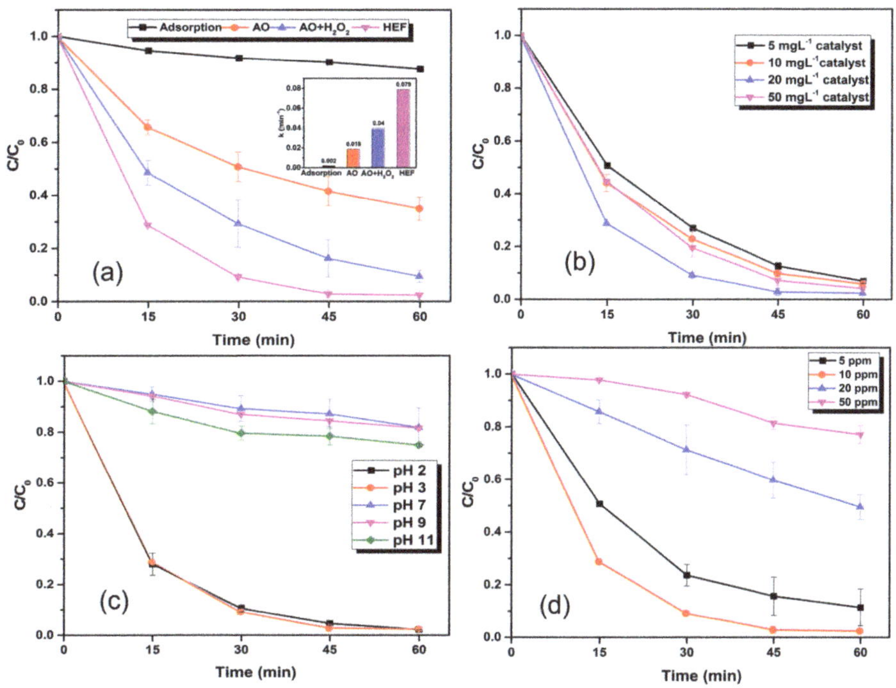

Figure 3. (**a**) Degradation of RhB with different process with rate of reaction in inset (**b**) Effect of catalyst dosage (**c**) Effect of pH (**d**) Effect of Initial dye concentration.

Similar results were obtained in HEF treatment of tannery wastewater [49]. The author used Fe_3O_4/Mn_3O_4/ZnO–rGO hybrid quaternary nano-catalyst. Graphite was used as the cathode and Ti/IrO_2/RuO_2 as the anode. The authors compared different processes of HEF, AO + H_2O_2, AO and adsorption. Higher degradation was obtained by HEF with 97.08% removal. A study by Jinisha et al. [36] compared electro-sorption, adsorption and HEF process. Higher degradation was shown by HEF (97.7%) process. A study conducted by Bedolla-Guzman and co-workers [50] compared the efficiency of AO + H_2O_2 and EF process in reactive yellow 160 dye with boron-doped diamond (BDD) as the anode and carbon polytetrafluorethylene (PTFE) as the cathode with Fe^{2+} as catalyst. The authors achieved higher efficiency with the EF process (91%). The study on the degradation of sunset yellow FCF azo dye [51] revealed that on comparison of AO$-H_2O_2$ and EF, the authors achieved higher degradation with EF. Only 88% dye was removed by AO$-H_2O_2$ in 360 min, whereas in EF around 50% dye was removed within 5 min, which fully decolorized in 45 min.

3.2.2. Effect of Catalyst Dosage

The catalyst dosage has a crucial role in the HEF process [52]. In order to optimize the catalyst dosage, HEF treatment was carried out by varying the catalyst dosage (5 mg L^{-1}, 10 mg L^{-1}, 20 mg L^{-1} and 50 mg L^{-1}). As illustrated in Figure 3b, on increasing the catalyst dosage from 5 mg L^{-1} to 10 mg L^{-1}, RhB removal slightly improved from 93.09% $\pm$ 0.0081 to 94.02% $\pm$ 0.0123. As, Fe^{2+} generation is less at lower catalyst concentrations, the production of hydroxyl radical is also less, which explains the lesser efficiency in degradation at 5 mg L^{-1} catalyst concentration [23,41,53]. When increasing catalyst dosage to 20 mg L^{-1}, a sharp increase in pollutant removal was evident, as RhB removal reached 97.51% $\pm$ 0.0002 within 60 min. Whereas, further increase in catalyst dosage lowered the RhB removal (reduced to 95.81% $\pm$ 0.0066).When excessive catalyst concentration is present in the system, the generated hydroxyl radicals could be scavenged and could lead to hydroperoxyl radical formation which has less oxidation potential than $^{\bullet}OH$ (Equations (5)–(7)) [23,53,54]. Thus, the optimized catalyst concentration is taken as 20 mg L^{-1} for all further experiments.

$$Fe^{2+} + HO^{\bullet} \rightarrow Fe^{3+} + OH^{-} \quad (5)$$

$$Fe^{2+} + H_2O_2 \rightarrow Fe - OOH^{2+} + H^{+} \quad (6)$$

$$Fe - OOH^{2+} \rightarrow HO_2^{\bullet} + Fe^{2+} \quad (7)$$

3.2.3. Effect of pH

The solution pH is a decisive parameter of the Fenton process [55]. An acidic pH, preferably pH 3, results in more radical generation, thus more activity. To determine the optimum pH for the HEF process, experiments were carried out under different pH of 2, 3, 7, 9, and 11. As Figure 3c indicates, higher RhB degradation is evident at acidic conditions (2 and 3). At pH 2, RhB removal of 97.72% $\pm$ 0.0021 and at pH 3, removal of 97.52% $\pm$ 0.0002, were observed. Whereas at neutral pH, only 18.08% $\pm$ 0.0765 of RhB was removed. At pH 9, 18.37% $\pm$ 0.0082 and at pH 11, 25.06% $\pm$ 0.0077 removal, were obtained.

Many studies reported the higher degradation of compounds at pH 3. Zheng et al. [38] observed a high degradation in RhB at pH of 3 on using $NiFe_2O_4/Fe_2O_3$ as the heterogeneous catalyst with graphite felt as the electrodes. Nidheesh et al. [54] studied the degradation of RhB using magnetite as the catalyst and graphite as the electrode. The authors also obtained higher efficiency at the pH of 3. Fayazi and Ghanei-Motlagh, [56] obtained higher degradation efficiency of methylene blue at pH 3 when sepiolite/pyrite composite was used as the catalyst, and graphite and platinum sheet was used as the electrodes. On the degradation of Ponceau SS dye using heterogeneous electro-Fenton process, dos Santos et al. [57] observed higher degradation at pH 3, when vermiculite was used as the catalyst.

For pH above 5, precipitation of ferric oxyhydroxide ($FeOOH^{2+}$) and ferric hydroxide ($Fe(OH)_3$) by ferric ions takes place terminating the Fenton reaction [58,59]. Formation of ferric hydroxides of iron species occurs at higher at higher pH values leading to the lowering of Fe^{2+}/Fe^{3+} ratio [23,60,61]. For the formation of H_2O_2, H^+ is needed (Equation (1)). H^+ is available only in acidic conditions. So, pH 3 is taken as optimum pH for further experiments.

3.2.4. Effect of Voltage

For achieving better process efficiency and for the formation of Fenton reagent, applied voltage or current density is a crucial parameter [12]. The effect of different applied voltages ranging from 1 V to 5 V were scrutinized to find out the optimum voltage. As the applied voltage was increased, the degradation also increased and after a certain range the degradation rate decreased. When the applied voltage was 1 V, the removal percentage was 22.15% ± 0.0414; on increasing the applied voltage to 2 V, degradation enhanced to 37.32% ± 0.0856. An upsurge degradation was achieved when the applied voltage was 3 V i.e., 97.52% ± 0.0002. On applying a voltage of 4 V, a similar removal percentage of 3 V was obtained (97.44% ± 0.0009) and was further decreased on applying 5 V (68.21% ± 0.0672). The RhB removal attained with different applied voltages is illustrated in Figure 4a.

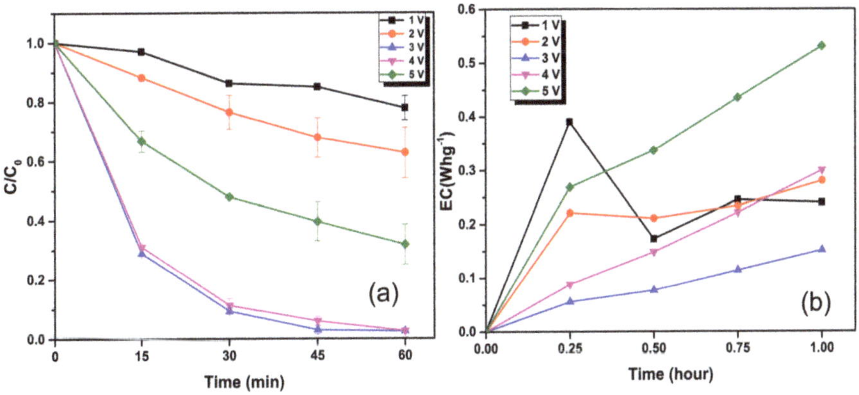

Figure 4. (**a**) Effect of voltage on RhB degradation. (**b**) Energy consumption with varying voltage.

The enhancement in degradation with increasing voltage is due to the increase in the production of •OH [8,54]. Reduction in oxidation at higher voltages after a certain level is due to the hydrogen gas evolution, decomposition of H_2O_2 (Equation (8)) and formation of H_2O.

$$2H_2O_2 \rightarrow 4H^+ + O_2 + 4e^- \quad (8)$$

The energy consumption for the RhB removal via HEF is calculated for varying applied voltages (Figure 4b). On comparing different voltages, 3 V showed significant reduction in RhB concentration by utilizing the lower energy. In both higher and lower voltage conditions, energy efficiency was less. Higher energy consumption will increase the cost of the process and thus a voltage of 3 V was selected for further experiments [54].

3.2.5. Effect of Initial Dye Concentration and Electrolysis Time

The effect of RhB concentration in HEF activity was evaluated by changing the initial pollutant concentration of 5 ppm, 10 ppm, 20 ppm and 50 ppm. The different removal rates with different concentrations of initial dye are explained in Figure 3d. For 5 ppm dye, 88.57% ± 0.0699 removal was observed which enhanced to 97.51% ± 0.0002 when the dye concentration was increased to 10 ppm. Further increase in concentration affected the efficiency drastically. When the initial dye concentration was 20 ppm, 50.36% ± 0.0472 removal was obtained, which lowered to 22.82% ± 0.0335 when the initial dye concentration was raised to 50 ppm.

Rate of removal of dye decreases with an increase in dye concentration [62]. The decrease in removal of dye with an increase in concentration can be ascribed to the lesser number of available •OH for the oxidation [25]. The lower pollutant removal in case of initial dye concentration below 10 ppm is due to the insufficient radical generation. This results in lower dye removal as collision between the particles is limited [54].

RhB degradation readily increases with electrolysis time. From Figure 3d, it is clear that a good amount of RhB particles have been degraded within the first 15 min of the electrolysis. As the time increased, around 97.51% ± 0.0002 was degraded from the system at the end of 60 min. Nidheesh et al. [54,63] reported the increasing RhB degradation with an increase in time. The decolorization rate was higher at the initial time intervals of electrolysis because of the higher formation of •OH. The degradation rate decreased as the time increased, because of the decrease in collision of RhB molecules with •OH [54,63].

3.2.6. Effect of Ions

The effect of different anions such as chlorides, sulphates and carbonates on RhB solution at optimized conditions was studied. Figure 5 illustrates the effect of various ions on the degradation of RhB. 50 mg L^{-1} and 100 mg L^{-1} of each NaCl, Na_2SO_4 and Na_2CO_3 were added to the RhB solution.

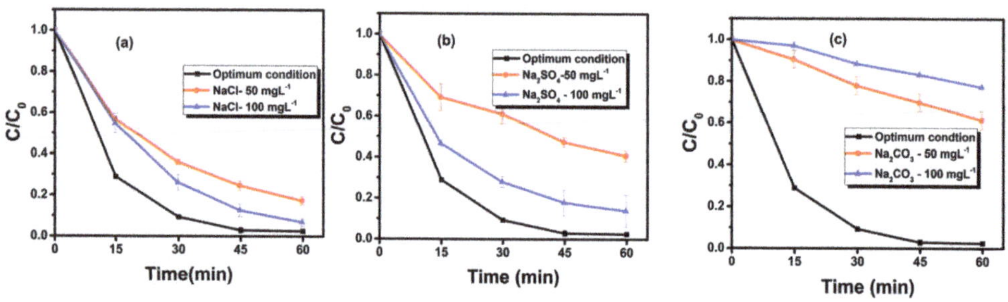

Figure 5. (**a**) Effect of NaCl on the degradation of RhB. (**b**) Effect of Na_2SO_4 on the degradation of RhB. (**c**) Effect of Na_2CO_3 on the degradation of RhB Experimental conditions: pH-3, Voltage–3, Catalyst dosage- 20 mg L^{-1}, Initial dye concentration-10 ppm.

When NaCl was added into the RhB solution, the removal percentage was 68.21% ± 0.0189 for 50 mg L^{-1} of NaCl and 93.1% ± 0.0184 for 100 mg L^{-1} of NaCl. When 50 mg L^{-1} of Na_2SO_4 was added to the RhB solution the degradation percentage obtained was 59.35% ± 0.0265, and on addition of 100 mg L^{-1} the degradation obtained was 86.1% ± 0.0782. When 50 mg L^{-1} of Na_2CO_3 was added, the degradation percentage obtained was 38.63% ± 0.0437, and 22.63% ± 0.0039 when 100 mg L^{-1} of Na_2CO_3 was added. Among the different anions added, the presence of carbonates has shown a strong influence in HEF activity. Whereas other ions such as chloride and sulphate inhibit HEF activity at lower concentration, and on higher dosage, can result in supporting the HEF activity via additional radical generation.

The presence of Cl^- in wastewaters have a suppressing effect on AOPs as they are said to have scavenging effect on •OH [64]. However, in this case, addition of chlorides as well as sulphates does not diminish the degradation but on the contrary increased the degradation with the increase of concentration. The upgrade in degradation with the increase of sulphate concentration was reported by Zhou et al. [65]. The authors explained the increase in degradation as the enhancement of current density with sulphate ions. The comparative higher degradation in chloride media may be due to the formation of active chlorine species [12]. The decrease in degradation in the presence of anions is due to the scavenging effect of hydroxyl radicals by sulphate (Equation (9)) and chloride (Equations (10) and (11) forming hydroxyl ions [54]. In the presence of carbonates, the degradation of RhB is suppressed by the formation of $FeCO_3$. Jinisha et al. [36] concluded that the effect of anions like sulphates and carbonates does not influence the production of H_2O_2.

$$HO^{\bullet} + SO_4^{2-} \rightarrow OH^- + SO_4^- \quad (9)$$

$$HO^{\bullet} + Cl^- \leftrightarrow HOCl^- \quad (10)$$

$$HOCl^{-\bullet} + Cl^- \rightarrow Cl_2^- + OH^- \quad (11)$$

3.2.7. Radical Scavenging Tests

To evaluate the contribution of various radicals such as hydroxyl radical on the surface ($^{\bullet}OH_{surf}$) and on bulk ($^{\bullet}OH_{bulk}$), super oxides ($O_2^{-\bullet}$) and singlet oxygen ($^1O_2$) scavengers such as t-butanol, KI, benzoquinone and sodium azide (NaN_3) were utilized. The scavenger t-butanol is known to have a good ability to scavenge $^{\bullet}OH$ [12]. Benzoquinone was used as superoxide radical scavenger [61]. The surface $^{\bullet}OH$ present on the catalyst surface are scavenged by KI [66]. The potential generation of singlet oxygen can be evaluated by sodium azide [67].

As shown in Figure 6a, the surface $^{\bullet}OH$ activity is insignificant, whereas the bulk $^{\bullet}OH$ radical strongly contributed to HEF activity. From Figure 6a, the non-radical pathway via singlet oxygen is the predominant pathway of HEF activity, and followed the trend, $^1O_2 > {^{\bullet}OH_{bulk}} > O_2^{-\bullet}$.

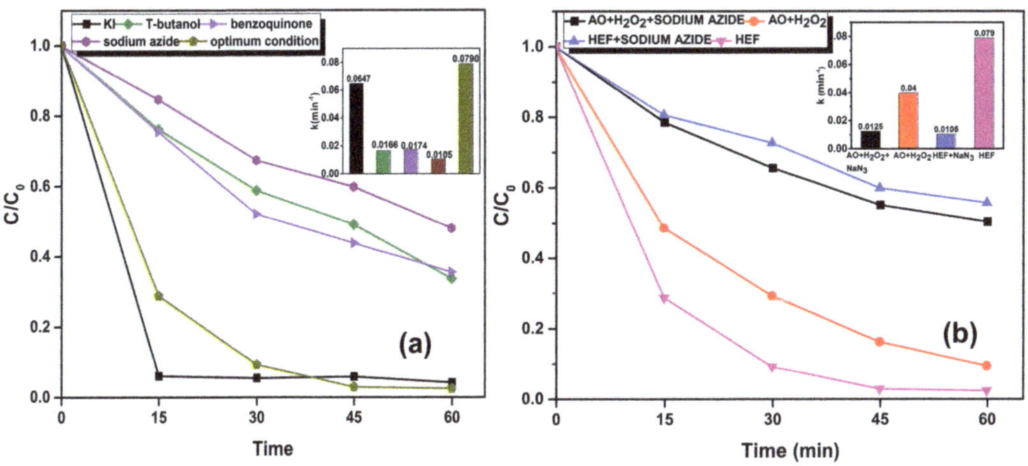

Figure 6. (**a**) Radical scavenging tests with rate of reactions; (**b**) sodium azide quenching experiments at different conditions.

To confirm the contribution of singlet oxygen in the system, quenching using NaN_3 was conducted in the presence and absence of the $MnFe_2O_4$–GO catalyst. As illustrated in Figure 6b, a strong presence of a non-radical pathway is evident in the AO + H_2O_2 system, where the NaN_3 lowered the RhB degradation rate to 0.0125 min^{-1}, comparable to the HEF with NaN_3 (0.0105 min^{-1}). This observation indicates the formation of $^1O_2$ is not solely from the catalyst.

Similarly, Lu et al. [68] reported the formation of singlet oxygen in the AO+H_2O_2 system in the presence of chloride (Cl^-). Herein, the chloride presence is evident, and the reasons might be (1) from the tap water used for making RhB-simulated water (37.7 mg L^{-1} of chloride), (2) Cl^- release from the RhB structure due to degradation. So, the singlet oxygen formation can be attributed to the singlet oxygen formation in chloride medium.

3.2.8. Role of Fe, Mn, GO and $MnFe_2O_4$

The individual contribution of Fe, Mn, GO and $MnFe_2O_4$ spinel structure in providing HEF activity were investigated using catalysts prepared under respective conditions by excluding other precursors involved. Because of the availability of numerous oxygen functional groups, large surface area, electrical conductivity and mechanical stability, graphene and GO are very promising support materials [69,70], and thereby enhance catalytic activity [71]. The enhancement in catalytic activity of the L–GO–ZnO composite on the photocatalytic activity of RhB was reported by Yaqoob et al. [10]. There are a few studies in the literature which report the catalytic activity of GO [72]. Thus, the role of GO alone is also examined to investigate the possibility of GO as a catalyst for HEF activity. Figure 7a depicts the different degradation rates of RhB obtained under each condition. From Figure 7a, it is clear that the catalyst works efficiently when it is in the form of spinel structure with GO support with a rate of RhB removal of 0.079 min^{-1}. In the case of GO, Mn,

Fe and $MnFe_2O_4$ without GO, the rates of RhB removal were 0.0449 min^{-1}, 0.0601 min^{-1}, 0.0610 min^{-1}, 0.0640 min^{-1}, respectively (inset of Figure 7a). The GO solely does not directly contribute to the HEF activity, as the RhB degradation in that case is similar to that of AO + H_2O_2 (0.04 min^{-1}). Similar results were obtained by Yao et al. [71]. In addition, this study observed an additional benefit of GO in the catalyst composite development, as the GO support contributed to lowering the solvothermal synthesis duration. In the absence of GO, $MnFe_2O_4$ development required a minimum of 9 h, whereas in the presence of GO, proper catalyst development was observed within 6 h of solvothermal treatment.

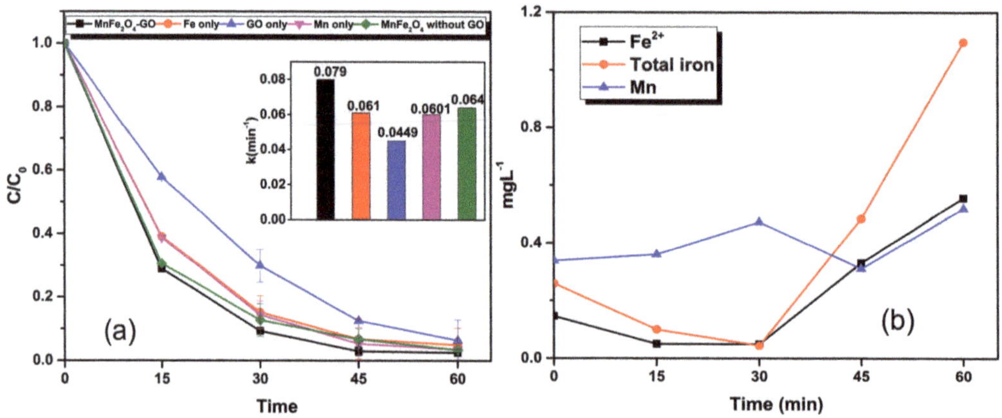

Figure 7. (**a**) Role of Fe, Mn and GO in catalyst with rate of reaction. (**b**) Leaching of Fe and Mn.

3.2.9. Leaching Studies

The stability of the as-synthesized catalyst was evaluated by analyzing the Fe and Mn leaching using ICP-OES and 1,10- phenanthroline method [39]. The amount of leached iron and Mn from the catalyst is illustrated in the Figure 7b. The amount of Fe^{2+} and total iron in the initial 30 min showed no significant leaching. After 30 min, there was a gradual increase in concentration, indicating the leaching of Fe^{2+} and total iron into the solution. The amount of Mn leached out showed fluctuating results which indicated that no gradual Mn release occurred.

3.2.10. Mechanism of Dye Removal

The mechanism involved in the HEF oxidation of RhB is given in Figure 8. The Fe^{2+} from the catalyst decomposes in-situ generated H_2O_2 to form hydroxyl radicals (Equation (5)) which degrade the RhB molecules into CO_2, H_2O and other byproducts. Mn metal present in $MnFe_2O_4$ strongly accelerated the H_2O_2 decomposition, as in Equations (12) and (13) [34,35]. The Fe^{2+} formation from Fe^{3+} occurred by electron loss (which is explained by Equations (1)–(3)). All the possible reductions of Fe^{3+} to Fe^{2+} and degradation occur at the cathode, as RhB is a cationic dye [54].

$$Mn^{2+} + H_2O_2 \rightarrow Mn^{3+} + HO^{\bullet} + OH^- \quad (12)$$

$$Mn^{3+} + H_2O_2 \rightarrow Mn^{2+} + HO_2^{\bullet} + H^+ \quad (13)$$

(RhB) degradation by reactive radicals is possible by ring-opening reaction and N-de-ethylation pathways [63]. As explained in Section 3.2.1, the slow decolorization of the RhB in AO + H_2O_2 system in the absence of catalyst may be facilitated by singlet oxygen. Singlet oxygen can be possibly formed by deprotonization, followed by electron rearrangement of oxygen molecules and disproportionation of H_2O_2 [73–75].

The results from this study compared with the efficiency of various catalysts in the removal of various dyes using HEF process, and given in Table 1. From Table 1, it is clear that the use of $MnFe_2O_4$–GO composite catalyst in the present study has significantly reduced the total electrolysis time, while keeping higher dye removal efficiency. The efficiency of the process is higher when compared with the conventional EF process [13].

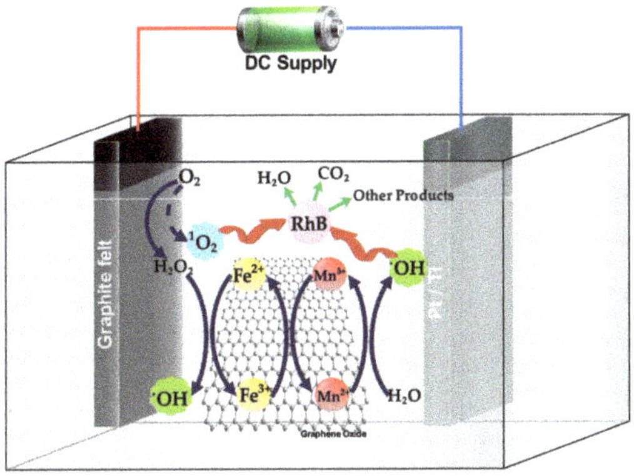

Figure 8. Mechanism of dye removal.

Table 1. Efficiency of various catalysts used for removal of dyes by electro-Fenton process.

Catalyst Used	Dye Removed	Efficiency of the Catalyst	Electrodes Used	Electrolysis Time	Reference
$MnFe_2O_4$-GO	Rhodamine B	97.51% colour removal	Pt/Ti and graphite felt	60 min	Present study
Fe^{2+}	Reactive yellow 160	91% colour removal	BDD and air diffusion cathode	360 min	[50]
Fe^{2+}	Alizarin red	>95% TOC removal	Graphite felt	210 min	[13]
Magnetite	Rhodamine B	97.3% colour removal	Graphite	180 min	[54]
Fe_3O_4 nanoparticles	Methyl orange	86.6% colour removal	RuO_2/Ti and C-PTFE	90 min	[76]
Modified sponge iron	Rhodamine B	99.71% colour removal	Pt sheet and GDE	120 min	[37]
Iron doped SBA-15 mesoporous silica	Rhodamine B	97.7% colour removal	Graphite	180 min	[36]
Fe_3O_4/rGO	Reactive red 195	93.34% colour removal	Stainless steel rods	60 min	[25]
Vermiculite	Ponceau SS	92.4% colour removal	BDD and carbon PTFE air diffusion electrode	360 min	[57]

3.3. HEF Activity in Real Textile Wastewater

The wastewater collected was slightly acidic in nature (pH of 5.34) and had a conductivity of 13.2 mS cm^{-1}. The optimized experimental conditions (pH-3, applied voltage of 3 V, catalyst concentration of 20 mg L^{-1}) were applied to the textile wastewater for its effective treatment. The colour change was examined at a wavelength of 555 nm. As illustrated in Figure 9, colour reduction of 61.4% $\pm$ 0.061 was obtained after HEF treatment. The decolorization is caused by the breakdown of the dyes belonging to the chromogen group into its by-products [77]. Similarly, Nidheesh et al. [63] found 97.5% colour removal in textile wastewater by the HEF process. Colour removal in real textile water was investigated in batch and continuous modes by Nidheesh and Gandhimathi [77]. In the batch study, there was 83% colour removal and in the continuous study, 68% colour removal was obtained.

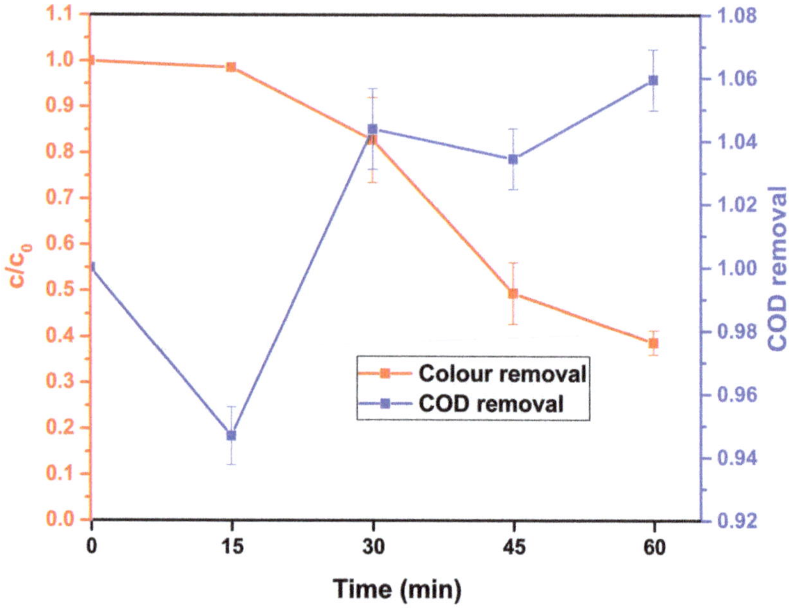

Figure 9. Colour removal and COD (mg L^{-1}) removal of real textile wastewater.

The COD were determined at different intervals to understand the degree of pollutant mineralization. The COD removal within 1 h is shown in Figure 9. Fluctuating COD percentage removal was observed. The abnormal result may be due to the mineralization of the compounds and intermediates present in the wastewater [78]. The degradation of complex organic pollutants, (non-detectable in COD analysis) to simpler organic pollutants detectable in COD analysis may result in rising COD values during the HEF process [77]. In general, the dyes present in textile wastewater are aromatic compounds with multiple aromatic rings and are not degradable by dichromate ions (used for COD analysis). However, these compounds are degradable by hydroxyl radicals, as they have higher oxidation potential than dichromate ions. However, the intermediate products and byproducts formed during the HEF process may not be as stable as that of dye and can be degraded by dichromate ions (thus, COD values will be provided by these compounds).

The TOC of the textile wastewater at different time intervals was determined, and the values before and after treatment are shown in Table 2. The initial TOC of 1551 mg L^{-1} reduced to 1478 mg L^{-1} after an electrolysis time of 60 min, with an increase in BOD from 631.58 to 1894.74 mg L^{-1} (Table 2). The conversion of complex organic structures to simpler biodegradable organic compounds, instead of total mineralization, could end up in lower TOC removal with improvement in BOD [79,80]. Generally, carboxylic acids such as oxalic acids are formed as intermediates in the degradation of dyes [81]. These compounds are unable to be degraded with $^{\bullet}OH$ due to their inefficiency [82], while microorganisms are found to be effective for the degradation of organic compounds [83]. This might have resulted in the increase in BOD.

Textile effluent characteristics before and after treatment were examined (Table 2). A slight increase in chloride and sulphate values was observed after the treatment. The increase in chloride and sulphate values may be due to the mineralization of chloride and sulphate containing complex aromatics present in the wastewater. Textile wastewaters are characterized by a high content of chloride in the range of 1600–2100 mg L^{-1}, as reported by the earlier studies [3]. Brillas et al. [5] suggested the increasing chances of chlorinated compounds in the wastewater after electrochemical processes. Studies have reported the increase in concentration of inorganics after electrochemical treatment [84].

For the treatment of textile effluents, hybrid systems are considered, as they will resolve the various limitations associated with the treatment using single systems. The increase in chloride and sulphate after HEF treatment may affect subsequent treatment. For example, AOPs followed by biological treatments are preferred because of the enhancement in biodegradability after the treatment [85]. However, the increase in chloride and sulphate can hamper the biological treatments [84]. Sodium and magnesium have been significantly decreased after treatment. The BOD/COD ratio shot up from 0.07 to 0.21, indicating the increase in biodegradability of the wastewater after treatment. Thus, the HEF process can be effectively used as a pre-treatment process for wastewater.

Table 2. Physicochemical characteristics of textile wastewater.

Parameters	Before Treatment	After Treatment
BOD (mg L^{-1})	631.58	1894.74
COD (mg L^{-1})	8521.81	9028.43
TOC (mg L^{-1})	1551	1478
Chloride (mg L^{-1})	3830.12	4080.74
Sulphate (mg L^{-1})	3936.71	5111.11
Sodium (mg L^{-1})	3939	3631
Magnesium (mg L^{-1})	121	14

4. Conclusions and Future Perspectives

Solvothermally synthesized $MnFe_2O_4$ supported on GO was used as the catalyst for the effective removal of RhB from aqueous solution by HEF process. The XRD, FTIR and TEM confirmed the formation of $MnFe_2O_4$ over GO. The optimized experimental conditions were found as pH 3, applied voltage of 3 V, catalyst concentration of 20 mg L^{-1} and initial pollutant concentration of 10 ppm. After an electrolysis time of 60 min, a removal efficiency of 97.51% could be achieved with these optimized conditions. Bulk $^{\bullet}OH$ and super-oxides were found out to be the superior oxidants controlling the process. The role of GO in catalyst was significant, as the time for the catalyst formation could be reduced. Mn leaching from the catalyst was found to be insignificant. However, the leaching of iron could be seen after an electrolysis time of 30 min. The optimized experimental conditions were applied to the real textile wastewater and an efficient colour removal of 61.23% at 555 nm could be observed. A TOC removal of 4.71% was obtained. Even though the COD results were fluctuating, an increase in BOD/COD ratio from 0.07 to 0.21 was observed after treatment, which indicated that the biodegradability of the wastewater was significantly enhanced. Thus, the HEF process could be used as an effective pre-treatment method for wastewaters.

The main challenge faced during the HEF process was the need for frequent replacement of the cathode, especially for the treatment of real wastewater. The efficiency of the dye removal greatly depends on the cathode, and passivation of the cathode during treatment will negatively affect its efficiency. Another challenge faced is associated with the variation in the behavioural properties of the catalyst during multiple synthesis.

The need for pilot-scale research in textile wastewater has to be mentioned. Further studies require an increase in the efficiency of textile wastewater in terms of colour and COD, as well as TOC. The degradation of textile wastewater can also be found as a factor of time, i.e., whether the optimized conditions will increase efficiency in extending the electrolysis time. Biological studies can be conducted to confirm the biodegradability of the textile wastewater after treatment.

Author Contributions: Conceptualization, P.V.N.; methodology, P.V.N. and J.S.; investigation, G.A. and J.S.; writing—original draft preparation, G.A.; writing—review and editing, P.V.N. and G.A.; supervision, P.V.N. All authors have read and agreed to the published version of the manuscript.

Funding: This research received no external funding.

Institutional Review Board Statement: Not applicable.

Informed Consent Statement: Not applicable.

Data Availability Statement: Data is contained within the article.

Acknowledgments: The authors are thankful to the Director, CSIR NEERI, for their support. The authors would also like to thank the Sophisticated Analytical Instrument Facility (SAIF), Indian Institute of Technology-Bombay, India; Indian Science Technology and Engineering Facilities Map (I-STEM), the Centre for Nano Science and Engineering, Indian Institute of Science, Bangalore; and the Institute Instrumentation Centre (IIC), the Indian Institute of Technology, Roorkee, India, for their assistance with various analyses.

Conflicts of Interest: The authors declare no conflict of interest.

References

1. Dung, N.T.; Duong, L.T.; Hoa, N.T.; Thao, V.D.; Ngan, L.V.; Huy, N.N. A comprehensive study on the heterogeneous electro-Fenton degradation of tartrazine in water using $CoFe_2O_4$/carbon felt cathode. *Chemosphere* **2022**, *287*, 132141. [CrossRef] [PubMed]
2. Yaqoob, A.A.; Parveen, T.; Umar, K.; Ibrahim, M.N.M. Role of nanomaterials in the treatment of wastewater: A review. *Water* **2020**, *12*, 495. [CrossRef]
3. GilPavas, E.; Correa-Sánchez, S. Optimization of the heterogeneous electro-Fenton process assisted by scrap zero-valent iron for treating textile wastewater: Assessment of toxicity and biodegradability. *J. Water Process Eng.* **2019**, *32*, 100924. [CrossRef]
4. Yavuz, Y.; Shahbazi, R.; Koparal, A.S.; Öğütveren, Ü.B. Treatment of Basic Red 29 dye solution using iron-aluminum electrode pairs by electrocoagulation and electro-Fenton methods. *Environ. Sci. Pollut. Res.* **2014**, *21*, 8603–8609. [CrossRef] [PubMed]
5. Brillas, E.; Martínez-Huitle, C.A. Decontamination of wastewaters containing synthetic organic dyes by electrochemical methods. An updated review. *Appl. Catal. B Environ.* **2015**, *166*, 603–643. [CrossRef]
6. Saigl, Z.M. Various adsorbents for removal of rhodamine b dye: A review. *Indones. J. Chem.* **2021**, *21*, 1039–1056. [CrossRef]
7. Berradi, M.; Hsissou, R.; Khudhair, M.; Assouag, M.; Cherkaoui, O.; El Bachiri, A.; El Harfi, A. Textile finishing dyes and their impact on aquatic environs. *Heliyon* **2019**, *5*, e02711. [CrossRef]
8. Kourdali, S.; Badis, A.; Boucherit, A. Degradation of direct yellow 9 by electro-Fenton: Process study and optimization and, monitoring of treated water toxicity using catalase. *Ecotoxicol. Environ. Saf.* **2014**, *110*, 110–120. [CrossRef]
9. Kaur, P.; Sangal, V.K.; Kushwaha, J.P. Parametric study of electro-Fenton treatment for real textile wastewater, disposal study and its cost analysis. *Int. J. Environ. Sci. Technol.* **2019**, *16*, 801–810. [CrossRef]
10. Yaqoob, A.A.; Noor, N.H.b.M.; Umar, K.; Adnan, R.; Ibrahim, M.N.M.; Rashid, M. Graphene oxide–ZnO nanocomposite: An efficient visible light photocatalyst for degradation of rhodamine B. *Appl. Nanosci.* **2021**, *11*, 1291–1302. [CrossRef]
11. Eslami, A.; Moradi, M.; Mehdipour, F. Decolorization and COD removal from real textile wastewater by chemical and electrochemical Fenton processes: A comparative study Background. *J. Environ. Health Sci. Eng.* **2013**, *11*, 31. [CrossRef] [PubMed]
12. Titchou, F.E.; Zazou, H.; Afanga, H.; El Gaayda, J.; Ait Akbour, R.; Hamdani, M.; Oturan, M.A. Electro-Fenton process for the removal of Direct Red 23 using BDD anode in chloride and sulfate media. *J. Electroanal. Chem.* **2021**, *897*, 115560. [CrossRef]
13. Panizza, M.; Oturan, M.A. Degradation of Alizarin Red by electro-Fenton process using a graphite-felt cathode. *Electrochim. Acta* **2011**, *56*, 7084–7087. [CrossRef]
14. Nidheesh, P.V.; Ganiyu, S.O.; Martínez-Huitle, C.A.; Mousset, E.; Olvera-Vargas, H.; Trellu, C.; Zhou, M.; Oturan, M.A. Recent advances in electro-Fenton process and its emerging applications. *Crit. Rev. Environ. Sci. Technol.* **2022**. [CrossRef]
15. Nidheesh, P.V.; Gandhimathi, R. Trends in electro-Fenton process for water and wastewater treatment: An overview. *Desalination* **2012**, *299*, 1–15. [CrossRef]
16. Brillas, E.; Sirés, I.; Oturan, M.A. Electro-Fenton Process and Related Electrochemical Technologies Based on Fenton's Reaction Chemistry. *Chem. Rev.* **2009**, *109*, 6570–6631. [CrossRef] [PubMed]
17. Ganiyu, S.O.; Zhou, M.; Martínez-Huitle, C.A. Heterogeneous electro-Fenton and photoelectro-Fenton processes: A critical review of fundamental principles and application for water/wastewater treatment. *Appl. Catal. B Environ.* **2018**, *235*, 103–129. [CrossRef]
18. Krishnan, S.; Martínez-Huitle, C.A.; Nidheesh, P.V. An overview of chelate modified electro-Fenton processes. *J. Environ. Chem. Eng.* **2022**, *10*, 107183. [CrossRef]
19. Nidheesh, P.V.; Zhou, M.; Oturan, M.A. An overview on the removal of synthetic dyes from water by electrochemical advanced oxidation processes. *Chemosphere* **2018**, *197*, 210–227. [CrossRef]
20. Florenza, X.; Solano, A.M.S.; Centellas, F.; Martínez-Huitle, C.A.; Brillas, E.; Garcia-Segura, S. Degradation of the azo dye Acid Red 1 by anodic oxidation and indirect electrochemical processes based on Fenton's reaction chemistry. Relationship between decolorization, mineralization and products. *Electrochim. Acta* **2014**, *142*, 276–288. [CrossRef]
21. Zhang, C.; Ren, G.; Wang, W.; Yu, X.; Yu, F.; Zhang, Q.; Zhou, M. A new type of continuous-flow heterogeneous electro-Fenton reactor for Tartrazine degradation. *Sep. Purif. Technol.* **2019**, *208*, 76–82. [CrossRef]
22. Garcia-Rodriguez, O.; Mousset, E.; Olvera-Vargas, H.; Lefebvre, O. Electrochemical treatment of highly concentrated wastewater: A review of experimental and modeling approaches from lab- to full-scale. *Crit. Rev. Environ. Sci. Technol.* **2020**, *52*, 240–309. [CrossRef]

23. Zhang, B.; Hou, Y.; Yu, Z.; Liu, Y.; Huang, J.; Qian, L.; Xiong, J. Three-dimensional electro-Fenton degradation of Rhodamine B with efficient Fe-Cu/kaolin particle electrodes: Electrodes optimization, kinetics, influencing factors and mechanism. *Sep. Purif. Technol.* **2019**, *210*, 60–68. [CrossRef]
24. Garcia-Rodriguez, O.; Lee, Y.Y.; Olvera-Vargas, H.; Deng, F.; Wang, Z.; Lefebvre, O. Mineralization of electronic wastewater by electro-Fenton with an enhanced graphene-based gas diffusion cathode. *Electrochim. Acta* **2018**, *276*, 12–20. [CrossRef]
25. Nazari, P.; Setayesh, S.R. Effective degradation of Reactive Red 195 via heterogeneous electro-Fenton treatment: Theoretical study and optimization. *Int. J. Environ. Sci. Technol.* **2019**, *16*, 6329–6346. [CrossRef]
26. Plakas, K.V.; Karabelas, A.J. Electro-Fenton applications in the water industry. *Handb. Environ. Chem.* **2018**, *61*, 343–378.
27. Ganiyu, S.O.; Martínez-Huitle, C.A.; Oturan, M.A.; Martinez-Hutile, C.A.; Oturan, M.A. Electrochemical advanced oxidation processes for wastewater treatment: Advances in formation and detection of reactive species and mechanisms. *Curr. Opin. Electrochem.* **2020**, *27*, 100678. [CrossRef]
28. Nidheesh, P.V.; Olvera-Vargas, H.; Oturan, N.; Oturan, M.A.A. Heterogeneous Electro-Fenton Process: Principles and Applications. In *Electro-Fenton Process: New Trends and Scale-Up*; Zhou, M., Oturan, M.A., Sirés, I., Eds.; Springer: Singapore, 2018; Volume 61, pp. 85–110.
29. Ben Hafaiedh, N.; Fourcade, F.; Bellakhal, N.; Amrane, A. Iron oxide nanoparticles as heterogeneous electro-Fenton catalysts for the removal of AR18 azo dye. *Environ. Technol.* **2020**, *41*, 2146–2153. [CrossRef]
30. Qin, H.; Cheng, H.; Li, H.; Wang, Y. Degradation of ofloxacin, amoxicillin and tetracycline antibiotics using magnetic core–shell $MnFe_2O_4$@C-NH_2 as a heterogeneous Fenton catalyst. *Chem. Eng. J.* **2020**, *396*, 125304. [CrossRef]
31. Kanwal, A.; Yaqoob, A.A.; Siddique, A.; Bhawani, S.A.; Ibrahim, M.N.M.; Umar, K. Hybrid Nanocomposites Based on Graphene and Its Derivatives: From Preparation to Applications. In *Graphene and Nanoparticles Hybrid Nanocomposites*; Springer: Singapore, 2021; pp. 261–281.
32. Kao, S.; Wu, J.; Wang, C.; Wang, Y. Effect of Fe^{2+} and Mn^{2+} catalysts on the performance of Bio-electro-Fenton microbial fuel cells. *J. Biostat. Biom. Appl.* **2015**, *1*, 103.
33. Dirany, A.; Sirés, I.; Oturan, N.; Oturan, M.A. Electrochemical abatement of the antibiotic sulfamethoxazole from water. *Chemosphere* **2010**, *81*, 594–602. [CrossRef] [PubMed]
34. Liang, X.; Zhong, Y.; He, H.; Yuan, P.; Zhu, J.; Zhu, S.; Jiang, Z. The application of chromium substituted magnetite as heterogeneous Fenton catalyst for the degradation of aqueous cationic and anionic dyes. *Chem. Eng. J.* **2012**, *191*, 177–184. [CrossRef]
35. Zhong, Y.; Liang, X.; He, Z.; Tan, W.; Zhu, J.; Yuan, P.; Zhu, R.; He, H. Environmental The constraints of transition metal substitutions (Ti, Cr, Mn, Co and Ni) in magnetite on its catalytic activity in heterogeneous Fenton and UV/Fenton reaction: From the perspective of hydroxyl radical generation. *Appl. Catal. B Environ.* **2014**, *150–151*, 612–618. [CrossRef]
36. Jinisha, R.; Gandhimathi, R.; Ramesh, S.T.; Nidheesh, P.V.; Velmathi, S. Removal of rhodamine B dye from aqueous solution by electro-Fenton process using iron-doped mesoporous silica as a heterogeneous catalyst. *Chemosphere* **2018**, *200*, 446–454. [CrossRef] [PubMed]
37. Tian, J.; Sharshar, M.M.; Yang, M.; Mu, T.; Xing, J. Degradation of Rhodamine B at neutral pH using modified sponge iron as a heterogeneous electro-Fenton catalyst. *Environ. Prog. Sustain. Energy* **2018**, *37*, 989–995. [CrossRef]
38. Zheng, Y.; Qiu, S.; Deng, F.; Zhu, Y.; Ma, F.; Li, G. A charcoal-shaped catalyst $NiFe_2O_4/Fe_2O_3$ in electro-Fenton: High activity, wide pH range and catalytic mechanism. *Environ. Technol.* **2021**, *42*, 1996–2008. [CrossRef]
39. Scaria, J.; Nidheesh, P.V. Magnetite–reduced graphene oxide nanocomposite as an efficient heterogeneous Fenton catalyst for the degradation of tetracycline antibiotics. *Environ. Sci. Water Res. Technol.* **2022**, *8*, 1261–1276. [CrossRef]
40. APHA. *Standard Methods for the Examination of Water and Wastewater*, 22nd ed.; Rice, E.W., Baird, R.B., Eaton, A.D., Clesceri, L.S., Eds.; American Public Health Association; American Water Works Association; Water Environment Federation: Washington, DC, USA, 2012.
41. Nidheesh, P.V.; Gandhimathi, R. Electro Fenton oxidation for the removal of Rhodamine B from aqueous solution in a bubble column reactor under continuous mode. *Desalin. Water Treat.* **2015**, *55*, 263–271. [CrossRef]
42. Zhang, W.; Hou, X.; Lin, Z.; Yao, L.; Wang, X.; Gao, Y.; Hu, S. Hollow microspheres and nanoparticles MnFe2O4 as superior anode materials for lithium ion batteries. *J. Mater. Sci. Mater. Electron.* **2015**, *26*, 9535–9545. [CrossRef]
43. Jacintha, A.M.; Manikandan, A.; Chinnaraj, K.; Antony, S.A.; Neeraja, P. Comparative studies of Spinel MnFe2O4 nanostructures: Structural, morphological, optical, magnetic and catalytic properties. *J. Nanosci. Nanotechnol.* **2015**, *15*, 9732–9740. [CrossRef]
44. Abdullah, S.I.; Ansari, M.N.M. Mechanical properties of graphene oxide (GO)/epoxy composites. *HBRC J.* **2015**, *11*, 151–156. [CrossRef]
45. Paul, R.K. Unique Preparation And Characterization Of Graphene Nano-Materials. 2017. Available online: http://lib.buet.ac.bd:8080/xmlui/handle/123456789/4635 (accessed on 17 October 2022).
46. Goodarz Naseri, M.; Saion, E.B.; Kamali, A. An Overview on Nanocrystalline $ZnFe_2O_4$, $MnFe_2O_4$, and $CoFe_2O_4$ Synthesized by a Thermal Treatment Method. *ISRN Nanotechnol.* **2012**, *2012*, 604241. [CrossRef]
47. Singh, G.; Chandra, S. Electrochemical performance of $MnFe_2O_4$ nano-ferrites synthesized using thermal decomposition method. *Int. J. Hydrogen Energy* **2018**, *43*, 4058–4066. [CrossRef]

48. Liu, W.; Qian, J.; Wang, K.; Xu, H.; Jiang, D.; Liu, Q.; Yang, X.; Li, H. Magnetically Separable Fe_3O_4 Nanoparticles-Decorated Reduced Graphene Oxide Nanocomposite for Catalytic Wet Hydrogen Peroxide Oxidation. *J. Inorg. Organomet. Polym. Mater.* **2013**, *23*, 907–916. [CrossRef]
49. Ozturk, D. Fe_3O_4/Mn_3O_4/ZnO-rGO hybrid quaternary nano-catalyst for effective treatment of tannery wastewater with the heterogeneous electro-Fenton process: Process optimization. *Sci. Total Environ.* **2022**, *828*, 154473. [CrossRef]
50. Bedolla-Guzman, A.; Sirés, I.; Thiam, A.; Peralta-Hernández, J.M.; Gutiérrez-Granados, S.; Brillas, E. Application of anodic oxidation, electro-Fenton and UVA photoelectro-Fenton to decolorize and mineralize acidic solutions of Reactive Yellow 160 azo dye. *Electrochim. Acta* **2016**, *206*, 307–316. [CrossRef]
51. Moreira, F.C.; Garcia-Segura, S.; Vilar, V.J.P.; Boaventura, R.A.R.; Brillas, E. Decolorization and mineralization of Sunset Yellow FCF azo dye by anodic oxidation, electro-Fenton, UVA photoelectro-Fenton and solar photoelectro-Fenton processes. *Appl. Catal. B Environ.* **2013**, *142–143*, 877–890. [CrossRef]
52. Nidheesh, P.V.; Gandhimathi, R. Comparative Removal of Rhodamine B from Aqueous Solution by Electro-Fenton and Electro-Fenton-Like Processes. *Clean—Soil Air Water* **2014**, *42*, 779–784. [CrossRef]
53. Lin, H.; Zhang, H.; Wang, X.; Wang, L.; Wu, J. Electro-fenton removal of Orange II in a divided cell: Reaction mechanism, degradation pathway and toxicity evolution. *Sep. Purif. Technol.* **2014**, *122*, 533–540. [CrossRef]
54. Nidheesh, P.V.; Gandhimathi, R.; Velmathi, S.; Sanjini, N.S. Magnetite as a heterogeneous electro Fenton catalyst for the removal of Rhodamine B from aqueous solution. *RSC Adv.* **2014**, *4*, 5698–5708. [CrossRef]
55. Oturan, N.; A. Oturan, M. Electro-Fenton Process: Background, New Developments, and Applications. In *Electrochemical Water and Wastewater Treatment*; Martinez-Huitle, C.A., Rodrigo, M.A., Scialdone, O., Eds.; Elsevier: Oxford, UK; Cambridge, MA, USA, 2018; pp. 193–221.
56. Fayazi, M.; Ghanei-Motlagh, M. Electrochemical mineralization of methylene blue dye using electro-Fenton oxidation catalyzed by a novel sepiolite/pyrite nanocomposite. *Int. J. Environ. Sci. Technol.* **2020**, *17*, 4541–4548. [CrossRef]
57. dos Santos, A.J.; Sirés, I.; Alves, A.P.M.; Martínez-Huitle, C.A.; Brillas, E. Vermiculite as heterogeneous catalyst in electrochemical Fenton-based processes: Application to the oxidation of Ponceau SS dye. *Chemosphere* **2020**, *240*, 124838. [CrossRef] [PubMed]
58. Malakootian, M.; Moridi, A. Efficiency of electro-Fenton process in removing Acid Red 18 dye from aqueous solutions. *Process Saf. Environ. Prot.* **2017**, *111*, 138–147. [CrossRef]
59. Babuponnusami, A.; Muthukumar, K. A review on Fenton and improvements to the Fenton process for wastewater treatment. *J. Environ. Chem. Eng.* **2014**, *2*, 557–572. [CrossRef]
60. Sirés, I.; Brillas, E. Electro-Fenton Process: Fundamentals and Reactivity. In *Electro-Fenton Process: New Trends and Scale-Up*; Zhou, M., Oturan, M.A., Sirés, I., Eds.; Springer Singapore: Singapore, 2018; pp. 1–28.
61. Teymori, M.; Khorsandi, H.; Aghapour, A.A.; Jafari, S.J.; Maleki, R. Electro-Fenton method for the removal of Malachite Green: Effect of operational parameters. *Appl. Water Sci.* **2020**, *10*, 39. [CrossRef]
62. Nidheesh, P.V.; Gandhimathi, R. Removal of Rhodamine B from aqueous solution using graphite–graphite electro-Fenton system. *Desalin. Water Treat.* **2014**, *52*, 1872–1877. [CrossRef]
63. Nidheesh, P.V.; Gandhimathi, R.; Sanjini, N.S. $NaHCO_3$ enhanced Rhodamine B removal from aqueous solution by graphite-graphite electro Fenton system. *Sep. Purif. Technol.* **2014**, *132*, 568–573. [CrossRef]
64. Gao, S.; Wang, Z.; Wang, H.; Jia, Y.; Xu, N.; Wang, X.; Wang, J.; Zhang, C.; Tian, T.; Shen, W. Peroxydisulfate activation using B-doped biochar for the degradation of oxytetracycline in water. *Appl. Surf. Sci.* **2022**, *599*, 153917. [CrossRef]
65. Zhou, M.; Yu, Q.; Lei, L.; Barton, G. Electro-Fenton method for the removal of methyl red in an efficient electrochemical system. *Sep. Purif. Technol.* **2007**, *57*, 380–387. [CrossRef]
66. Du, X.; Fu, W.; Su, P.; Cai, J.; Zhou, M. Internal-micro-electrolysis-enhanced heterogeneous electro-Fenton process catalyzed by Fe/Fe_3C@PC core–shell hybrid for sulfamethazine degradation. *Chem. Eng. J.* **2020**, *398*, 125681. [CrossRef]
67. Alizadeh Fard, M.; Barkdoll, B. Effects of oxalate and persulfate addition to Electrofenton and Electrofenton-Fenton processes for oxidation of Ketoprofen: Determination of reactive species and mass balance analysis. *Electrochim. Acta* **2018**, *265*, 209–220. [CrossRef]
68. Lu, X.; Zhou, X.; Qiu, W.; Wang, Z.; Cheng, H.; Zhang, H.; Yu, J.; Wang, D.; Ma, J. Singlet oxygen involved electrochemical disinfection by anodic oxidation of H_2O_2 in the presence of Cl^-. *Chem. Eng. J.* **2022**, *446*, 136871. [CrossRef]
69. Divyapriya, G.; Nidheesh, P.V. Importance of Graphene in the Electro-Fenton Process. *ACS Omega* **2020**, *5*, 4725–4732. [CrossRef] [PubMed]
70. Gopinath, A.; Pisharody, L.; Popat, A.; Nidheesh, P.V. Supported catalysts for heterogeneous electro-Fenton processes: Recent trends and future directions. *Curr. Opin. Solid State Mater. Sci.* **2022**, *26*, 100981. [CrossRef]
71. Yao, Y.; Cai, Y.; Lu, F.; Wei, F.; Wang, X.; Wang, S. Magnetic recoverable $MnFe_2O_4$ and $MnFe_2O_4$-graphene hybrid as heterogeneous catalysts of peroxymonosulfate activation for efficient degradation of aqueous organic pollutants. *J. Hazard. Mater.* **2014**, *270*, 61–70. [CrossRef]
72. Scaria, J.; Gopinath, A.; Ranjith, N.; Ravindran, V.; Ummar, S.; Nidheesh, P.V.; Kumar, M.S. Carbonaceous materials as effective adsorbents and catalysts for the removal of emerging contaminants from water. *J. Clean. Prod.* **2022**, *350*, 131319. [CrossRef]
73. Yu, D.; Wu, F.; He, J.; Bai, L.; Zheng, Y.; Wang, Z.; Zhang, J. Tuned layered double hydroxide-based catalysts inducing singlet oxygen evolution: Reactive oxygen species evolution mechanism exploration, norfloxacin degradation and catalysts screen based on machine learning. *Appl. Catal. B Environ.* **2023**, *320*, 121880. [CrossRef]

74. Cai, H.; Zou, J.; Lin, J.; Li, J.; Huang, Y.; Zhang, S.; Yuan, B.; Ma, J. Sodium hydroxide-enhanced acetaminophen elimination in heat/peroxymonosulfate system: Production of singlet oxygen and hydroxyl radical. *Chem. Eng. J.* **2022**, *429*, 132438. [CrossRef]
75. Li, J.; Zou, J.; Zhang, S.; Cai, H.; Huang, Y.; Lin, J.; Li, Q.; Yuan, B.; Ma, J. Sodium tetraborate simultaneously enhances the degradation of acetaminophen and reduces the formation potential of chlorinated by-products with heat-activated peroxymonosulfate oxidation. *Water Res.* **2022**, *1*, 119095. [CrossRef]
76. Jiang, H.; Sun, Y.; Feng, J.; Wang, J. Heterogeneous electro-Fenton oxidation of azo dye methyl orange catalyzed by magnetic Fe3O4 nanoparticles. *Water Sci. Technol.* **2016**, *74*, 1116–1126. [CrossRef]
77. Nidheesh, P.V.; Gandhimathi, R. Textile Wastewater Treatment by Electro-Fenton Process in Batch and Continuous Modes. *J. Hazardous, Toxic, Radioact. Waste* **2015**, *19*, 04014038. [CrossRef]
78. Hafaiedh, N.B.; Bellakhal, N. Mineralization of Synthetic and Industrial Food Effluent Containing Acid Red18 by Electro-Fenton Process using a Graphite-Felt Cathode. *Int. J. Sci. Res.* **2015**, *4*, 2643–2649.
79. Basha, C.A.; Selvakumar, K.V.; Prabhu, H.J.; Sivashanmugam, P.; Lee, C.W. Degradation studies for textile reactive dye by combined electrochemical, microbial and photocatalytic methods. *Sep. Purif. Technol.* **2011**, *79*, 303–309. [CrossRef]
80. Nidheesh, P.V.; Gandhimathi, R. Effect of solution pH on the performance of three electrolytic advanced oxidation processes for the treatment of textile wastewater and sludge characteristics. *RSC Adv.* **2014**, *4*, 27946–27954. [CrossRef]
81. Olvera-Vargas, H.; Oturan, N.; Aravindakumar, C.T.; Paul, M.M.S.; Sharma, V.K.; Oturan, M.A. Electro-oxidation of the dye azure B: Kinetics, mechanism, and by-products. *Environ. Sci. Pollut. Res.* **2014**, *21*, 8379–8386. [CrossRef] [PubMed]
82. Ruiz, E.J.; Hernández-Ramírez, A.; Peralta-Hernández, J.M.; Arias, C.; Brillas, E. Application of solar photoelectro-Fenton technology to azo dyes mineralization: Effect of current density, Fe^{2+} and dye concentrations. *Chem. Eng. J.* **2011**, *171*, 385–392. [CrossRef]
83. Yao, X.; Wang, K.; Zhang, S.; Liang, S.; Li, K.; Wang, C.; Zhang, T.; Li, H.; Wang, J.; Dong, L.; et al. Degradation of the mixture of ethyl formate, propionic aldehyde, and acetone by Aeromonas salmonicida: A novel microorganism screened from biomass generated in the citric acid fermentation industry. *Chemosphere* **2020**, *258*, 127320. [CrossRef]
84. Periyasamy, S.; Lin, X.; Ganiyu, S.O.; Kamaraj, S.-K.; Thiam, A.; Liu, D. Insight into BDD electrochemical oxidation of florfenicol in water: Kinetics, reaction mechanism, and toxicity. *Chemosphere* **2022**, *288*, 132433. [CrossRef]
85. Liu, Y.; Li, K.; Xu, W.; Du, B.; Wei, Q.; Liu, B.; Wei, D. GO/PEDOT:NaPSS modified cathode as heterogeneous electro-Fenton pretreatment and subsequently aerobic granular sludge biological degradation for dye wastewater treatment. *Sci. Total Environ.* **2020**, *700*, 134536. [CrossRef]

water
MDPI

Article

Optimized Removal of Azo Dyes from Simulated Wastewater through Advanced Plasma Technique with Novel Reactor

Yang Liu [1,*], Jia-Wei Song [1], Jia Bao [1,*], Xin-Jun Shen [1], Cheng-Long Li [1], Xin Wang [1] and Li-Xin Shao [2]

[1] School of Environmental and Chemical Engineering, Shenyang University of Technology, Shenyang 110870, China
[2] School of Mechanical Engineering, Shenyang University of Technology, Shenyang 110870, China
* Correspondence: liuyang@sut.edu.cn (Y.L.); baojia@sut.edu.cn (J.B.)

Abstract: Increasing attention has been paid to removal of aqueous contaminations resulting from azo dyes in water by plasma technology. However, the influence factors and removal mechanism of plasma technology were still obscure, moreover, energy consumption and oxidized degradation efficiency of plasma reactor were also inferior. In the present study, a comparative analysis was performed using 100 mg/L of Methyl Orange (MO) in the simulated wastewater with a novel plasma reactor to achieve the ideal parameters involving voltage, discharge gap, and discharge needle numbers. Therefore, the optimal removal rate for MO could be up to 95.1% and the energy consumption was only 0.26 kWh/g after the plasma treatment for 60 min, when the voltage was set as 15 kV, the discharge gap was 20 mm, and the discharge needle numbers was 5. Based upon the response surface methodology (RSM), the removal rate of MO was predicted as 99.3% by massive optimization values in software, and the optimum conditions were confirmed with the plasma treatment period of 60 min, the voltage of 14.8 kV, the discharge gap of 20 mm, and the discharge needles of 5. Plasma associated with catalysts systems including plasma, plasma/Fe^{2+}, plasma/PS, and plasma/PS/Fe^{2+} were further investigated, and the best removal rate for MO reached 99.2% at 60 min under the plasma/PS/Fe^{2+} system due to simultaneously synergistic reactions of $HO^{\bullet}$ and $SO_4^{\bullet-}$. Moreover, it was also revealed that –N=N– bond was attacked and broken by active species like $HO^{\bullet}$, and the oxidized by-products of benzenesulfonic acid and phenolsulfonic acid might be generated, via the analysis of the variation in the absorbances through UV-Vis spectrophotometry during the plasma treatment. As a result, the advanced plasma technique in this study presented excellent efficacy for MO removal from simulated wastewater with low energy consumption.

Keywords: azo dyes; Methyl Orange; novel plasma reactor; oxidized degradation; plasma technology; response surface methodology (RSM)

Citation: Liu, Y.; Song, J.-W.; Bao, J.; Shen, X.-J.; Li, C.-L.; Wang, X.; Shao, L.-X. Optimized Removal of Azo Dyes from Simulated Wastewater through Advanced Plasma Technique with Novel Reactor. *Water* **2022**, *14*, 3152. https://doi.org/10.3390/w14193152

Academic Editor: Zacharias Frontistis

Received: 14 September 2022
Accepted: 4 October 2022
Published: 6 October 2022

1. Introduction

In 1862, the inartificial dyes were initially obtained from natural sources, including leaves, roots, and branches, etc. [1], which usually limited the range of colors [2]. Thereafter, various kinds of artificial dyes were explored to satisfy daily demands. According to their chemical structures and chromophores, dyes are classified as azo (mono-azo, di-azo, tri-azo, poly-azo), anthraquinone, phthalocyanine, diarylmethane, triarylmethane, indigo, azine, oxazine, thiazine, xanthene, nitro, nitroso, methine, thiazole, indamine, indophenol, lactone, amino ketone, hydroxyl ketone stibene, and sulfur dyes [3]. So far, there are over two thousand types of dyes available for commercial purposes in the market with quantities exceeding 7×10^5 tons worldwide annually [4,5], of which over 50% were azo dyes [4]. Azo dyes are a class of organic compounds with azo linkages (–N=N–) connecting aryl groups, which were widely used in textile, printing, paper and other fields [1,4].

Azo dyes possessed strong physical and chemical stability, extraordinary resistance to high temperature, photolysis, biological and oxidative degradation [6]. Therefore, great quantities utilizing and inappropriate discharge of azo dyes have resulted in water pollutions and accumulations in the water cycle, threatening the safety of drinking water. Textile-processing units discharged extremely colored water effluents with dye content in the range of 10–200 mg/L [3]. For instance, the outflow concentration of the dye Acid Orange 10 from the final clarifier of a textile factory in India was 45 mg/L [7]. The outflow concentrations of azo dye from 14 Ramadan textile industries in Iraq ranged between 20 mg/L and 50 mg/L [7]. Human exposure to these azo dyes and their intermediates would cause mutagenic, carcinogenic, nervous system disorders, and dermatitis [8], as well as severe and long-lasting impact on the liver, kidney, brain, and reproduction system [9]. Therefore, increasing efforts would be warranted to implement the elimination of azo dyes from waters.

Azo dyes usually contain a strong stable benzene ring structure and azo bonds that prevent mineralization, and even dissolved functional groups of sulfonate, hydroxyl, or carboxyl, leading to difficult water treatment [10–13]. Extensive research has been done on utilizing physical and chemical techniques for the removal of azo dyes, including adsorption, chemical oxidation, biological techniques, and plasma technology [14–16], etc. The adsorption method could adsorb azo dyes effectively, but only by transferring the contaminants from water to the solid phase, without achieving the degradation effect [17], and the adsorbent regeneration was an intractable problem [18]. The chemical oxidation process could decolorize azo dyes, but the mineralization rates were usually inferior, and the addition of chemical agents was inclined to cause secondary pollution [16,19]. The biological method was used for azo dyes degradation due to their high activity and strong adaptability [20], but high levels of azo dyes could inhibit the activity of bacteria [16]. Therefore, an environmentally friendly and high efficiency degradation technique for azo dyes was required for developing to overcome the above-mentioned problems.

Plasma technology is considered as the advanced oxidation process (AOP); the plasma discharge could break molecular structure efficiently by its generation of strong active species ($HO^{\bullet}$, $H^{\bullet}O^{\bullet}$, $HO_2{}^{\bullet}$, $NO_2{}^{\bullet}$, etc.) and other physiochemical effects (UV irradiation, shockwaves, local high temperature, etc.) in situ [17,21,22]. The water molecule of solution in the discharge process leads to produce $HO^{\bullet}$ and $H^{\bullet}$ by its dissociation, ionization, and vibrational/rotational excitation, as shown in Equations (1)–(4) [15]. $H^{\bullet}$ and $O^{\bullet}$ could be generated by vibrationally/rotationally excited water molecules that are released into a lower energetic state, as shown in Equations (5)–(7) [15]. Subsequently, the enduring H_2O_2 could be formed by dimerization of $HO^{\bullet}$ (Equation (8)) [23]. In addition, plasma containing water has UV light emission as a result of exited species relaxation to lower energetic states, which are generated from the collisions between electrons and neutral molecules [15]. In the plasma treatment process, the organic molecules (M) absorb the radiation and transfer into an excited state (M*) by ultraviolet radiation, whereafter the excited M* could be decomposed into a new product in the transversion of M* goes back to the ground state immediately due to short lifetime (10^{-9}–10^{-8} s), as shown in the Equation (9) [15]. Thus, this technique was suitable for the removal of degradation-resistant organic compounds [16]. Ma et al. have used Dielectric Barrier Discharge (DBD) plasma technology to degrade 44.36 g/L of methylene blue (MB), the degradation rate could reach up to 98.3% after 3 min [24]. Sarangapani et al., have used Box-Benkhen Design (BBD) model and Response Surface Methodology (RSM) to optimize the experimental parameters of the DBD plasma technique for Methyl Orange (MO) degradation, which determined the optimal experimental conditions with voltage at 70 kV, treatment time as 120 s, and MO concentration as 100 mg/L [25].

$$\text{Dissociation}: \ H_2O + e \rightarrow HO^{\bullet} + H^{\bullet} + e \tag{1}$$

$$\text{Ionization}: \ H_2O + e \rightarrow 2e + H_2O^{+} \tag{2}$$

$$H_2O^{+} + H_2O \rightarrow HO^{\bullet} + H_3O^{+} \tag{3}$$

$$\text{Vibrational/rotational excitation}: \ H_2O + e \rightarrow H_2O^{*} + e \tag{4}$$

$$H_2O^*+H_2O \rightarrow H_2O + H^\bullet+HO^\bullet \quad (5)$$

$$H_2O^*+H_2O \rightarrow H_2+O^\bullet+H_2O \quad (6)$$

$$H_2O^*+H_2O \rightarrow 2H^\bullet+O^\bullet+H_2O \quad (7)$$

$$HO^\bullet+HO^\bullet \rightarrow H_2O_2 \quad (8)$$

$$M + hv \rightarrow M^* \rightarrow \text{product} \quad (9)$$

However, the influence factors and degradation mechanism of plasma technology on azo dyes at the molecular level were still obscure, as well as usually followed with high energy consumption and inferior degradation efficiency due to low mass transfer efficiency between gas and liquid phase in plasma reactor [26,27]. In general, active species including $HO^\bullet$, $H^\bullet$, $O^\bullet$, etc. generated from plasma could penetrate into liquid, but with shallow penetration depth and attain some dozens of microns [28].

Therefore, the establishment of the interaction between active species and organic contaminants would enhance the removal efficacy. In this study, a novel reactor with an improved plasma area was designed through regulating the grounded plate and the plasma discharge area, which could increase the removal efficiency by the intensive reaction between active species and MO contaminants on the interface of gas-liquid. Moreover, the addition of typical catalysts such as persulfate (PS) and Fe^{2+} would present excellent degradation efficacy due to the high oxidizability of generated active species [29–32].

In order to solve the above-said problems, the plasma technique for the azo dyes was implemented in the present study to achieve the objectives as follows: (1) confirm the optimal influence parameters by comparing the plasma removal performance on the target azo dye of MO under different conditions, involving the reaction time, voltage, discharge gap, and discharge needle numbers; (2) improve MO removal efficiency and reduce energy consumption using the novel plasma reactor through the optimal response surface methodology (RSM); (3) investigate the removal efficiency of plasma associated with catalyst system including PS/Fe^{2+}; and (4) determine the removal mechanism via the UV-Vis analysis of MO during the plasma process.

2. Materials and Methods

2.1. Chemicals and Reagents

Methyl Orange (MO, 99.9%) was acquired from Fluka (Steinheim, Germany). Sodium hydroxide (NaOH, 99.9%), ferrous sulfate ($FeSO_4$, 99.9%), and sodium persulphate ($Na_2S_2O_8$, 99.9%) were purchased from Acros Organics (Geel, Belgium). All solutions were prepared using ultrapure water with a conductance of 18.2 MΩ/cm (Millipore, Bedford, MA, USA).

2.2. Experimental Setup and Procedure

A novel reactor was designed as a cylinder with well appropriate between the plasma discharge area and the grounded electrode in the bottom of the reactor. The novel reactor could increase the effective discharge area for active species formation during the plasma process. The fully propagated plasma occupied a roughly circular region on the liquid surface with visually uniform leader distribution [33]. As shown in Figure 1, the novel plasma reactor was an organic glass reactor with a 50 mL effective volume for MO removal from simulated wastewater. In detail, the bottom radius was 50 mm, the height was 70 mm, and the wall thickness was 5 mm. Two air vents with a radius of 4 mm were set on both sides, which were 20 mm away from the top of the reactor. A sample collection hole with a 4 mm diameter was installed at 12 mm away from the bottom of reactor. The mass transfer efficiency of active species in liquid could be improved by enhancing the air tightness of the reactor, all vents were blocked with an airtight polymer cap during the operation. The circular grounded metal electrode plate with a 45 mm radius and 2 mm thickness was immersed in the MO solution, the same size as a perforated metal plate fixed with needle-like electrodes was used as high voltage discharge electrode plate. Needle-like electrodes were stainless steels which widely used due to their mechanical and anti-corrosion properties [34]. The nylon columns were used to support the high voltage electrode and the grounded electrode, and the high voltage electrode could be adjusted by the distance above the solution.

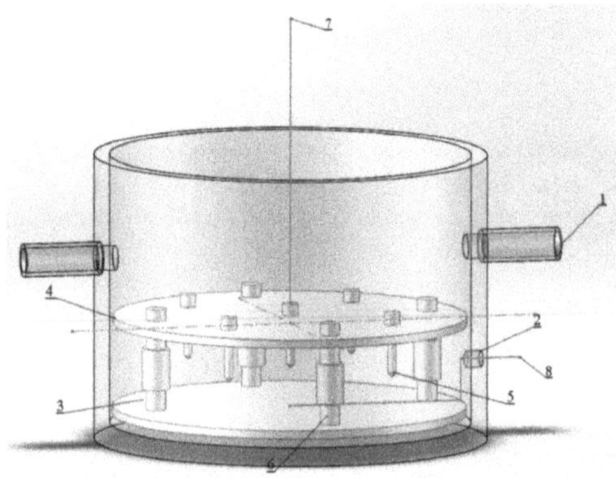

Figure 1. The sketch map of a novel plasma reactor. (1. Air vent 2. Sample collection hole 3. Grounded electrode plate 4. Discharge electrode plate 5. Discharge needle 6. Nylon column 7. Power lead 8. Ground lead).

The high voltage direct current (DC) power supply (LYZGF, Zhi-Cheng Company, China) was used to generate plasma, characterized with the range of 0–60 kV for discharge voltage and 0–5 mA for output current. Based upon the important influence parameters, the discharge voltages were in the range between 11 kV and 15 kV, the discharge gap was between 19 mm and 21 mm, and discharge needle numbers were 3, 4, and 5, together with the initial MO concentration of 100 mg/L in simulated solution [35–37]. The added catalysts of Fe^{2+} and PS were set as 0.02 mmol/L under the optimal experimental conditions after the experimental factors were optimized by the RSM in the experiment of MO removal with plasma.

2.3. Sample Preparation and Analysis

The experimental duration of each plasma process was 60 min. Samples were taken at 0 min, 10 min, 20 min, 30 min, and 60 min, the reactor effluent from MO wastewater treatment was analyzed after filtration of glass fiber membrane (0.45 μm). The samples were scanned using an ultraviolet and visible (UV-Vis) spectrophotometer (DR 5000, Hach Company, Loveland, CO, USA) [38]. The absorbance was compared with the standard curve ($Y = 0.7563x - 0.00831$ $R^2 = 0.998$) established at the maximum absorption wavelength of MO 462 nm [39], subsequently the concentration and removal rate of MO was calculated separately. The TOC values of collected samples were determined by the TOC analyzer (multi NC 3100, Analytik Jena AG Company, Jena, Germany) via direct nonpurgeable organic carbon (NPOC) measurement method to confirm the reduction of organic substances in solutions [24].

2.4. Data Analysis

1. Removal rate of MO (η) %

$$\eta = \frac{C_0 - C_t}{C_0}$$

where η was the removal rate of MO in the solution, C_0 and C_t were the concentration of MO solution at times of 0 and t, respectively.

2. Energy consumption (W) kWh/g

$$W = \frac{IUt}{\Delta m}$$

where W was the energy consumption of removing organic matter per unit weight, I was the output current, U was the output voltage, t was the discharge time, Δm was the removal weight of organic matter.

3. Mineralization rate (X_{TOC}) %

$$X_{TOC} = \frac{TOC_0 - TOC_t}{TOC_0}$$

where X_{TOC} was the mineralization rate of MO, TOC_0 and TOC_t were total organic carbon at 0 min and t min, respectively.

2.5. RSM Experimental Design

Design-experimental V8.0.6.1 statistical software (Stat-Ease, Minneapolis, MN, USA) was applied to carry out the experimental design, analysis of variance (ANOVA), mathematical modeling, and 3D response surface [13]. Plackett-Burman, Central-Composite, and Box-Behnkhen Design (BDD) were widely used methods for RSM experiments, in which BDD followed the least number of quadratic model fitting experiments at three levels. As shown in Table 1, four significant operating parameters of a novel plasma reactor involving time (min), voltage (kV), discharge gap (mm), and discharge needle numbers were optimized using BBD in this study, and the removal rate of MO was the factor of response surface analysis [40]. In this study, four parameters presented three variation levels equally, which were codified as −1, 0, and 1 [13].

Table 1. Response surface factor level table in novel plasma reactor.

Parameter	Horizontal Extent		
	−1	0	1
Time (min)	10	35	60
Voltage (kV)	11	13	15
Discharge gap (mm)	19	20	21
Discharge needle numbers	3	4	5

3. Results and Discussion

3.1. The Influence of Voltage

Discharge plasma originated by high voltage on the gas-liquid surface would generate various active species in the solution [35]. Thus, the discharge voltage was a significant factor on the removal of azo dye during the plasma treatment. To evaluate the effect of discharge voltage on MO removal, systematic experiments were implemented on plasma treatment with different applied voltages of 11 kV, 13 kV, and 15 kV. The other experiment parameters of the discharge needle numbers were set as 5, and the discharge gap was 20 mm. As shown from the contour plot of the response surface in Figure 2, the colors changed gradually from purple to red, which demonstrated that the removal rates of MO were enhanced over time. In addition, there was an increase at the same altitude, the removal rates of MO were improved with the voltage increasing in a certain applied voltage range. The interaction of voltage and treatment time showed a positive effect for the removal rate of MO. The removal rates of MO were 85.0%, 90.3%, and 95.1% at 60 min when each applied voltage were 11 kV, 13 kV, and 15 kV, respectively. This phenomenon was consistent with supplementary free radicals generated by high applied voltage [41]. Because additional energy electrons could be produced with the application of increasing voltages, which could intensify the collision chance between energetic electrons and air molecules effectively, thereafter producing extra excited oxygen atoms that could react with water molecules to generate extra $HO^{\bullet}$ [35,42]. On the other hand, the plasma treatment process could generate ultraviolet radiation that improved the removal rate of MO, and the intensity of ultraviolet radiation could be enhanced at relatively high applied voltage [35].

However, high applied voltage would cost additional energy consumption and reduce energy utilization [43]. Therefore, the applied voltage was selected as 15 kV.

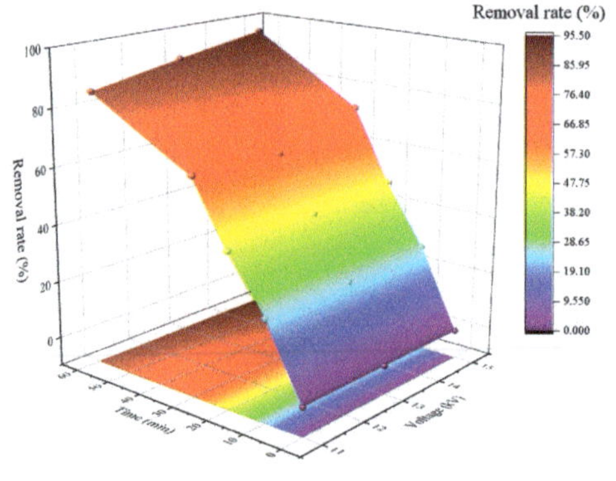

Figure 2. The influence of voltage on the removal rate of MO.

3.2. *The Influence of Discharge Gap*

Suitable discharge gap is important for MO removal in plasma, which influences the electric filed in the gap [36]. In this study, systematic experiments were conducted in plasma treatment with various discharge gaps of 19 mm, 20 mm, and 21 mm. The other parameters of the experiment including the applied voltage of 15 kV, and the number of discharge needles of 5. As shown from the contour plot of the response surface in Figure 3, the optimum removal rates of MO could achieve 89.1%, 96.7%, and 79.2% when the discharge gap was 19 mm, 20 mm, and 21 mm, respectively. The MO removal rates improved firstly when the discharge gap between 19 mm and 20 mm, but decreased subsequently when the discharge gap increased to 21 mm. It was because that corona discharge would change into unstable and noisy spark discharge when the discharge gap was low excessively, the generated spark discharge was not conductive to the active species due to the gap breakdown [44], whereas the removal rate of MO decreased to 79.2% when the discharge gap amplified to 21 mm. The discharge gap was related to electric filed intensity which could influence the movement speed of gas molecules in the discharge area [45], thereafter, the increased discharge gap would weaken the electric field intensity which declined the movement speed of gas molecules in the air under the same input energy and then generate scarcer active species [44,45]. Therefore, 20 mm was chosen as the optimal discharge gap for the plasma treatment.

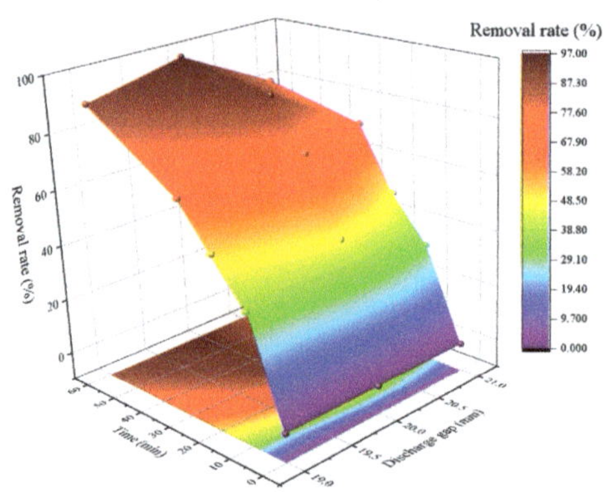

Figure 3. The influence of discharge gap on removal rate of MO.

3.3. The Influence of Discharge Needle Numbers

The numbers of discharge needle electrodes are related to the plasma channels between the discharge needle and liquid surface [37]; thus, it is important to get a suitable discharge needle numbers to enhance the removal rate of MO. To estimate the influence of discharge needle numbers on MO removal, different discharge needle numbers of 3, 4, and 5 were conducted in this plasma treatment. The other parameters were related to the applied voltage of 15 kV, and the discharge gap of 20 mm. As shown in the contour plot of the response surface in Figure 4, the removal rates of MO were improved with the growth of the treatment period and discharge needles number, subsequently the removal rate of MO was up to 95.1% at 60 min when the discharge needles number was 5. The results verified that the increase in discharge needle numbers expanded the discharge region to promote the chemical reactions between active species and MO [46]. However, the removal rates of MO were decreased to 82.5% and 82.8% individually at 60 min when the discharge needle numbers were 3 and 4, respectively. In general, needle-like electrodes generated fully propagated plasma channels at the liquid surface, and all types of active species were produced near plasma channels, thereafter the removal of MO mainly occurred in this circular area [33,47]. With the decline of discharge needle numbers, the plasma channels decreased, so the removal rates of MO were inferior due to depressed reaction region of active species with MO [37]. Therefore, 5 were chosen as the optimal discharge needles numbers in this study.

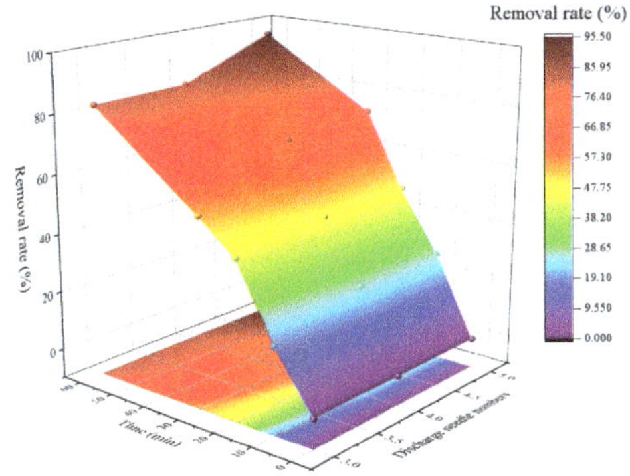

Figure 4. The influence of discharge needles number on removal rate of MO.

3.4. Analysis of RSM

RSM refers to a mathematical and statistical approach used to evaluate various aspects, including designing experiments, developing models, examining many independent variables, and assessing the optimum conditions for responses [48]. In the present study, four significant influencing parameters were selected and optimized at three levels through the ANOVA and mathematical modeling, for which, the removal rates of MO in the plasma treatment were applied as the response value. The test results were listed in Table 2.

The significance of experimental factors was judged by F-value and p-value in ANOVA analysis, F-value was the accuracy of generated quadratic polynomial equation computed statistically, and the experimental factor was significant when the p-value < 0.05. The ANOVA results for response surface of the quadratic model were listed in Table 3. MO independent variable of time (A and A^2), voltage (B and B^2), discharge gap (C^2), discharge needle numbers (D), and the interaction of AB were significant for the removal efficiency. However, the interaction of AC, AD, BC, BD, and CD were insignificant, and the independent variable of C and D^2 were also insignificant. Therefore, the influence parameters for the MO by the plasma technique followed the order of time (A) $\approx$ voltage (B) > discharge needle numbers (D) > discharge gap (C). The Predicted R^2 = 0.7119 was in reasonable agreement with the Adjusted R^2 = 0.9000 because the difference < 0.2. The signal of noise ratio about Adep Precision = 16.954 > 4 demonstrating that the signal was acceptable due to among the desirable region. According to the parameters of p-value < 0.0001, R^2 = 0.9500, which demonstrated the prediction model was appropriate for describing the plasma experimental data. As shown in Figure 5a, all residuals spread along a straight line that demonstrated a normal probability distribution plot of MO obtained removal rates which were employed to validate the normality of studentized residuals. Furthermore, Figure 5b correspondingly exhibited the predict and actual values of removal rates as another confirmation.

Table 2. Results of the removal experiment of MO.

The Experimental Number	Time (min)	Voltage (kV)	Discharge Gap (mm)	Discharge Needle Numbers	Removal Rate (%)
1	60.00	13.00	20.00	3.00	71.52
2	10.00	13.00	21.00	4.00	9.54
3	60.00	13.00	20.00	5.00	90.25
4	35.00	15.00	20.00	5.00	72.73
5	35.00	13.00	19.00	5.00	64.4
6	10.00	15.00	20.00	4.00	26.84
7	60.00	11.00	20.00	4.00	43.77
8	10.00	13.00	20.00	3.00	25.26
9	35.00	11.00	20.00	5.00	64.53
10	60.00	13.00	21.00	4.00	80.59
11	35.00	15.00	20.00	3.00	54.08
12	60.00	13.00	19.00	4.00	55.58
13	35.00	11.00	19.00	4.00	20.88
14	35.00	13.00	20.00	4.00	59.24
15	10.00	11.00	20.00	4.00	20.76
16	10.00	13.00	19.00	4.00	9.41
17	35.00	15.00	21.00	4.00	47.34
18	35.00	13.00	20.00	4.00	59.24
19	10.00	13.00	20.00	5.00	25.49
20	35.00	13.00	19.00	3.00	42.35
21	35.00	13.00	20.00	4.00	59.24
22	35.00	11.00	21.00	4.00	27.24
23	35.00	15.00	19.00	4.00	58.45
24	35.00	13.00	21.00	3.00	42.78
25	35.00	13.00	20.00	4.00	59.24
26	35.00	11.00	20.00	3.00	22.48
27	60.00	15.00	20.00	4.00	82.83
28	35.00	13.00	21.00	5.00	51.7

Table 3. MO ANOVA results for response surface of quadratic model.

Source	Sum of Squares	df	Mean Square	F-Value	p-Value	
Model	13,001.19	14	928.66	18.99	<0.0001	significant
A-Time	7866.37	1	7866.37	160.89	<0.0001	
B-Voltage	1694.80	1	1694.80	34.66	<0.0001	
C-discharge gap	5.49	1	5.49	0.11	0.7424	
D-discharge needle numbers	1019.92	1	1019.92	20.86	0.0004	
AB	271.92	1	271.92	5.56	0.0334	
AC	154.75	1	154.75	3.17	0.0969	
AD	85.56	1	85.56	1.75	0.2071	
BC	76.30	1	76.30	1.56	0.2321	
BD	136.89	1	136.89	2.80	0.1165	
CD	43.10	1	43.10	0.88	0.3637	
A^2	443.32	1	443.32	8.86	0.0100	
B^2	432.13	1	432.13	8.84	0.0101	
C^2	952.96	1	952.96	19.49	0.0006	

Table 3. *Cont.*

Source	Sum of Squares	df	Mean Square	F-Value	*p*-Value
D^2	41.94	1	41.94	0.86	0.3700
Residual	684.49	14	48.89		
Lack of Fit	684.49	10	68.45		
Pure Error	0.000	4	0.000		
Cor Total	13,685.68	28			

R^2 = 0.9500 Adjusted R^2 = 0.9000 Predicted R^2 = 0.7119 Adeq Presion = 16.954.

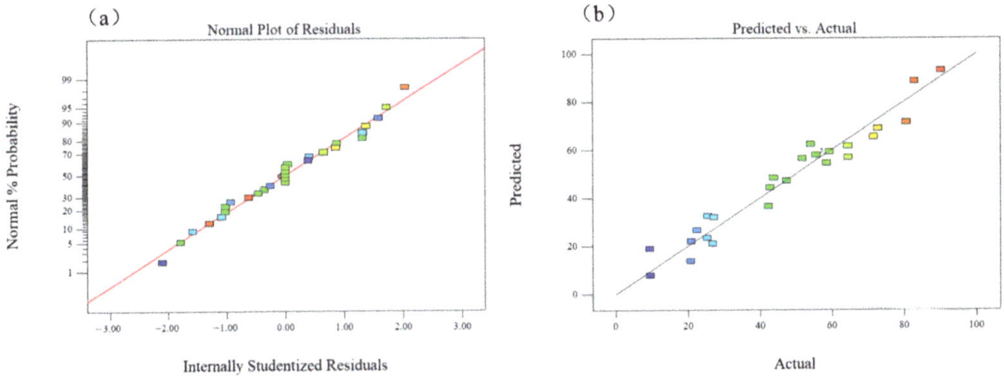

Figure 5. The normal probability plot of studentized residuals of MO of the model for the removal rates (**a**) and the experimental response values versus the predicted response value of MO removal rates (**b**).

The RSM for the removal rates of MO according to the four important parameters achieved from the plasma experiment were presented in Figure 6a–f. Four parameters were demonstrated in 3D surface plots. The interactions among the voltage, time, discharge gap, as well as discharge needle numbers were correlated to the ANOVA results for response surface of quadratic model. In addition, when the removal rate of MO was predicted as 99.3% by massive optimization values in software, correspondingly, the optimum conditions including the voltage of 14.8 kV, the plasma treatment time of 60 min, the discharge gap of 20 mm, discharge needles of 5 were confirmed. This conclusion demonstrated that the results of RSM experiment conditions were consistent with the results of previous single-factor experiment conditions. Moreover, the regression model could be conformed as a second-order response surface by the fitting analysis of multi-linear regression, as shown in the following Equation (10).

$$Y = 59.24 + 25.60A + 11.88B + 0.68C + 9.22D + 8.24AB + 6.22AC + 4.63A - D - 4.37BC - 5.85BD - 3.28CD - 8.17A^2 - 8.16B^2 - 12.12C^2 + 2.54D^2 \quad (10)$$

3.5. Advantages of Removal Effect of MO in Novel Plasma Reactor

In order to verify the advantages of novel designed plasma reactor, the comparison of MO removal efficiencies applied with the same voltage of 15 kV was implemented between the novel and conventional plasma reactors (Figure S1). As shown in the Figure 7, the optimum removal rate of the novel plasma reactor could reach 95.1% at 60 min, while that of the conventional plasma reactor was only 40.5%. In addition, the energy consumption of the novel reactor for MO removal was only 0.26 kWh/g, which was much lower than that of the conventional plasma reactor with 1.1 kWh/g significantly. This might be related to the increased plasma area between the diameters of the grounded plate and the plasma discharge area, contributing to fully propagated plasma that occupied a roughly circular region on the gas-liquid surface with uniform leader distribution [33], furthermore, increasing the mass transfer efficiency and interaction between the active species and MO contaminant. Therefore, the novel designed plasma reactor could show both advantages of higher removal efficiency and lower energy consumption during the process of MO removal treatment.

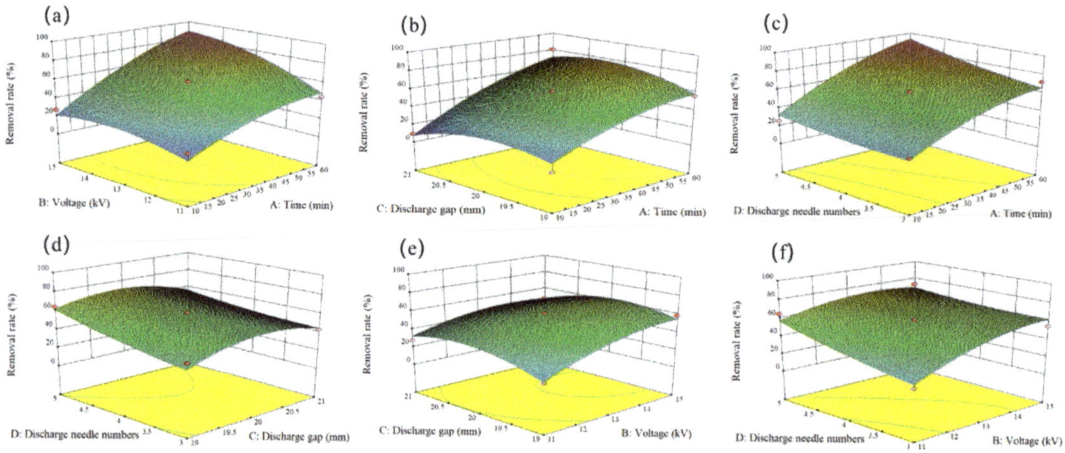

Figure 6. RSM for degradation efficiency of MO as a function of voltage and time (**a**), discharge gap and time (**b**), discharge needle numbers and time (**c**), discharge needle numbers and discharge gap (**d**), discharge gap and voltage (**e**), discharge needle numbers and voltage (**f**).

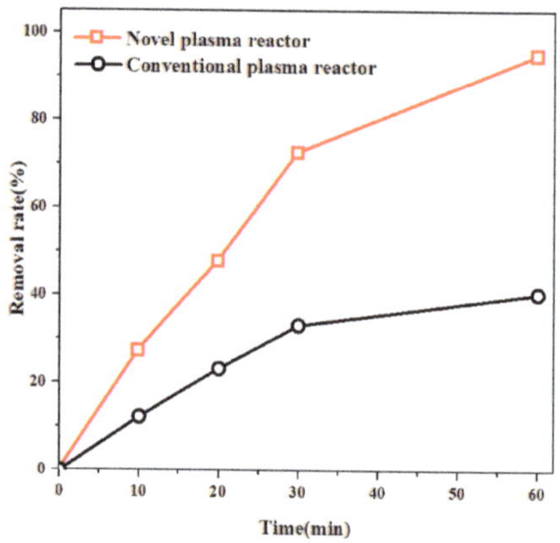

Figure 7. The comparison of novel and conventional reactor for the removal rates of MO (the novel plasma reactor (V = 50 mL) and the conventional plasma reactor (V = 150 mL) under the same conditions of initial concentration = 100 mg/L, voltage = 15 kV, discharge gap = 20 mm, and discharge needle numbers = 5).

3.6. *The Influence of Catalysts on MO*

The MO removal efficiency could be further improved by adding catalysts due to the rapid interaction between active species and organic contaminants. For instance, Fe^{2+} could react with H_2O_2 in a solution, similarly to the Fenton reaction, to produce supplementary $HO^{\bullet}$ [49], which was beneficial to improve the removal of organic matter greatly. PS could be easily activated by UV and heated by plasma, resulting in oxidizing to $SO_4^{\bullet-}$ strongly [50]. $SO_4^{\bullet-}$ showed a commendable oxidization effect on the improvement of degradation efficiency due to the advantages of strong stability [51]. Therefore, it is necessary to explore the influence of catalysts of PS and Fe^{2+} in the same experimental conditions optimized by RSM. As shown in Figure 8, four plasma system involving plasma, plasma/Fe^{2+}, plasma/PS, and plasma/Fe^{2+}/PS altogether performed remarkable treatment efficiency, and plasma/PS/Fe^{2+} exhibited the maximum removal rate for MO up to 99.2% at 60 min. The removal rates of MO could reach 97.0% and 97.3% under the plasma/Fe^{2+} system and plasma/PS system, respectively, when the concentration of Fe^{2+} and PS was both 0.02 mmol/L. Compared with the single plasma system, their removal rates of MO were enhanced slightly. This phenomenon might be related to the advanced oxidation of catalysts addition. For one thing, Fe^{2+} could efficiently catalyze H_2O_2 into advanced oxidation radical of $HO^{\bullet}$ by the Fenton reaction, whereafter this active $HO^{\bullet}$ could strengthen the removal efficiency of MO, as shown in Equations (11) and (12) [52]. For the other thing, as shown in Equation (13), PS could be activated by plasma to produce $SO_4^{\bullet-}$ radical [16,50], thus the removal rate of MO was enhanced by the $SO_4^{\bullet-}$ radical attack on the MO molecules efficiently. However, compared with the plasma/PS system, the plasma/Fe^{2+} system presented higher removal velocity. It was because that Fe^{2+} could catalyze H_2O_2 into a large number of active $HO^{\bullet}$ radicals in a short time [29]. In addition, $HO^{\bullet}$ was a nonselective radical with the oxidation potential of E_0 = 1.8–2.7 V, which inclined to breakdown –N=N– bond of MO with low bond energy into a single structure resulting in decoloring initially. Whereas, $SO_4^{\bullet-}$ radical was a strong one-electron oxidant that would degrade aromatics selectively due to its higher oxidation potential of E_0 = 2.5–3.1 V [24,53,54]. Moreover, X_{TOC} of plasma/Fe^{2+} and plasma/PS system at 60 min were 16.1% and 20.7%, respectively. Higher X_{TOC} presented by plasma/PS carried out further efforts to mineralize MO resulting in lower removal velocity. This also could explain the optimum removal efficiency of plasma/PS/Fe^{2+} system was attributable to the simultaneously synergistic reaction of $HO^{\bullet}$ and $SO_4^{\bullet-}$.

$$Fe^{2+}+H_2O_2 \rightarrow Fe^{3+}+HO^{\bullet}+OH^- \quad (11)$$

$$Fe^{3+}+H_2O_2 \rightarrow Fe^{2+}+HO_2^{\bullet} \quad (12)$$

$$S_2O_8^{2-} \xrightarrow{plasma} 2SO_4^{\bullet-} \quad (13)$$

$$SO_4^{\bullet-}+OH^- \rightarrow HO^{\bullet}+SO_4^{2-} \quad (14)$$

$$SO_4^{\bullet-}+H_2O \rightarrow HSO_4^-+HO^{\bullet} \quad (15)$$

$$SO_4^{\bullet-}+SO_4^{\bullet-} \rightarrow S_2O_8^{2-} \quad (16)$$

$$Fe^{2+}+S_2O_8^{2-} \rightarrow Fe^{3+}+SO_4^{\bullet-}+SO_4^{2-} \quad (17)$$

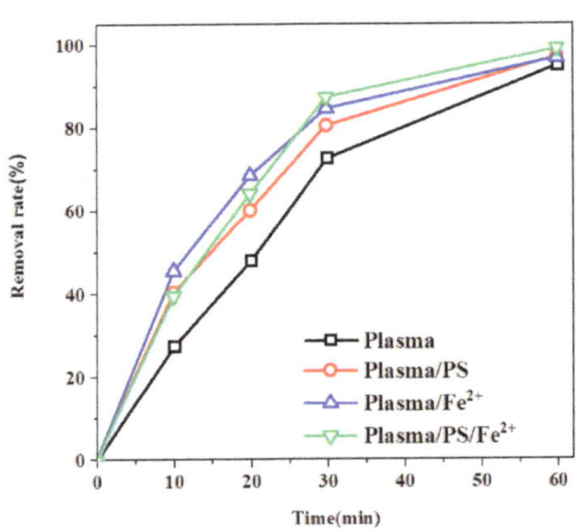

Figure 8. The removal rates of MO under plasma, plasma/PS, plasma/Fe^{2+} and plasma/PS/Fe^{2+} systems.

3.7. Removal Mechanism of Plasma for MO Removal

The MO removal mechanism was proposed by the UV-Vis spectrophotometry analysis between 200 and 900 nm in the optimal experiment conditions during the plasma treatment process. As shown in Figure 9, the initial MO solution presented two absorption peaks of 268 nm and 463 nm. The maximum absorption at 463 nm corresponded to the π-π conjugated chain of the –N=N– structure, and the other one at 268 nm corresponded to the characteristic absorption peak of the benzene ring [13].

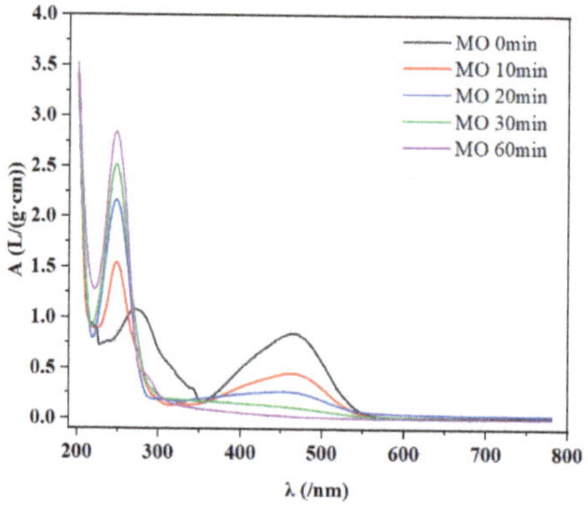

Figure 9. Absorbance spectra of MO solution over the reaction time.

With the reaction proceeding, the wavelength of MO at 463 nm was diminished gradually, which demonstrated the decrease of MO concentration. However, the wavelength of the benzene ring monomer at 268 nm was amplified progressively, possibly because the strong π-π conjugated system between –N=N–might be destroyed by $HO^\bullet$ and $O^\bullet$ in the plasma treatment process, thereafter MO was degraded into SO_3 and phenylsulfinate ions [24]. Moreover, the peak of MO at 268 nm occurred blue shift slightly, demonstrating the obvious characteristic absorption peak of benzene ring. This phenomenon might be related to the occurrence of new productions involving monomer compounds of benzenesulfonic acid and phenolsulfonic acid [55], due to the MO oxidized by large amounts of free radicals generated during the plasma treatment.

4. Conclusions

In summary, the removal of MO in the simulated wastewater by novel plasma technique could be feasible. Intensive studies on simulated aqueous solutions containing MO were implemented on the conditions of removal treatment involving the voltage, the numbers of discharge needle, and the distances of discharge gap. The optimal removal rates for MO could achieve 95.1% with novel plasma reactor, when the treatment duration was 60 min, the voltage was set at 15 kV, discharge gap as 20 mm, and discharge needle numbers were 5, respectively. An accurate predicted method of RSM was established to optimize the plasma treatment process for the removal of MO from simulated wastewater with the four important influencing parameters above involved. The influence factors for MO removal followed the order of time and voltage > discharge needle numbers > discharge gap. The removal rate of MO was predicted as 99.3% by optimization values in software, the optimum conditions were confirmed as the voltage of 14.8 kV, the plasma treatment time of 60 min, the discharge gap of 20 mm, and discharge needle numbers of 5. Compared with the conventional reactor, the novel plasma reactor showed the advantages of high removal efficiency and low energy consumption for MO removal. The investigations of plasma associated with catalysts systems revealed the optimal removal rate for MO achieved 99.2% at 60 min by plasma/PS/Fe^{2+} system due to simultaneous synergistic reaction of $HO^\bullet$ and $SO_4^{\bullet-}$. Furthermore, through analyzing the variation in the absorbances with UV-Vis spectrophotometry during plasma process for the MO removal, it was found that –N=N– bond was initially attacked and broken by active species like HO•, and the oxidized intermediates of benzenesulfonic acid and phenolsulfonic acid might be generated. As a result, the present study demonstrated the excellent and efficient removal for MO through novel plasma treatment technique, and extensive investigations about novel plasma reactor design to improve mineralization rate should be further implemented in the future.

Supplementary Materials: The following supporting information can be downloaded at: https://www.mdpi.com/article/10.3390/w14193152/s1, Figure S1: The sketch map of conventional plasma reactor.

Author Contributions: Conceptualization, Y.L. and J.B.; Data curation, Y.L., J.-W.S. and C.-L.L.; Formal analysis, J.-W.S., C.-L.L. and L.-X.S.; Funding acquisition, J.B.; Investigation, J.-W.S., X.-J.S. and X.W.; Methodology, Y.L. and J.B.; Project administration, Y.L. and J.B.; Resources, Y.L. and J.B.; Software, J.-W.S. and L.-X.S.; Supervision, Y.L., J.B., X.-J.S. and X.W.; Validation, J.-W.S. and X.-J.S.; Visualization, L.-X.S.; Writing—original draft, Y.L. and J.-W.S.; Writing—review & editing, Y.L. and J.B. All authors have read and agreed to the published version of the manuscript.

Funding: This research was funded by the National Natural Science Foundation of China (No. 21976124 and No. 21507092), the Natural Science Foundation of Liaoning Province of China (No. 2019-ZD-0217), and Liaoning Revitalization Talents Program (No. XLYC2007195).

Data Availability Statement: Not applicable.

Conflicts of Interest: The authors declare no conflict of interest.

References

1. Jamee, R.; Siddique, R. Biodegradation of Synthetic Dyes of Textile Effluent by Microorganisms: An Environmentally and Economically Sustainable Approach. *Eur. Microbiol. Immu.* **2019**, *9*, 114–118. [CrossRef] [PubMed]
2. Ajaz, M.; Shakeel, S.; Rehman, A. Microbial use for azo dye degradation-a strategy for dye bioremediation. *Int. Microbiol.* **2020**, *23*, 149–159. [CrossRef]
3. Selvaraj, V.; Karthika, T.S.; Mansiy, C.; Alagar, M. An over review on recently developed techniques, mechanisms and intermediate involved in the advanced azo dye degradation for industrial applications. *J. Mol.Struct.* **2021**, *1224*, 129195. [CrossRef]
4. Fatima, M.; Farooq, R.; Lindstrom, R.W.; Saeed, M. A review on biocatalytic decomposition of azo dyes and electrons recovery. *J. Mol. Liq.* **2017**, *246*, 275–281. [CrossRef]
5. Chung, K.T. Azo dyes and human health: A review. *J. Environ. Sci. Health C* **2016**, *4*, 233–261. [CrossRef] [PubMed]
6. Pirkarami, A.; Olya, M.E.; Limaee, N.Y. Decolorization of azo dyes by photo electro adsorption process using polyaniline coated electrode. *Prog. Org. Coat.* **2013**, *76*, 682–688. [CrossRef]
7. Yaseen, D.A.; Scholz, M. Textile dye wastewater characteristics and constituents of synthetic effluents: A critical review. *Int. J. Environ. Sci. Technol.* **2019**, *16*, 1193–1226. [CrossRef]
8. Dahiya, D.; Nigam, P.S. Waste Management by Biological Approach Employing Natural Substrates and Microbial Agents for the Remediation of Dyes' Wastewater. *Appl. Sci.* **2020**, *10*, 2958. [CrossRef]
9. Donkadokula, N.Y.; Kola, A.K.; Naz, I.; Saroj, D. A review on advanced physico-chemical and biological textile dye wastewater treatment techniques. *Rev. Environ. Sci. Biotechnol.* **2020**, *19*, 543–560. [CrossRef]
10. Li, C.; Zhang, M.H.; Song, C.W.; Tao, P.; Sun, M.H.; Shao, M.H.; Wang, T.H. Enhanced Treatment Ability of Membrane Technology by Integrating an Electric Field for Dye Wastewater Treatment: A Review. *J. Aoac. Int.* **2018**, *101*, 1341–1352. [CrossRef] [PubMed]
11. Katheresan, V.; Kanseto, J.; Lau, S.Y. Efficiency of various recent wastewater dye removal methods: A review. *J. Environ. Chem. Eng.* **2018**, *6*, 4676–4697. [CrossRef]
12. Perez, A.; Poznyak, T.; Chairez, I.; Guzman-Zavaleta, Z.J.; Alfaro-Ponce, M. Influence of Sodium Sulfate on the Direct Red 28 Degradation by Ozone in a Wastewater Recycling Process: A Stoichiometric and Novel Image Analysis. *Mater. Lett.* **2019**, *42*, 428–438. [CrossRef]
13. Liu, Y.; Li, C.L.; Bao, J.; Wang, X.; Yu, W.J.; Shao, L.X. Degradation of Azo Dyes with Different Functional Groups in Simulated Wastewater by Electrocoagulation. *Water* **2022**, *14*, 123. [CrossRef]
14. Shanker, U.; Rani, M.; Jassal, V. Degradation of hazardous organic dyes in water by nanomaterials. *Environ. Chem. Lett.* **2017**, *15*, 623–642. [CrossRef]
15. Jiang, B.; Zheng, J.T.; Qiu, S.; Wu, M.B.; Zhang, Q.H.; Yan, Z.F.; Xue, Q.Z. Review on electrical discharge plasma technology for wastewater remediation. *J. Colloid Interface Sci.* **2014**, *236*, 348–368. [CrossRef]
16. Fan, J.W.; Wu, H.X.; Liu, R.Y.; Meng, L.Y.; Sun, Y.J. Review on the treatment of organic wastewater by discharge plasma combined with oxidants and catalysts. *Environ. Sci. Pollut. R.* **2021**, *28*, 2522–2548. [CrossRef] [PubMed]

17. Shang, K.F.; Li, J.; Morent, R. Hybrid electric discharge plasma technologies for water decontamination: A short review. *Plasma. Sci. Technol.* **2019**, *21*, 043001. [CrossRef]
18. Cui, M.H.; Liu, W.Z.; Cui, D. Recent advancements in azo dye decolorization in bio-electrochemical systems (BESs): Insights into decolorization mechanism and practical application. *Water Res.* **2021**, *203*, 117512. [CrossRef] [PubMed]
19. Goncharuk, W.; Klishchenko, R.E.; Kornienko, I.V. Destruction of GT Azo Active Orange Dye in the Flow-Through Plasma-Chemical Reactor. *J. Water Chem. Technol.* **2018**, *40*, 185–189. [CrossRef]
20. Sen, S.K.; Raut, S.; Bandyopadhyay, P.; Raut, S. Fungal decolouration and degradation of azo dyes: A review. *Fungal. Biol. Rev.* **2016**, *30*, 112–133. [CrossRef]
21. Topolovec, B.; Skoro, N.; Puac, N.; Petrovic, M. Pathways of organic micropollutants degradation in atmospheric pressure plasma processing -A review. *Chemosphere* **2022**, *294*, 133606. [CrossRef]
22. Russo, M.; Lervolino, G.; Vaiano, V.; Palma, V. Non-Thermal Plasma Coupled with Catalyst for the Degradation of Water Pollutants: A Review. *Catalysts* **2020**, *10*, 1438. [CrossRef]
23. Magureanu, M.; Bradu, C.; Parvulescu, V.I. Plasma processes for the treatment of water contaminated with harmful organic compounds. *J. Phys. D Appl. Phys.* **2018**, *51*, 313002. [CrossRef]
24. Ma, S.; Lee, S.; Kim, K.; Im, J.; Jeon, H. Purification of organic pollutants in cationic thiazine and azo dye solutions using plasma-based advanced oxidation process via submerged multi-hole dielectric barrier discharge. *Sep. Purif. Technol.* **2021**, *255*, 117715. [CrossRef]
25. Sarangapani, C.; Dixit, Y.; Milosavljevic, V.; Bourke, P.; Sullivan, C.; Cullen, P.J. Optimization of atmospheric air plasma for degradation of organic dyes in wastewater. *Water Sci. Technol.* **2017**, *75*, 207–219. [CrossRef]
26. Wu, L.H.; Xie, Q.L.; Lv, Y.B.; Wu, Z.Y.; Liang, X.J.; Lu, M.Z.; Nie, Y. Degradation of Methylene Blue via Dielectric Barrier Discharge Plasma Treatment. *Water* **2019**, *11*, 1818. [CrossRef]
27. Koppenol, W.H. Names for inorganic radicals (IUPAC recommendations 2000). *Pure. Appl. Chem.* **2000**, *72*, 437–446. [CrossRef]
28. Essiptchouk, A.; Petraconi, G.; Miranda, F.; Saraiva, A.C.V.; Charakhovski, L. Glycerine degradation by submerged plasma. *J. Phys. D Appl. Phys.* **2019**, *52*, 465201. [CrossRef]
29. Shang, K.F.; Li, W.F.; Wang, X.J.; Lu, N.; Jiang, N.; Li, J. Degradation of p-nitrophenol by DBD plasma/Fe^{2+}/persulfate oxidation process. *Sep. Purif. Technol.* **2019**, *218*, 106–112. [CrossRef]
30. Chen, W.M.; Gu, Z.P.; Guo, S.P.; Li, Q.B. Microwave-assisted Fe-0-activated persulfate process for treating explosives in production wastewater. *Chem. Eng. J.* **2020**, *391*, 123497. [CrossRef]
31. Peng, J.B.; Wang, Z.X.; Wang, S.Y.; Liu, J.; Zhang, Y.Z.; Wang, B.J.; Gong, Z.M.; Wang, M.J.; Dong, H.; Shi, J.L. Enhanced removal of methylparaben mediated by cobalt/carbon nanotubes (Co/CNTs) activated peroxymonosulfate in chloride-containing water: Reaction kinetics, mechanisms and pathways. *Chem. Eng. J.* **2021**, *409*, 128176. [CrossRef]
32. Qi, C.D.; Wen, Y.N.; Zhao, Y.J.; Dai, Y.H.; Li, Y.P.; Xu, C.M.; Yang, S.G.; He, H. Enhanced degradation of organic contaminants by Fe(III)/peroxymonosulfate process with L-cysteine. *Chin. Chem. Lett.* **2022**, *33*, 2125–2128. [CrossRef]
33. Stratton, G.R.; Bellona, C.L.; Dai, F.; Holsen, T.M.; Thagard, S.M. Plasma-based water treatment: Conception and application of a new general principle for reactor design. *Chem. Eng. J.* **2015**, *273*, 543–550. [CrossRef]
34. Oke, S.R.; Ige, O.O.; Falodun, O.E.; Okoro, A.M.; Mphahlele, M.R.; Olubambi, P.A. Powder metallurgy of stainless steels and composites: A review of mechanical alloying and spark plasma sintering. *Int. J. Adv. Manuf. Tech.* **2019**, *102*, 3271–3290. [CrossRef]
35. Zhang, C.X.; Sun, Y.B.; Yu, Z.Q.; Zhang, G.Y.; Feng, J.W. Simultaneous removal of Cr(VI) and acid orange 7 from water solution by dielectric barrier discharge plasma. *Chemosphere* **2018**, *191*, 527–536. [CrossRef]
36. Zhang, C.; Qiu, J.T.; Kong, F.; Hou, X.M.; Fang, Z.; Yin, Y.; Shao, T. Plasma surface treatment of Cu by nanosecond-pulse diffuse discharges in atmospheric air. *Plasma. Sci. Technol.* **2018**, *20*, 014001. [CrossRef]
37. Sun, B.; Aye, N.N.; Gao, Z.Y.; Lv, D.; Zhu, X.M.; Sato, M. Characteristics of gas-liquid pulsed discharge plasma reactor and dye decoloration efficiency. *J. Environ. Sci.* **2012**, *24*, 840–845. [CrossRef]
38. Kakhki, R.M.; Rahni, S.Y.; Karimian, A. Removal of Methyl Orange from aqueous solutions by a novel, high efficient and low cost copper-modified nanoalum. *Inorg. Nano-Met. Chem.* **2021**, *51*, 1291–1296.
39. Ko, S.J.; Yamaguchi, T.; Salles, F.; Oh, J.M. Systematic utilization of layered double hydroxide nanosheets for effective removal of Methyl Orange from an aqueous system by pi-pi stacking-induced nanoconfinement. *J. Environ. Manag.* **2021**, *277*, 111455. [CrossRef]
40. Bagheri, F.; Chaibakhsh, N. Efficient visible-light photocatalytic ozonation for dye degradation using Fe_2O_3/MoS_2 nanocomposite. *Sep. Sci. Technol.* **2021**, *56*, 3022–3032. [CrossRef]
41. Meiyazhagan, S.; Yugeswaran, S.; Ananthapadmanabhan, P.V.; Suresh, K. Process and kinetics of dye degradation using microplasma and its feasibility in textile effluent detoxification. *J. Water Process Eng.* **2020**, *37*, 101519. [CrossRef]
42. Hafeez, A.; Javed, F.; Fazal, T.; Shezaed, N.; Amjad, U.E.S.; Rehman, M.S.U. Intensification of ozone generation and degradation of azo dye in non-thermal hybrid corona-DBD plasma micro-reactor. *Chem. Eng. Processing* **2021**, *159*, 108205. [CrossRef]
43. Takahashi, K.; Takeda, M.; Konno, R.; Takaki, K.; Satta, N. Influence of Electric Parameters on Hydroxyl Radical Production by Positive Pulsed Discharge Inside of a Bubble in Water. *IEEE Trans. Plasma Sci.* **2019**, *47*, 1105–1113. [CrossRef]
44. Guo, H.; Wang, Y.W.; Liao, L.N.; Li, Z.; Pan, S.J.; Puyang, C.D.; Su, Y.Y.; Zhang, Y.; Wang, T.C.; Ren, J.Y. Review on remediation of organic-contaminated soil by discharge plasma: Plasma types, impact factors, plasma-assisted catalysis, and indexes for remediation. *Chem. Eng. J.* **2022**, *436*, 135239. [CrossRef]

45. Hu, X.Y.; Wang, B.W. Removal of pefloxacin from wastewater by dielectric barrier discharge plasma: Mechanism and degradation pathways. *J. Environ. Chem. Eng.* **2021**, *9*, 105720. [CrossRef]
46. Lee, H.D.; Kim, J.O.; Chung, J.W. Degradation of Methyl Orange by pulsed corona discharge process in water. *Desalin. Water Treat.* **2015**, *53*, 2767–2773. [CrossRef]
47. Xiang, H.J.; Lei, B.; Yuan, X.C.; Lv, Q.A.; Zhang, Q. Design and Simulation of New Type Reactor in the Wastewater Treatment System Based on Discharge Plasma. *IEEE Trans. Plasma Sci.* **2019**, *47*, 952–957. [CrossRef]
48. Jadaa, W.; Prakash, A.; Ray, A.K. Modeling of Degradation of Diazo Dye in Swirl-Flow Photocatalytic Reactor: Response Surface Approach. *Catalysts* **2020**, *10*, 1418. [CrossRef]
49. Wang, J.K.; Yao, Z.P.; Wang, Y.J.; Xia, Q.X.; Chu, H.Y.; Jiang, Z.H. Preparation of immobilized coating Fenton-like catalyst for high efficient degradation of phenol. *Environ. Pollut.* **2017**, *224*, 552–558. [CrossRef]
50. Liu, Y.; Qu, G.Z.; Sun, Q.H.; Jia, H.Z.; Wang, T.C.; Zhu, L.Y. Endogenously activated persulfate by non-thermal plasma for Cu(II)-EDTA decomplexation: Synergistic effect and mechanisms. *Chem. Eng. J.* **2021**, *406*, 126774. [CrossRef]
51. Yang, D.Z.; Zhou, X.F.; Liang, J.P.; Xu, Q.N.; Wang, H.L.; Yang, K.; Wang, B.; Wang, W.C. Degradation of methylene blue in liquid using high-voltage pulsed discharge plasma synergizing iron-based catalyst-activated persulfate. *J. Phys. D Appl. Phys.* **2021**, *54*, 244002. [CrossRef]
52. Li, H.; Song, R.Y.; Wang, Y.Y.; Zhong, R.W.; Zhang, Y.; Zhou, J.; Wang, T.C.; Zhu, L.Y. Simultaneous removal of antibiotic-resistant bacteria and its resistance genes in water by plasma oxidation: Highlights the effects of inorganic ions. *Sep. Purif. Technol.* **2022**, *218*, 119672. [CrossRef]
53. Yuan, R.X.; Ramjaun, S.N.; Wang, Z.H.; Liu, J.S. Effects of chloride ion on degradation of Acid Orange 7 by sulfate radical-based advanced oxidation process: Implications for formation of chlorinated aromatic compounds. *J. Hazard. Mater.* **2011**, *196*, 173–179. [CrossRef] [PubMed]
54. Das, S.; Kamat, P.V.; Padmaja, S.; Au, V.; Madison, S.A. Free radical induced oxidation of the azo dye Acid Yellow 9. *J. Chem. Soc. Perkin Trans. 2* **1999**, *6*, 1219–1223. [CrossRef]
55. Wang, B.W.; Xu, M.; Chi, C.M.; Chao, W.; Dajun, M.J. Degradation of Methyl Orange using dielectric barrier discharge water falling film reactor. *J. Adv. Oxid. Technol.* **2017**, *20*, 20170021. [CrossRef]

water MDPI

Article

Fenton Process for Treating Acrylic Manufacturing Wastewater: Parameter Optimization, Performance Evaluation, Degradation Mechanism

Zhiwei Lin, Chunhui Zhang *, Peidong Su, Wenjing Lu, Zhao Zhang, Xinling Wang and Wanyue Hu

College of Chemistry and Environmental Engineering, China University of Mining and Technology (Beijing), Beijing 100083, China
* Correspondence: ZCHcumtb@hotmail.com; Tel.: +86-139-1017-6209

Abstract: Acrylic manufacturing wastewater is characterized by high toxicity, poor biodegradability, high chemical oxygen demand (COD) and ammonia nitrogen. Herein, we exploited traditional Fenton technology to treat acrylic fiber manufacturing wastewater. The impacts of key operating variables including the initial concentration of H_2O_2 (C_{H2O2}), the initial concentration of Fe^{2+} (D_{Fe2+}), and solution pH (pH) on the COD removal rate (R_{COD}) were explored and the treatment process was optimized by Response Surface Methodology (RSM). The results indicated that the optimum parameters are determined as pH 3.0, 7.44 mmol/L of Fe^{2+} and 60.90 mmol/L of H_2O_2 during Fenton process. For the actual acrylic manufacturing wastewater treatment shows that the removal rates for COD, TOC, NH_4^+-N and TN are 61.45%~66.51%, 67.82%~70.99%, 55.67%~60.97% and 56.45%~61.03%, respectively. It can meet the textile dyeing and finishing industry water pollutant discharge standard (GB4287-2012). During the Fenton reaction, the effective degradation and removal of organic matter is mainly achieved by HO• oxidation, supplemented by flocculation and sedimentation of Fe^{3+} complexes. This study will provide useful implications in the process parameters for the practical application of Fenton method in acrylic acid production wastewater.

Keywords: acrylic fiber manufacturing wastewater; Fenton reaction; advanced oxidation method; degradation mechanism

Citation: Lin, Z.; Zhang, C.; Su, P.; Lu, W.; Zhang, Z.; Wang, X.; Hu, W. Fenton Process for Treating Acrylic Manufacturing Wastewater: Parameter Optimization, Performance Evaluation, Degradation Mechanism. *Water* **2022**, *14*, 2913. https://doi.org/10.3390/w14182913

Academic Editor: Christos S. Akratos

Received: 19 August 2022
Accepted: 14 September 2022
Published: 17 September 2022

1. Introduction

Acrylic fibers are one kinds of the significant manufacturing raw materials for the textile industry [1]. During the manufacture of acrylic fibers, a large volume of wastewater is inevitably generated. Owing to its high toxicity, poor biodegradability, high COD, and ammonia nitrogen, acrylic fiber manufacturing wastewater has been recognized as one of the problematic organic wastewaters [2]. In general, conventional biological methods have been used in the degradation of organic matters in acrylic fiber manufacturing wastewater. However, the effluent water quality treated by conventional biological processes such as A/O and A^2/O alone usually cannot meet the national discharge standards [3]. Therefore, it is imperative to develop stable, effective, economical combined processes to treat acrylic manufacturing wastewater.

At present, the main approaches for the treatment of acrylic fiber manufacturing wastewater include but not limit to bio-enhanced treatment [4,5], membrane technology [6,7], advanced oxidation processes (AOP) [8,9], and coupled treatment processes [10]. C. Gong et al. [11] used an electro-coagulation (EC) process to pretreat acrylic fiber manufacturing wastewater. Under optimal conditions, the removal rate of total organic carbon was 44%, and the (Biochemical Oxygen Demand) BOD_5/COD ratio was increased to 0.35. J. Wei et al. [12] combined Fenton-sequencing batch membrane bioreactor process for the treatment of acrylic fiber manufacturing wastewater. After Fenton oxidation treatment, the COD removal rate was 47.0%, and the ratio of BOD_5/COD increased from 0.35 to

0.69. Microbubble-ozonation was used to treat refractory wet-spun acrylic fiber manufacturing wastewater with 42%, 21%, and 42% removal rates of COD, NH_3-N, and UV_{254}, respectively [13]. The degradation of refractory organics such as alkane aromatic compounds in wastewater was enhanced by microbubble ozone treatment, which resulted in improved biodegradability of the wastewater. J. Wei et al. [14] utilized the Fered-Fenton process to treat acrylic fiber manufacturing wastewater, resulting in an increase in the BOD_5/COD ratio from 0.29 to over 0.68 after 180 min of treatment in which Ti was used as the cathode and RuO_2/Ti as the anode. T. Zheng et al. [15] fabricated a three-dimensional electrochemical oxidation reactor to treat wet-spun acrylic fiber manufacturing wastewater, which significantly elevated the treatment efficiency of COD, NH_3-N, TOC, and UV_{254} by 44.5%, 38.8%, 27.2%, and 10.9%, respectively. X. Xu and Z. Shao et al. [16] developed highly efficient peroxy monosulfate activated catalysts (e.g., $LaSrCo_{0.8}Fe_{0.2}O_4$) that showed excellent performance in catalytic advanced oxidation applications of difficult to degrade organic pollutants. In terms of this, the enhanced pretreatment of acrylic fiber wastewater using advanced oxidation processes can achieve satisfactory treatment results.

Fenton advanced oxidation process, which is based on the generation of hydroxyl radicals (HO•), has the advantages of high oxidation and strong electron affinity [17]. The Fenton process can decompose refractory organic matter, and its by-product, Fe^{3+}, have a favorable flocculation effect, which can simultaneously complete the degradation-coagulation-precipitation removal of organic matter. Therefore, the Fenton method is considered an effective method for treating refractory organic wastewater [18]. However, there are many intermediate products and side reactions in the Fenton reaction system. Once the reaction conditions are not adequately controlled, the utilization rate of H_2O_2 will decrease, resulting in insufficient mineralization of organic matter and a severe waste of raw materials [19]. Therefore, during the application of the Fenton method, the use of reagents, the conversion trend of H_2O_2, and the influence of operating parameters on the reaction system should be investigated to achieve the efficient utilization of the Fenton process and the economic degradation of pollutants. As mentioned above, this study adopts the response surface methodology to explore the effect of factors such as the concentration of H_2O_2, the amount of Fe^{2+} added, and the initial pH value on the degradation of organic matter. In addition, we discussed the interaction between various factors, the removal effect of refractory organic matter, and the mechanism of the reaction process. This research provides technical reference and a theoretical basis for applying the Fenton catalytic oxidation method in the treatment of acrylic fiber manufacturing wastewater.

2. Materials and Methods

2.1. Wastewater

The experimental acrylic fiber wastewater was obtained from the effluent of the secondary sedimentation tank after A/O process treatment in a sewage treatment plant in Jilin Province. The experimental wastewater was pretreated with concentrated sulfuric acid within 12 h after sampling. The main purpose of pretreatment of acrylic acid wastewater with concentrated sulfuric acid was to lower the pH of the solution to about 2. The storage time of acrylic wastewater was prolonged by lowering the pH of the water sample to slow down the microbial activity and inhibit the hydrolysis of ammonia-containing compounds. It is worth noting that the water samples must be used within 48 h. The physical and chemical indicators of wastewater are shown in Table 1.

Table 1. Water quality indexes of water samples.

Parameter	COD $mg \cdot L^{-1}$	BOD_5 $mg \cdot L^{-1}$	TOC $mg \cdot L^{-1}$	NH_3-N $mg \cdot L^{-1}$	TN $mg \cdot L^{-1}$	BOD_5/COD	pH
Amount	249–270	8–10	90–100	60–66	78–95	0.030–0.040	5.4–5.8

2.2. Chemicals

All chemicals are of analytical grade. NaOH and H_2SO_4 (98%) were purchased from Beijing Chemical Reagent Co., Beijing, China. $FeSO_4 \cdot 7H_2O$ and H_2O_2 (30%) were purchased from Beijing Lanyi Chemical Co., Beijing, China.

2.3. Experimental Method

Since the degradation efficiency of COD by Fenton process was usually affected by the initial pH of the solution, the $H_2O_2/FeSO_4$ ratio, the concentration of H_2O_2, $FeSO_4$, reaction time, and temperature [20,21]. The initial concentrations of Fe^{2+} and H_2O_2 as well as the pH determined the amount of HO• production in the Fenton reaction, which was the active substance for the direct oxidation of COD [2,22–24]. The initial concentration of H_2O_2 (code A), the initial concentration of Fe^{2+} (code B), and the pH value (code C) was chosen as the primary factors. Firstly, single-factor experiments were conducted to explore the effect of COD degradation under different conditions. After that, response surface methodology was adopted to optimize conditions and analyze the interaction between factors to optimize the Fenton process. Taking the removal rate of COD as the response value, a three-factor and three-level Box-Behnken response surface design was carried out. The experimental factors, level codes, design schemes, and response values are shown in Table 2.

Table 2. Coded levels and corresponding values for test factors in RSM experimental design.

Run	Factor						Response, R (%)
	A: [H_2O_2]		B: [Fe^{2+}]		C: pH		
	Coded Level	Corresponding Value ($mmol \cdot L^{-1}$)	Coded Level	Corresponding Value ($mmol \cdot L^{-1}$)	Coded Level	-	
S1	−1	56.25	−1	4.16	0	3	51.92 ± 1.25
S2	1	93.75	−1	4.16	0	3	43.73 ± 1.02
S3	−1	56.25	1	9.38	0	3	54.91 ± 1.28
S4	1	93.75	1	9.38	0	3	45.57 ± 1.06
S5	−1	56.25	0	6.25	−1	2	44.72 ± 0.98
S6	1	93.75	0	6.25	−1	2	39.66 ± 0.89
S7	−1	56.25	0	6.25	1	4	50.63 ± 1.05
S8	1	93.75	0	6.25	1	4	42.17 ± 1.07
S9	0	75.00	−1	4.16	−1	2	34.95 ± 0.67
S10	0	75.00	1	9.38	−1	2	49.71 ± 1.09
S11	0	75.00	−1	4.16	1	4	38.16 ± 0.88
S12	0	75.00	1	9.38	1	4	41.42 ± 0.87
S13	0	75.00	0	6.25	0	3	57.04 ± 1.35
S14	0	75.00	0	6.25	0	3	57.88 ± 1.39
S15	0	75.00	0	6.25	0	3	56.89 ± 1.38
S16	0	75.00	0	6.25	0	3	57.28 ± 1.42
S17	0	75.00	0	6.25	0	3	57.54 ± 1.36

The experimental setup and procedure are shown in Figure 1. Firstly, 300.0 mL of acrylic fiber wastewater was added to the 500.0 mL beaker. The pH adjustment was achieved by titrating 0.1 or 1 mol/L of H_2SO_4 (or NaOH) solution under the condition of stirring intensity around 140 rpm. In which the pH of the solution was monitored by (LEICI PHB-4) portable pH meter to achieve precise regulation of the solution pH. Secondly, $FeSO_4 \cdot 7H_2O$ was added to the solution and stirred at 140.0 rpm. In addition, H_2O_2 (30% in mass fraction) was added to the solution to initiate the Fenton oxidation reaction. Finally, samples were taken at intervals to test the concentrations of Fe^{2+}, COD, and H_2O_2.

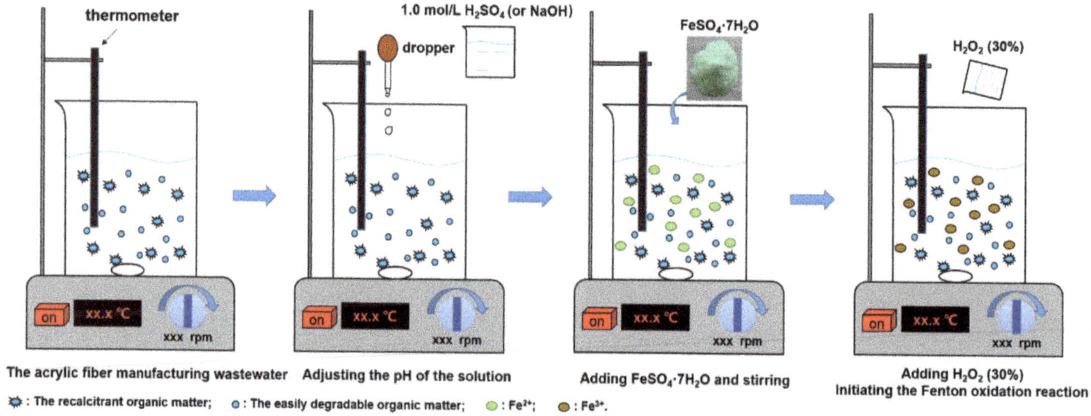

Figure 1. Fenton oxidation experiment.

2.4. Analytical Methods

The method of COD detection refers to the standard method "Water quality-Determination of the chemical oxygen demand-Dichromate method, HJ 828-2017" published by the Ministry of Ecology and Environment of the People's Republic of China [25]. The detection method of ammonia nitrogen refers to the standard method "Water quali-ty-Determination of ammonia nitrogen-Nessler's reagent spectrophotometry, HJ535-2009" published by the Ministry of Ecology and Environment of the People's Republic of China [26]. The determination method of iron content refers to the standard method "Water quality-Determination of Iron-phenanthroline spectrophotometry, HJ/T 345-2007" published by the Ministry of Ecology and Environment of the People's Republic of China [27]. The concentration of H_2O_2 was determined spectrophotometrically with potassium titanium oxalate. 5 mL of water sample was added to a 25 mL cuvette. 5 mL of potassium titanium oxalate solution (0.02 mol/L) was added. Deionized water was used to dilute to the scale. After standing for 8 min, the absorbance was determined at 385 nm. Deionized water was used as a reference during the determination [28,29].

GC-MS (Gas chromatograph-mass spectrometer) was used to detect the type (or concentration) of organics. The supernatant before and after the Fenton reaction was first filtered through a 0.45 mm membrane. Then the aqueous samples were then subjected to solid phase extraction, elution, dehydration and concentration procedures to complete the aqueous sample pretreatment. Samples were analyzed by GC-MS (GC (7890)-MS (5975), Agilent Technologies Inc.) equipped with a separation column (DB-5MS 30 m × 0.25 mm × 0.25 μm) with column template at 40 °C for 2 min. The inlet temperature was 290 °C for 4 min. Carrier gas conditions: high-purity nitrogen with a flow rate of 1.0 mL/min. MS conditions: ion source temperature 280 °C, interface temperature 280 °C, solvent delay time 5.0 min.

3. Results and Discussion

3.1. Examination of Main Factors That Affect the Removal of COD

In the Fenton reaction system, the initial concentration of Fe^{2+} and H_2O_2 as well as the pH value play a crucial role in the degradation of organic pollutants. [30]. Since they determine the amount of HO• production in the Fenton reaction, which is the active species for the direct oxidation of COD. Typically, the initial concentration of Fe^{2+} and pH was set to be 2~10 mmol/L and 3.0, respectively [31].

3.1.1. Initial Concentration of H_2O_2

Under this condition: (i) the initial concentration of Fe^{2+} was 6.25 mmol/L; (ii) the initial pH was 3.0; The effect of the initial concentration of H_2O_2 on the COD removal rate of acrylic fiber wastewater is shown in Figure 2a; The changes of Fe^{2+} and H_2O_2 concentration in the wastewater after the reaction are shown in Figure 2b,c.

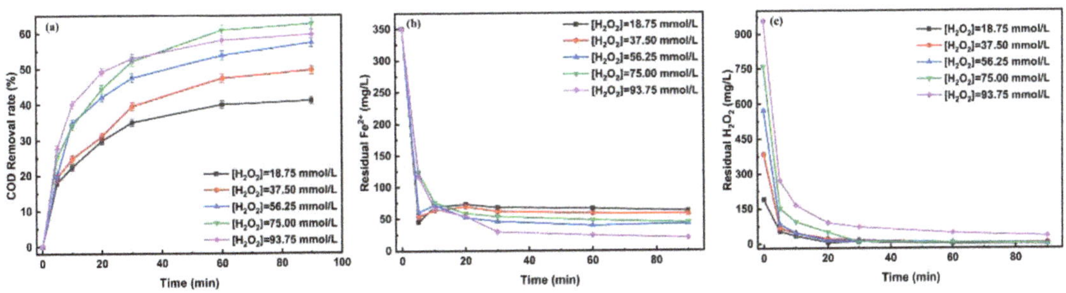

Figure 2. Influence of initial concentration of H_2O_2 on COD removal rate (**a**), the residual concentration of Fe^{2+} (**b**), and the residual concentration of H_2O_2 (**c**).

As shown in Figure 2a, the COD removal efficiency increased from 41.03% to 62.39% as the initial concentration of H_2O_2 increased from 18.75 to 75.00 mmol/L. However, when the initial concentration of H_2O_2 increased to 93.75 mmol/L, the COD removal rate decreased slightly. The reason is that the increased H_2O_2 concentration can accelerate the reaction between H_2O_2 and Fe^{2+}, leading to the generation of more HO• in the solution, thereby improving the COD removal rate. However, excess H_2O_2 will be quenched by HO•, causing the consumption of the HO• and the production of hydrogen peroxide radicals (HOO•) Equation (1) [32]. Simultaneously, the excess H_2O_2 will be self-decomposed to produce H_2O and O_2, reducing the utilization rate Equation (2).

$$Fe^{2+} + H_2O_2 \rightarrow Fe^{3+} + OH^- + HO\bullet \quad (1)$$

$$H_2O_2 + HO\bullet \rightarrow HOO\bullet + H_2O \quad (2)$$

The higher the concentration of H_2O_2 is, the lower the residual concentration of Fe^{2+} will be (Figure 2b). This trend was attributed to the increased H_2O_2 concentration promoting the conversion of Fe^{2+} to Fe^{3+}. Notably, it can be observed that when the H_2O_2 concentration was added to 93.75 mmol/L, the residual amount of H_2O_2 in the solution at 30 min was 71.13 mg/L (or 2.09 mmol/L) (Figure 2c). This can be ascribed to the that the low Fe^{2+} concentration is hard to trigger the H_2O_2 reaction [33]. Therefore, it can be concluded that the initial concentration of H_2O_2 was not as high as possible, and the best initial concentration is 75.00 mmol/L.

3.1.2. The Initial Concentration of Fe^{2+}

As shown in Figure 3a, the effect of Fe^{2+} concentration on the removal rate of COD at pH 3.0 and H_2O_2 concentration of 75 mmol/L. Figure 3b,c shows the changes of Fe^{2+} and H_2O_2 concentrations in the aqueous solution after the reaction.

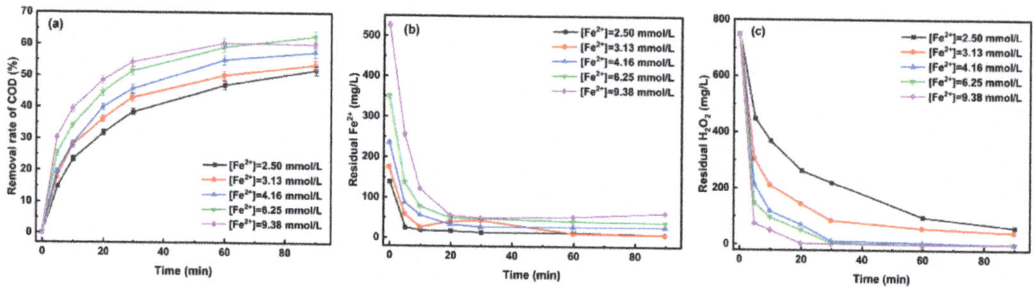

Figure 3. Influence of initial concentration of Fe^{2+} on COD removal rate (**a**), the residual concentration of Fe^{2+} (**b**), and the residual concentration of H_2O_2 (**c**).

As shown in Figure 3a, with the increase of initial concentration of Fe^{2+}, the COD removal rate was increased gradually. The COD removal rate increased significantly from 51% to 65% when the initial concentration of Fe^{2+} increased from 2.50 mmol/L to 6.25 mmol/L. It can be observed that when the initial concentration of Fe^{2+} increases to 9.38 mmol/L, the COD removal rate decreases slightly. The trend of changes in the concentration of Fe^{2+} and H_2O_2 after the Fenton reaction was shown in Figure 3b,c. Within the reaction time of 0~5 min (Figure 3b), it was clearly observed that the Fe^{2+} concentration decreased rapidly, and then the curve gradually stabilizes. By comparing the slope of the reaction curve, it can be proved that the increased Fe^{2+} concentration can facilitate the reaction rate. Furthermore, the consumption rate of H_2O_2 increased with the increase of $^+$ the initial concentration of Fe^2 (Figure 3c). Typically, at a low initial concentration of Fe^{2+}, the rate of the reaction Equation (1) was slow and the amount of HO• produced was less than sufficient to oxidize the organic matter in water [17]. Meanwhile, the excess H_2O_2 reacted with the generated HO• Equation (2), which consumes H_2O_2 in the water that had not been catalytically decomposed, resulting in a lower COD removal rate [34]. However, when the initial concentration of Fe^{2+} was too high, too much HO• was produced at the beginning of the reaction. At the same time, a side reaction Equation (3) occurred, consuming HO• that had not yet participated in the oxidation reaction [35]. In addition, a large amount of HO• would react with each other to form H_2O and O_2 Equation (4) [35], which leaded to a decrease in COD removal rate. At a dose of 6.25 mmol/L of Fe^{2+}, the H_2O_2 reacted completely within 30 min and the best COD removal was achieved. Therefore, the optimal initial concentration of Fe^{2+} was 6.25 mmol/L.

$$Fe^{2+} + HO\bullet \rightarrow Fe^{3+} + OH^- \quad (3)$$

$$4HO\bullet \rightarrow 2H_2O + O_2 \quad (4)$$

3.1.3. Initial pH

The effect of initial pH on the COD removal rate of acrylic fiber wastewater is shown in Figure 4a. The change curve of the concentration of Fe^{2+} and H_2O_2 after the oxidation-reduction reaction is shown in Figure 4b,c. the initial concentration of Fe^{2+} was 6.25 mmol/L, and the initial concentration of H_2O_2 was 75 mmol/L.

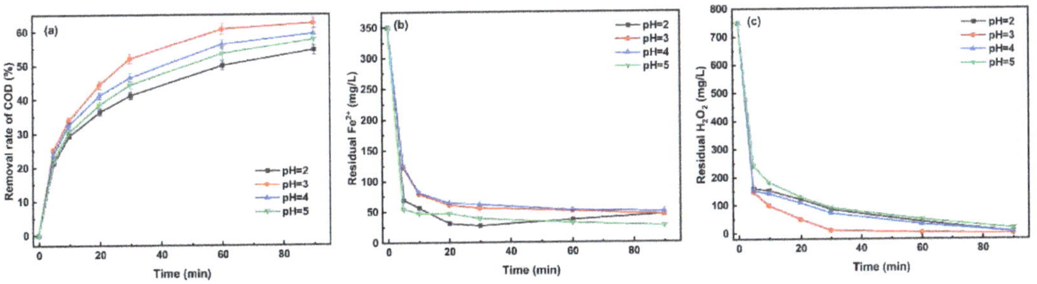

Figure 4. Influence of initial pH on COD removal rate (**a**), the residual concentration of Fe^{2+} (**b**), and the residual concentration of H_2O_2 (**c**).

The results presented in Figure 4a showed that it can be obtained the highest COD removal rate of 65.79% at a pH of 3.0, then the removal rate decreased slightly as the pH increased. According to the reaction mechanism of the Fenton reagent, the initial pH could directly interfere the complex equilibrium system of Fe^{2+} and Fe^{3+} in the solution, thereby affecting the oxidation ability of the Fenton reagent. As exhibited in Figure 4b,c, when the pH value was 2, the concentration of Fe^{2+} in the solution was relatively low while the concentration of H_2O_2 was relatively high within 30 min before the reaction. Due to the higher concentration of $[H^+]$, the reaction Equation (5) was in a reverse reaction state. Fe^{3+} reduction to Fe^{2+} was inhibited in this state, and the catalytic reaction could not proceed smoothly [36]. As a result, the generation rate of HO• was slowed down, and the oxidation capacity was reduced. When the pH value was > 3, it was obvious that the removal rate of COD gradually decreased, mainly due to the increase of OH^- concentration, which leads to the decrease of reaction rate Equation (1), and thus the amount of HO• decreases. As shown in Figure 4b,c, most of the Fe^{2+} and H_2O_2 in the system reacted quickly and were consumed during the first 10 min. When the reaction time was between 10 and 30 min, the Fe^{2+} concentration gradually decreased and tended to be stable. When the pH was 2, 4, and 5, the concentration of H_2O_2 did not stabilize until 90 min. Notably, the highest utilization and reaction rate of H_2O_2 was achieved at pH 3, and the reaction equilibrium could be reached within 30 min. Therefore, the optimal pH value of this study was 3.0, which was consistent with the results of other research.

$$Fe^{3+} + H_2O_2 \rightarrow Fe^{2+} + H^+ + HOO\bullet \quad (5)$$

3.2. Response Surface Analysis

3.2.1. Regression Model and Analysis of Variance

The results of the RSM model were presented in Table 3, giving the analysis of variance (ANOVA). It should be noted that the F value (Fisher variation ratio) and p value (probability value) in the analysis of the variance table were the leading indicators, showing the significance and adequacy of the model used. p value less than 0.05 means that the model was significant, while a value greater than 0.10 was usually regarded as a less critical factor [37]. The ANOVA of this model showed that the F value of lack of fit was 99.01, and the very low p value of lack of fit was 0.0003 (<0.05), indicating that these parameters in the model were highly significant. The F value of the regression model was 14.49, and $p < 0.05$, which also proved that the regression model was highly significant, and the experiment was reliable. As can be seen from Table 3, the R-Squared (R^2) and Adjusted R-Squared (Raj^2) of the model, respectively, were 0.95 and 0.89, indicating that the regression equation was highly reliable, which can explain 89% of the response value changes. Since p value was less than 0.05, A, B, B^2, C^2 can be regarded as vital terms. In addition, it was clear that C, A × B, and A × C were considered irrelevant terms. Therefore, the author obtained the equation relationship between the response and the variable, which was represented by the second-order polynomial equation fitting based on the coding factors (A, B, C):

$$R = 57.33 - 3.88\,A + 2.86\,B + 0.42\,C - 0.29\,A \times B - 0.85\,A \times C - 2.87\,B \times C - 2.53\,A^2 - 5.76\,B^2 - 10.50\,C^2 \quad (6)$$

where R is the removal of COD, A is the coded value of H_2O_2 concentration, B is the coded value of Fe^{2+} concentration, C is the coded value for pH.

Table 3. Covariance analysis of linear, quadratic and interaction variables of the regression model.

Source	Sum of Squares (SS)	Degree of Freedom (df)	Mean Square (MS)	F-Value (F)	*p*-Value (*p*)
Model	906.22	9	100.69	14.94	0.0009
A-[H_2O_2]	120.51	1	120.51	17.89	0.0039
B-[Fe^{2+}]	65.27	1	65.27	9.69	0.0170
C-pH	1.39	1	1.39	0.21	0.6629
A × B	0.33	1	0.33	0.049	0.8310
A × C	2.89	1	2.89	0.43	0.5334
B × C	33.06	1	33.06	4.91	0.0623
A^2	26.94	1	26.94	4.00	0.0857
B^2	139.90	1	139.90	20.76	0.0026
C^2	464.37	1	464.37	68.92	<0.0001
Residual	47.16	7	6.74		
Lack of Fit	46.54	3	15.51	99.01	0.0003
Pure Error	0.63	4	0.16		
Cor Total	953.38	16			
R^2	0.95				
Raj^2	0.89				

The above regression equation can be used to predict the R-value within the range of the factors in this study. The value and sign of the regression coefficient indicated the influence of each item on the response. It can be seen from Equation (6) that the item with the highest regression coefficient in the interaction term was B*C, which indicated that the interaction term occupies a dominant position in the overall response. The positive sign of the coefficient indicated a synergistic effect, and the negative sign indicated an antagonistic effect.

3.2.2. Response Surface Optimization

According to Table 2, the *p* value of A × B, A × C, and B × C were 0.831, 0.533, and 0.0623, respectively. To further determine the interaction between variables, the model Equation (1) was used for fitting, and the result was shown in Figure 5. It was clear that the removal rate of COD increased from 29.20% to 59.30% with the changes of A and B, when the pH was 3 (code value was 0). A view of Figure 5a,d, at the position of the center point, the range of A was near the −0.61 level and the range of B was near the 0.25 level. From Figure 5b,e, at the position of the center point, the range of A was near the −0.65 level and the range of pH was near the 0.10 level. As described in Figure 5c,f, at the position of the center point, the range of B was near the 0.25 level and the range of C was near the 0.1 level. The larger the difference between the coded value of the Center Point and the 0.00 level, the greater the interaction between the two factors [38]. As explained previously, the order of the influence of each variable on COD degradation was: A-[H_2O_2] > B-[Fe^{2+}] > C-pH. In addition, it was apparent that the interaction between B-[Fe^{2+}] and C-pH had a more substantial impact.

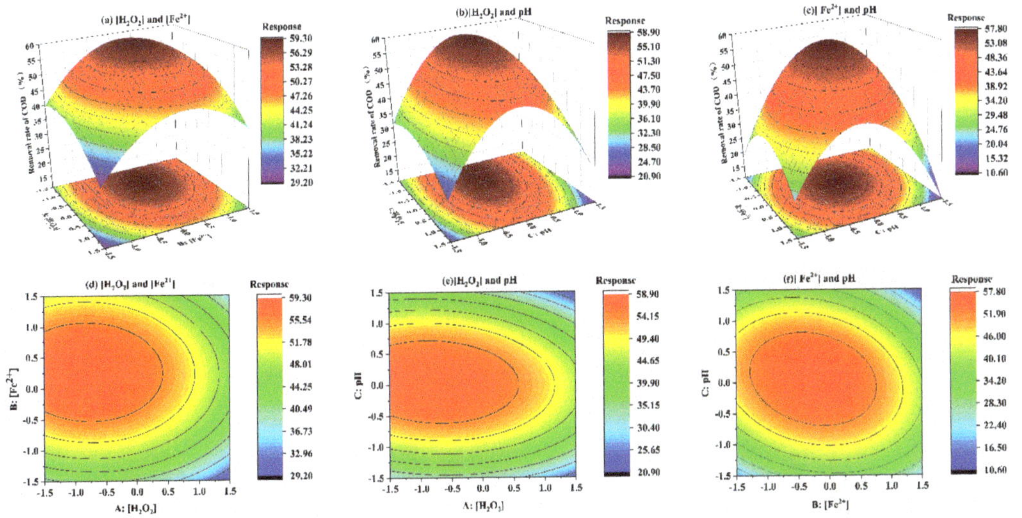

Figure 5. A three-dimensional response surface for the response of COD removal rate in terms of coded A and B (**a**), coded A and c (**b**), coded B and C (**c**); The corresponding contour plots for the response of COD removal rate in terms of coded A and B (**d**), coded A and c (**e**), coded B and C (**f**).

3.2.3. Verification of Optimal COD Degradation Conditions

As shown in Figure 6, after linear fitting between the predicted and actual values, the R^2 was 0.95, which showed that the predicted result had a high degree of credibility [39]. With the use of the Fenton process to degrade COD in acrylic fiber wastewater, and through response surface optimization analysis, the optimal degradation conditions were: (i)the initial concentration of H_2O_2 was 60.30 mmol/L; (ii) the initial concentration of Fe^{2+} was 7.46 mmol/L; (iii) the pH was 3; (iv) COD removal rate was 59.22%. The experiment was repeated three times under the above-optimized conditions, and the results were presented in Table 4. Obviously, it can be observed that after treatment, the acrylic fiber wastewater effluent can meet the secondary discharge requirements of the "Integrated Wastewater Discharge Standard" (GB4287-2012) [40].). The COD removal rate's actual and predicted values were 63.2% and 59.8%, and the error was less than 5%. Therefore, the confidence of the obtained model was strong.

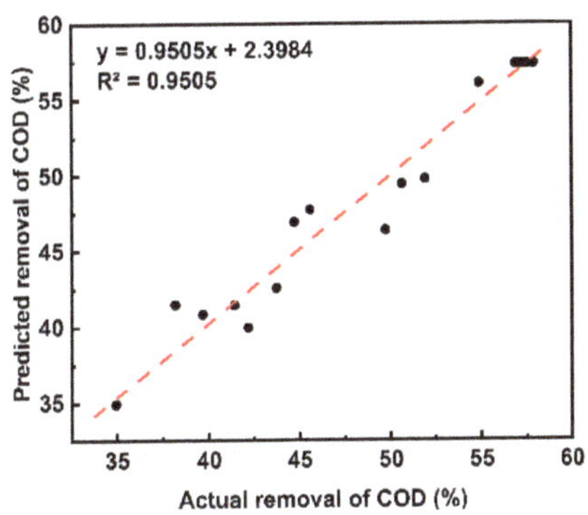

Figure 6. The removal of COD: Comparison of actual value and predicted value.

Table 4. Removing effect of acrylic fiber wastewater by Fenton process.

Water Quality Index	Influent	Effluent	Emission Limit
pH	5.5 ± 0.2	6.4 ± 0.2	6~9
COD/(mg·L^{-1})	259.6 ± 8.5	93.3 ± 3.5	200
TOC/(mg·L^{-1})	92.3 ± 2.5	28.2 ± 0.7	30
NH_4^+-N/(mg·L^{-1})	63.5 ± 2.6	26.4 ± 0.6	25
TN/(mg·L^{-1})	83.8 ± 3.2	34.5 ± 0.6	50

3.3. Removal Effect of Refractory Organic Matter

In order to evaluate the degradation effect of toxic and difficult-to-degrade organic pollutants in acrylic fiber wastewater, gas chromatography-mass spectrometry (GC-MS) was used to detect toxic and hazardous substances in the wastewater. Based on the optimal process conditions (The initial concentration of H_2O_2 was 60.30 mmol/L, the initial concentration of Fe^{2+} was 7.46 mmol/L, the pH was 3), the actual acrylic manufacturing wastewater was treated by the Fenton method. The results were presented in Figure 7. It was evident that the chromatographic peaks of most organics in the water after Fenton treatment were weakened. Moreover, the raw water contained 16 primary organic pollutants, mainly including aromatic hydrocarbons and long-chain alkanes. It can be seen that the concentration of organic pollutants had been significantly reduced after the Fenton process, demonstrating that the Fenton process can convert refractory organic matter into small molecular organics or inorganics [41]. In addition, as shown in Table 5, Undecane, Diisobutyl phthalate and Dibutyl phthalate still existed in the effluent. We speculated that Undecane might be the final product of the degradation of long-chain alkanes. In the Fenton oxidation process, HO• preferentially oxidized small molecular organic compounds, followed by other large molecular organic compounds [42]. Since Diisobutyl phthalate and Dibutyl phthalate were macromolecular organic compounds, and the concentration in the influent was relatively high. As a result, Diisobutyl phthalate and Dibutyl phthalate in the effluent was not wholly removed. Notably, the Fenton process can remove most of the refractory organic matter in acrylic fiber wastewater.

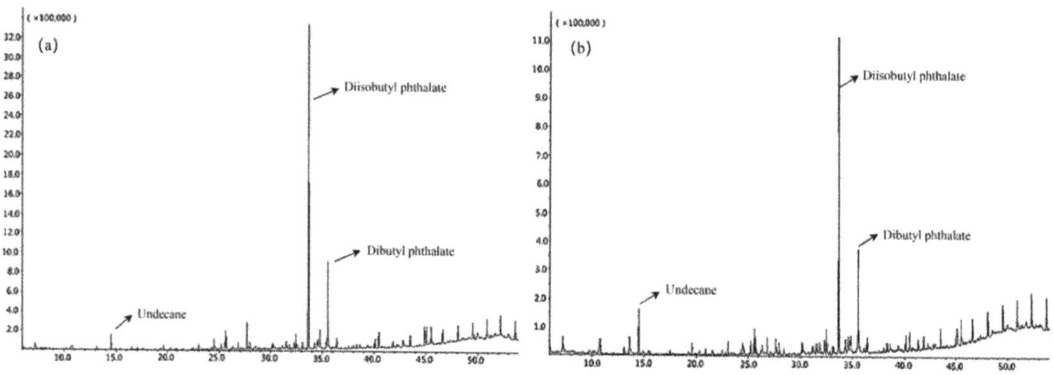

Figure 7. GC-MS patterns: (**a**) raw water, (**b**) water samples treated by Fenton process.

Table 5. The main organic pollutants.

Number	Peak Time (min)	Organic Matter	Structural Formula	Molecular Formula	The Degree of Match (%)	Occasion	
						Influent	Effluent
1	14.532	Undecane		$C_{11}H_{24}$	94	√	√
2	24.509	1,4,5-Trimethylnaphthalene	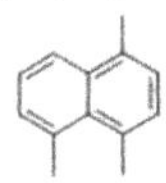	$C_{13}H_{14}$	62	√	
3	25.643	2,6-Di-tert-butyl-4-methylphenol	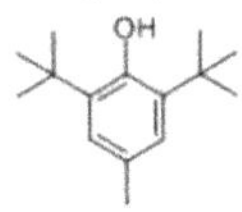	$C_{15}H_{24}O$	96	√	
4	25.732	2,4-di- tert -butylphenol	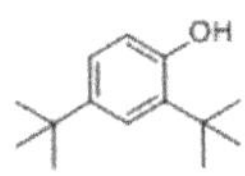	$C_{14}H_{22}O$	94	√	
5	27.657	2,2,4-trimethyl-1,3-pentanediol diisobutyrate	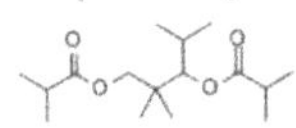	$C_{16}H_{30}O_4$	59	√	
6	32.404	Octadecane		$C_{18}H_{38}$	99	√	
7	33.656	Diisobutyl phthalate	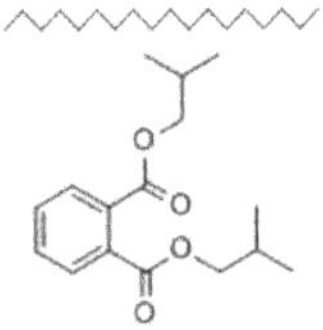	$C_{16}H_{22}O_4$	90	√	√
8	35.535	Dibutyl phthalate	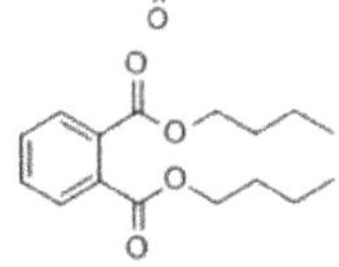	$C_{16}H_{22}O_4$	94	√	√
9	40.446	2,6-diphenylpyridine	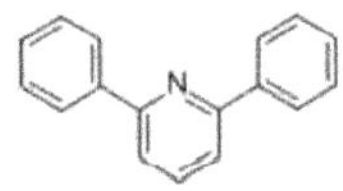	$C_{17}H_{13}N$	97	√	
10	43.488	Tetracosane		$C_{24}H_{50}$	98	√	
11	44.882	Benzo(H)quinoline	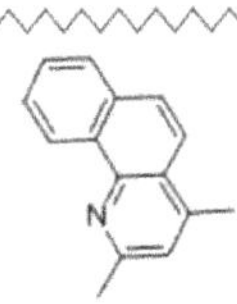	$C_{15}H_{13}N$	50	√	
12	45.073	Heneicosane		$C_{21}H_{44}$	93	√	
13	45.51	DEHP, di-(2-ethylhexyl) phthalate.	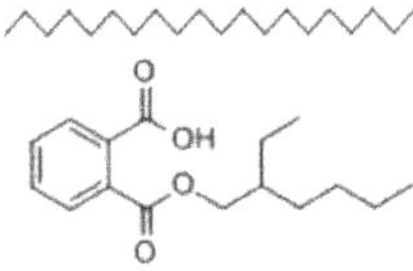	$C_{16}H_{22}O_4$	90	√	
14	46.617 48.100 49.528	Hexacosane		$C_{26}H_{54}$	98	√	
15	50.907 53.703	Octacosane		$C_{28}H_{58}$	98	√	
16	52.240	Eicosane	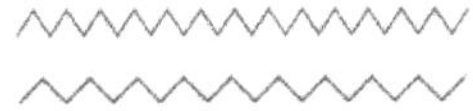	$C_{20}H_{42}$	98	√	

3.4. Mechanism Analysis

In brief, the degradation of acrylic fiber wastewater by the Fenton process was based on the HO•, which can degrade organic pollutants through oxidation, generated during the electron transfer process between H_2O_2 and a homogeneous metal catalyst (Fe^{2+}) [43]. Equations (1)–(5), (7)–(9) described the essential reactions in the Fenton process [2,44–47]. The production of HO• represented the beginning of the process, which the interaction of

Fe^{2+} with H_2O_2 Equation (1). Meanwhile, the generated Fe^{3+}, which were reduced to Fe^{2+} by H_2O_2, undergo a Fenton-like reaction Equation (5). When there were too many Fe^{2+}, it would react with HO• to produce Fe^{3+} Equation (3). Therefore, Equations (1) and (3) were the primary sources of Fe^{3+}. Excessive H_2O_2 reacted with HO• to generate HOO•, which oxidation activity was lower than HO•.

A large amount of HOO• was consumed through reacting with Fe^{3+}, Fe^{2+}, and HO• Equations (7)–(9), enabling an effective cyclic mechanism of Fe^{3+} and Fe^{2+}.

$$Fe^{3+} + HOO\bullet \rightarrow Fe^{2+} + H^+ + O_2 \quad (7)$$

$$Fe^{2+} + HOO\bullet \rightarrow Fe^{3+} + HO_2^- \quad (8)$$

$$HOO\bullet + HO\bullet \rightarrow H_2O + O_2 \quad (9)$$

In addition, a great deal of HO•, which had high oxidation, was produced through the above reaction. In order to demonstrate that the degradation of acrylic acid wastewater is based on the oxidation of HO•, HO• scavenging experiments based on the degradation of acrylic acid wastewater by Fenton system were carried out. It has been shown that isopropyl alcohol reacts rapidly with HO- and competes with the target substrate, thus effectively inhibiting the reaction of the target substrate with HO• [48]. Therefore, in this experiment isopropyl alcohol (IPA) was used to remove the active species associated with the degradation of organic matter. Figure 8 shows the effect of using different concentrations (0%, 2%, 6%, 10%) of isopropanol on the degradation of organic matter in acrylic acid wastewater based on the optimal fenton process conditions (mentioned in Section 3.2.3). As can be seen from Figure 8, the inhibition effect of isopropanol is obvious, with 2%, 6%, and 10% isopropanol almost completely inhibiting the degradation of organics. It indicates that HO• is the main active species in this Fenton system for the degradation of organic matter. Therefore, the mechanism in the degradation process of organic matter in acrylic wastewater was speculated as follows. In the presence of organic molecules (RH) from acrylic manufacturing wastewater, the generated HO• attacked the organic molecules by absorbing protons, producing highly reactive organic radicals (R•), which could be further oxidized to generate ROH Equations (10) and (11) [20,43,49]. Further, ROH was oxidized by HO• to form R_1OOH and R_2H to achieve the conversion of organic macromolecules to organic small molecules Equations (10) and (11). In addition, R_1OOH was easily oxidized by HO• to form smaller molecular weight organic matter (R_3H), H_2O and CO_2 Equation (12). The efficient degradation of organic matter in acrylic wastewater was achieved through the oxidation reactions of Equations (10)–(13) (where R, R_1, R_2, and R_3 represent one or more products in the degradation stage). Furthermore, H_2O_2 was generated via a reaction among HO• Equation (14).

$$HO\bullet + RH \rightarrow R\bullet + H_2O \quad (10)$$

$$HO\bullet + R\bullet \rightarrow ROH \quad (11)$$

$$HO\bullet + ROH \rightarrow R_1OOH + R_2H \quad (12)$$

$$HO\bullet + R_1OOH \rightarrow R_3H + CO_2 + H_2O \quad (13)$$

$$HO\bullet + HO\bullet \rightarrow H_2O_2 \quad (14)$$

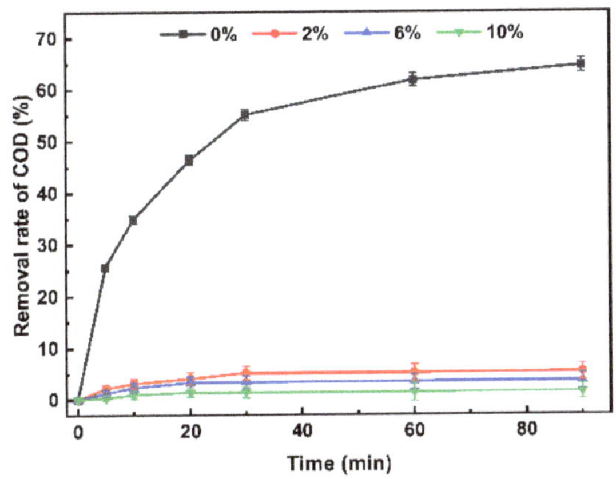

Figure 8. Effect of HO• scavenger on the degradation of organic matter in acrylic acid wastewater.

The mechanism for the degradation of acrylic manufacturing wastewater by the Fenton process is explained in Figure 9. The chain reaction can continue until the H_2O_2 was completely consumed because Fe^{2+} acted as a catalyst and transmitter in the reaction. Various free radicals and intermediate substances, such as HOO•, Fe^{3+} and HO•, were generated during the reaction as the nodes of the chemical reaction chain. In addition, Fe^{3+} was readily hydrolyzed and forms complexes. The flocculation and precipitation processes of these complexes also played an important role in removing organic matter. The degradation of organic matter was achieved through a series of reactions as described above.

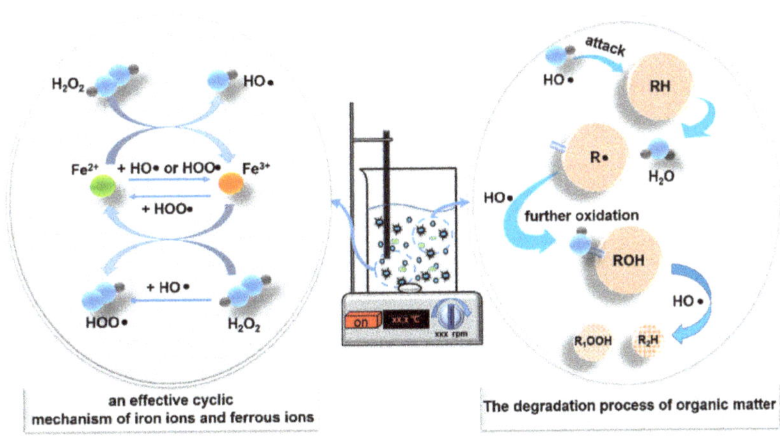

Figure 9. Schematic diagram of Fenton reaction process.

4. Conclusions

(1) The optimization analysis combined with the response surface method showed that the optimal degradation conditions for acrylic fiber wastewater using the Fenton method are: (i) initial H_2O_2 concentration of 60.90 mmol/L; (ii) initial Fe^{2+} concentration of 7.44 mmol/L; (iii) pH of 3. The predicted degradation efficiency of the model equation was 59.8%, and the actual COD degradation rate was 63.2%.

(2) The deviation between the actual and fitted model values was less than 5%, indicating that the model equation had a high degree of credibility. According to the analysis of the influence of factors and variables, it can be seen that the influence order of the

three factors was $[H_2O_2] > [Fe^{2+}] > pH$. In addition, the interaction between $[Fe^{2+}]$ and pH had the most significant impact on the degradation of COD.

(3) For the actual acrylic fiber wastewater treatment, the removal rate of COD, TOC, NH_4^+-N, TN is 61.45%~66.51%, 67.82%~70.99%, 55.67%~60.97%, 56.45%~61.03%, respectively. The effluent met the textile dyeing and finishing industry water pollutant discharge standard "GB4287-2012". The COD, TOC, NH_4^+-N, and TN were decreased to 93.3 ± 3.5 mg/L, 28.2 ± 0.7 mg/L, 26.4 ± 0.6 mg/L, 34.5 ± 0.6 mg/L, respectively.

(4) HO• generated during electron transfer between H_2O_2 and Fe^{2+} effectively decompose organic pollutants in acrylic production wastewater. In this case, 13 kinds of aromatic hydrocarbons and long-chain alkanes in acrylic fiber wastewater had been effectively removed.

Author Contributions: Conceptualization, Z.L. and C.Z.; methodology, Z.L.; software, Z.L., and P.S.; validation, Z.L., C.Z. and P.S.; formal analysis, Z.L.; investigation, Z.L.; resources, Z.L.; data curation, Z.L.; writing—original draft preparation, Z.L.; writing—review and editing, P.S. and Z.L.; visualization, W.L. and Z.Z.; supervision, Z.Z., X.W. and W.H.; project administration, C.Z.; funding acquisition, C.Z. All authors have read and agreed to the published version of the manuscript.

Funding: This research received no external funding.

Institutional Review Board Statement: "Not applicable" for studies not involving humans or animals.

Informed Consent Statement: "Not applicable." for studies not involving humans.

Data Availability Statement: Data are contained within the article.

Acknowledgments: This research is supported by the National Research Foundation China, and China University of Mining and Technology (Beijing) under the laboratory of the Institute for Total Process Pollution Control and Circular Economy.

Conflicts of Interest: The authors declare no conflict of interest.

References

1. Pattnaik, P.; Dangayach, G.S.; Bhardwaj, A.K. A review on the sustainability of textile industries wastewater with and without treatment methodologies. *Rev. Environ. Health* **2018**, *33*, 163–203. [CrossRef] [PubMed]
2. Li, J.; Luan, Z.; Yu, L.; Ji, Z. Pretreatment of acrylic fiber manufacturing wastewater by the Fenton process. *Desalination* **2012**, *284*, 62–65. [CrossRef]
3. Holkar, C.R.; Jadhav, A.J.; Pinjari, D.V.; Mahamuni, N.M.; Pandit, A.B. A critical review on textile wastewater treatments: Possible approaches. *J. Environ. Manag.* **2016**, *182*, 351–366. [CrossRef] [PubMed]
4. Hai, F.I.; Yamamoto, K.; Nakajima, F.; Fukushi, K. Bioaugmented membrane bioreactor (MBR) with a GAC-packed zone for high rate textile wastewater treatment. *Water Res.* **2011**, *45*, 2199–2206. [CrossRef]
5. Qin, L.; Zhang, G.; Meng, Q.; Xu, L.; Lv, B. Enhanced MBR by internal micro-electrolysis for degradation of anthraquinone dye wastewater. *Chem. Eng. J.* **2012**, *210*, 575–584. [CrossRef]
6. Cinperi, N.C.; Ozturk, E.; Yigit, N.O.; Kitis, M. Treatment of woolen textile wastewater using membrane bioreactor, nanofiltration and reverse osmosis for reuse in production processes. *J. Clean. Prod.* **2019**, *223*, 837–848. [CrossRef]
7. Deowan, S.A.; Galiano, F.; Hoinkis, J.; Johnson, D.; Altinkaya, S.A.; Gabriele, B.; Hilal, N.; Drioli, E.; Figoli, A. Novel low-fouling membrane bioreactor (MBR) for industrial wastewater treatment. *J. Membr. Sci.* **2016**, *510*, 524–532. [CrossRef]
8. GilPavas, E.; Dobrosz-Gómez, I.; Gómez-García, M.Á. Coagulation-flocculation sequential with Fenton or Photo-Fenton processes as an alternative for the industrial textile wastewater treatment. *J. Environ. Manag.* **2017**, *191*, 189–197. [CrossRef]
9. Giannakis, S.; Lin, K.-Y.A.; Ghanbari, F. A review of the recent advances on the treatment of industrial wastewaters by Sulfate Radical-based Advanced Oxidation Processes (SR-AOPs). *Chem. Eng. J.* **2021**, *406*, 127083. [CrossRef]
10. Paździor, K.; Bilińska, L.; Ledakowicz, S. A review of the existing and emerging technologies in the combination of AOPs and biological processes in industrial textile wastewater treatment. *Chem. Eng. J.* **2019**, *376*, 120597. [CrossRef]
11. Gong, C.; Zhang, Z.; Li, H.; Li, D.; Wu, B.; Sun, Y.; Cheng, Y. Electrocoagulation pretreatment of wet-spun acrylic fibers manufacturing wastewater to improve its biodegradability. *J. Hazard. Mater.* **2014**, *274*, 465–472. [CrossRef] [PubMed]
12. Wei, J.; Song, Y.; Meng, X.; Pic, J.-S. Combination of Fenton oxidation and sequencing batch membrane bioreactor for treatment of dry-spun acrylic fiber wastewater. *Environ. Earth Sci.* **2015**, *73*, 4911–4921. [CrossRef]
13. Zheng, T.; Wang, Q.; Zhang, T.; Shi, Z.; Tian, Y.; Shi, S.; Smale, N.; Wang, J. Microbubble enhanced ozonation process for advanced treatment of wastewater produced in acrylic fiber manufacturing industry. *J. Hazard. Mater.* **2015**, *287*, 412–420. [CrossRef] [PubMed]

14. Wei, J.; Song, Y.; Meng, X.; Tu, X.; Pic, J.-S. Transformation characteristics of organic pollutants in Fered-Fenton process for dry-spun acrylic fiber wastewater treatment. *Water Sci. Technol.* **2014**, *70*, 1976–1982. [CrossRef]
15. Zheng, T.; Wang, Q.; Shi, Z.; Fang, Y.; Shi, S.; Wang, J.; Wu, C. Advanced treatment of wet-spun acrylic fiber manufacturing wastewater using three-dimensional electrochemical oxidation. *J. Environ. Sci.* **2016**, *50*, 21–31. [CrossRef] [PubMed]
16. Yang, L.; Jiao, Y.; Xu, X.; Pan, Y.; Su, C.; Duan, X.; Sun, H.; Liu, S.; Wang, S.; Shao, Z. Superstructures with Atomic-Level Arranged Perovskite and Oxide Layers for Advanced Oxidation with an Enhanced Non-Free Radical Pathway. *ACS Sustain. Chem. Eng.* **2022**, *10*, 1899–1909. [CrossRef]
17. Han, X.; Lu, H.; Gao, Y.; Chen, X.; Yang, M. The role of in situ Fenton coagulation on the removal of benzoic acid. *Chemosphere* **2020**, *238*, 124632. [CrossRef]
18. Pignatello, J.J.; Oliveros, E.; MacKay, A. Advanced Oxidation Processes for Organic Contaminant Destruction Based on the Fenton Reaction and Related Chemistry. *Crit. Rev. Environ. Sci. Technol.* **2006**, *37*, 273–275. [CrossRef]
19. He, D.-Q.; Zhang, Y.-J.; Pei, D.-N.; Huang, G.-X.; Liu, C.; Li, J.; Yu, H.-Q. Degradation of benzoic acid in an advanced oxidation process: The effects of reducing agents. *J. Hazard. Mater.* **2020**, *382*, 121090. [CrossRef]
20. Neyens, E.; Baeyens, J. A review of classic Fenton's peroxidation as an advanced oxidation technique. *J. Hazard. Mater.* **2003**, *98*, 33–50. [CrossRef]
21. Brillas, E.; Sirés, I.; Oturan, M.A. Electro-Fenton Process and Related Electrochemical Technologies Based on Fenton's Reaction Chemistry. *Chem. Rev.* **2009**, *109*, 6570–6631. [CrossRef] [PubMed]
22. Walha, K.; Amar, R.B.; Quemeneur, F.; Jaouen, P. Treatment by nanofiltration and reverse osmosis of high salinity drilling water for seafood washing and processing Abstract. *Desalination* **2008**, *219*, 231–239. [CrossRef]
23. Wei, J.; Song, Y.; Tu, X.; Zhao, L.; Zhi, E. Pretreatment of dry-spun acrylic fiber manufacturing wastewater by Fenton process: Optimization, kinetics and mechanisms. *Chem. Eng. J.* **2013**, *218*, 319–326. [CrossRef]
24. Pérez, M.; Torrades, F.; Domènech, X.; Peral, J. Fenton and photo-Fenton oxidation of textile effluents. *Water Res.* **2002**, *36*, 2703–2710. [CrossRef]
25. NEPA. Water Quality-Determination of the Chemical Oxygen Demand-Dichromate Method. China National Standard HJ828-2017; China National Environmental Protection Administration(NEPA), 2017. Available online: https://www.mee.gov.cn/ywgz/fgbz/bz/bzwb/jcffbz/201704/W020170606398873416325.pdf (accessed on 1 May 2017).
26. NEPA. Water Quality-Determination of Ammonia Nitrogen-Nessler's Reagent Spectrophotometry. China National Standard HJ535-2009; China National Environmental Protection Administration(NEPA), 2009. Available online: https://www.mee.gov.cn/ywgz/fgbz/bz/bzwb/jcffbz/201001/W020180319535685015821.pdf (accessed on 1 April 2010).
27. NEPA. Water Quality—Determination of Iron—Phenanthroline Spectrophotometry. China National Standard HJ/T 345—2007; China National Environmental Protection Administration(NEPA), 2007. Available online: https://www.mee.gov.cn/ywgz/fgbz/bz/bzwb/jcffbz/200703/W020120104561152753262.pdf (accessed on 1 May 2007).
28. Zhao, H.; Dong, M.; Wang, Z.; Wang, H.; Qi, H. Roles of free radicals in NO oxidation by Fenton system and the enhancement on NO oxidation and H2O2 utilization efficiency. *Environ. Technol.* **2020**, *41*, 109–116. [CrossRef]
29. LU, P. Spectrophotometric determination of hydrogen peroxide in Fenton advanced oxidation systems by potassium titanyl oxalate. *Archit. Eng. Technol. Des.* **2014**, *582–582*, 517. [CrossRef]
30. Tekbaş, M.; Yatmaz, H.C.; Bektaş, N. Heterogeneous photo-Fenton oxidation of reactive azo dye solutions using iron exchanged zeolite as a catalyst. *Microporous Mesoporous Mater.* **2008**, *115*, 594–602. [CrossRef]
31. Teel, A.L.; Warberg, C.R.; Atkinson, D.A.; Watts, R.J. Comparison of mineral and soluble iron Fenton's catalysts for the treatment of trichloroethylene. *Water Res.* **2001**, *35*, 977–984. [CrossRef]
32. Verma, M.; Haritash, A.K. Degradation of amoxicillin by Fenton and Fenton-integrated hybrid oxidation processes. *J. Environ. Chem. Eng.* **2019**, *7*, 102886. [CrossRef]
33. Zazouli, M.A.; Yousefi, Z.; Eslami, A.; Ardebilian, M.B. Municipal solid waste landfill leachate treatment by fenton, photo-fenton and fenton-like processes: Effect of some variables. *Iran. J. Environ. Health Sci. Eng.* **2012**, *9*, 3. [CrossRef]
34. Pera-Titus, M.; García-Molina, V.; Baños, M.A.; Giménez, J.; Esplugas, S. Degradation of chlorophenols by means of advanced oxidation processes: A general review. *Appl. Catal. B: Environ.* **2004**, *47*, 219–256. [CrossRef]
35. Rott, E.; Minke, R.; Bali, U.; Steinmetz, H. Removal of phosphonates from industrial wastewater with UV/FeII, Fenton and UV/Fenton treatment. *Water Res.* **2017**, *122*, 345–354. [CrossRef] [PubMed]
36. Yap, C.L.; Gan, S.; Ng, H.K. Fenton based remediation of polycyclic aromatic hydrocarbons-contaminated soils. *Chemosphere* **2011**, *83*, 1414–1430. [CrossRef] [PubMed]
37. Hou, P.; Cannon, F.S.; Nieto-Delgado, C.; Brown, N.R.; Gu, X. Effect of preparation protocol on anchoring quaternary ammonium/epoxide-forming compound into granular activated carbon for perchlorate adsorption: Enhancement by Response Surface Methodology. *Chem. Eng. J.* **2013**, *223*, 309–317. [CrossRef]
38. Luo, D. Optimization of total polysaccharide extraction from Dioscorea nipponica Makino using response surface methodology and uniform design. *Carbohydr. Polym.* **2012**, *90*, 284–288. [CrossRef]
39. Behera, S.K.; Meena, H.; Chakraborty, S.; Meikap, B.C. Application of response surface methodology (RSM) for optimization of leaching parameters for ash reduction from low-grade coal. *Int. J. Min. Sci. Technol.* **2018**, *28*, 621–629. [CrossRef]

40. NEPA. Discharge Standards of Water Pollutants for Dyeing and Finishing of Textile Industry. China National Standard GB4287-2012. China National Environmental Protection Administration(NEPA), 2012. Available online: https://www.doc88.com/p-998284323481.html?r=1 (accessed on 1 January 2013).
41. Gogate, P.R.; Pandit, A.B. A review of imperative technologies for wastewater treatment I: Oxidation technologies at ambient conditions. *Adv. Environ. Res.* **2004**, *8*, 501–551. [CrossRef]
42. Oller, I.; Malato, S.; Sánchez-Pérez, J.A. Combination of Advanced Oxidation Processes and biological treatments for wastewater decontamination—A review. *Sci. Total Environ.* **2011**, *409*, 4141–4166. [CrossRef]
43. Cai, Q.Q.; Lee, B.C.Y.; Ong, S.L.; Hu, J.Y. Fluidized-bed Fenton technologies for recalcitrant industrial wastewater treatment–Recent advances, challenges and perspective. *Water Res.* **2021**, *190*, 116692. [CrossRef]
44. Miklos, D.B.; Remy, C.; Jekel, M.; Linden, K.G.; Drewes, J.E.; Hübner, U. Evaluation of advanced oxidation processes for water and wastewater treatment—A critical review. *Water Res.* **2018**, *139*, 118–131. [CrossRef]
45. Oturan, M.A.; Aaron, J.-J. Advanced Oxidation Processes in Water/Wastewater Treatment: Principles and Applications. A Review. *Crit. Rev. Environ. Sci. Technol.* **2014**, *44*, 2577–2641. [CrossRef]
46. Andreozzi, R.; Caprio, V.; Insola, A.; Marotta, R. Advanced oxidation processes (AOP) for water purification and recovery. *Catal. Today* **1999**, *53*, 51–59. [CrossRef]
47. Babuponnusami, A.; Muthukumar, K. Advanced oxidation of phenol: A comparison between Fenton, electro-Fenton, sono-electro-Fenton and photo-electro-Fenton processes. *Chem. Eng. J.* **2012**, *183*, 1–9. [CrossRef]
48. Gu, X.; Qin, N.; Zhang, P.; Hu, Y.; Zhang, Y.-N.; Zhao, G. In-situ synthesis of {111}TiO_2/Ti photoelectrode to boost efficient removal of dimethyl phthalate based on a bi-functional interface. *Chem. Eng. J.* **2021**, *422*, 129980. [CrossRef]
49. Liang, D.; Li, N.; An, J.; Ma, J.; Wu, Y.; Liu, H. Fenton-based technologies as efficient advanced oxidation processes for microcystin-LR degradation. *Sci. Total Environ.* **2021**, *753*, 141809. [CrossRef] [PubMed]

water

MDPI

Article

Achieving Sustainable Development Goal 6 Electrochemical-Based Solution for Treating Groundwater Polluted by Fuel Station

Júlio César Oliveira da Silva [1], Aline Maria Sales Solano [1,2,*], Inalmar D. Barbosa Segundo [1,*], Elisama Vieira dos Santos [1,3], Carlos A. Martínez-Huitle [1,3,4,*] and Djalma Ribeiro da Silva [1]

1 Institute of Chemistry, Federal University of Rio Grande do Norte, Lagoa Nova, Natal 59078-970, RN, Brazil
2 Federal Institute of Education, Science, and Technology of Rio Grande do Norte/Nova Cruz, Av. José Rodrigues de Aquino Filho, Nova Cruz 59215-000, RN, Brazil
3 National Institute for Alternative Technologies of Detection, Toxicological Evaluation and Removal of Micropollutants and Radioactives (INCT-DATREM), Institute of Chemistry, Universidad Estatal Paulista Júlio de Mesquita Filho, Araraquara 14800-900, SP, Brazil
4 Department Chemie, Johannes Gutenberg-Universität Mainz Duesbergweg 10-14, 55128 Mainz, Germany
* Correspondence: lininhamss@hotmail.com (A.M.S.S.); idbsegundo@gmail.com (I.D.B.S.); carlosmh@quimica.ufrn.br (C.A.M.-H.)

Abstract: Oil leakage occurs at fuel service stations due to improper storage, which pollutes soil and, subsequently, can reach the groundwater. Many compounds of petroleum-derived fuels pose hazards to aquatic systems, and so must be treated to guarantee clean and safe consumption, which is a right proposed by the United Nations in their Sustainable Development Goal 6 (SDG 6: Clean Water and Sanitation). In this study, contaminated groundwater with emerging pollutants by petroleum-derived fuel was electrochemically treated in constantly mixed 0.5 L samples using three different anodes: Ni/BDD, Ti/Pt, Ti/RuO_2. Parameters were investigated according to chemical oxygen demand (COD), energy consumption analysis, by applying different electrodes, current densities (*j*), time, and the use of Na_2SO_4 as an electrolyte. Despite a similar COD decrease, better degradation was achieved after 240 min of electrochemical treatment at Ti/RuO_2 system (almost 70%) by applying 30 mA cm^{-2}, even without electrolyte. Furthermore, energy consumption was lower with the RuO_2 anode, and greater when 0.5 M of Na_2SO_4 was added; while the order, when compared with the other electrocatalytic materials, was Ti/RuO_2 > Ti/Pt > Ni/BDD. Thereafter, aiming to verify the viability of treatment at a large scale, a pilot flow plant with a capacity of 5 L was used, with a double-sided Ti/RuO_2 as the anode, and two stainless steel cathodes. The optimal conditions for the effective treatment of the polluted water were a *j* of 30 mA cm^{-2}, and 0.5 M of Na_2SO_4, resulting in 68% degradation after 300 min, with almost complete removal of BTEX compounds (benzene, toluene, ethyl-benzene, and xylene, which are found in emerging pollutants) from the water and other toxic compounds. These significant results proved that the technology used here could be an effective SDG 6 electrochemical-based solution for the treatment of groundwater, seeking to improve the quality of water, removing contaminants, and focusing on Brazilian environmental legislations and, consequently, converting pollutants into effluent that can be returned to the water cycle.

Keywords: anodic oxidation; BTEX; groundwater pollution; electrochemical oxidation; sustainable development goals

Citation: da Silva, J.C.O.; Solano, A.M.S.; Barbosa Segundo, I.D.; dos Santos, E.V.; Martínez-Huitle, C.A.; da Silva, D.R. Achieving Sustainable Development Goal 6 Electrochemical-Based Solution for Treating Groundwater Polluted by Fuel Station. *Water* **2022**, *14*, 2911. https://doi.org/10.3390/w14182911

Academic Editors: Dionysios (Dion) Demetriou Dionysiou, Yujue Wang and Huijiao Wang

Received: 6 August 2022
Accepted: 13 September 2022
Published: 17 September 2022

1. Introduction

Groundwater refers to any amount of water located below the earth's surface, constituting an important water reserve for consumption, mainly for regions where there are waterbodies unable to meet the needs of the population [1]. The importance of preserving and/or recovering these water systems is essential for both the current society and future generations in order to accomplish the Sustainable Development Goals (SDGs) proposed by the United Nations (UN). SDG 6 regards ensuring availability and sustainable management of water and sanitation for all, seeking to attend not only drinking water, but also water harvesting, water efficiency, desalination, wastewater treatment, recycling, and reuse technologies [2]. According to the UN, the way to promote this goal by 2030 is through the implementation of the following targets: (i) to achieve safe and affordable drinking water; (ii) sanitation and hygiene and end open defecation; (iii) improve water quality, wastewater treatment, and safe reuse; (iv) increase water-use efficiency and ensure freshwater supplies; (v) implement integrated water resources management; and (vi) protect and restore water-related ecosystems [3].

However, several industrial and human activities (e.g., population growth, the increasing demand for energy consumption, the exploitation of natural resources, and the inappropriate disposal or storage of products harmful to human health) have caused significant pollution problems, such as the contamination of rivers and groundwater, subsequently provoking negative environmental impacts that can affect the whole ecosystem [4]. Some recalcitrant and hazardous contaminants such as polycyclic aromatic hydrocarbon (PAH), polychlorinated biphenyl (PCB) [5], BTEX compounds [6], and trichloroethylene [7] are frequently identified in polluted groundwater.

Conventional water treatment methods (wet-air oxidation, Fenton type processes, super critical water oxidation, photocatalysis, adsorption, etc.) attend to the problems related to SDG 6 in several sectors, but sometimes these approaches can be operationally intensive, chemically severe, and energetically expensive. Therefore, the design and development of environmentally innovative SDG 6 solutions [8,9] have been significantly stimulated. In the last years, electrochemical advanced oxidation technologies have received great attention because these processes are considered feasible alternatives to prevent or solve many environmental problems related to water pollution [10,11]. At the same time, the use of renewable technologies as the main electrical energy source, instead of the conventional electrical supply, has extended the applicability of these approaches in different water-security sectors [12,13].

This type of SDG 6 technology efficiently eliminates a great number of persistent organic pollutants [14–16] and microorganisms by promoting the in-situ electrogeneration of oxidizing species. Among the SDG 6 electrochemical-based solutions, anodic oxidation (AO) or electrooxidation (EO) is the most popular approach to be considered since it is a robust, easy, safe, and eco-friendly water treatment solution [15,16]. Based on their electrocatalytic fundamentals, AO can follow two ways on the oxidation of organic pollutants at the anodic surface (M) by direct electron transfer and/or more quickly by the reaction with the $^{\bullet}$OH which were formed from the oxidation of water (Equation (1)) [15,17,18]:

$$M + H_2O \rightarrow M(^{\bullet}OH) + H^+ + e^- \quad (1)$$

The amount of the electrogenerated heterogenous free $^{\bullet}$OH depends on the electrocatalytic material used, like active (e.g., Pt, IrO_2, and RuO_2) and non-active (e.g., PbO_2, SnO_2, and boron-doped diamond (BDD)), favoring the conversion of organics into by-products (due to lower amounts of free $^{\bullet}$OH, occurring direct oxidation) or promoting the complete incineration of pollutants to CO_2 and water (due to the higher amount of free $^{\bullet}$OH, once indirect oxidation takes place) [10,19,20]. Although non-active anodes are frequently used because they are more effective to degrade different pollutants (ranging from $\mu g\ L^{-1}$ to $g\ L^{-1}$ [16,21]) in various water matrices [22–26], significant advances in strategies for preparing/synthesizing novel electrocatalytic materials have been achieved to replace non-active anodes [26,27] as well as to develop/design innovative electrochemical reactors aiming to successfully translate from laboratory cells to pilot scale-up [28,29].

Therefore, this work aims to evaluate the efficiency of the electrochemical degradation of groundwater contaminated by fuel leak underneath a gas station at different anodic materials (Ti/RuO_2, Ti/Pt, and NB/BDD), in a batch stirring cell, applying 10, 30, and 60 mA cm^{-2} at 25 °C. After that, an electrocatalytic material was selected to scale up the electrochemical system aiming to study the effective assessment of the water quality with an innovative SDG 6 electrochemical-based solution to treat 5 L of fuel polluted-groundwater using a reactor with 286 cm^2 of geometrical area. The scaling up of the electrochemical device intends to prove that this technology could be an effective solution for water treatment, fulfilling SDG 6 via the improvement of the quality of water and returning it to the water cycle.

2. Materials and Methods

2.1. Characterization of Groundwater Effluent

The yellowish groundwater sample was collected in areas of fuel recovery stations in Natal city, in the northeastern Brazilian region, by trained technicians of the Universidade Federal do Rio Grande do Norte, approximately 10 m below the water level, and removing the supernatant oil. Different chemical compounds in the collected sample were identified and quantified by gas chromatography—mass spectroscopy (GC—MS) technique using commercially available standard compounds. EPA standard method 8021 was used to evaluate the sample by GC with a serial detector CG-PID/FID, column OV624 (gas flow: 3.0 mL min^{-1}), with initial temperature of 40.15 °C/1.0 min, after 10 °C/min to 150 °C, with a total analysis time of 32 min. Split-less injector temperature was 250 °C, lamp temperature (PID) was 280 °C, and detector FID was 300 °C. Before the GC—MS analysis, the samples were treated by solid-phase micro-extraction (SPME) with an 85 µm film of poly-(dimethylsiloxane) fiber. Once the extraction was completed (10 min), the fiber was withdrawn back inside the fiber holder, removed from the reaction sample, and then injected by a 1:25 split method. The temperature of the transfer line was 240 °C, the ions trap was 170 °C, and the desorption time in the split-less mode at 260 °C was 3 min. Other physical-chemical characteristics were also measured (Table 1).

2.2. EO Experiments

Two electrochemical reactors were used to perform the bulk-electrolysis: (i) a conventional laboratory cell with a capacity of 0.5 L (Figure 1a) and (ii) an electrochemical pilot flow cell with a capacity of 5 L (Figure 1b). In the former, AO experiments were carried out under galvanostatic conditions using a MINIPA MPL-3305M power supply with Ti/Pt, Ti/RuO_2, or Nb/BDD as anodes, and a titanium plate as the cathode (each one of them with a geometrical area of 13.5 cm^2; and an electrode-gap of 1.5 cm) by applying 10, 30, and 60 mA cm^{-2} at 25 °C; while an electrochemical flow cell with a double-sided RuO_2 anode and two stainless steel plates as the cathodes (each one of them with a geometrical area of about 286 cm^{-2} and electrode-gaps of 3 cm) were used for electrochemically treating polluted-groundwater effluent, which was recirculated by the pilot plant using a pump with a flow rate of 153 L h^{-1}, by applying 30 mA cm^{-2} at 25 °C (experimental conditions were established by the preliminary experiments with batch cell) in the latter.

2.3. Analytical Methods

Polluted-groundwater decontamination was monitored from its chemical oxygen demand (COD) and total organic carbon (TOC) reductions, determined on a Hanna Instruments HI 83399 multiparameter photometer and TOC analyzer (Analytik Jena model multi-N/C 3100, by injecting 30 mL aliquots), respectively. On the one hand, COD measurements were determined by using 2 mL of effluent sample in pre-prepared kit HANNA solutions which remained 2 h under the digestion process at 150 °C into a thermo-reactor (HANNA model HI 839800). On the other hand, BOD_5 was measured using an OxiTop® system (manometric respirometry). Another method for organic analysis is according to the specific ultraviolet absorbance at 254 nm ($SUVA_{254}$, L mg^{-1} m^{-1}), which can be used as an indicator of aromaticity and chemical reactivity for aquatic organic matter samples from a wide range of water sources. It is obtained by dividing the UV absorbance at 254 nm (m^{-1}) of samples by TOC values (mg L^{-1}) [30]. Finally, pH conditions were monitored using a Nova Instruments pH-meter HANNA and the conductivity was measured with an Electrical Conductivity meter model HI4321.

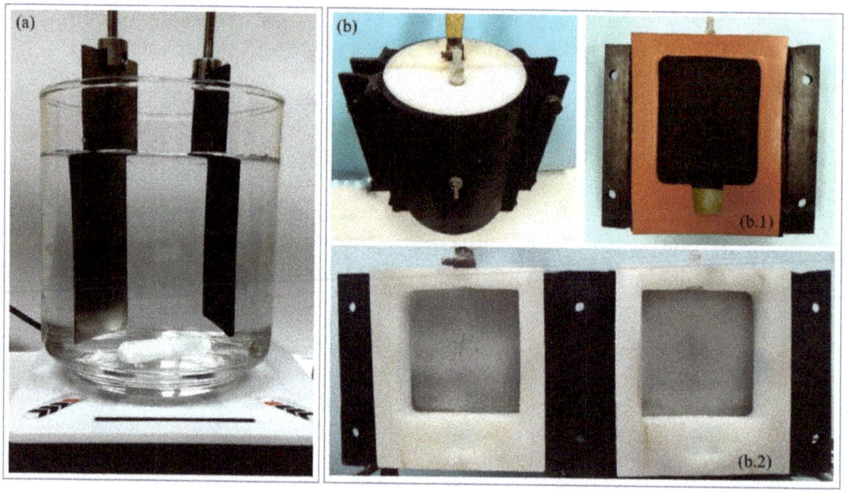

Figure 1. Electrochemical reactors: batch mode (**a**) and the flow cell (**b**) with the Ti/RuO$_2$ anode placed in the middle of the reactor (**b.1**), and the opened reactor detailing the two stainless steel plates as cathodes.

The discoloration was determined by measuring the absorbance reduction with a Shimadzu UV 1800 spectrophotometer during the AO tests of groundwater effluent at different applied current densities. Wavelength peaks at 436, 525, and 620 nm were taken from the UV-Vis spectrum between 200–800 nm, and then DFZy calculation was made according to Equation (2) following the method DIN EN 7884:2012 [31].

$$\mathrm{DFZ_y} = 100 \times \left(\frac{\mathrm{E_y}}{\mathrm{d}}\right) \tag{2}$$

where $\mathrm{E_y}$ is the absorbance at a y wavelength and d is the cell path length in cm.

2.4. Electrochemical Flow Reactor Characterization

The electrochemical flow reactor was hydrodynamically characterized by the limiting diffusion current technique. Potassium ferricyanide ($K_4Fe(CN_6)$) solutions, ranging from 20 to 80 mmol L^{-1} in 0.5 mol L^{-1} NaOH, were electrolyzed with a flow rate of 153 L h^{-1}, and the potential-current profiles for the oxidation-reduction reactions (Equations (3) and (4)) were determined by increasing the voltage stepwise.

$$\mathrm{Fe(CN)_6}^{3-} + \mathrm{e}^- \rightarrow \mathrm{Fe(CN)_6}^{4-} \tag{3}$$

$$\mathrm{Fe(CN)_6}^{4-} \rightarrow \mathrm{Fe(CN)_6}^{3-} + \mathrm{e}^- \tag{4}$$

Then, polarization curves were obtained and plotted to estimate the limiting currents at each one of the ferricyanide ($K_4Fe(CN_6)$) solutions at different concentrations [32]. From the slope value obtained by the profile limiting current vs. redox couple concentration, the mass-transfer coefficient of the reactor was estimated by Equation (5).

$$k_m = \frac{\mathrm{i_L}}{\mathrm{zFAC_\infty}} \tag{5}$$

k_m is the mass-transfer coefficient (m s^{-1}), i_L is the electrolysis limiting current (A), z is the electrons transfer in the redox reaction (for this redox pair, 1), F is the Faraday constant (96,487 C mol^{-1}), A is the electrode surface area (m^2), and C_∞ is the bulk species concentration (mol dm^{-3}).

3. Results

3.1. Effluent Characterization

GC—MS results (see Table 1) showed that the concentrations of benzene, toluene, ethylbenzene, xylene (called BTEX), and phenol in the groundwater collected-sample were higher than the maximum allowed for human consumption, as well as 230 mg L^{-1} of COD. Furthermore, other physical characteristics were measured and reported in Table 1.

3.2. Influence of the Applied *j* and the Effect of Different Anode Materials Using the Batch Reactor

Electrocatalytic material and current density (j) are two of the main operating parameters that determine the effectiveness of the decontamination of water matrices by electrochemical advanced oxidation processes (EAOPs). A higher concentration of oxidizing species can be electrochemically generated, and higher pollutant degradation can be attained, depending on the anode and j selected. Therefore, it is essential to study the anodic material and j effects in order to determine the degradation efficiency of the SDG 6 electrochemical-based solution proposed here, since there are electrodes and j values that promote undesirable parasitic reactions, leading to similar or even lower pollutant removal [33]. To do that, Ti/RuO_2, Ti/Pt, or Nb/BDD anode were used in a batch reactor to investigate the effect of the j on the EO of groundwater by applying 10, 30, and 60 mA cm^{-2} at 25 °C.

Table 1. Physical-chemical characterization before EO treatment of groundwater collected sample.

Parameter	Before Treatment	MAV [a]
pH	6.59	-
Conductivity (mS cm^{-1})	0.292	-
Phenol content (μg L^{-1})	6	3
Benzene (μg L^{-1})	96.6	5
Toluene (μg L^{-1})	2441	170
Ethylbenzene (μg L^{-1})	925.5	200
Xylene (*o*-, *p*- and *m*-) (μg L^{-1})	5435.5	300
BOD (mg L^{-1})	80.5	-
COD (mg L^{-1})	230	-
TOC (mg L^{-1})	91.5	-
Total petroleum hydrocarbons (TPHs) (mg L^{-1})	4.66	-
Color (DFZ at 436, 525, and 620 nm, respectively)	9.5, 6.5, 4.9	-
Absorbance at 254 nm (AU)	0.714	-
$SUVA_{254}$	0.78	-
Turbidity (NTU)	7.9	-

Note: [a] Maximum allowed value for groundwater (human consumption) based on current legislation in Brazil [34].

Figures 2 and 3 show that both factors (type of electrode and j) significantly influence the COD decay, using a batch stirring reactor with 0.5 L of effluent (the lines represent the trend of the symbols, which are the real values). As can be observed, no complete organic matter removal was attained independently of the anodic material and *j*, respectively. By applying 30 mA cm^{-2}, 78.3%, 81%, and 78.2% of COD removals were achieved at Ti/RuO$_2$, Ti/Pt, and Nb/BDD anodes, respectively. Meanwhile, COD removals of about 75.6% and 77% were reached for Ti/RuO$_2$ and Ti/Pt, respectively, by applying 60 mA cm^{-2}. This decrease in efficacy was associated with the promotion of the oxygen evolution reaction over organics oxidation at higher *j* due to the anodic electrode used, such as Ti/RuO$_2$ and Ti/Pt, which are considered active anodes. Conversely, ineffective COD removals were achieved at 10 mA cm^{-2} because a lower concentration of free $^{\bullet}$OH radicals is produced at Ti/RuO$_2$ and Ti/Pt electrodes. COD results preliminarily evidenced that the process is potentially controlled by mass transport conditions at all electrocatalytic materials because of the concentration of pollutants, type of electrocatalytic material, undesired reactions at higher *j* (e.g., oxygen evolution reaction), and the flow-hydrodynamic cell conditions. Nevertheless, the mass transport conditions affecting the process at Ti/RuO$_2$ and Ti/Pt electrodes seem to be more significant than diamond electrode. In fact, COD removal increased approximately 4.2% and 10.1% by applying 60 mA cm^{-2} from the removal efficiencies achieved at 30 and 10 mA cm^{-2}, respectively, when the BDD electrode was used. This behavior is due to the physisorbed $^{\bullet}$OH, which are efficiently electrogenerated via water discharge (Equation (1)) at the non-active diamond surface, and confined close to the anode surface in the reaction cage, favoring the degradation of organic matter and, consequently, achieving the best oxidation performances compared with the others [35,36].

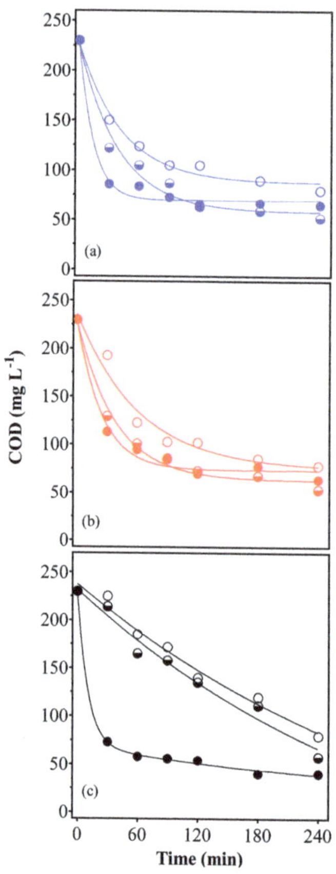

Figure 2. Influence of 10, 30, and 60 mA cm^{-2} (empty, half-, and full-colored, respectively), as a function of time on the COD removal during the EO: (**a**) Ti/RuO$_2$, (**b**) Ti/Pt, and (**c**) Nb/BDD.

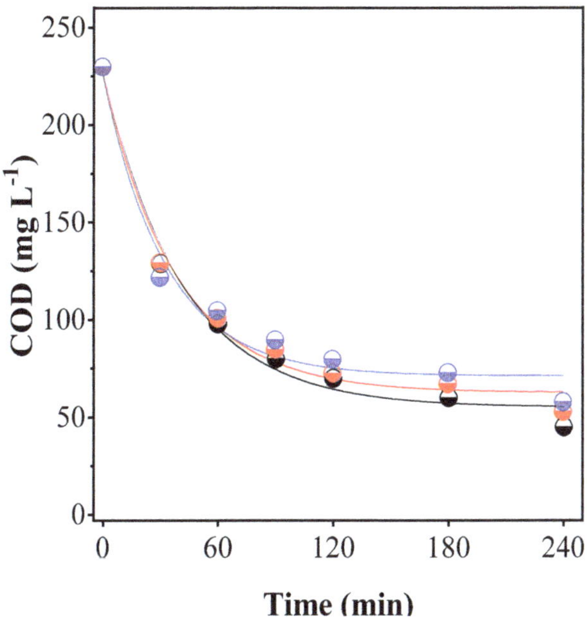

Figure 3. Influence of the electrocatalytic material (Ti/RuO_2 ; Ti/Pt ; Nb/BDD) on COD decay as a function of time, when Na_2SO_4 was added to the real water matrix by applying 30 mA cm^{-2} at 25 °C.

The results have indicated that the degradation of the pollutants present in the effluent was mainly controlled by mass transport in several cases (at Ti/RuO_2, Ti/Pt, and Nb/BDD anodes) because, independently of the *j* and the exponential, COD decayed at the beginning of the electrolysis; it remained almost stable in the last 150 min of the treatment. This behavior was also confirmed by the estimation of the limiting current (I_{lim}) value, considering the initial COD as well as the hydrodynamic conditions of the batch cell (Equation (5)), as already reported in our previous work.

$$I_{lim}(t) = 4FAk_m COD(t) \quad (6)$$

where $I_{lim(t)}$ is the limiting current (A) at a given time *t*, 4 is the number of exchanged electrons, *A* is the electrode area (m^2), *F* is the Faraday's constant, k_m is the average mass transport coefficient in the electrochemical reactor (m s^{-1}), and COD(t) is the chemical oxygen demand (mol O_2 m^{-3}) at a given time *t*.

An I_{lim} of 0.11 A was estimated, considering a k_m of 1.9×10^{-5} m s^{-1}. This current value is lower than all the currents applied in this work (0.14, 0.41, and 0.82 A for 10, 30, and 60 mA cm^{-2}, respectively), confirming that the AO technology, under these conditions, is taking place under mass transport control. On the other hand, I_{lim} of 0.11 A is significantly lower than 0.81 (value estimated for the *j* of 60 mA cm^{-2}), which explains the fact that there was not a significant enhancement at higher *j*, as a consequence of a higher charge which was consumed in parasite non-oxidizing reactions such as oxygen evolution. These assumptions, treating a real water matrix, are in agreement with various studies [37–39] about the EO of real wastewater. Based on the results obtained, a *j* of 30 mA cm^{-2} was selected as the optimum value to be applied in a new set of experiments by adding Na_2SO_4 as a supporting electrolyte where the effects of the reduction of COD were studied with Ti/RuO_2, Ti/Pt, and Nb/BDD anodes.

3.3. Comparative Groundwater EO Using Different Anodes Adding Sulfate as Supporting Electrolyte

In accordance with other studies [40–42], the addition of salts such as NaCl or Na_2SO_4 in real effluent favors an increase in conductivity, having a positive effect on the degradation of organic matter. On the one hand, the water discharge is favored (Equation (1)), producing $^{\bullet}OH$; while, on the other hand, the electrogeneration of additional oxidizing species is promoted by the oxidation of salt precursors, such as NaCl or Na_2SO_4 [43], enhancing the elimination of pollutants from water matrix. In this sense, a known amount of Na_2SO_4 was added to 0.5 L achieving a concentration of about 0.5 M.

In the case of Ti/RuO_2 and Ti/Pt, $^{\bullet}OH$ are strongly adsorbed on their surfaces, higher oxides are formed (Equation (7)), which participate as a mediator (Equation (8)) in the selective oxidation of organics, R (so-called electrochemical conversion), which occurs in concomitance with oxygen evolution (Equation (9)), as a side reaction, affecting the efficiency of the EO process.

$$M(^{\bullet}OH) \rightarrow MO + H^+ + e^- \quad (7)$$

$$MO + R \rightarrow M + RO \quad (8)$$

$$MO \rightarrow M + \frac{1}{2} O_2 \quad (9)$$

In this sense, direct electrochemical oxidation at Ti/RuO_2 and Ti/Pt surfaces is preferentially favored (direct electron transfer from the organics to the anode surface), limiting the effectiveness of the process. Conversely, when the Nb/BBD anode is used, the $^{\bullet}OH$ are weakly adsorbed on its surface, rapidly reacting with the organic molecules, consequently leading to electrochemical incineration (CO_2, and water (Equation (10)).

$$M(^{\bullet}OH) + R \rightarrow M + m\, CO_2 + n\, H_2O + x\, H^+ + y\, e^- \quad (10)$$

This reaction also competes with the side reaction of O_2 evolution (Equation (11)), which can significantly affect the efficiency of the electrolytic treatment.

$$M(^{\bullet}OH) \rightarrow M + \frac{1}{2} O_2 + H^+ + e^- \quad (11)$$

Considering the information above and the results in Figure 4, the partial elimination of COD in all electrodes is a consequence of the formation of byproducts during the electrochemical treatment of the real water matrix by applying 30 mA cm^{-2}, even when the effluent conductivity was increased due to the addition of Na_2SO_4. Additionally, the promotion of the oxygen evolution reaction (Equation (9)), as an undesired reaction, can be favored at Ti/RuO_2 and Ti/Pt electrodes when the conductivity is increased, consequently limiting the elimination of organics. Meanwhile, the slight organics removal improvement of the effluent when BDD was used and Na_2SO_4 added could be associated with the electrochemical production of sulfate-based oxidants, such as $SO_4^{\bullet-}$ and $S_2O_8^{2-}$. Their formation, via direct or indirect electrosynthetic routes, depends on the BDD surface properties, electrolyte medium, its concentration, and the current or potential applied [44–47] (Equations (12)–(16)).

$$2SO_4^{2-} \rightarrow S_2O_8^{2-} + 2e^- \quad (12)$$

$$SO_4^{2-} + {}^{\bullet}OH \rightarrow SO_4^{-\bullet} + H^+ + e^- \quad (13)$$

$$HSO_4^- + {}^{\bullet}OH \rightarrow SO_4^{-\bullet} + H_2O \quad (14)$$

$$H_2SO_4 + {}^{\bullet}OH \rightarrow SO_4^{-\bullet} + H_3O^+ \quad (15)$$

$$BDD(SO_4^{-\bullet}) + SO_4^{-\bullet} \rightarrow S_2O_8^{2-} \quad (16)$$

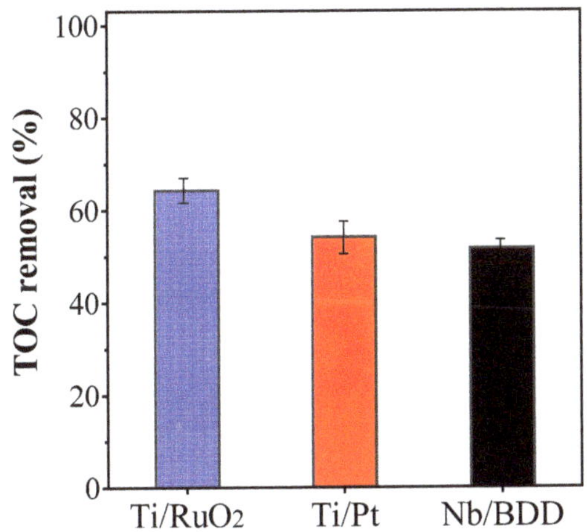

Figure 4. TOC removals, as a function of electrocatalytic material, when no Na_2SO_4 was added to real water matrix by applying 30 mA cm^{-2} at 25 °C.

This mix of oxidants remains at the reaction cage (BDD-$^{\bullet}OH/SO_4^{\bullet-}/S_2O_8^{2-}$), oxidizing several organic pollutants in different water matrices [44,48], deactivating microorganisms [47,49], and depolluting soil matrices (by using soil washing, electrokinetic soil remediation, or soil flushing) [50,51]. However, ionic species in the water matrices could act as scavengers, trapping the oxidants and, consequently, reducing their oxidative action [24,52], like Cl^-, HCO_3^{2-}, and others. Therefore, the effectiveness of the process was also evaluated by the organic matter degradation (COD) and mineralization (TOC).

In terms of COD removal, as can be observed in Figure 3, no significant removals were achieved (ranging from 75–80%) by applying 30 mA cm^{-2} when sulfate was added in the effluent. Comparable efficiencies were achieved without the addition of sulfate in the effluent (78.26%, 81%, and 78.2% of COD removals were achieved at Ti/RuO_2, Ti/Pt, and Nb/BDD anodes, respectively), under similar experimental conditions. Nevertheless, some differences were observed when the mineralization removal was determined by TOC measurements (Figure 4). As can be observed, Ti/RuO_2 anode efficiently mineralized the organic matter, reaching approximately 64.61%. Meanwhile, Ti/Pt and Nb/BDD anodes achieved about 54.20 and 51.74% of TOC removals, respectively. This behavior can be associated with the electrode deactivation of surface and oxidants scavengers. In the former, Pt surface can be deactivated due to the adsorption of BTEX compounds or their oxidation byproducts [42]; while the oxidants electrogenerated at BDD could be trapped by the scavenger species in the water matrix, in the latter. Conversely, the enhancement on the mineralization at Ti/RuO_2 can be related to the occurrence of direct electrochemical oxidation of organics and byproducts at its surface (without surface deactivation), in concomitance with the oxygen evolution reaction.

Thus, with these results, and aiming to understand the energy requirements needed in terms of kWh $gCOD^{-1}$ to evaluate the feasibility of the process, energy consumption was estimated by using Equation (17).

$$EC\ (kWh\ (gCOD)^{-1}) = (E_{cell} \times I \times t)/(\Delta(COD)_{exp} \times V_s) \quad (17)$$

where V_s is the solution volume (L), I refers to the current intensity (A), $\Delta(COD)_{exp}$ (g O_2 L^{-1}) refers to the difference between initial COD and COD at time t, t to the electrolysis time in hours, and E_{cell} to the average potential difference of the cell (V). Table 2 reports the EC values for all experimental conditions, in the absence and presence of sulfate in the effluent.

Table 2. Energy consumption values (kWh $gCOD^{-1}$), as a function of j and electrocatalytic material, after 240 min of electrochemical treatment, adding or not adding Na_2SO_4 in the effluent, and considering approximately between 75–80% of COD removal.

	j (mA cm^{-2})			
Electrode	10 [a]	30 [a]	60 [a]	30 [b]
Ti/RuO_2	0.023	0.068	0.174	0.050
Ti/Pt	0.020	0.073	0.195	0.068
Nb/BDD	0.028	0.098	0.272	0.085

Note: [a] without Na_2SO_4; [b] with Na_2SO_4.

It was possible to observe that, as the j was increased, the energy consumption significantly increased. After 240 min of EO, the tests with Ti/RuO_2 adding Na_2SO_4 (COD removal = 75%) and without Na_2SO_4 (COD removal = 78%) lead to lower energy consumptions of 0.068 and 0.050 kWh $kgCOD^{-1}$, respectively, by applying 30 mA cm^{-2}. Meanwhile, for the experiments using Ti/Pt and Nb/BDD, in the presence and absence of sulfate, more substantial electrical requirements are needed to achieve at least 78% of organic matter removal. Although the EO experiments with BDD seem to be optimal in terms of organic matter removal, the use of oxide-based electrodes could be more eco-technologically suitable.

Another feature that should be considered is that the two-electrode cell operating under galvanostatic mode serves to provide only general information about the electrical requirements of each EO assay and, consequently, the economic costs. Although this cell option has specific benefits, it could offer apparent experimental conditions that are not reproducible and not completely fixed; consequently, the EO behavior is difficult to be scaled. In this sense, an electrochemical flow reactor is the most appropriate strategy to scale up the water SDG 6 solution because it allows for improving the electrode size, batch recirculation mode operation, the use of more economic electrode materials and shapes, and their components are suitable to be produced in large amounts. For this reason, Ti/RuO_2 electrode was chosen for upgrading the EO process.

3.4. EO of Groundwater at the Pre-Pilot Plant

The mass transfer characterization of the pre-pilot plant was firstly performed (Figure 1). Then, the polarization curves (current response vs. voltage) were registered during electrolysis, in the electrochemical flow reactor, of different solutions of ferricyanide (i.e., 20–80 mol m^{-3}) in 0.5 mol L^{-1} NaOH at a constant flow rate of 4.25×10^{-5} m^3 s^{-1}. As can be observed from Figure 5, well-defined limiting current-profiles were achieved at all polarization curves. It is clear that the limiting current increased when an increase in the concentration of ferricyanide solution was attained. This behavior indicated that the mass transfer rate at the electrode surface depends on the concentration of ions in the electrolyte. By plotting the limiting current profiles, as a function of the ferricyanide concentration, a linear relationship was reached (inset in Figure 5) and, consequently, the experimental k_m was estimated by Equation (5).

Mass-transfer rate was approximately 8.8×10^{-6} m s^{-1}; this shows that the design allowed a good migration of ferricyanide ions towards the electrode surface during electrolysis due to its design and hydrodynamic conditions. However, the value of k_m obtained with the batch cell was slightly higher than the pre-pilot plant (1.9×10^{-5} and 8.8×10^{-6} m s^{-1}, for batch and flow cells, respectively), reaching better mass transfer in the former configuration. This behavior is completely associated with the residence-solution time on the reaction compartment, which is lesser at the pre-pilot plant than that probably attained at the batch cell. It is important to also state here that the treated-effluent volume is different at both cells, and it significantly influences the k_m as well as the hydrodynamic reactor configuration. In the electrochemical flow cell, the variation of k_m may not necessarily result in significant treatment efficiency for complex organics dissolved in the groundwater polluted by the fuel station.

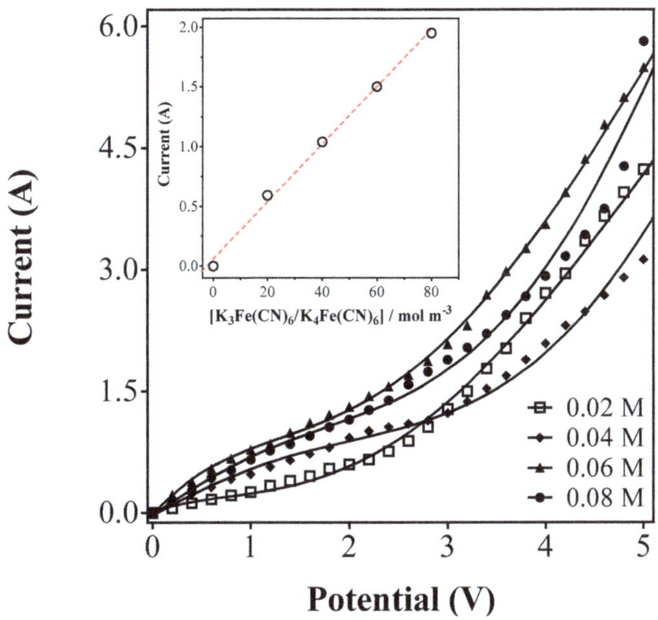

Figure 5. Polarization curves for the mass-transfer characterization of the electrochemical flow cell using 20–80 mol m^{-3} ferricyanide redox couple in 0.5 M NaOH. Inset: variation of limiting current as a function of ferricyanide concentration.

In this frame, the electrochemical treatment was extended to 5 L of polluted-groundwater effluent with Na_2SO_4 ($\approx$0.5 M) using a pilot plant with Ti/RuO_2 double-side anode by applying 30 mA cm^{-2} at 25 °C to verify its viability for an industrial scale. As cathodes, two stainless steel plates were used. Figure 6 shows that the degradation of organic matter in the real effluent reached about 68% after 300 min of treatment. This behavior is mainly due to the reaction of the organics in the effluent with $^{\bullet}OH$ at the RuO_2 surface as well as the hydrodynamic conditions of the reactor and the increase in the conductivity by adding Na_2SO_4. By determining COD removal, 87% was achieved, evidencing the effectiveness of SDG 6 electrochemical-based solution to eliminate dissolved organics in groundwater. The concentration of some pollutants after the treatment was efficiently reduced (see Table 3). Discoloration was also achieved by using the pre-pilot plant (Figure 6 and Table 3), giving general information about the elimination of organics from the effluent, especially when dealing with real matrices and/or advanced oxidation/reduction technologies. The chromophore groups, which are part of the molecules responsible for the effluent color, are fragmented during the electrochemical treatment via the participation of oxidizing species, consequently promoting a reduction of the color. However, spectral interferences can be attained during the spectrophotometric measurements due to the formation of intermediates and water matrix components (e.g., natural organic matter, inorganic species), which may absorb radiation at the wavelength of the target organic pollutants. An average of 65% of color abatement is achieved with this methodology of treatment (Table 3), which means that the yellowish color of the wastewater disappears during the 5 h oxidation. However, an increment of $SUVA_{254}$ value occurs, probably due to the difficulty in the removal of recalcitrant aromatic compounds, evidenced by the maintenance of TOC values even with a large COD decrease. This behavior is normal in real wastewater matrices [39,53]. Finally, considering the E_{cell}, electrolysis time (300 min), and COD removal, 2.53 kWh kg COD^{-1} was necessary to efficiently reduce the concentration of main pollutants and other dissolved organics (see Table 3).

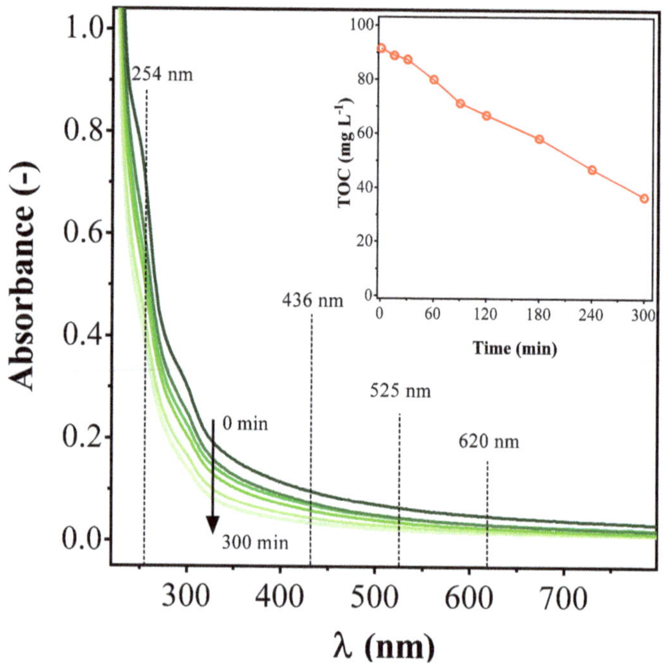

Figure 6. Wavelength spectrum and TOC decay (inset), as a function of time, of 5 L of polluted groundwater effluent by applying 30 mA cm^{-2} at 25 °C in presence of Na_2SO_4 (≈0.5 M), with the pilot plant with Ti/RuO_2 double-side anode.

Table 3. Physical-chemical characterization after EO treatment at the pre-pilot plant with a Ti/RuO_2 double-sided anode.

Parameter	Before Treatment	After Treatment	MAV [a]
COD (mg O_2 L^{-1})	230	30	-
TOC (mg C L^{-1})	91.5	36.7	-
pH	6.59	6.72	-
Phenol content (μg L^{-1})	6	<1	3
Benzene (μg L^{-1})	96.6	<1.5	5
Toluene (μg L^{-1})	2441	<1.5	170
Ethylbenzene (μg L^{-1})	925.5	<1.5	200
Xylene (*o*-, *p*- and *m*-) (μg L^{-1})	5435.5	500	300
PAHs (mg L^{-1})	4.66	4	-
Color (DFZ at 436, 525, and 620 nm)	9.5; 6.5; 4.9	3.5; 2.2; 1.7	-
Absorbance at 254 nm (AU)	0.714	0.402	
$SUVA_{254}$	0.78	1.09	-

Note: [a] Maximum Allowed Value for groundwater (human consumption) based on current legislation in Brazil [34].

4. Conclusions

The electrochemical oxidation treatment of polluted groundwater proposed in this work is proper for the efficient removal of color and organics of major concern from the environment, which is in agreement with SDG 6 (i.e., depollution of water, sanitation, disinfection, water treatment, as well as their specialized use for monitoring pollutants) [8,54]. The comparative analysis of organics removal of effluent was performed by employing three different anodes (Ti, Ru, and BDD) while varying the current density, in addition to the use of Na_2SO_4 as an electrolyte. Less energy consumption was identified with the Ti/RuO_2 anode, besides higher mineralization. In this way, the viability of industrial-scale treatment of water with the same characteristics as shown in this work is demonstrated by employing a pre-pilot with a Ti/RuO_2 double-sided anode, 0.5 M of Na_2SO_4 as the electrolyte, and at 30 mA cm^{-2} (25 °C, 5 h treatment). With these characteristics, 87% of COD abatement was achieved, mineralization of almost 40% through TOC analysis, and abatement of 65% of color. The BTEX compounds concentration was lowered to values authorized by the Brazilian environmental legislations, with exception of xylene, which decreased by 90% but requires a longer reaction time for fulfillment of the normative legislation for human consumption.

Author Contributions: Conceptualization, D.R.d.S.; E.V.d.S.; C.A.M.-H.; methodology, J.C.O.d.S.; A.M.S.S.; I.D.B.S.; formal analysis, J.C.O.d.S.; A.M.S.S.; E.V.d.S. and C.A.M.-H.; investigation J.C.O.d.S.; A.M.S.S.; D.R.d.S.; E.V.d.S.; C.A.M.-H.; resources, C.A.M.-H.; data curation, J.C.O.d.S.; E.V.d.S.; D.R.d.S. and C.A.M.-H.; writing—original draft preparation, J.C.O.d.S.; A.M.S.S.; I.D.B.S.; E.V.d.S. and C.A.M.-H.; writing—review and editing, J.C.O.d.S.; A.M.S.S.; E.V.d.S. and C.A.M.-H.; funding acquisition, E.V.d.S. and C.A.M.-H. All authors have read and agreed to the published version of the manuscript.

Funding: Conselho Nacional de Desenvolvimento Científico e Tecnológico (CNPq, Brazil) (306323/2018-4, 312595/2019-0, 439344/2018-2, 150933/2021-5). Fundação de Amparo à Pesquisa do Estado de São Paulo (Brazil), FAPESP 2014/50945-4 and 2019/13113-4. Alexander von Humboldt Foundation (Germany) and CAPES (Brazil) 88881.136108/2017-01.

Institutional Review Board Statement: Not applicable.

Informed Consent Statement: Not applicable.

Data Availability Statement: Not applicable.

Acknowledgments: Financial supports from Conselho Nacional de Desenvolvimento Científico e Tecnológico (CNPq, Brazil) (306323/2018-4, 312595/2019-0, 439344/2018-2), and from Fundação de Amparo à Pesquisa do Estado de São Paulo (Brazil), FAPESP 2014/50945-4 and 2019/13113-4, are gratefully acknowledged. Carlos A. Martínez-Huitle acknowledges the funding provided by the Alexander von Humboldt Foundation (Germany) and CAPES (Brazil) as a Humboldt fellowship for Experienced Researcher (88881.136108/2017-01) at the Johannes Gutenberg-Universität Mainz, Germany. Inalmar D. Barbosa Segundo gratefully acknowledges the financial support under post-doctoral grant awarded (CNPq, 150933/2021-5).

Conflicts of Interest: The authors declare no conflict of interest.

References

1. Foster, S.; Chilton, J. Policy experience with groundwater protection from diffuse pollution—A review. *Curr. Opin. Environ. Sci. Health* **2021**, *23*, 100288. [CrossRef]
2. Yang, Y.; Cheng, Y. Evaluating the ability of transformed urban agglomerations to achieve Sustainable Development Goal 6 from the perspective of the water planet boundary: Evidence from Guanzhong in China. *J. Clean. Prod.* **2021**, *314*, 128038. [CrossRef]
3. United Nations. *Sustainable Development Goal 6 Synthesis Report 2018 on Water and Sanitation*; United Nations: New York, NY, USA, 2018.
4. Chamanehpour, E.; Sayadi, M.H.; Yousefi, E. The potential evaluation of groundwater pollution based on the intrinsic and the specific vulnerability index. *Groundw. Sustain. Dev.* **2019**, *10*, 100313. [CrossRef]
5. Ibigbami, O.A.; Adeyeye, E.I.; Adelodun, A.A. Polychlorinated Biphenyls and Polycyclic Aromatic Hydrocarbons in Groundwater of Fuel-Impacted Areas in Ado-Ekiti, Nigeria. *Polycycl. Aromat. Compd.* **2020**, *42*, 2433–2446. [CrossRef]
6. Liu, S.-H.; Lai, C.-Y.; Ye, J.-W.; Lin, C.-W. Increasing removal of benzene from groundwater using stacked tubular air-cathode microbial fuel cells. *J. Clean. Prod.* **2018**, *194*, 78–84. [CrossRef]
7. Lien, P.; Yang, Z.; Chang, Y.; Tu, Y.; Kao, C. Enhanced bioremediation of TCE-contaminated groundwater with coexistence of fuel oil: Effectiveness and mechanism study. *Chem. Eng. J.* **2016**, *289*, 525–536. [CrossRef]
8. Misra, A.K. Climate change and challenges of water and food security. *Int. J. Sustain. Built Environ.* **2014**, *3*, 153–165. [CrossRef]

9. Mishra, B.K.; Chakraborty, S.; Kumar, P.; Saraswat, C. Correction to: Sustainable Solutions for Urban Water Security. In *Sustainable Solutions for Urban Water Security: Innovative Studies*; Mishra, B.K., Chakraborty, S., Kumar, P., Saraswat, C., Eds.; Springer International Publishing: Cham, Switzerland, 2020; p. C1. ISBN 978-3-030-53110-2.
10. Martínez-Huitle, C.A.; Panizza, M. Electrochemical oxidation of organic pollutants for wastewater treatment. *Curr. Opin. Electrochem.* **2018**, *11*, 62–71. [CrossRef]
11. Martínez-Huitle, C.A.; Sirés, I.; Rodrigo, M.A. Editorial overview: Electrochemical technologies for wastewater treatment with a bright future in the forthcoming years to benefit of our society. *Curr. Opin. Electrochem.* **2021**, *30*, 100905. [CrossRef]
12. Ganiyu, S.O.; Martínez-Huitle, C.A. The use of renewable energies driving electrochemical technologies for environmental applications. *Curr. Opin. Electrochem.* **2020**, *22*, 211–220. [CrossRef]
13. Henrique, J.M.; Monteiro, M.K.; Cardozo, J.C.; Martínez-Huitle, C.A.; da Silva, D.R.; dos Santos, E.V. Integrated-electrochemical approaches powered by photovoltaic energy for detecting and treating paracetamol in water. *J. Electroanal. Chem.* **2020**, *876*, 114734. [CrossRef]
14. Martínez-Huitle, C.A.; Rodrigo, M.A.; Sirés, I.; Scialdone, O. Single and Coupled Electrochemical Processes and Reactors for the Abatement of Organic Water Pollutants: A Critical Review. *Chem. Rev.* **2015**, *115*, 13362–13407. [CrossRef] [PubMed]
15. Panizza, M.; Cerisola, G. Direct And Mediated Anodic Oxidation of Organic Pollutants. *Chem. Rev.* **2009**, *109*, 6541–6569. [CrossRef] [PubMed]
16. Brillas, E.; Martínez-Huitle, C.A. Decontamination of wastewaters containing synthetic organic dyes by electrochemical methods. An updated review. *Appl. Catal. B Environ.* **2015**, *166–167*, 603–643. [CrossRef]
17. Tavares, M.G.; da Silva, L.V.; Solano, A.M.S.; Tonholo, J.; Martínez-Huitle, C.A.; Zanta, C.L. Electrochemical oxidation of Methyl Red using Ti/Ru0.3Ti0.7O2 and Ti/Pt anodes. *Chem. Eng. J.* **2012**, *204–206*, 141–150. [CrossRef]
18. Solano, A.M.S.; Martínez-Huitle, C.A.; Garcia-Segura, S.; El-Ghenymy, A.; Brillas, E. Application of electrochemical advanced oxidation processes with a boron-doped diamond anode to degrade acidic solutions of Reactive Blue 15 (Turqueoise Blue) dye. *Electrochim. Acta* **2016**, *197*, 210–220. [CrossRef]
19. Martínez-Huitle, C.A.; Brillas, E. A critical review over the electrochemical disinfection of bacteria in synthetic and real wastewaters using a boron-doped diamond anode. *Curr. Opin. Solid State Mater. Sci.* **2021**, *25*, 100926. [CrossRef]
20. Ganiyu, S.O.; Martínez-Huitle, C.A.; Oturan, M.A. Electrochemical advanced oxidation processes for wastewater treatment: Advances in formation and detection of reactive species and mechanisms. *Curr. Opin. Electrochem.* **2020**, *27*, 100678. [CrossRef]
21. Bergmann, H. Electrochemical disinfection—State of the art and tendencies. *Curr. Opin. Electrochem.* **2021**, *28*, 100694. [CrossRef]
22. Escalona-Durán, F.; da Silva, D.R.; Martínez-Huitle, C.A.; Villegas-Guzman, P. The synergic persulfate-sodium dodecyl sulfate effect during the electro-oxidation of caffeine using active and non-active anodes. *Chemosphere* **2020**, *253*, 126599. [CrossRef]
23. Chaplin, B.P. The Prospect of Electrochemical Technologies Advancing Worldwide Water Treatment. *Accounts Chem. Res.* **2019**, *52*, 596–604. [CrossRef] [PubMed]
24. Ganiyu, S.O.; Martínez-Huitle, C.A. Nature, Mechanisms and Reactivity of Electrogenerated Reactive Species at Thin-Film Boron-Doped Diamond (BDD) Electrodes During Electrochemical Wastewater Treatment. *ChemElectroChem* **2019**, *6*, 2379–2392. [CrossRef]
25. Nidheesh, P.V.; Divyapriya, G.; Oturan, N.; Trellu, C.; Oturan, M.A. Environmental Applications of Boron-Doped Diamond Electrodes: 1. Applications in Water and Wastewater Treatment. *ChemElectroChem* **2019**, *6*, 2124–2142. [CrossRef]
26. He, Y.; Lin, H.; Guo, Z.; Zhang, W.; Li, H.; Huang, W. Recent developments and advances in boron-doped diamond electrodes for electrochemical oxidation of organic pollutants. *Sep. Purif. Technol.* **2018**, *212*, 802–821. [CrossRef]
27. Du, X.; Oturan, M.A.; Zhou, M.; Belkessa, N.; Su, P.; Cai, J.; Trellu, C.; Mousset, E. Nanostructured electrodes for electrocatalytic advanced oxidation processes: From materials preparation to mechanisms understanding and wastewater treatment applications. *Appl. Catal. B Environ.* **2021**, *296*, 120332. [CrossRef]
28. Pérez, J.; Llanos, J.; Sáez, C.; López, C.; Cañizares, P.; Rodrigo, M. Development of an innovative approach for low-impact wastewater treatment: A microfluidic flow-through electrochemical reactor. *Chem. Eng. J.* **2018**, *351*, 766–772. [CrossRef]
29. Yang, N.; Yu, S.; Macpherson, J.V.; Einaga, Y.; Zhao, H.; Zhao, G.; Swain, G.M.; Jiang, X. Conductive diamond: Synthesis, properties, and electrochemical applications. *Chem. Soc. Rev.* **2018**, *48*, 157–204. [CrossRef]
30. Manan, T.S.B.A.; Khan, T.; Mohtar, W.H.M.W.; Beddu, S.; Kamal, N.L.M.; Yavari, S.; Jusoh, H.; Qazi, S.; Supaat, S.K.B.I.; Adnan, F.; et al. Dataset on specific UV absorbances (SUVA254) at stretch components of Perak River basin. *Data Brief* **2020**, *30*, 105518. [CrossRef]
31. German Institute for Standardization. Water Quality - Examination and Determination of Colour (ISO 7887:2012-04), 2012. Available online: https://shop.standards.ie/en-ie/standards/din-en-iso-7887-2012-04-399902_saig_din_din_875612/ (accessed on 9 September 2022).
32. Martínez-Huitle, C.A.; Quiroz, M.A.; Comninellis, C.; Ferro, S.; De Battisti, A. Electrochemical incineration of chloranilic acid using Ti/IrO_2, Pb/PbO_2 and Si/BDD electrodes. *Electrochim. Acta* **2004**, *50*, 949–956. [CrossRef]
33. Moreira, F.C.; Boaventura, R.A.R.; Brillas, E.; Vilar, V.J.P. Electrochemical advanced oxidation processes: A review on their application to synthetic and real wastewaters. *Appl. Catal. B Environ.* **2017**, *202*, 217–261. [CrossRef]
34. Brazilian National Council of the Environment. Resolution Number 396, 3 April 2008. Available online: https://www.braziliannr.com/brazilian-environmental-legislation/conama-resolution-39608/ (accessed on 9 September 2022).

35. Cardozo, J.C.; da Silva, D.R.; Martínez-Huitle, C.A.; Quiroz, M.A.; dos Santos, E.V. The versatile behavior of diamond electrodes—Electrochemical examination of the anti-psychotic drug olanzapine (OL) oxidation as a model organic aqueous solution. *Electrochim. Acta* **2022**, *411*, 140063. [CrossRef]
36. Ganiyu, S.O.; dos Santos, E.V.; Martínez-Huitle, C.A.; Waldvogel, S.R. Opportunities and challenges of thin-film boron-doped diamond electrochemistry for valuable resources recovery from waste: Organic, inorganic, and volatile product electrosynthesis. *Curr. Opin. Electrochem.* **2021**, *32*, 100903. [CrossRef]
37. Panizza, M.; Cerisola, G. Applicability of electrochemical methods to carwash wastewaters for reuse. Part 1: Anodic oxidation with diamond and lead dioxide anodes. *J. Electroanal. Chem.* **2010**, *638*, 28–32. [CrossRef]
38. Ganiyu, S.O.; dos Santos, E.V.; Costa, E.C.T.D.A.; Martínez-Huitle, C.A. Electrochemical advanced oxidation processes (EAOPs) as alternative treatment techniques for carwash wastewater reclamation. *Chemosphere* **2018**, *211*, 998–1006. [CrossRef]
39. Segundo, I.D.B.; Martins, R.J.; Boaventura, R.A.; Silva, T.F.; Moreira, F.C.; Vilar, V.J. Finding a suitable treatment solution for a leachate from a non-hazardous industrial solid waste landfill. *J. Environ. Chem. Eng.* **2021**, *9*, 105168. [CrossRef]
40. Brito, L.R.; Ganiyu, S.O.; dos Santos, E.V.; Oturan, M.A.; Martínez-Huitle, C.A. Removal of antibiotic rifampicin from aqueous media by advanced electrochemical oxidation: Role of electrode materials, electrolytes and real water matrices. *Electrochim. Acta* **2021**, *396*, 139254. [CrossRef]
41. Solano, A.M.S.; de Araújo, C.K.C.; de Melo, J.V.; Peralta-Hernandez, J.M.; da Silva, D.R.; Martínez-Huitle, C.A. Decontamination of real textile industrial effluent by strong oxidant species electrogenerated on diamond electrode: Viability and disadvantages of this electrochemical technology. *Appl. Catal. B Environ.* **2012**, *130–131*, 112–120. [CrossRef]
42. de Moura, D.C.; Brito, C.D.N.; Quiroz, M.A.; Pergher, S.; Martinez-Huitle, C.A. Cl-mediated electrochemical oxidation for treating an effluent using platinum and diamond anodes. *J. Water Process Eng.* **2015**, *8*, e31–e36. [CrossRef]
43. Sirés, I.; Brillas, E.; Oturan, M.A.; Rodrigo, M.A.; Panizza, M. Electrochemical Advanced Oxidation Processes: Today and Tomorrow. A Review. *Environ. Sci. Pollut. Res.* **2014**, *21*, 8336–8367. [CrossRef]
44. Divyapriya, G.; Nidheesh, P. Electrochemically generated sulfate radicals by boron doped diamond and its environmental applications. *Curr. Opin. Solid State Mater. Sci.* **2021**, *25*, 100921. [CrossRef]
45. Barreto, J.P.D.P.; Araújo, K.; de Araújo, D.M.; Martinez-Huitle, C.A. Effect of sp3/sp2 Ratio on Boron Doped Diamond Films for Producing Persulfate. *ECS Electrochem. Lett.* **2015**, *4*, E9–E11. [CrossRef]
46. Araújo, K.C.D.F.; da Silva, D.R.; dos Santos, E.V.; Varela, H.; Martínez-Huitle, C.A. Investigation of persulfate production on BDD anode by understanding the impact of water concentration. *J. Electroanal. Chem.* **2020**, *860*, 113927. [CrossRef]
47. Srivastava, V.; Kumar, M.S.; Nidheesh, P.V.; Martínez-Huitle, C.A. Electro catalytic generation of reactive species at diamond electrodes and applications in microbial inactivation. *Curr. Opin. Electrochem.* **2021**, *30*, 100849. [CrossRef]
48. Serrano, K.G. A critical review on the electrochemical production and use of peroxo-compounds. *Curr. Opin. Electrochem.* **2020**, *27*, 100679. [CrossRef]
49. Scaria, J.; Nidheesh, P.V. Comparison of hydroxyl-radical-based advanced oxidation processes with sulfate radical-based advanced oxidation processes. *Curr. Opin. Chem. Eng.* **2022**, *36*, 100830. [CrossRef]
50. Divyapriya, G.; Singh, S.; Martínez-Huitle, C.A.; Scaria, J.; Karim, A.V.; Nidheesh, P. Treatment of real wastewater by photoelectrochemical methods: An overview. *Chemosphere* **2021**, *276*, 130188. [CrossRef]
51. Ferreira, M.B.; Solano, A.M.S.; Dos Santos, E.V.; Martínez-Huitle, C.A.; Ganiyu, S.O. Coupling of Anodic Oxidation and Soil Remediation Processes: A Review. *Materials* **2020**, *13*, 4309. [CrossRef]
52. Ganiyu, S.O.; El-Din, M.G. Insight into in-situ radical and non-radical oxidative degradation of organic compounds in complex real matrix during electrooxidation with boron doped diamond electrode: A case study of oil sands process water treatment. *Appl. Catal. B Environ.* **2020**, *279*, 119366. [CrossRef]
53. Segundo, I.D.B.; Moreira, F.C.; Silva, T.F.; Webler, A.D.; Boaventura, R.A.; Vilar, V.J. Development of a treatment train for the remediation of a hazardous industrial waste landfill leachate: A big challenge. *Sci. Total Environ.* **2020**, *741*, 140165. [CrossRef]
54. Mishra, B.K.; Chakraborty, S.; Kumar, P.; Saraswat, C. *Sustainable Solutions for Urban Water Security*; Springer International Publishing: New York, NY, USA, 2020. [CrossRef]

water

MDPI

Article

Insights into the Kinetics, Theoretical Model and Mechanism of Free Radical Synergistic Degradation of Micropollutants in UV/Peroxydisulfate Process

Zhixiong Liu [1], Wenlei Qin [2], Lei Sun [1], Huiyu Dong [2], Xiangjuan Yuan [1,3,*], Fei Pan [1] and Dongsheng Xia [1,3,*]

1 School of Environmental Engineering, Wuhan Textile University, Wuhan 430073, China
2 Research Center for Eco-Environmental Sciences, University of Chinese Academy of Sciences, 18 Shuang-qing Road, Beijing 100085, China
3 Engineering Research Center for Clean Production of Textile Dyeing and Printing, Ministry of Education, Wuhan 430073, China
* Correspondence: yuanxiangjuan1986@outlook.com (X.Y.); dongsheng_xia@wtu.edu.cn (D.X.); Tel.: +86-027-5936-7685 (X.Y.); Fax: +86-027-5936-7338 (X.Y.)

Abstract: The degradation of acyclovir (ACY) and atenolol (ATL) in the UV/peroxydisulfate (UV/PDS) process has been systematically considered, focusing on the degradation kinetics, theoretical models, and reaction pathways via applying a microfluidic UV reaction system. The removal efficiencies of ACY and ATL were >94.8%, and the apparent degradation rate constants (k_{obs}) were 0.0931 and 0.1938 min^{-1} at pH 6.0 in the UV/PDS system. The sulfate radical ($SO_4^{\bullet-}$) and hydroxyl radical ($^{\bullet}OH$) were identified as the major reactive radicals. The pH-dependent reaction rate constants of ACY and ATL with $^{\bullet}OH$ and $SO_4^{\bullet-}$ were measured via the competing kinetics. Meanwhile, the contributions of $^{\bullet}OH$ and $SO_4^{\bullet-}$ for ACY and ATL degradation were calculated by the radical steady-state hypothesis, and the results revealed that $SO_4^{\bullet-}$ occupied a decisive position (>84.5%) for the elimination of ACY and ATL. The contribution of $^{\bullet}OH$ became more significant with the increasing pH, while $SO_4^{\bullet-}$ was still dominant. Moreover, ACY and ATL degradation performance were systematically evaluated via the experiments and Kintecus model under different operational parameters (Cl^-, Br^-, HCO_3^-, NOM, etc.) in the UV/PDS process. Furthermore, the plausible reaction pathways of ACY and ATL were elucidated based on the Fukui function theory and ultra-performance liquid chromatography-tandem quadrupole time-of-flight mass spectrometry (UPLC-QTOF-MS) analysis. The UV/PDS process has been demonstrated to be an efficient and potential application for micropollutants mitigation.

Keywords: UV/PDS; micropollutants; radical contribution rates; kinetic models; degradation pathways

Citation: Liu, Z.; Qin, W.; Sun, L.; Dong, H.; Yuan, X.; Pan, F.; Xia, D. Insights into the Kinetics, Theoretical Model and Mechanism of Free Radical Synergistic Degradation of Micropollutants in UV/Peroxydisulfate Process. *Water* **2022**, *14*, 2811. https://doi.org/10.3390/w14182811

Academic Editors: Dionysios (Dion) Demetriou Dionysiou, Yujue Wang and Huijiao Wang

Received: 18 August 2022
Accepted: 6 September 2022
Published: 9 September 2022

1. Introduction

Acyclovir (ACY) and atenolol (ATL), as the typical and most widely used antiviral and β-blocker agents, are frequently adopted against cardiovascular diseases and two common virus infections (herpes simplex and varicella-zoster) [1,2]. ACY and ATL have been frequently detected in different aquatic environments (from ng L^{-1} to μg L^{-1}) owing to their scarce biodegradability and widespread usage during the last decade [2–4]. ACY and ATL could not be effectively eliminated by the conventional treatment processes, including precipitation, biodegradation, and flocculation; hence, wastewater treatment plant effluents are regarded as one of the most important pollution resources. Therefore, in order to effectively respond to Green China and Carbon Neutrality [5,6], it is urgent to explore efficient and eco-friendly technologies for the diminution of ACY and ATL in water treatment.

In this sense, the UV/peroxydisulfate process (UV/PDS) has received considerable attention recently due to the in situ generation of sulfate radicals ($SO_4^{\bullet-}$, 2.5–3.1 V) and hydroxyl radicals (${}^{\bullet}OH$, 1.8–2.7 V), relatively high stability, the wider operational pH range, and its high mineralization efficiency of organic pollutants [7–11]. To date, many studies have been conducted to investigate the removal efficiencies of diverse pollutants, the effects of main influential parameters, and the identification of reactive species, etc., in the UV/PDS process [12–15]. Liu et al. (2013) compared the degradation performance of ATL in UV/hydrogen peroxide (UV/H_2O_2) and UV/PDS processes and observed that UV/PDS was superior to UV/H_2O_2 for efficient ATL degradation due to the longer half-life time of $SO_4^{\bullet-}$ and its higher reactivity with ATL than ${}^{\bullet}OH$ [3]. Although the removal efficiencies of diverse micropollutants in the UV/PDS process have been widely investigated, limited reports have taken note of the contributions and reaction mechanisms of $SO_4^{\bullet-}$ and ${}^{\bullet}OH$ under different operational conditions.

On the basis of previous studies, the degradation of pollutants in the UV/PDS process is mainly attributed to ${}^{\bullet}OH$, $SO_4^{\bullet-}$, and UV direct photolysis [9,12,16]. In order to accurately investigate the contributions and reaction mechanisms of $SO_4^{\bullet-}$ and ${}^{\bullet}OH$ in the UV/PDS system, the microfluid-based UV reaction system (MVPS) was deployed in this study. The MVPS system, with the advantages of a small total volume, minimum waste liquid generation, and controllable reaction rate via radiation flux, could explore the dynamics process in detail and in-depth [17]. Moreover, both the steady-state assumption and the Kintecus model were applied to compare the radical contribution rates. Furthermore, the elucidation of the degradation pathways of ACY and ATL based on density functional theory (DFT) has rarely been reported in the UV/PDS system.

Therefore, the aims of the present study are (1) to explore the degradation performance of ACY and ATL and the effects of influencing factors in the UV/PDS process; (2) to determine the pH-dependent reaction rate constants of ${}^{\bullet}OH$ and $SO_4^{\bullet-}$ with ACY and ATL; (3) to calculate the contributions of ${}^{\bullet}OH$ and $SO_4^{\bullet-}$ at different operational factors via steady-state assumption and the Kintecus software; (4) to elucidate the degradation pathways of ACY and ATL based on DFT and ultra-performance liquid chromatography-tandem quadrupole time-of-flight mass spectrometry (UPLC-QTOF-MS) analysis in the UV/PDS process. It is worth noting that this research would provide some theoretical support and preliminary exploration, especially for the applicability in wastewater for micropollutants abatement replying on the UV/PDS process.

2. Material and Methods

2.1. Chemicals

Sinopharm Chemical Reagent (Shanghai, China) provided the majority of the commonly used experimental drugs. The PDS, humic acid sodium salt, ACY (>99.0%), ATL (>98%), and 5,5-dimethyl-1-pyrrolidine N-oxide (DMPO, >95%) were obtained from Sigma-Aldrich (St. Louis, MO, USA) and Aladdin (Shanghai, China). Methanol (MeOH) and acetonitrile (ACN) were chromatographically pure and supplied by Fisher Scientific (Geel, Belgium). Ultrapure water (>18.2 MΩ cm), obtained from an A10 Milli-Q system (Millipore, Boston, MA, USA), was used in all of the experiments.

2.2. Experimental Procedures

The batch experiments were conducted in the MVPS equipped with a micro-fluorescent silica detector (MFSD) in the present study (Figure S1). The detailed structure and operational parameters were described in the study by Li et al. [17]. The photon fluence rates at 254 nm ($F_{p,o,UV}$) in the MVPS was determined and calculated to be 1.12×10^{-3} Einstein m^{-2} s^{-1} in this study, based on the degradation of uridine (0.01 mM) (Text S1, Figure S2) [18].

Before the batch experiments in MVPS, the lamp was warmed for 30 min until the MFSD signal was stabilized. A mixture solution of moderate PDS reagent, ACY and ATL solution, and phosphate buffer (5 mM) was pumped through the UV tube to receive irradiation in MVPS, and the reaction was started. All of the reactions were performed at 20 °C. The samples were withdrawn from the MVPS system at interval times, immediately quenched with $Na_2S_2O_3$ solution, and stored at 4 °C for subsequent analysis.

2.3. Analytical Methods

The e2695 high-performance liquid chromatography (HPLC) system equipped with a Sunfire C18 column (150 mm × 4.6 mm, 5μm) and a 2489 Vis-UV detector (Waters, Milford, MA, USA) was used to measure the concentrations of ACY and ATL during the photolysis reaction. MeOH and water (v:v = 40:60), as the mobile phase with a flow rate of 0.55 mL min^{-1}, were deployed for ACY detection, while phosphate and ACN (v:v = 95:5) at a flow rate of 1.0 mL min^{-1} was used for ATL analysis. DMPO, a spin-trapping reagent, was employed for the identification of radicals for the electron paramagnetic resonance test (EPR, Bruker A-300 spectrometer, Karlsruhe, Baden Wuerttemberg, Germany). Moreover, the UPLC-QTOF-MS system (Agilent, Wilmington, DE, USA) was applied to determine the intermediates of ACY and ATL in the UV/PDS process. The detection parameters of the UPLC-QTOF-MS system are Supplemented in Text S2 in detail.

2.4. Determination of the Second-Order Reaction Rate Constants

The second-order reaction rate constants of ACY and ATL with $^\bullet$OH ($k_{^\bullet OH,ACY}$ and $k_{^\bullet OH,ATL}$) and $SO_4^{\bullet-}$ ($k_{SO_4^{\bullet-},ACY}$ and $k_{SO_4^{\bullet-},ATL}$) at different pH values were determined using the competitive kinetics method. The p-chlorobenzoic acid (pCBA) ($k_{^\bullet OH,pCBA}$ = 5×10^9 M^{-1} s^{-1}) and benzoic acid (BA) ($k_{SO_4^{\bullet-},BA}$ = 1.2×10^9 M^{-1} s^{-1}) were chosen as the competitive chemicals in the UV/H_2O_2 and UV/PDS processes, respectively [19]. The calculation steps and results are depicted in Text S3 and Figures S3–S7.

2.5. Determination of the Relative Contributions of $^\bullet$OH and $SO_4^{\bullet-}$

The degradation of ACY and ATL in the UV/PDS system could be represented as Equations (1) and (2).

$$-\ln\frac{[\mathrm{ACY}]_T}{[\mathrm{ACY}]_0} = k_{^\bullet OH,ACY}\int[^\bullet\mathrm{OH}]dt + k_{SO_4^{-},ACY}\int[\mathrm{SO_4^{\bullet-}}]dt + k_{PDS,ACY}\int[\mathrm{PDS}]dt + k_{UV,ACY}t \quad (1)$$

$$-\ln\frac{[\mathrm{ATL}]_T}{[\mathrm{ATL}]_0} = k_{^\bullet OH,ATL}\int[^\bullet\mathrm{OH}]dt + k_{SO_4^{-},ATL}\int[\mathrm{SO_4^{\bullet-}}]dt + k_{PDS,ATL}\int[\mathrm{PDS}]dt + k_{UV,ATL}t \quad (2)$$

The PDS oxidation and UV direct photolysis are negligible (Figures S8 and S9). Meanwhile, Equations (1) and (2) can be simplified to Equations (3) and (4) on account of the steady-state assumption of the radicals.

$$\begin{aligned}-\ln\frac{[\mathrm{ACY}]_T}{[\mathrm{ACY}]_0} &\approx k_{^\bullet OH,ACY}\int[^\bullet\mathrm{OH}]dt + k_{SO_4^{\bullet-},ACY}\int[\mathrm{SO_4^{\bullet-}}]dt \\ &= (k_{^\bullet OH,ACY}[^\bullet\mathrm{OH}]_{ss} + k_{SO_4^{\bullet-},ACY}[\mathrm{SO_4^{\bullet-}}]_{ss})t\end{aligned} \quad (3)$$

$$\begin{aligned}-\ln\frac{[\mathrm{ATL}]_T}{[\mathrm{ATL}]_0} &\approx k_{^\bullet OH,ATL}\int[^\bullet\mathrm{OH}]dt + k_{SO_4^{\bullet-},ATL}\int[\mathrm{SO_4^{\bullet-}}]dt \\ &= (k_{^\bullet OH,ATL}[^\bullet\mathrm{OH}]_{ss} + k_{SO_4^{\bullet-},ATL}[\mathrm{SO_4^{\bullet-}}]_{ss})t\end{aligned} \quad (4)$$

where $[^\bullet OH]_{ss}$ and $[SO_4^{\bullet-}]_{ss}$ represent the steady-state concentrations of $^\bullet$OH and $SO_4^{\bullet-}$, respectively. In addition, the apparent degradation rate constants of ACY and ATL (k_{obs}) would be further simplified to Equations (5)–(7).

$$k_{obs} = k_{^\bullet OH,ACY}[^\bullet\mathrm{OH}]_{ss} + k_{SO_4^{\bullet-},ACY}[\mathrm{SO_4^{\bullet-}}]_{ss} \quad (5)$$

$$k_{obs} = k_{^\bullet OH,ATL}[^\bullet\mathrm{OH}]_{ss} + k_{SO_4^{\bullet-},ATL}[\mathrm{SO_4^{\bullet-}}]_{ss} \quad (6)$$

$$k_{obs} = k_{exp,^\bullet OH} + k_{exp,SO_4^{\bullet-}} \quad (7)$$

Nitrobenzene (NB), a typical probe compound of $^{\bullet}$OH ($k_{^{\bullet}OH,NB}$ = 3.0 × 10^9 M^{-1} s^{-1}, $k_{SO_4^{\bullet-},NB}$ < 10^6 M^{-1} s^{-1}), was used to determine the [$^{\bullet}$OH]$_{ss}$ [20,21] in the UV/PDS process. The [$^{\bullet}$OH]$_{ss}$ during the reaction could be further expressed as Equation (8):

$$k_{obs,NB} = k_{^{\bullet}OH,NB}[^{\bullet}\mathrm{OH}]_{ss} + k_{UV,NB} \quad (8)$$

where $k_{obs,NB}$ and $k_{UV,NB}$ indicate the pseudo-first-order rate constant and the single UV photolysis rate constant (min^{-1}) of NB (Figure S10). $k_{^{\bullet}OH,NB}$ represents the reaction rate constants of NB with $^{\bullet}$OH.

2.6. Kinetic Model

In this study, a reaction kinetic model was performed using Kintecus 6.80 software (James C. Ianni, Albuquerqua, NM, USA) [22] to simulate the concentrations of ACY, ATL, and reactive radicals in the UV/PDS process on the basis of 150 elementary reactions (Table S1). All of the rate constants of the above reactions were acquired from our determinations, literature, or estimations of similar reactions. The Kintecus model, with its high predictability and accuracy, has been widely used for simulating chemical reaction kinetics in the UV/chlorine [23], UV/H_2O_2 [24,25], and UV/PDS processes [12,25].

2.7. Quantum Chemistry Calculation

The molecular regioselectivity of ACY and ATL was investigated by the quantum chemical calculation based on DFT. All of the geometry optimizations were implemented at the B3LYP/6–31 G (d) level using the Gaussian 09 program [26]. The wavefunction analysis was carried out at the B3LYP/6–31G (d) level using Multiwfn 3.8 (T. Lu, Beijing, China) [27], based on the structure configuration from the Gaussian optimization. The condensed Fukui function (f^0) was employed as a popular and powerful tool in this study to predict the regioselectivity of ACY and ATL to radical attack at the atomic level. All of the calculation details of f^0 are Supplemented in Text S4.

3. Results and Discussion

3.1. Degradation Efficiencies of ACY and ATL in Different Processes

Figure 1 reveals that the removal of ACY and ATL was significantly promoted in the UV/PDS process with the k_{obs} values of 0.0931 and 0.1938 min^{-1} ($R^2 \geq 0.99$), respectively. However, the ACY and ATL degradation via direct PDS oxidation and UV photolysis can be neglected (Figure 1). The UV/PDS process can produce both $SO_4^{\bullet-}$ and $^{\bullet}$OH due to UV activation of peroxide bond and sulfate conversion [19] (Figure S11) (Equations (9) and (10)) [12,28], resulting in distinct improvement for the elimination of quinolone drugs [29], dyestuff [30], nonsteroidal anti-inflammatory drugs [10], etc.

$$S_2O_8^{\ 2-} + hv_{254} \rightarrow 2SO_4^{\bullet-} \quad (9)$$

$$SO_4^{\bullet-} + OH^- \rightarrow {}^{\bullet}OH + SO_4^{\ 2-} \quad (10)$$

To verify the roles of $SO_4^{\bullet-}$ and $^{\bullet}$OH for the enhanced ACY and ATL removal, the scavenger experiments were performed in the UV/PDS process (Figure 1). Tertiary butanol (TBA) was used to quench $^{\bullet}$OH ($k_{^{\bullet}OH,TBA}$= (3.8–7.6) × 10^8 M^{-1} s^{-1} and $k_{SO_4^{\bullet-},TBA}$= (4.0−9.1) × 10^5 M^{-1} s^{-1}) [31–33], and MeOH can quench both $SO_4^{\bullet-}$ and $^{\bullet}$OH ($k_{^{\bullet}OH,MeOH}$ = 9.7 × 10^8 M^{-1} s^{-1} and $k_{SO_4^{\bullet-},MeOH}$ = 2.5 × 10^7 M^{-1} s^{-1}) [34–36]. It is notable from Figure 1 that in the presence of TBA and MeOH, the k_{obs} values of ACY degradation reduced by 7.53% and 57.0%, while they declined by 6.73% and 42.0% for ATL, respectively. It is noteworthy from Figure 1 that $SO_4^{\bullet-}$ instead of $^{\bullet}$OH played a vital role in both ATL and ACY degradation in the UV/PDS process. Therefore, it is essential to evaluate the specific contribution rates of $SO_4^{\bullet-}$ and $^{\bullet}$OH for pollutant removal under different parameters in the UV/PDS process.

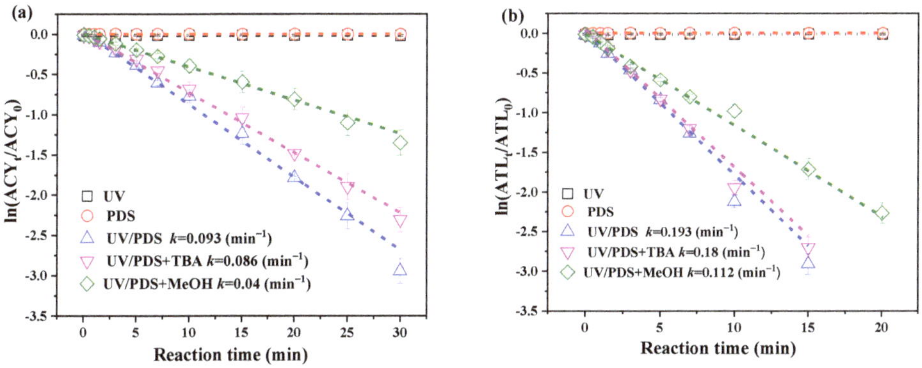

(**a**) (**b**)

Figure 1. ACY (**a**) and ATL (**b**) degradation by UV, PDS, UV/PDS, UV/PDS + TBA, and UV/PDS + MeOH processes. Conditions: $[ACY]_0$ = 0.022 mM, $[ACY]_0$ = 0.019 mM, $[PDS]_0$ = 0.6 mM, $[TBA]_0$ = $[MeOH]_0$ = 10 mM, pH = 6.0.

3.2. Effects of PDS Dosage

The effects of PDS dosage on the degradation of ACY and ATL were explored and modeled by the Kintecus 6.80 software(James C. Ianni, Albuquerqua, NM, USA). As depicted in Figure 2, with the increment in the PDS dosage from 0.1 to 1.0 mM, the degradation of both ACY and ATL was significantly strengthened, and the k_{obs} values of ACY and ATL degradation were promoted from 0.0104 to 0.1213 min^{-1} and 0.0432 to 0.3031 min^{-1} (R^2 > 0.99), respectively. The specific rate constants of $^{\bullet}OH$ and $SO_4^{\bullet-}$ for ACY and ATL degradation ($k_{exp,\bullet OH}$ and $k_{exp,\ SO_4^{\bullet-}}$) calculated by the steady-state assumption are also presented in Figure 2. The modeled degradation rates of ACY and ATL ($k_{obs,mod}$) highly agreed with the experimental values (k_{obs}) in the UV/PDS process within a credible range (<25%).

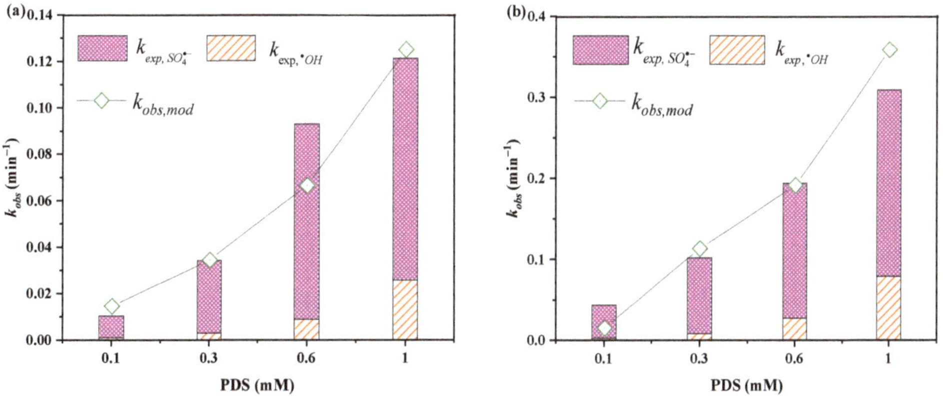

(**a**) (**b**)

Figure 2. Effects of PDS dosage on the radical contributions and k_{obs} of ACY (**a**) and ATL (**b**) in UV/PDS process. Conditions: $[ACY]_0$ = 0.022 mM, $[ATL]_0$ = 0.019 mM, $[PDS]_0$ = 0.6 mM, pH = 6.0.

In addition, it can be noted from Table 1 that the $[^{\bullet}OH]_{ss}$ and $[SO_4^{\bullet-}]_{ss}$ in the UV/PDS process gradually increased from 1.00×10^{-14} to 28.72×10^{-14} M and from 1.46

to 8.38 × 10^{-13} M as the PDS dosage rose from 0.1 to 1.0 mM. Due to the rapid photodecomposition rate of PDS (Figure S12), the yields of both $[^{\bullet}OH]_{ss}$ and $[SO_4^{\bullet-}]_{ss}$ in the UV/PDS system were evidently expedited (Table 1) [37], resulting in the gradual increment of the $k_{exp,\bullet OH}$ and $k_{exp,\ SO_4^{\bullet-}}$ values of ACY and ATL (Figure 2). The higher $[SO_4^{\bullet-}]_{ss}$ led to the higher contribution of $SO_4^{\bullet-}$ to the degradation of ACY and ATL, manifesting that $SO_4^{\bullet-}$ rather than $^{\bullet}OH$ played a major role in pollutant removal. It can be obviously seen from Figure 2 that the modeled results ($k_{obs,mod}$) derived from the kinetic model were in good agreement with the experimental values (k_{obs}) in most cases.

Table 1. $[^{\bullet}OH]_{ss}$ and $[SO_4^{\bullet-}]_{ss}$ in UV/PDS process at different PDS dosages and pHs.

Radical Species (M)	PDS (mM)				pH			
	0.1	0.3	0.6	1.0	6.0	9.0	10.0	11.0
$[^{\bullet}OH]_{ss} \times 10^{-14}$	1.00	3.00	9.83	28.72	9.44	12.27	17.72	17.86
$[SO_4^{\bullet-}]_{ss} \times 10^{-13}$	1.46	3.39	6.03	8.38	6.13	3.83	2.66	1.79

3.3. Effects of Solution pH

The $k_{\bullet OH,ACY}$,$k_{SO_4^{\bullet-},ACY}$, $k_{\bullet OH,ATL}$, and $k_{SO_4^{\bullet-},ACY}$ values at pH 6,0–11.0 were determined based on the competition kinetics, and the results are depicted in Figure 3. The $k_{\bullet OH,ACY}$ and $k_{SO_4^{\bullet-},ACY}$ values declined from 1.50 × 10^9 to 1.30 × 10^9 M^{-1} s^{-1} and from 1.58 × 10^9 to 1.00 × 10^9 M^{-1} s^{-1}, respectively, with the pH value rising from 6.0 to 11.0 (Figure 3a,b). Russo et al. reported the $k_{\bullet OH,ACY}$ to be 2.3 × 10^9 M^{-1} s^{-1} at pH 6.0 [2], and the $k_{\bullet OH,ACY}$ and $k_{SO_4^{\bullet-},ACY}$ values at different pHs have been seldom investigated. For ATL (Figure 3c,d), with the increment of pH (6.0–11.0), the $k_{\bullet OH,ATL}$ decreased from 4.6 × 10^9 to 4.11 × 10^9 M^{-1} s^{-1}, while the $k_{SO_4^{\bullet-},ATL}$ increased from 4.56 × 10^9 to 5.64 × 10^9 M^{-1} s^{-1}. The pH dependency of the second-order reaction rate constant could be ascribed to a combination of various effects due to the reaction of free radicals ($^{\bullet}OH$ and $SO_4^{\bullet-}$) with dissimilar pollutant species, which are the significant parameters of the following kinetic study.

As shown in Figure 4, with the increase in the solution pH from 6.0 to 11.0, the k_{obs} of ACY and ATL distinctly declined from 0.0931 to 0.071 min^{-1} and 0.1938 to 0.101 min^{-1}, respectively, and the $k_{obs,mod}$ were consistent with the experimental results. Additionally, as the pH increased from 6.0 to 11.0, the $k_{exp,\bullet OH}$ slightly enhanced from 0.00878 to 0.01378 min^{-1} for ACY degradation and from 0.02607 to 0.04357 min^{-1} for ATL degradation, respectively. However, the overt reduction in $k_{exp,\ SO_4^{\bullet-}}$ for ACY and ATL degradation was observed. The $k_{exp,\ SO_4^{\bullet-}}$ of ACY and ATL decreased by 30.59% and 63.79%, respectively (Figure 4).

Moreover, the $[^{\bullet}OH]_{ss}$ and $[SO_4^{\bullet-}]_{ss}$ exhibited similar change patterns with the $k_{exp,\ \bullet OH}$ and $k_{exp,\ SO_4^{\bullet-}}$ (Table 1). It is notable that the $[^{\bullet}OH]_{ss}$ obviously increased from 9.44 × 10^{-14} to 17.86 × 10^{-14} M, while the $[SO_4^{\bullet-}]_{ss}$ reduced from 6.13 to 1.79 × 10^{-13} M with the increasing pH from 6.0 to 11.0 in the UV/PDS processes. Guan et al. (2011) [38] discovered that when the solution pH > 9.3, the distinct fast conversion of $SO_4^{\bullet-}$ to $^{\bullet}OH$ caused the declining formation rate of $SO_4^{\bullet-}$ from photolysis of peroxymonosulfate (PMS) and the reduction in the oxidative power, which were in agreement with the changes of $[^{\bullet}OH]_{ss}$ and $[SO_4^{\bullet-}]_{ss}$ in the current study. Liu et al. 2013 [3] investigated the elimination of ATL in the UV/PDS process with a pH range of 3.0–11.0 and discovered the declined degradation rate of ATL with the pH decreasing from 7.0 to 9.0. Furthermore, the reaction of OH^- with $SO_4^{\bullet-}$ by Equation (10) at high pH, could also result in the decreasing trend of $SO_4^{\bullet-}$ [29].

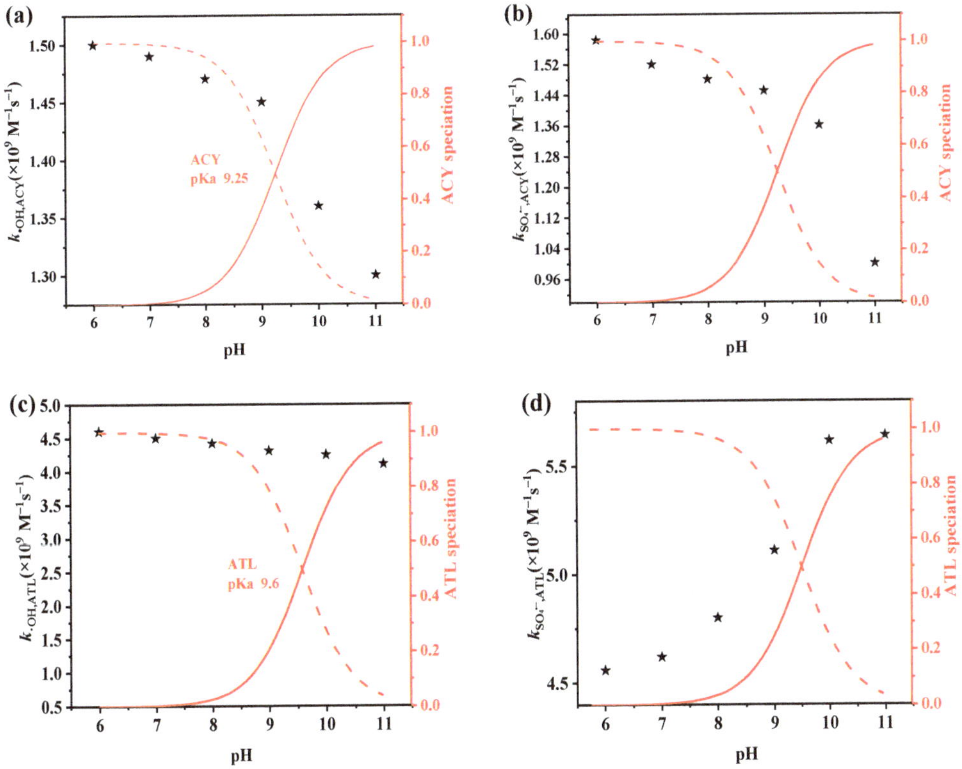

Figure 3. The second-order rate constants of $k_{\bullet OH,ACY}$(**a**) $k_{SO_4^{\bullet-},ACY}$(**b**), $k_{\bullet OH,ATL}$(**c**), and $k_{SO_4^{\bullet-},ATL}$ (**d**) under different pHs. (black star, measured k; red lines, speciation). Conditions: $[ACY]_0 = 0.022$ mM, $[ATL]_0 = 0.019$ mM, $[PDS]_0 = 0.6$ mM, pH = 6.0.

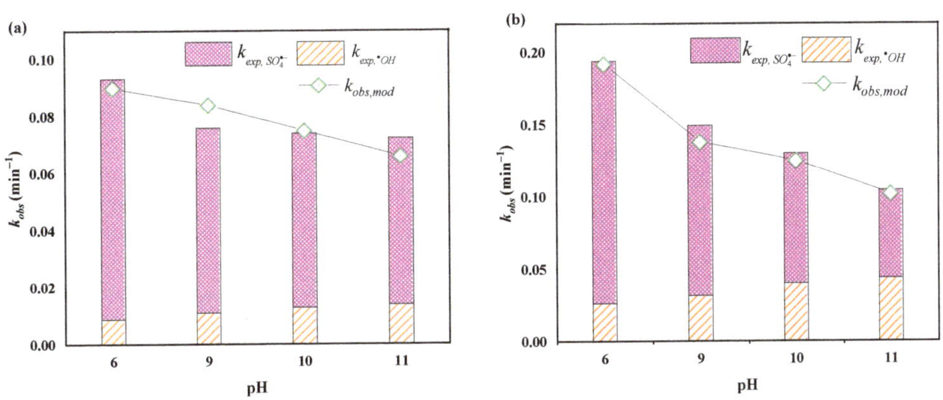

Figure 4. Effects of solution pHs on the radical contributions and k_{obs} of ACY (**a**) and ATL (**b**) in UV/PDS process. Conditions: $[ACY]_0 = 0.022$ mM, $[ATL]_0 = 0.019$ mM, $[PDS]_0 = 0.6$ mM.

3.4. Effects of Operation Parameters in UV/PDS Process

3.4.1. Chloride

The existence of Cl^- significantly affected the removal efficiencies of ACY and ATL in the UV/PDS process (Figure 5). As shown in Figure 5a,b, the k_{obs} of ACY and ATL decreased by 36.56% and 53.36%, with augmenting Cl^- concentration from 0 to 3 mM. Additionally, the $k_{mod,{}^\bullet OH}$ increased by 62.0–344.4% and $k_{mod,SO_4^{\bullet-}}$ declined by 82.7–85.1% in the presence of 3 mM Cl^-, respectively. With the augmentation of the Cl^- concentration from 0.5 to 3 mM, the $[{}^\bullet OH]_{ss}$ gradually increased from 6.62×10^{-14} to 32.24×10^{-14} M, while the $[SO_4^{\bullet-}]_{ss}$ decreased from 4.52×10^{-13} to 1.38×10^{-13} M for ACY and ATL degradation (Table S2). Lutze et al. [23] reported that $SO_4^{\bullet-}$ would be converted to ${}^\bullet OH$ in the presence of Cl^- at pH $\geq$ 5.0 (Equations (11)–(15)). Cl^- can react quickly with ${}^\bullet OH$ to form $ClOH^{\bullet-}$ with a rate constant of 4.3×10^9 M^{-1} s^{-1}(Equation (13), reverse). In addition, the extremely rapid decomposition rate constant of $ClOH^{\bullet-}$ (6.1×10^9 M^{-1} s^{-1}, Equation (13)) can obviously suppress the reaction between Cl^- and ${}^\bullet OH$. However, only at pH < 3.0 did the yield of $Cl^\bullet$ from the reaction of $ClOH^{\bullet-}$ with H^+ become important (Equation (14)) [39,40]. Noteworthily, the reactions between Cl^- and ${}^\bullet OH/SO_4^{\bullet-}$ could also produce several reactive chlorine species ($Cl^\bullet$, $Cl_2^{\bullet-}$, $ClO^\bullet$, etc.).

$$SO_4^{\bullet-} + Cl^- \rightarrow SO_4^{2-} + Cl^\bullet \; k = 3.0 \times 10^8 \quad M^{-1}s^{-1} \tag{11}$$

$$Cl^\bullet + H_2O \rightarrow ClOH^{\bullet-} + H^+ \; k = 2.5 \times 10^5 \quad M^{-1}s^{-1} \tag{12}$$

$$ClOH^{\bullet-} \leftrightarrow {}^\bullet OH + Cl^- \; k_{for} = 6.1 \times 10^9 \; M^{-1}s^{-1}, k_{rev} = 4.3 \times 10^9 \; M^{-1}s^{-1} \tag{13}$$

$$ClOH^{\bullet-} + H^+ \rightarrow Cl^\bullet + H_2O \; k = 2.1 \times 10^{10} \; M^{-1}s^{-1} \tag{14}$$

$$Cl^\bullet + Cl^- \rightarrow Cl_2^{\bullet-} \; k = 8.5 \times 10^9 \quad M^{-1}s^{-1} \tag{15}$$

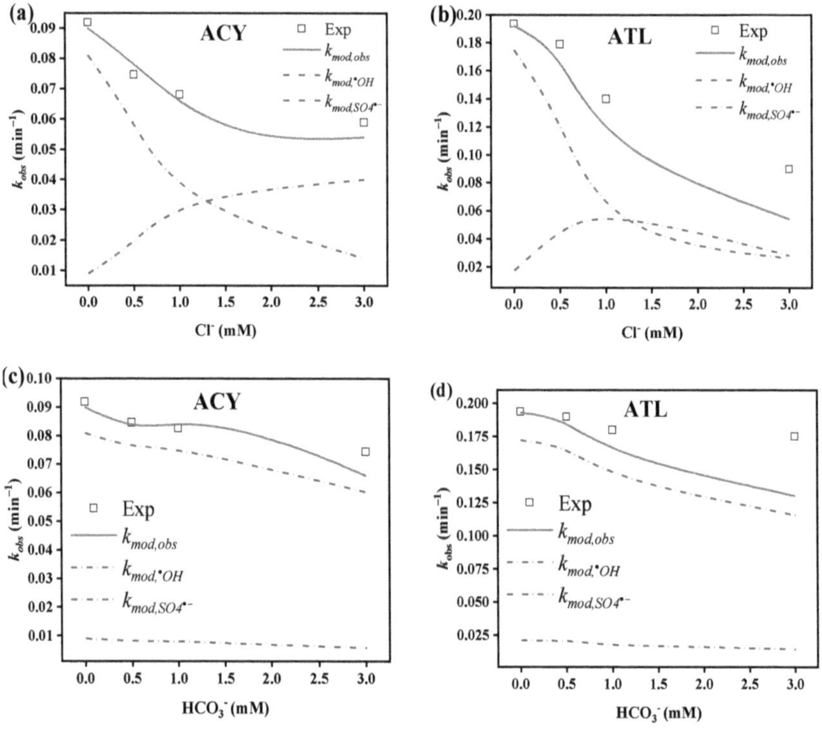

Figure 5. Effects of Cl^- and HCO_3^- dosages on the k_{obs} of ACY (**a**,**c**) and ATL (**b**,**d**) degradation and specific rates of ${}^\bullet OH$ and $SO_4^{\bullet-}$ in UV/PDS process. Conditions: $[ACY]_0$ = 0.022 mM, $[ATL]_0$ = 0.019mM, $[PDS]_0$ = 0.6 mM, pH = 6.0.

3.4.2. Bicarbonate

It can be seen from Figure 5c,d that the k_{obs} for the abatement of ACY and ATL declined by 8.8–11.82% with the addition of 3 mM HCO_3^- in the UV/PDS system. HCO_3^- can react quickly with $^{\bullet}OH$ (8.5×10^6 M^{-1} s^{-1}) and $SO_4^{\bullet-}$ (1.6×10^6 M^{-1} s^{-1}) to form secondary free radical $CO_3^{\bullet-}$ [6,41], leading to the scavenging effects of $^{\bullet}OH$ and $SO_4^{\bullet-}$. Furthermore, the presence of 3 mM HCO_3^- caused $k_{mod,^{\bullet}OH}$, and $k_{mod,SO_4^{\bullet-}}$ to decrease by 32.6–36.6% and 25.5–32.6%, respectively. Moreover, at pH 6.0, both H_2CO_3 and HCO_3^- (pK_{a1} = 6.3, pK_{a2} = 10.3) reacted with $SO_4^{\bullet-}$ in an analogous rate constant [42–44]. According to the simulation results (Table S2), $[^{\bullet}OH]_{ss}$ and $[SO_4^{\bullet-}]_{ss}$ distinctly decreased by 6.67%–5.69% and 7.08%–3.77% as the HCO_3^- concentration rose from 0.5 to 3 mM, respectively, resulting in the obvious reduction in the degradation rates of ACY and ATL.

3.4.3. Bromide

It can be observed from Figure 6a,b that the addition of Br^- inhibited the removal efficiencies of ACY and ATL in the UV/PDS process. The k_{obs} for ACY and ATL degradation decreased by 37.7 and 61.14%, respectively, as the Br^- concentration increased from 0 to 1.0 mM. Furthermore, $k_{mod,^{\bullet}OH}$ and $k_{mod,SO_4^{\bullet-}}$ decreased by 2.6–66.1% and 51.5–75.1% with the addition of 1.0 mM Br^-. Because Br^- exhibits higher rate constants reacting with $SO_4^{\bullet-}$ and $^{\bullet}OH$ (Equations (16) and (17)) than ACY and ATL do, it can be considered a scavenger of $SO_4^{\bullet-}$ and $^{\bullet}OH$, which may be the main reason for the decreased removal efficiencies of ACY and ATL. In addition, Lu et al. 2016 [45] also observed that the presence of Br^- notably suppressed the removal of o-phthalic acid (PA) in $SO_4^{\bullet-}$-based advanced oxidation systems, and PA probably did not react with the generated reactive bromide species directly.

$$SO_4^{\bullet-} + Br^- \rightarrow SO_4^{2-} + Br^{\bullet} \quad k = 3.5 \times 10^9 M^{-1}s^{-1} \tag{16}$$

$$^{\bullet}OH + Br^- \rightarrow BrOH^{\bullet-} \quad k = 1.0 \times 10^{10} M^{-1}s^{-1} \tag{17}$$

3.4.4. NOM

It can be observed from Figure 6c,d that the natural organic matter (NOM) markedly inhibited the removal efficiencies of ACY and ATL in the UV/PDS process. It is worth noting that the k_{obs} for ACY and ATL significantly dropped by 36.6% and 50.9% with the addition of 3.0 mgC L^{-1} NOM. Meanwhile, the $k_{mod,^{\bullet}OH}$, and $k_{mod,SO_4^{\bullet-}}$ decreased by 35.3–57.8% and 25.7–56.3%, respectively, under the same conditions (3 mgC L^{-1} NOM). It is noted from Figure 6 that the experimental k_{obs} for the abatement of ACY and ATL were highly compliant with the simulated results. It is well known that NOM can react with $^{\bullet}OH$ (1.4×10^4 L mgC^{-1} s^{-1}) and $SO_4^{\bullet-}$ (6.8×10^3 L mgC^{-1} s^{-1}), resulting in their reduction [23]. Furthermore, NOM could act as an inner filter of UV light affecting the yields of $^{\bullet}OH$ and $SO_4^{\bullet-}$ derived from direct photolysis of PDS [44,46].

3.4.5. Sulfate and Nitrate

It can be seen in Figure S13 that the abatement of ACY and ATL in the UV/PDS process was markedly restrained by the addition of SO_4^{2-}. The k_{obs} for ACY and ATL decreased by 15.1% and 13.8%, respectively, when the concentration of SO_4^{2-} increased to 3 mM. In addition, the $k_{mod,^{\bullet}OH}$ and $k_{mod,SO_4^{\bullet-}}$ decreased by 18.6–35.3% and 19.2–33.1% in the presence of 3 mM SO_4^{2-}. Furthermore, we can intuitively find that the experimental values are consistent with the simulated values. The higher concentration of SO_4^{2-} could induce the reductions of the redox potential of $SO_4^{\bullet-}/SO_4^{2-}$ and the oxidation capacity of the UV/PDS process on the basis of the Nernst equation (Equations (18) and (19)) [47,48]. Whereas $SO_4^{\bullet-}$ could also be formed through the reaction of $^{\bullet}OH$ with SO_4^{2-}(Equation (20)). However, the lower reaction rate constants of $SO_4^{\bullet-}$ with ACY/ATL than $^\bullet OH$ would lead to suppressive effects on the k_{obs} of ACY and ATL degradation.

$$SO_4^{\bullet -} + e^- \rightarrow SO_4^{2-} \tag{18}$$

$$E_{(SO_4^{\bullet -}/SO_4^{2-})} = E^{\theta}_{(SO_4^{\bullet -}/SO_4^{2-})} + \frac{RT}{zF} \ln \frac{[SO_4^{\bullet -}]}{[SO_4^{2-}]} \tag{19}$$

$$SO_4^{2-} + {}^\bullet OH \rightarrow SO_4^{\bullet -} + OH^- \tag{20}$$

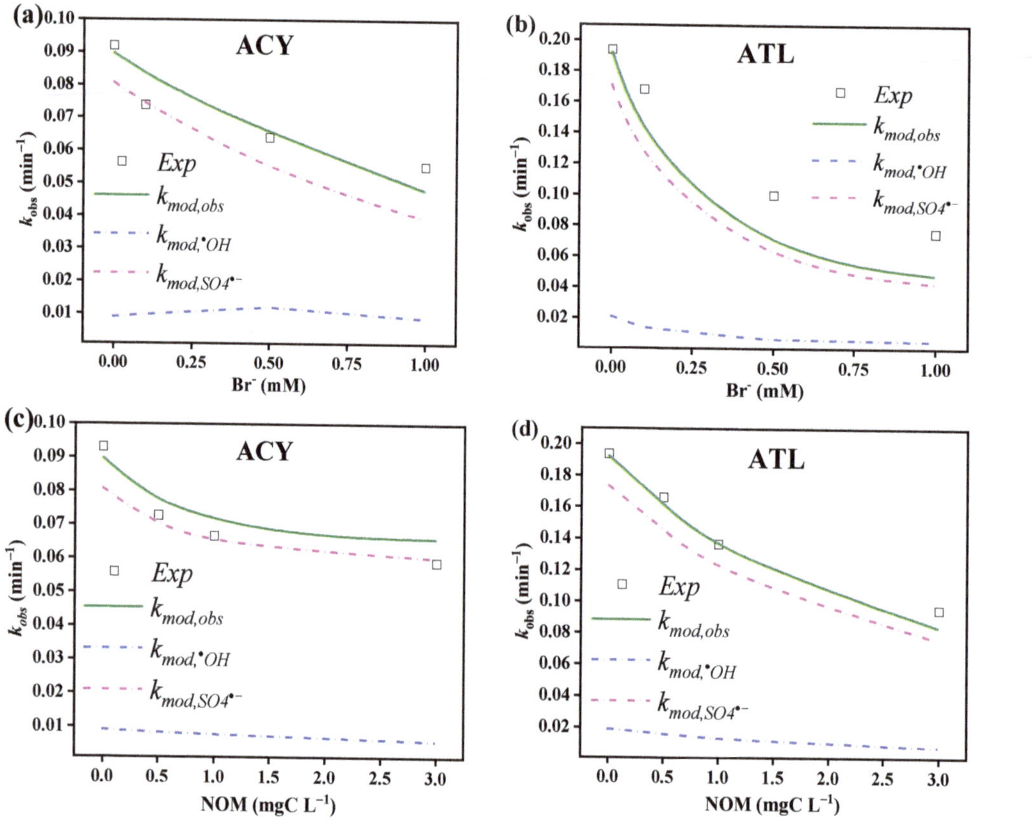

Figure 6. Effect of Br^- dosage and NOM on the k_{obs} of ACY (**a**,**c**) and ATL (**b**,**d**) degradation and specific rates of $^\bullet OH$ and $SO_4^{\bullet-}$ in UV/PDS process. Conditions: $[ACY]_0$ = 0.022 mM, $[ATL]_0$ = 0.019 mM, $[PDS]_0$ = 0.6 mM, pH = 6.0.

NO_3^- (0.5–3 mM) visibly inhibited the removal efficiencies of ACY and ATL in the UV/PDS system (Figure S14). The k_{obs} for ACY and ATL decreased by 26.0% and 27.7%, as the NO_3^- concentration rose to 3 mM. The $k_{mod,^\bullet OH}$ and $k_{mod,SO_4^{\bullet-}}$ decreased by 18.6–37.1% and 19.2–36.3% in the presence of 3 mM NO_3^-, respectively. NO_3^- can be used as a photosensitizer, consuming UV_{254} during the reaction. Wang et al. demonstrated the strong inhibition of NO_3^- on the degradation of TAP in the UV/PDS system [49]. Lin et al. revealed that NO_3^- can significantly reduce the radiation intensity received by PMS, thereby slowing down the degradation of BPA [21].

3.5. Oxidation Mechanisms and Degradation Pathways Speculation

The f^0 of ACY and ATL based on the Hirshfeld charge was calculated using Multiwfn software [46,50,51], and the results are represented in Tables S3 and S4 and Figures 7 and 8. Moreover, Tables S5 and S6 tabulated the reasonable transformation products (TPs) of ACY and ATL [52,53], which were detected by UPLC-QTOF-MS.

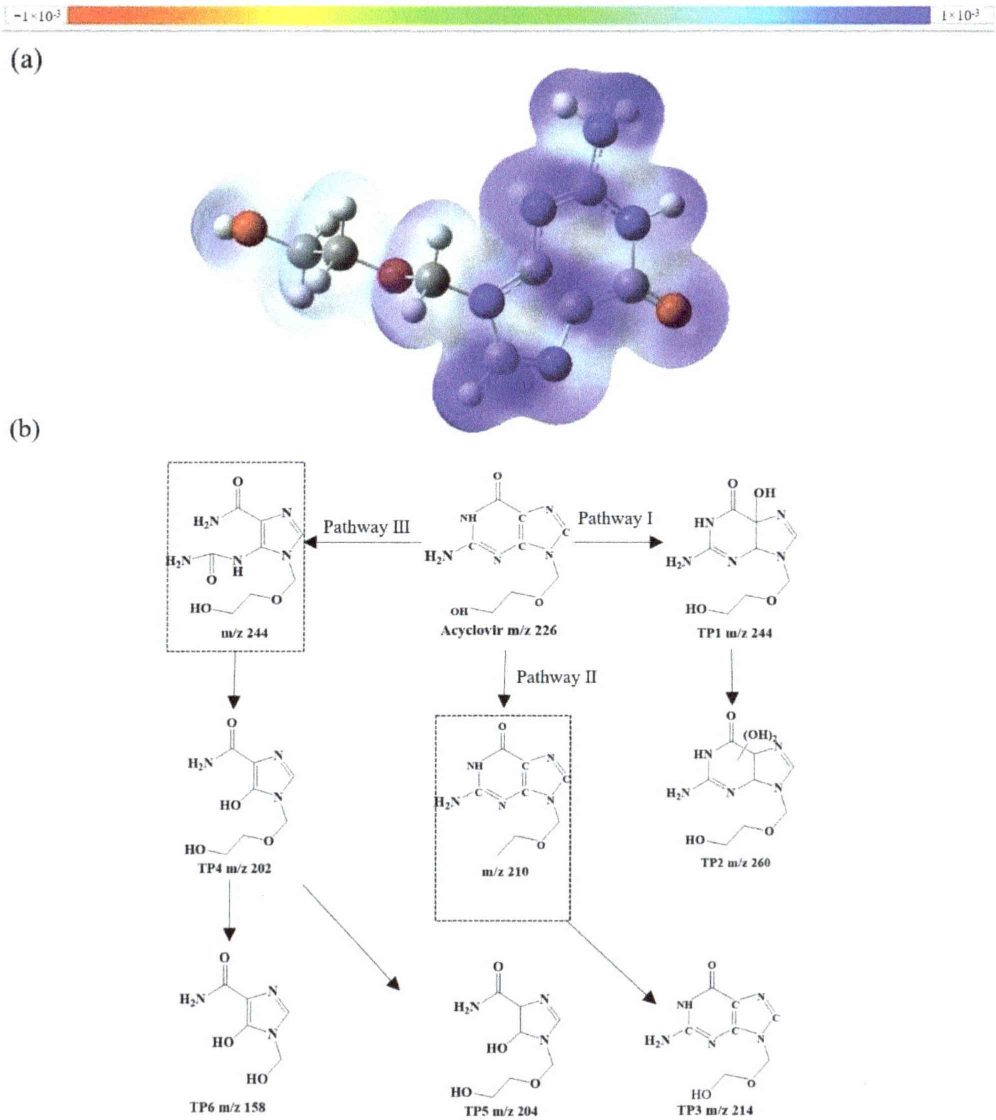

Figure 7. The contour surfaces of Fukui function f^0 for ACY with isovalue of 0.01 (**a**) and the plausible transformation pathways of ACY in UV/PDS process (**b**). Conditions: $[ACY]_0 = 0.022$ mM, $[PDS]_0 = 0.6$ mM, pH = 6.0.

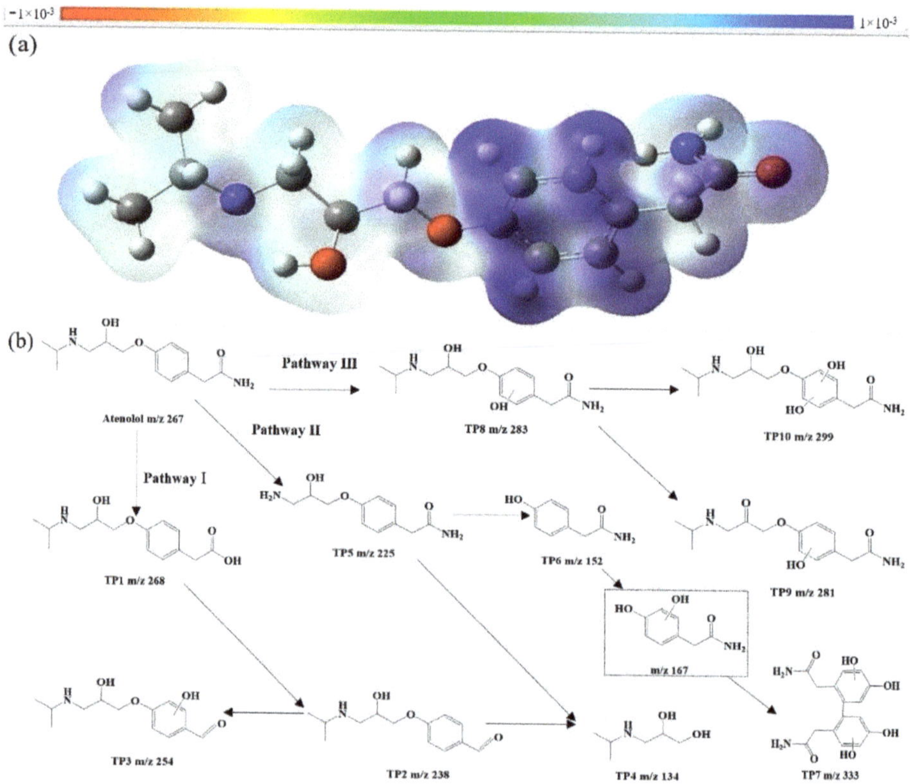

Figure 8. The contour surfaces of Fukui function f^0 for ATL with isovalue of 0.01 (**a**) and the plausible transformation pathways of ATL in UV/PDS process (**b**). Conditions: $[ATL]_0 = 0.022$ mM, $[PDS]_0 = 0.6$ mM, pH = 6.0.

According to the results in Table S3 and Figure 7a, the highest values of f^0 (blue area in Figure 7a) were achieved in the C1, C2, C5, C8, N6, N16, and O15 positions of ACY (Figure S15a), which were much easier to be attacked by $^{\bullet}OH$ and/or $SO_4^{\bullet-}$. An et al. [54] also found similar degradation patterns for ACY, and the dominating reaction pathway of ACY with $^{\bullet}OH$ initiated by the addition onto the ortho site of methoxy substituent on the benzene ring (i.e., C1 and C2 positions in this study). As shown in Figure 7b, combining the six detected transformation intermediates (Table S5) and the results of the Fukui function calculations, three degradation pathways for ACY in the UV/PDS process were proposed. Pathway I underwent the electrophilic addition reaction through $^{\bullet}OH$ attacking the ACY N-heterocycle, resulting in the formation of the hydroxylated products TP1 and TP2 through dihydroxylation. Pathway II: ACY may also transform to TP3 through a de-alcoholic reaction and subsequent hydroxylation. Pathway III: The C5 and N6 could be further attacked by $^{\bullet}OH$ and $SO_4^{\bullet-}$, further causing ring opening and the production of TP4 through further hydroxylation. Then, TP4 could generate TP5 through H addition or TP6 by de-alcoholic reaction and hydroxylation.

It can be observed from Table S4 and Figure 8a that the $^{\bullet}OH/SO_4^{\bullet-}$ radicals most favorably attack the C5, C6, C7, C18, C19, O1, and O9 positions of ATL (Figure S15b). In addition, Miao et al. [55] also reported that $^{\bullet}OH$ would most likely attack the C3 position of ATL (i.e., the C6 position in this study) to produce hydroxylation products. In addition, Miao et al. [55] also reported that the atom with a higher $2FED^2_{HOMO}$ value in the ATL structure was the O11 position (i.e., O9 position in this study), indicating that the breaking of the ether bond is reasonable. Herein, three plausible transformation pathways of ATL (Figure 8b) on the basis of the 10 transformation products (Table S6) were as follows. Pathway I: the oxidation of the primary amide group in the main side chain would lead to the formation of TP1 with m/z 268. Subsequently, the addition of oxygen to the alkyl group and hydrogen abstraction by $^{\bullet}OH$ contributed to the generation of TP2 and was further hydroxylated to produce TP3. $^{\bullet}OH/SO_4^{\bullet-}$ can attack ether linkage and undergo cleavage of ether linkage to form TP4 with m/z 134. Pathway II: The secondary amine structure of ATL could be attacked by $^{\bullet}OH/SO_4^{\bullet-}$, causing the cleavage of the C−N bond to produce TP5. Afterward, TP4 and TP6 could also be attributed to the split of the ether linkage caused by the attack of $^{\bullet}OH/SO_4^{\bullet-}$. After that, the reaction initiated at the secondary amine moieties (–NH–) and dimerization would lead to the yield of TP7. Pathway III revealed the electrophilic addition reaction through $^{\bullet}OH$ attacking the ATL aromatic ring, producing the hydroxylated product TP8, which was then carbonylated to generate TP9 and hydroxylated to TP10.

4. Conclusions

In summary, the degradation of ACY and ATL derived from $SO_4^{\bullet-}$ and $^{\bullet}OH$ in the UV/PDS process was comprehensively investigated concerning kinetic simulations and degradation mechanisms. The second-order rate constants at 6.0−11.0 for ACY and ATL reacting with $^{\bullet}OH$ were estimated to be 1.30−1.50 and 3.45−4.60 $M^{-1}s^{-1}$, and those for ACY and ATL with $SO_4^{\bullet-}$ were 1.00−1.58 and 4.56−5.64 $M^{-1}s^{-1}$, respectively. In addition, the radical quenching experiments and the steady-state assumption calculation demonstrated that $SO_4^{\bullet-}$ instead of $^{\bullet}OH$ was the predominant radical for pollutant elimination in the UV/PDS process. Additionally, the kinetic model deployed in the current study could well predict the degradation rates of ACY and ATL under different operational parameters. The degradation rates of $^{\bullet}OH$ and $SO_4^{\bullet-}$ for ACY and ATL degradation were highly pH-dependent. Based on the prediction of the active sites via the Fukui function and the intermediates of ACY and ATL identified by UPLC-QTOF-MS, the plausible transformation pathways of ACY and ATL in the UV/PDS process have been proposed. This study could provide some theoretical support and preliminary exploration for pollutant removal in the UV/PDS system. Furthermore, the toxicity of the target pollutants and their intermediates in the UV/PDS system could be considered in future studies.

Supplementary Materials: The following supporting information can be downloaded at: https://www.mdpi.com/article/10.3390/w14182811/s1, Text S1. Determination of $F_{p,o,UV}$ in the MVPS; Text S2. UPLC-QTOF-MS method; Text S3. Determination of the second-order reaction rate constants; Text S4. Calculation of condensed Fukui function; Table S1: Principal reactions and rate constants in the UV/PDS system; Table S2: Modeled steady-state concentrations of $^{\bullet}OH$ and $SO_4^{\bullet-}$ in different water matrices; Table S3: Calculated condensed Fukui function and dual descriptor of ACY; Table S4: Calculated condensed Fukui function and dual descriptor of ATL; Table S5: Major intermediates of ACY detected in UV/PDS process; Table S6: Major intermediates of ATL detected in UV/PDS process; Figure S1: Schematic diagram of the MVPS; Figure S2: The degradation of uridine under direct UV photolysis; Figure S3: Determination of the second-order rate constant for the reaction of ACY with $^{\bullet}OH$ at pH = 6.0 (a), pH = 9.0 (b), and pH = 11.0 (c); Figure S4: Determination of the second-order rate constant for the reaction of ACY with $SO_4^{\bullet-}$ at pH = 6.0 (a), pH = 9.0 (b), and pH = 11.0 (c); Figure S5: Determination of the second-order rate constant for the reaction of ATL with $^{\bullet}OH$ at pH = 6.0 (a), pH = 9.0 (b), and pH = 11.0 (c); Figure S6: Determination of the second-order rate constant for the reaction of ATL with $SO_4^{\bullet-}$ pH = 6.0 (a), pH = 9.0 (b), and pH = 11.0 (c); Figure S7: The existence forms of competing substrates pCBA(a) and BA(b) at different pH; Figure S8: Determination of ACY by UV irradiation (a) and PDS oxidation (b) at different pH; Figure S9: Determination of ATL by UV irradiation (a) and PDS oxidation (b) at different pH; Figure S10: The apparent degradation rate constants of NB in direct UV photolysis and UV/PDS processes at different pHs; Figure S11: EPR spectra of $^{\bullet}OH$ and $SO_4^{\bullet-}$ adducts in UV/PDS process; Figure S12: Attenuation of PDS in UV/PDS process for the degradation of ACY (a) and ATL (b); Figure S13: Effect of SO_4^{2-} dosage on the kobs of ACY (a) and ATL (b) degradation and specific rates of •OH and $SO_4^{•-}$ in the UV/PDS process; Figure S14: Effect of NO_3^- dosage on the k_{obs} of ACY (a) and ATL (b) degradation and specific rates of •OH and $SO_4^{•-}$ in UV/PDS process; Figure S15: Structural formula of ACY(a) and ATL (b) [56–74].

Author Contributions: Conceptualization, X.Y. and D.X.; methodology, Z.L.; software, H.D.; validation, H.D. and X.Y.; formal analysis, L.S.; investigation, Z.L. and W.Q.; writing—original draft preparation, Z.L.; writing—review and editing, X.Y. and F.P.; supervision, X.Y. and D.X.; funding acquisition, X.Y. All authors have read and agreed to the published version of the manuscript.

Funding: This work was financially supported by the National Natural Science Foundation of China (51808412).

Data Availability Statement: FigShare [https://doi.org/10.6084/m9.figshare.21067627 (accessed on 17 August 2022)]; FigShare [https://doi.org/10.6084/m9.figshare.21067618 (accessed on 17 August 2022)].

Acknowledgments: We are very grateful to Jun Men from the analysis and testing center of the Institute of Hydrobiology, Chinese Academy of Sciences with the great help for the identification and analysis of intermediates in this study.

Conflicts of Interest: The authors declare no conflict of interest.

References

1. Yu, X.P.; Qin, W.L.; Yuan, X.J.; Sun, L.; Pan, F.; Xia, D.S. Synergistic mechanism and degradation kinetics for atenolol elimination via integrated UV/ozone/peroxymonosulfate process. *J. Hazard. Mater.* **2021**, *407*, 124393. [CrossRef] [PubMed]
2. Russo, D.; Siciliano, A.; Guida, M.; Galdiero, E.; Amoresano, A.; Andreozzi, R.; Reis, N.M.; Puma, G.L.; Marotta, R. Photodegradation and ecotoxicology of acyclovir in water under UV254 and UV254/H_2O_2 processes. *Water Res.* **2017**, *122*, 591–602. [CrossRef] [PubMed]
3. Liu, X.; Fang, L.; Zhou, Y.C.; Zhang, T.Q.; Shao, Y. Comparison of UV/PDS and UV/H_2O_2 processes for the degradation of atenolol in water. *J. Environ. Sci.* **2013**, *25*, 1519–1528. [CrossRef]
4. Liang, C.J.; Wang, Z.S.; Mohanty, N. Influences of carbonate and chloride ions on persulfate oxidation of trichloroethylene at 20 °C. *Sci. Total. Environ.* **2006**, *370*, 271–277. [CrossRef] [PubMed]
5. Kranina, E.I. China on the way to achieving carbon neutrality. *Financ. J.* **2021**, *13*, 51–61. [CrossRef]
6. Bushukina, V.I. Specific features of renewable energy development in the world and russia. *Financ. J.* **2021**, *13*, 93–107. [CrossRef]
7. Ma, X.; Chen, H.; Chen, R.; Hu, X. Direct and activated chlorine dioxide oxidation for micropollutant abatement: A review on kinetics, reactive sites, and degradation pathway. *Water* **2022**, *14*, 2028. [CrossRef]
8. Parolini, M.; Pedriali, A.; Binelli, A. Application of a biomarker response index for ranking the toxicity of five pharmaceutical and personal care products (PPCPs) to the bivalve Dreissena polymorpha, Arch. Environ. *Contam.Toxicol.* **2013**, *64*, 439–447. [CrossRef]
9. Lai, F.; Tian, F.X.; Xu, B.; Ye, W.K.; Gao, Y.Q.; Chen, C.; Xing, H.B.; Wang, B.; Xie, M.J.; Hu, X.J. A comparative study on the degradation of phenylurea herbicides by UV/persulfate process: Kinetics, mechanisms, energy demand and toxicity evaluation associated with DBPs. *Chem. Eng. J.* **2022**, *428*, 132088. [CrossRef]
10. Fu, Y.Y.; Gao, X.S.; Geng, J.J.; Li, S.L.; Wu, G.; Ren, H.Q. Degradation of three nonsteroidal anti-inflammatory drugs by UV/persulfate: Degradation mechanisms, efficiency in effluents disposal. *Chem. Eng. J.* **2019**, *356*, 1032–1041. [CrossRef]
11. Hou, S.D.; Ling, L.; Shang, C.; Guan, Y.H.; Fang, J.Y. Degradation kinetics and pathways of haloacetonitriles by the UV/persulfate process. *Chem. Eng. J.* **2017**, *320*, 478–484. [CrossRef]
12. Fu, Y.Y.; Wu, G.; Geng, J.J.; Li, J.C.; Li, S.N.; Ren, H.Q. Kinetics and modeling of artificial sweeteners degradation in wastewater by the UV/persulfate process. *Water Res.* **2019**, *150*, 12–20. [CrossRef] [PubMed]
13. Wang, Z.Y.; An, N.; Shao, Y.S.; Gao, N.Y.; Du, E.; Xu, B. Experimental and simulation investigations of UV/persulfate treatment in presence of bromide: Effects on degradation kinetics, formation of brominated disinfection byproducts and bromate. *Sep. Purif. Technol.* **2020**, *242*, 116767. [CrossRef]
14. Acero, J.L.; Benítez, F.J.; Real, F.J.; Rodríguez, E. Degradation of selected emerging contaminants by UV-activated persulfate: Kinetics and influence of matrix constituents. *Sep. Purif. Technol.* **2018**, *201*, 41–50. [CrossRef]
15. Li, Q.; Wang, L.F.; Fang, X.H.; Zhang, L.; Li, J.J.; Xie, H.Y. Synergistic effect of photocatalytic degradation of hexabromocyclododecane in water by UV/TiO_2/persulfate. *Catalysts* **2019**, *9*, 189. [CrossRef]
16. Zhang, Y.Q.; Xiao, Y.J.; Zhong, Y.; Lim, T.T. Comparison of amoxicillin photodegradation in the UV/H_2O_2 and UV/persulfate systems: Reaction kinetics, degradation pathways, and antibacterial activity. *Chem. Eng. J.* **2019**, *372*, 420–428. [CrossRef]
17. Li, M.K.; Wang, C.; Yau, M.L.; Bolton, J.R.; Qiang, Z.M. Sulfamethazine degradation in water by the VUV/UV process: Kinetics, mechanism and antibacterial activity determination based on a mini-fluidic VUV/UV photoreaction system. *Water Res.* **2017**, *108*, 348–355. [CrossRef]

18. Li, M.; Li, W.T.; Bolton, J.R.; Blatchley, E.R.; Qiang, Z.M. Organic pollutant degradation in water by the vacuum-ultraviolet/ultraviolet/H_2O_2 process: Inhibition and enhancement roles of H_2O_2. *Environ. Sci. Technol.* **2019**, *53*, 912–918. [CrossRef]
19. Neghi, N.; Krishnan, N.R.; Kumar, M. Analysis of metronidazole removal and micro-toxicity in photolytic systems: Effects of persulfate dosage, anions and reactor operation-mode. *J. Environ. Chem. Eng.* **2018**, *6*, 754–761. [CrossRef]
20. Lin, Z.; Qin, W.L.; Sun, L.; Yuan, X.J.; Xia, D.S. Kinetics and mechanism of sulfate radical- and hydroxyl radical-induced degradation of Bisphenol A in VUV/UV/peroxymonosulfate system. *J. Environ. Chem. Eng.* **2020**, *38*, 101636. [CrossRef]
21. Xiao, Y.J.; Zhang, L.F.; Zhang, W.; Lim, K.Y.; Webster, R.D.; Lim, T.T. Comparative evaluation of iodoacids removal by UV/persulfate and UV/H_2O_2 processes. *Water Res.* **2016**, *102*, 629–639. [CrossRef] [PubMed]
22. Lutze, H.V.; Bircher, S.; Rapp, I.; Kerlin, N.; Bakkour, R.; Geisler, M.; von Sonntag, C.; Schmidt, T.C. Degradation of chlorotriazine pesticides by sulfate radicals and the influence of organic matter. *Environ. Sci. Technol.* **2015**, *49*, 1673–1680. [CrossRef] [PubMed]
23. Zhao, X.; Jiang, J.; Pang, S.Y.; Guan, C.T.; Li, J.; Wang, Z.; Ma, J.; Luo, C.W. Degradation of iopamidol by three UV-based oxidation processes: Kinetics, pathways, and formation of iodinated disinfection byproducts. *Chemosphere* **2019**, *221*, 270–277. [CrossRef] [PubMed]
24. Cong, J.; Wen, G.; Huang, T.L.; Deng, L.Y.; Ma, J. Study on enhanced ozonation degradation of para-chlorobenzoic acid by peroxymonosulfate in aqueous solution. *Chem. Eng. J.* **2015**, *264*, 399–403. [CrossRef]
25. Li, M.X.; Sun, J.F.; Han, D.D.; Wei, B.; Mei, Q.; An, Z.X.; Wang, X.Y.; Cao, H.J.; Xie, J.; He, M.X. Theoretical investigation on the contribution of $^\bullet OH$, $SO_4^{\bullet-}$ and $CO_3^{\bullet-}$ radicals to the degradation of phenacetin in water: Mechanisms, kinetics, and toxicity evaluation. *Ecotoxicol. Environ. Saf.* **2020**, *204*, 110977. [CrossRef]
26. Lee, M.; Zimmermann-Steffens, S.G.; Arey, J.S.; Fenner, K.; Gunten, U.V. Development of prediction models for the reactivity of organic compounds with ozone in aqueous solution by quantum chemical calculations: The role of delocalized and localized molecular orbitals. *Environ. Sci. Technol.* **2015**, *49*, 9925–9935. [CrossRef] [PubMed]
27. Lu, T.; Chen, F.W. Multiwfn: A multifunctional wavefunction analyzer. *J. Comput. Chem.* **2012**, *33*, 580–592. [CrossRef]
28. Lee, M.Y.; Wang, W.L.; Xu, Z.B.; Ye, B.; Wu, Q.Y.; Hu, H.Y. The application of UV/PS oxidation for removal of a quaternary ammonium compound of dodecyl trimethyl ammonium chloride (DTAC): The kinetics and mechanism. *Sci. Total. Environ.* **2019**, *655*, 1261–1269. [CrossRef]
29. Liu, X.H.; Liu, Y.; Lu, S.Y.; Wang, Z.; Wang, Y.Q.; Zhang, G.D.; Guo, X.C.; Guo, W.; Zhang, T.T.; Xi, B.D. Degradation difference of ofloxacin and levofloxacin by UV/H_2O_2 and UV/PS (persulfate): Efficiency, factors and mechanism. *Chem. Eng. J.* **2020**, *385*, 123987. [CrossRef]
30. Ding, X.X.; Gutierrez, L.; Croue, J.P.; Li, M.; Wang, L.J.; Wang, Y.R. Hydroxyl and sulfate radical-based oxidation of RhB dye in UV/H_2O_2 and UV/persulfate systems: Kinetics, mechanisms, and comparison. *Chemosphere* **2020**, *253*, 126655. [CrossRef]
31. Yasmeen, H.; Zada, A.; Liu, S.X. Surface plasmon resonance electron channeled through amorphous aluminum oxide bridged ZnO coupled g-C_3N_4 significantly promotes charge separation for pollutants degradation under visible light. *J. Photochem. Photobiol. A Chem.* **2020**, *400*, 112681.
32. Yasmeen, H.; Zada, A.; Liu, S.X. Dye loaded MnO_2 and chlorine intercalated g-C_3N_4 coupling impart enhanced visible light photoactivities for pollutants degradation. *J. Photochem. Photobiol. A Chem.* **2019**, *380*, 111867. [CrossRef]
33. Zhang, Z.H.; Zada, A.; Cui, N.; Liu, N.W.; Liu, M.H.; Yang, Y.Z.; Jiang, D.L.; Jiang, J.H.; Liu, S.Y. Synthesis of Ag loaded ZnO/BiOCl with high photocatalytic performance for the removal of antibiotic pollutants. *Crystals* **2021**, *11*, 981. [CrossRef]
34. Anipsitakis, G.P.; Dionysiou, D.D. Transition metal/UV-based advanced oxidation technologies for water decontamination. *Appl. Catal. B Environ.* **2004**, *54*, 155–163. [CrossRef]
35. Lescano, M.R.; Lopez, A.O.; Romero, R.L.; Zalazar, C.S. Degradation of chlorpyrifos formulation in water by the UV/H_2O_2 process: Lumped kinetic modelling of total organic carbon removal. *J. Photoch. Photobio. A* **2021**, *404*, 112924. [CrossRef]
36. Yoon, S.H.; Jeong, S.; Lee, S. Oxidation of bisphenol A by UV/$S_2O_8^{2-}$: Comparison with UV/H_2O_2. *Environ. Technol.* **2012**, *33*, 123–128. [CrossRef]
37. He, X.X.; Mezyk, S.P.; Michael, I.; Fatta-Kassinos, D.; Dionysiou, D.D. Degradation kinetics and mechanism of beta-lactam antibiotics by the activation of H_2O_2 and $Na_2S_2O_8$ under UV-254nm irradiation. *J. Hazard. Mater.* **2014**, *279*, 375–383. [CrossRef]
38. Guan, Y.H.; Ma, J.; Li, X.C.; Fang, J.Y.; Chen, L.W. Influence of pH on the formation of sulfate and hydroxyl radicals in the UV/peroxymonosulfate system. *Environ. Sci. Technol.* **2011**, *45*, 9308–9314. [CrossRef]
39. Yang, Y.; Pignatello, J.J.; Ma, J.; Mitch, W.A. Comparison of halide impacts on the efficiency of contaminant degradation by sulfate and hydroxyl radical-based advanced oxidation processes (AOPs). *Environ. Sci. Technol.* **2014**, *48*, 2344–2351. [CrossRef]
40. Gunten, U.V. Ozonation of drinking water: Part II. Disinfection and by-product formation in presence of bromide, iodide or chlorine. *Water Res.* **2003**, *37*, 1469–1487. [CrossRef]
41. Lu, X.; Shao, Y.S.; Gao, N.Y.; Chen, J.X.; Zhang, Y.S.; Xiang, H.M.; Guo, Y.L. Degradation of diclofenac by UV-activated persulfate process: Kinetic studies, degradation pathways and toxicity assessments. *Ecotoxicol. Environ. Saf.* **2017**, *141*, 139–147. [CrossRef]
42. Luo, C.W.; Jiang, J.; Ma, J.; Pang, S.Y.; Liu, Y.Z.; Song, Y.; Guan, C.T.; Li, J.; Jin, Y.X.; Wu, D.J. Oxidation of the odorous compound 2,4,6-trichloroanisole by UV activated persulfate: Kinetics, products, and pathways. *Water Res.* **2016**, *96*, 12–21. [CrossRef] [PubMed]

43. Qin, W.L.; Lin, Z.; Dong, H.Y.; Yuan, X.J.; Qiang, Z.M.; Liu, S.; Xia, D.S. Kinetic and mechanistic insights into the abatement of clofibric acid by integrated UV/ozone/peroxydisulfate process: A modeling and theoretical study. *Water Res.* **2020**, *186*, 116336. [CrossRef] [PubMed]
44. Fang, J.Y.; Fu, Y.; Shang, C. The roles of reactive species in micropollutant degradation in the UV/free chlorine system. *Environ. Sci. Technol.* **2014**, *48*, 1859–1868. [CrossRef] [PubMed]
45. Lu, J.H.; Dong, W.; Ji, Y.F.; Kong, D.Y.; Huang, Q.G. Natural organic matter exposed to sulfate radicals increases its potential to form halogenated disinfection byproducts. *Environ. Sci. Technol.* **2016**, *50*, 5060–5067. [CrossRef]
46. Wu, Z.H.; Chen, C.Y.; Zhu, B.Z.; Huang, C.H.; An, T.C.; Meng, F.A.; Fang, J.Y. Reactive nitrogen species are also involved in the transformation of micropollutants by the UV/Monochloramine process. *Environ. Sci. Technol.* **2019**, *53*, 11142–11152. [CrossRef] [PubMed]
47. Wang, Y.R.; Chu, W. Degradation of a xanthene dye by Fe(II)-mediated activation of oxone process. *J. Hazard. Mater.* **2011**, *186*, 1455–1461. [CrossRef]
48. Jaafarzadeh, N.; Ghanbari, F.; Zahedi, A. Coupling electrooxidation and oxone for degradation of 2,4-Dichlorophenoxyacetic acid (2,4-D) from aqueous solutions. *J. Water Process Eng.* **2018**, *22*, 203–209. [CrossRef]
49. Wang, F.G.; Wang, W.J.; Yuan, S.J.; Wang, W.; Hu, Z.H. Comparison of UV/H_2O_2 and UV/PS processes for the degradation of thiamphenicol in aqueous solution. *J. Photoch. Photobio. A* **2017**, *348*, 79–88. [CrossRef]
50. Zhou, Z.; Liu, X.T.; Sun, K.; Lin, C.Y.; Ma, J.; He, M.C.; Ouyang, W. Persulfate-based advanced oxidation processes (AOPs) for organic-contaminated soil remediation: A review. *Chem. Eng. J.* **2019**, *372*, 836–851. [CrossRef]
51. Lu, T.; Chen, Q. Realization of conceptual density functional theory and information-theoretic approach in Multiwfn program. *Concept. Density Funct. Theory* **2022**, *2*, 631–647.
52. Guo, H.G.; Ke, T.L.; Gao, N.Y.; Liu, Y.; Cheng, X. Enhanced degradation of aqueous norfloxacin and enrofloxacin by UV-activated persulfate: Kinetics, pathways and deactivation. *Chem. Eng. J.* **2017**, *316*, 471–480. [CrossRef]
53. Wacławek, S.; Lutze, H.V.; Grübel, K.; Padil, V.V.T.; Černík, M.; Dionysiou, D.D. Chemistry of persulfates in water and wastewater treatment: A review. *Chem. Eng. J.* **2017**, *330*, 44–62. [CrossRef]
54. An, T.C.; An, J.B.; Gao, Y.P.; Li, G.Y.; Fang, H.S.; Song, W.H. Photocatalytic degradation and mineralization mechanism and toxicity assessment of antivirus drug acyclovir: Experimental and theoretical studies. *Appl. Catal. B Environ.* **2015**, *164*, 279–287. [CrossRef]
55. Miao, D.; Peng, J.B.; Zhou, X.H.; Qian, L.; Wang, M.J.; Zhai, L.; Gao, S.X. Oxidative degradation of atenolol by heat-activated persulfate: Kinetics, degradation pathways and distribution of transformation intermediates. *Chemosphere* **2018**, *207*, 174–182. [CrossRef]
56. Li, M.K.; Qiang, Z.M.; Bolton, J.R.; Qu, J.H.; Li, W.T. A mini-fluidic UV photoreaction system for bench-scale photochemical studies. *Environ. Sci. Technol. Lett.* **2015**, *2*, 297–301. [CrossRef]
57. Scholes, M.L.; Schuchmann, M.N.; Sonntag, C.V. Enhancement of radiation-induced base release from nucleosides in alkaline solution: Essential role of the $O^{\bullet-}$ radical. *Int. J. Radiat. Biol.* **1992**, *61*, 443–449. [CrossRef]
58. Yang, Y.; Pignatello, J.J.; Ma, J.; Mitch, W.A. Effect of matrix components on UV/H_2O_2 and UV/$S_2O_8^{2-}$ advanced oxidation processes for trace organic degradation in reverse osmosis brines from municipal wastewater reuse facilities. *Water Res.* **2016**, *89*, 192–200. [CrossRef]
59. Grebel, J.E.; Pignatello, J.J.; Mitch, W.A. Effect of halide ions and carbonates on organic contaminant degradation by hydroxyl radical-based advanced oxidation processes in saline waters. *Environ. Sci. Technol.* **2010**, *44*, 6822–6828. [CrossRef]
60. Yu, X.Y.; Bao, Z.C.; Barker, J.R. Free radical reactions involving $Cl^{\bullet}$, $Cl_2^{\bullet-}$, and $SO_4^{\bullet-}$ in the 248 nm Photolysis of Aqueous Solutions Containing $S_2O_8^{2-}$ and Cl^-. *J. Phys. Chem.* **2004**, *108*, 295–308. [CrossRef]
61. Peyton, G.R. The free-radical chemistry of persulfate-based total organic carbon analyzers. *Mar. Chem.* **1993**, *41*, 91–103. [CrossRef]
62. Klaning, U.K. Laser flash photolysis of HClO, ClO^-, HBrO, and BrO^- in aqueous solution. Reactions of Cl^- and Br^- atoms. *Ber. Bunsenges. Phys. Chem.* **1985**, *89*, 243–245. [CrossRef]
63. Zehavi, D.; Rabani, J. Oxidation of aqueous bromide ions by hydroxyl radicals. Pulse radiolytic investigation. *J. Phys. Chem.* **1972**, *3*, 76. [CrossRef]
64. Lin, M.Z.; Archirel, P.; Van-Oanh, N.T.; Muroya, Y.; Fu, H.; Yan, Y.; Nagaishi, R.; Kumagai, Y.; Katsumura, Y.; Mostafavi, M. Temperature dependent absorption spectra of Br^-, $Br_2^{\bullet-}$, and Br_3^- in aqueous solutions. *J. Phys. Chem. A.* **2011**, *115*, 4241–4247. [CrossRef] [PubMed]
65. Wagner, I.; Strehlow, H. On the flash-photolysis of bromide ions in aqueous-solutions. *Ber. Bunsenges. Phys. Chem.* **1987**, *91*, 1317–1321. [CrossRef]
66. Gunten, U.V.; Oliveras, Y. Advanced oxidation of bromide-containing waters: Bromate formation mechanisms. *Environ. Sci. Technol.* **1998**, *32*, 63–70. [CrossRef]
67. Matthew, B.M.; Anastasio, C. A chemical probe technique for the determination of reactive halogen species in aqueous solution Part 1-bromide solutions. *Atmos. Chem. Phys.* **2006**, *6*, 2423–2437. [CrossRef]
68. Neta, P.; Madhavan, V.; Zemel, H.; Fessenden, R.W. Rate constants and mechanisms of reaction of $SO_4^{\bullet-}$,with aromatic compounds. *J. Am. Chem. Soc.* **1977**, *5*, 163–164. [CrossRef]
69. Sutton, H.C.; Downes, M.T. Reactions of the HO2 radical in aqueous solution with bromine and related compounds. *J. Chem. Soc. Faraday Trans 1 Phys. Chem. Condens. Phases* **1972**, *68*, 1498–1507. [CrossRef]

70. Heeb, M.B.; Criquet, J.; Zimmermann-Steffens, S.G.; Gunten, U.V. Oxidative treatment of bromide-containing waters: Formation of bromine and its reactions with inorganic and organic compounds—A critical review. *Water Res.* **2014**, *48*, 15–42. [CrossRef]
71. Beckwith, R.C.; Wang, T.X.; Margerum, D.W. Equilibrium and kinetics of bromine hydrolysis. *Inorg. Chem.* **1996**, *35*, 995–1000. [CrossRef] [PubMed]
72. Liu, Y.Z.; Yang, Y.; Pang, S.Y.; Zhang, L.Q.; Ma, J.; Luo, C.W.; Guan, C.T.; Jiang, J. Mechanistic insight into suppression of bromate formation by dissolved organic matters in sulfate radical-based advanced oxidation processes. *Chem. Eng. J.* **2018**, *333*, 200–205. [CrossRef]
73. Schwarz, H.A.; Bielski, B.H.J. Reactions of hydroperoxo and superoxide with iodine and bromine and the iodide (I_2^-) and iodine atom reduction potentials. *J. Phys. Chem.* **1986**, *90*, 1445–1448. [CrossRef]
74. Buxton, G.V.; Dainton, F.S. The radiolysis of aqueous solutions of oxybromine compounds; the spectra and reactions of BrO and BrO_2. *Math. Phys. Sci.* **1968**, *304*, 427–439.

water

Article

Using Electrochemical Oxidation to Remove PFAS in Simulated Investigation-Derived Waste (IDW): Laboratory and Pilot-Scale Experiments

Amy Yanagida [1], Elise Webb [2], Clifford E. Harris [3], Mark Christenson [2] and Steve Comfort [1,*]

1 School of Natural Resources, University of Nebraska, Lincoln, NE 68583, USA
2 AirLift Environmental, LLC, 5900 N. 58th, Suite 5, Lincoln, NE 68583, USA
3 Department of Chemistry, Albion College, Albion, MI 49224, USA
* Correspondence: scomfort@unl.edu; Tel.: +1-402-472-1502

Abstract: Repeated use of aqueous firefighting foams at military aircraft training centers has contaminated groundwater with per and polyfluorinated alkyl substances (PFAS). To delineate the extent of PFAS contamination, numerous site investigations have occurred, which have generated large quantities of investigation-derived wastes (IDW). The commonly used treatment of incinerating PFAS-tainted IDW is costly, and was recently suspended by the Department of Defense. Given long term IDW storage in warehouses is not sustainable, our objective was to use electrochemical oxidation to degrade PFAS in contaminated water and then scale the technology toward IDW treatment. This was accomplished by conducting a series of laboratory and pilot-scale experiments that electrochemically oxidized PFAS using direct current with boron-doped diamond (BDD) electrodes. To improve destruction efficiency, and understand factors influencing degradation rates, we quantified the treatment effects of current density, pH, electrolyte and PFAS chain length. By using ^{14}C-labeled perfluorooctanoic acid (PFOA) and tracking temporal changes in both ^{14}C-activity and fluoride concentrations, we showed that oxidation of the carboxylic head (-$^{14}COOH \rightarrow {}^{14}CO_2$) was possible and up to 60% of the bonded fluorine was released into solution. We also reported the efficacy of a low-cost, 3D printed, four-electrode BDD reactor that was used to treat 189 L of PFOA and PFOS-contaminated water ($C_o \leq 10$ μg L^{-1}). Temporal monitoring of PFAS with LC/MS/MS in this pilot study showed that PFOS concentrations decreased from 9.62 μg L^{-1} to non-detectable (<0.05 μg L^{-1}) while PFOA dropped from a concentration of 8.16 to 0.114 μg L^{-1}. Efforts to improve reaction kinetics are ongoing, but current laboratory and pilot-scale results support electrochemical oxidation with BDD electrodes as a potential treatment for PFAS-tainted IDW.

Keywords: boron-doped diamond electrodes; per and polyfluorinated alkyl substances; chemical oxidation; ^{14}C-labeled PFOA

Citation: Yanagida, A.; Webb, E.; Harris, C.E.; Christenson, M.; Comfort, S. Using Electrochemical Oxidation to Remove PFAS in Simulated Investigation-Derived Waste (IDW): Laboratory and Pilot-Scale Experiments. *Water* **2022**, *14*, 2708. https://doi.org/10.3390/w14172708

Academic Editors: Yujue Wang, Dionysios (Dion) Demetriou Dionysiou and Huijiao Wang

Received: 22 July 2022
Accepted: 20 August 2022
Published: 31 August 2022

1. Introduction

Per- and polyfluoroalkyl substances (PFAS) are a family of synthesized chemicals that have been heavily used in manufacturing, often without adequate disposal. Commonly used to improve the quality of commercial products by making them resistant to heat, oil, and stains, PFAS are found in many household goods and industrial products such as surfactants, emulsifiers, and aqueous film forming foams (AFFFs). Due to their unique ability to resist heat and block oxygen, PFAS-containing firefighting foams were used at aircraft service centers where fire-fighting training operations were routinely performed. With more than three decades of repeated use at hundreds of military facilities across the U.S.A., PFAS have been detected in 61% of the groundwater samples taken around Department of Defense (DoD) facilities with concentrations ranging from μg L^{-1} (ppb) to low mg L^{-1} (ppm) concentrations [1–4].

Chemically, PFAS contain both a fluorinated carbon tail and a hydrophilic ionic head (-COOH, -SO_3H), and it is this unique structure that give PFAS their biphasic or surfactant characteristics. The PFAS molecule may be either fully (per–) or partially (poly–) fluorinated and can contain between two and eighteen carbon atoms [5]. Examples of the most common PFAS include perfluorooctanoic acid (PFOA; $C_8F_{15}O_2H$) and perfluorooctane sulfonic acid (PFOS; $C_8F_{17}SO_3H$). Although PFOA and PFOS are the two most frequently detected PFAS in contaminated aquifers, there are literally hundreds of PFAS precursors that have been released into the environment, which are also causes for concern.

Unfortunately, the chemical properties that make PFAS useful from an industrial perspective also make them mobile and recalcitrant once released into the environment. As true xenobiotics, PFAS have no known natural decomposition processes and the stability of the C-F bond makes them almost un-degradable by natural attenuation [6,7].

Now that toxicological studies have revealed the bio-accumulative, neurotoxic, and potentially carcinogenic nature of PFAS exposure [8–11], identifying and delineating the extent of PFAS contamination is paramount for protecting human and ecological health. Given the United States Environmental Protection Agency (USEPA) recently decreased its 2016 health advisor level (HAL) of 70 ng L^{-1} (70 parts per trillion, 70 ppt) for the singular or combined concentration of PFOA and PFOS, to 0.004 ng L^{-1} for PFOA and 0.02 ng L^{-1} for PFOS [12], the need to further delineate the extent of PFAS contamination in aquifers will only intensify.

The federal government has an environmental responsibility to delineate the extent of contamination at sites where firefighting foams were routinely used. While PFAS-contaminated groundwater is the primary concern, site investigations create a secondary problem that is continuing to grow, namely, the generation of investigation-derived waste (IDW). IDW is the water, soil and drill cuttings generated during well installations and sampling activities performed during contaminated site investigations. The potential risks and liabilities associated with PFAS exposure has previously caused IDW generators to take the conservative approach of incinerating their PFAS waste. While energy intensive and costly, incineration offered a workable solution. The temporary moratorium on incinerating PFAS-containing substances by the Department of Defense [13] has now changed the way IDW will be handled, and complicates disposal options. Given that PFAS site investigations will continue and long-term storage of containerized IDW in warehouses is not a sustainable solution, alternative treatment options are needed. In 2020, the EPA established the PFAS Innovative Treatment Team (PITT) to address the disposal and destruction of PFAS-containing media and waste. This PITT team identified four technologies with the potential to destroy PFAS-contaminated media and waste [14]; one of these technologies was electrochemical oxidation (EC).

As an alternative to incinerating PFAS-tainted IDW, we report herein on a series of laboratory and pilot-scale experiments that electrochemically oxidized PFAS with boron-doped diamond electrodes (EC-BDD) with direct current. The reason BDD anodes can degrade per- and polyfluoroalkyl substances is because of the greater overpotential they require to oxidize water to oxygen (relative to other anodes). The oxygen overpotential for gold, platinum, or glassy carbon anodes ranges between 1.7 and 2.2 V, but up to 2.5 V for BDD anodes [15]. This allows BDD electrodes to produce direct anodic oxidation of the PFAS and offer a "chemical-free" treatment for PFAS-contaminated water [16]. Our objectives were to quantify the effects of current density, pH, electrolyte and PFAS chain length on degradation kinetics and defluorination. We also report on the efficacy of a custom designed, 3D printed, electrode chamber where the number of BDD electrodes (2 vs. 4), power sources (1 vs. 2) and DC polarity switching, were varied. We then show the results from treating a 208 L (55 gal) container of PFOA- and PFOA-tainted water (C_o = <10 μg L^{-1}).

2. Materials and Methods

2.1. Chemicals

Chemicals were purchased from a variety of vendors and used as received. These chemicals included: perfluorooctanoic acid (PFOA; $C_8F_{15}O_2H$; >95% purity) and perfluorooctane sulfonate (PFOS; $C_8F_{17}SO_3$; >95% purity) (Fisher, Fair Lawn, NJ, USA). Shorter chain PFAS included: perfluorohexanoic acid, perfluorobutyric acid; perfluorohexanesulfonic acid, and perfluorobutanesulfonic acid (Fisher, Fair Lawn, NJ; >98% purity) (Table 1). We also used ^{14}C-labeled perfluorooctanoic acid (^{14}C-PFOA; 55 *m*Ci mmol^{-1}; American Radiolabeled Chemicals, St. Louis, MO, USA); sodium sulfate, ammonium fluoride, and H_2SO_4 (Fisher, Fair Lawn, NJ, U.S.A.). All solutions were prepared with Millipore water (18.2 MΩ cm^{-1} resistivity, 25 °C) from a Nanopure Barnstead E-pure system (Thermo Scientific, Waltham, MA, USA).

Table 1. Compound name, acronym, formula, and structure of PFAS used in laboratory and pilot-scale experiments.

Compound (C#)	Acronym	Formula	Structure
Perfluorooctanoic Acid (C8)	PFOA	$F(CF_2)_7COOH$	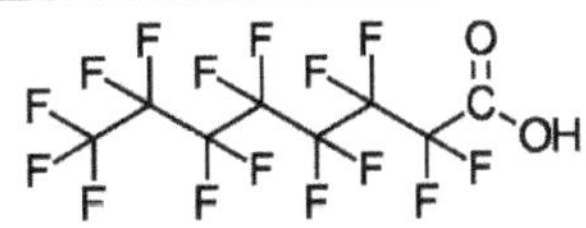
Perfluorohexanoic Acid (C6)	PFHxA	$F(CF_2)_5COOH$	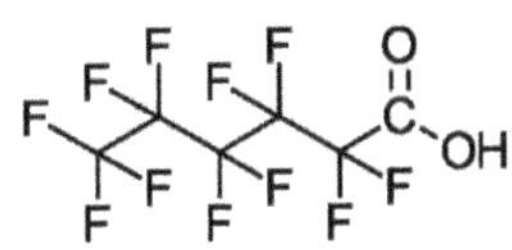
Perfluorobutyric Acid (C4)	PFBA	$F(CF_2)_3COOH$	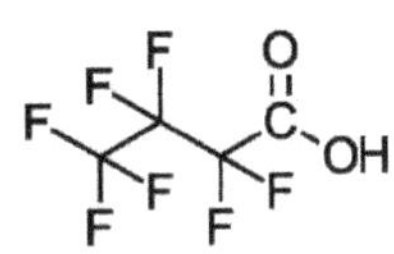
Perfluoropropionic Acid (C3)	PFPrA	$F(CF_2)_2COOH$	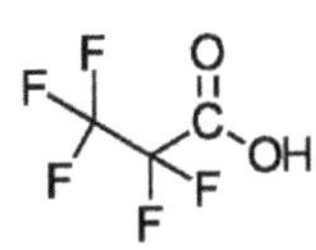
Perfluorooctane Sulfonic Acid (C8)	PFOS	$F(CF_2)_8SO_3H$	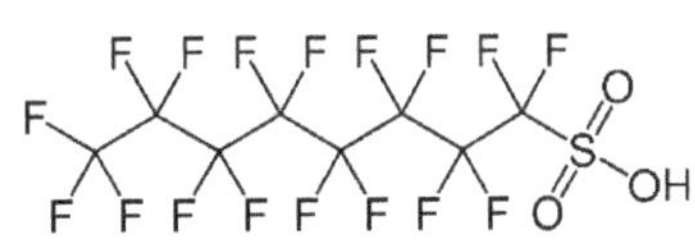

2.2. Chemical Analysis

Carbon-14 activity (^{14}C-activity) was determined by removing 1 mL subsamples from the batch reactors and mixing with 6 mL of Ultima Gold liquid scintillation cocktail (Packard, Meriden, CT, USA). Samples were then mixed and allowed to sit overnight in the dark before analyzing on a Packard 1900TR liquid scintillation counter (LSC; Packard Instrument, Downers Grove, IL, USA). A blank consisting of 6 mL Ultima Gold liquid scintillation cocktail was analyzed prior to running the samples and used to correct sample activity values (dpms).

Treating PFAS by electrochemical oxidation removes fluoride from the alkyl chain. Thus, temporal release of fluoride into solution can be representative of PFAS degradation. In select electrochemical experiments, samples were taken during treatment for fluoride analysis and run on a Dionex DX-120 ion chromatograph (IC). Standard solutions of fluoride were first prepared and injected into the IC to create a calibration curve.

Standards were made using ammonium fluoride. The fluoride standards used varied between 1 and 75 mg L^{-1}, depending on the concentration of PFAS treated. Choosing which standards to use for data analysis was based on the potential mass of fluoride that could be released into solution by breaking the C-F bonds. For example, defluorination of 10 mg L^{-1} PFOA could produce approximately 6.8 mg L^{-1} of fluoride; 100 mg L^{-1} PFOA could yield 68 mg L^{-1} fluoride, if defluorination was complete. All fluoride standards were run through the IC, and peak areas were used to generate a calibration curve through linear regression.

When ^{14}C-PFOA was not used in experiments, unlabeled PFOA and PFOS concentrations were determined at the University of Nebraska Water Center using a Waters Xevo TQ-S Micro triple quadrupole mass spectrometer with a 2D ultrahigh pressure liquid chromatography interface. Using EPA Method 533 [17], linear calibration curves were obtained for the concentration range of 0–20 μg L^{-1} using isotopically labelled internal standards. Specific standard concentrations included: 0, 0.1. 0.5, 1.0, 5, 10 and 20 μg L^{-1}. Limit of detection (LOD) and limit of quantification (LOQ) for the analytical method were determined by the following: duplicate analyses of standards; results from samples fortified at a concentration near the low calibration standard signal sensitivity; and the Student's t-test value for the number of replicates analyzed.

2.3. Electrochemical Experiments

2.3.1. Generalized Setup of Electrochemical Experiments

Most electrochemical experiments were run under similar stirred, single batch conditions. The typical experimental setup consisted of a 600 mL Erlenmeyer flask filled with 500 mL of a PFAS solution (8 μg L^{-1} to 100 mg L^{-1}). Enough electrolyte salt (Na_2SO_4) was added to bring the concentration to 10 mM Na_2SO_4. The solution was then spiked with 1 to 3 mL of stock ^{14}C-PFOA (55 mCi $mmol^{-1}$), which brought the solution's ^{14}C-activity to approximately 1500 to 3000 dpms mL^{-1} and 0.6 mL of diluted H_2SO_4 acid (10% v/v with H_2O) to decrease the solution pH (2.5). The experimental unit received a magnetic stir bar, was placed on a stir plate, and then mixed at a stir speed of 700 rpm. We used either one boron-doped diamond (BDD) anode and a platinum coated titanium (Pt/Ti) wire cathode, or two boron-doped diamond electrodes (NeoCoat®, La Chaux-de Fonds, Switzerland). The NeoCoat® electrodes consisted of a polycrystalline boron-doped diamond coating (5 μm coating, 2500 ppm B) deposited on both sides of a mesh niobium substrate. The dimensions of the mesh BDD electrodes were 25 × 100 × 1.4 mm. A plastic holder was fabricated that maintained an electrode spacing of 5 mm. These electrodes were connected to a DC power supply (30 V/20 A, Extech instruments, Nashua, NH, USA) and suspended in the solution so that they were submerged as fully as possible without the electrical alligator clips touching the solution. Electrical current was set at either 1 or 0.4 A, unless specified otherwise in presented graphs. Assuming an estimated electrode surface area of 25 cm^2, current densities ranged from 8 to 40 mA cm^{-2}. The supplied voltage varied with current density and electrolyte concentration and is reported on resulting graphs.

2.3.2. Effect of Electrical Current

Using a BDD anode and cathode, we quantified the effects of electrical current on a 100 mg L^{-1} perfluorooctanoic acid (0.24 mM PFOA) solution. Three separate experiments were run with the electrical current set at 0.2, 0.4 and 1 A (8, 16, 40 mA cm^{-2}). The experiments ran for 6 h, and 1.5 mL samples were periodically removed from the stirred reactor to quantify temporal decreases in ^{14}C-activity. One milliliter samples were mixed with LSC cocktail and analyzed in a liquid scintillation counter; ^{14}C concentrations were then fit to a first-order rate expression ($C = C_o e^{-kt}$) and compared.

Preliminary experiments were also run with low concentrations of PFOS (17 μg L^{-1}) and PFOA (4.4 μg L^{-1}), utilizing the BDD anode and cathodes and 8 mA cm^{-2}. These experiments were analyzed with a Micromass Quattro Micro Triple quadrupole liquid chromatograph/mass spectrometer with methods and results presented in the Supplementary Materials (SM).

2.3.3. Effect of Acid Addition (pH)

The effect of pH on PFOA degradation was determined by conducting parallel experiments using either two BDD electrodes, or a BDD/Pf/Ti electrode set, both run at an electrical current of 1 A (40 mA cm^{-2}). The PFAS test solution was a 100 mg L^{-1} perfluorooctanoic acid (0.24 *mM* PFOA) solution, spiked with ^{14}C-PFOA. We acidified the PFOA solution with diluted H_2SO_4 to a pH of 2.5 and left one companion solution untreated (no acid, pH 6–7). Batch reactors were run under the same conditions as above with volts ranging from 16 to 20 V.

2.3.4. Effect of Electrolyte

Using the generalized electrochemical procedures and two BDD electrodes as anode and cathode, degradation of PFOA was compared in a 10 *m*M Na_2SO_4 electrolyte salt versus a 100 *m*M KH_2PO_4 salt concentration. The pH of both solutions was adjusted to 2.5 before starting electrochemical oxidation.

2.3.5. The Effects of Reseeding PFOA and Fluoride on Degradation

Following the generalized procedures, ^{14}C-PFOA was reseeded into the stirred batch reactor and treated with two BDD electrodes at pH 2.5. Reseeding occurred at 2 h and 4 h after the initial (T = 0 h) experiment was started. Temporal samples were taken every 15 min, mixed with the cocktail, and analyzed by LSC.

To determine how fluoride released into solution was possibly influencing BDD performance, we conducted an experiment where generalized procedures were used to treat a 40 mg L^{-1} fluoride solution for 2 h before spiking with ^{14}C-PFOA and tracking changes in ^{14}C activity. A similar experiment was performed where 40 mg F^{-} L^{-1} was treated for 2 h and then spiked with a higher PFOA concentration (100 mg L^{-1} PFOA) so that defluorination could also be monitored.

2.3.6. Measuring Defluorination

To measure electrochemical oxidation and defluorination together, we used a 100 mg L^{-1} PFOA solution and spiked it with ^{14}C-PFOA. Using two BDD electrodes and direct current, we removed solution samples from the batch reactor every 30 to 60 min for 6 h. ^{14}C activity was determined by mixing 1 mL of sample with 6 mL of cocktail and counting on an LSC. Fluoride was measured directly with a Dionex DX-120 ion chromatograph using conductivity detection.

To determine how initial concentration influenced defluorination, we treated PFOA concentrations of 5, 10, 50 and 100 mg L^{-1} and ran them under standard EC conditions using an electrical current of 0.4 A (8 mA cm^{-2}; 12–14 V). Similarly, we determined how fluorinated alkyl chain length influenced the rate of defluorination. We used perfluorinated carbonates with carbon chains lengths of 8, 6, 4, and 3 carbons; specifically, these compounds included perfluorooctanoic acid (C8), perfluorohexanoic acid (C6), perfluorobutyric acid (C4), and perfluoropropionic acid (C3) (Table 1). The starting concentration of each compound was 0.24 mM, and each compound was treated with two BDD electrodes and run under standard conditions and 0.4 A.

2.3.7. Pilot-Scale Experiments

A 208 L barrel was filled with 189 L of tap water and spiked to ~10 μg L^{-1} PFOA, ~10 μg L^{-1} PFOS with a 10 *m*M Na_2SO_4 background matrix. We used a custom 3D printed four-electrode, flow-through reactor that allowed the electrodes to fit perfectly into a 3D mold and force contact with the PFAS molecules. The custom reactor was designed with the 3D CAD software SOILDWORKS (Dassault Systèmes Solidworks, Co., Waltham, MA, USA) and then printed on MakerBot Replicator 3D printer (Brooklyn, NY, USA) using MakerBot filament plastic. To further seal the reactor, flex sealant was used around the periphery of the angular joints and hose connections while a silicone adhesive sealant was used around each electrode. Reactor pieces and electrodes were assembled with 6 mm foam gaskets inserted between each section; electrodes were spaced 1 cm apart. PVC hose adaptors for inlet and outlets were connected to 3D printed material using PVC primer and glue. The final reactor design was fabricated to be semi-permanent (i.e., leak free), but still allowed for disassembly and electrode replacement.

Once constructed, the BDD reactor was connected to a custom designed direct current system (3E, Windsor Heights, IA, USA) that had a 500 W rheostat capable of adjusting current (0–16 A), 4 solid state relays for switching polarities, and two timers for adjusting positive and negative cycles. Pilot-scale experiments were run at 1 A with polarity reversed every 30 s. Tubing adaptors and a submersible pump were used to continuously circulate water through the reactor at a rate of 2.33 L min^{-1}. This flow rate allowed the 189 L of simulated IDW to cycle through the BDD reactor every 81 min. Temporal 1.5 mL samples were collected over 450 h and analyzed via LC/MS/MS.

Before treating the pilot-scale 208 L drum of PFAS-contaminated water, the assembled reactor was used to treat multiple batches of 2 L of 10 μg L^{-1} of PFOA in a 10 *m*M Na_2SO_4 matrix at a flow rate of 1.7 L min^{-1}. Variables tested with the 3D printed reactor were the number of BDD electrodes (2 vs. 4), power sources (1 vs. 2) and polarity (constant vs. switching).

3. Results and Discussion

3.1. Electrochemical Oxidation of PFAS

A series of laboratory batch experiments were undertaken using a boron-doped diamond (BDD) anode to electrochemically (EC) oxidize PFAS compounds. The advantage electrochemical oxidation has over other technologies, is that BDD anodes can initiate an electron removal from both the ionic heads of PFOA and PFOS-like structures [18–22] (Figure 1). The key factor for enhanced oxidation via EC-BDD treatment is the interaction the EC-generated hydroxyl radicals have with the electrode surface. In general, high oxidation power anodes are characterized by weak electrode–hydroxyl radical interactions resulting in a high current efficiency for organic oxidation and a low electrochemical activity for oxygen evolution. Based on this, boron-doped diamond can be considered as one of the ideal anode materials for electrochemical mineralization of organic contaminants [23].

While the exact mechanisms for electrochemical oxidation of PFAS are complex and still being investigated, there is general agreement that the rate-limiting step is the direct electron transfer at the anode, which results in cleavage of the head group (-COOH vs. $-SO_3H$) to produce the corresponding perfluoroalkyl radical (Figure 1); the perfluoroalkyl radical can then quickly react with •OH, O_2, or H_2O and degrade to the one-carbon shorter perfluoroheptanecarboxylate (i.e., perfluoro alkyl carbonate, Figure 1). The newly produced, 1-carbon shorter carboxylate then undergoes the same degradation cycle as the original PFOA, sequentially converting the carboxylic acid head to carbon dioxide, the fluorine atoms to hydrogen fluoride, and the CF_2 to another carboxylic acid group.

As discussed by Radjenovic et al. [16], the use of high PFAS concentrations (i.e., 10–100 mg/L) in laboratory experiments likely overestimates EC oxidation performance and makes extrapolating results to environmentally relevant matrices difficult. While we used high PFOA concentrations when fluoride analysis was also performed, the use of ^{14}C-PFOA provided the advantage of making analysis simple, quick, and negated any PFAS cross contamination interferences. Moreover, by measuring the ^{14}C activity of the solution (dpms mL^{-1}), converting to μCi, and then using the specific activity of the PFOA label (55,000 μCi $mmol^{-1}$), the PFOA concentration could be calculated. In most batch experiments that used only ^{14}C-PFOA, the typical starting concentrations (Co) were 6.8 to 10 μg PFOA L^{-1}, which would be environmentally relevant for many contaminated IDW matrices.

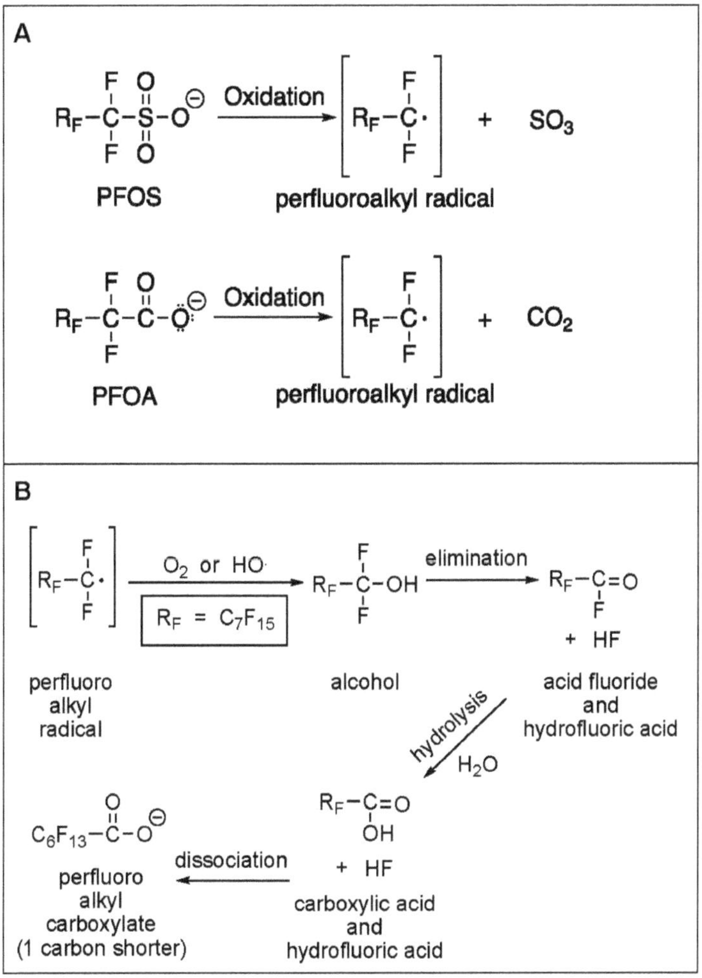

Figure 1. (**A**) Rate-limiting step for the treatment of PFOS versus PFOA to the perfluoroalkyl radical; (**B**) subsequent degradation of perfluoroalkyl radical.

3.2. *Effect of Electrical Current*

Potential is the major driving force for electrochemical oxidation of PFAS. Applied current density is a key experimental parameter that affects PFAS degradation and defluorination efficiency because it regulates the capability of the electron transfer rate and OH radical generation on the electrode surface. In the PFOA decomposition process, Ochiai et al. [24] found that when the current density was higher than 0.6 mA/cm^2, the direct electron oxidation was no longer the main mechanism. Using DC settings that would produce 0.6 mA/cm^2, we observed no change in PFOA concentrations. By using higher electrical current, however, we observed both near linear (zero) and first-order PFAS degradation rates (Figure 2). Based on previous research [25], we believe these observations can be explained by creating kinetic conditions that are either "current-controlled" or "mass-transport controlled". Under current-controlled conditions, the rate-limiting condition is electron transfer at the BDD anode to form the perfluoroalkyl radical (Figure 1). When PFAS is in abundance (i.e., not limiting), the rate of degradation is constant and controlled by the electron transfer at the BDD interface (i.e., zero-order, Figure 2). Under mass-transport controlled, the PFAS becomes limiting, and the degradation rate is controlled by the transfer of the PFAS to the BDD electrode (Figure 2). In other words, the reaction becomes concentration-dependent or first-order. By increasing the amperage from 0.2 to 0.4 and 1.0 A, the reaction rate changed from current-controlled to concentration-dependent (Figure 2).

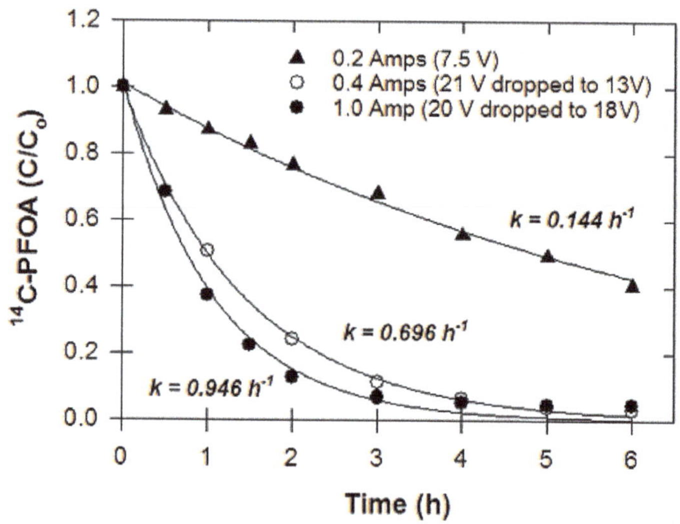

Figure 2. Temporal changes in PFOA concentrations under varying electrical currents (0.2, 0.4, and 1.0 A) and current density (8, 16, 40 mA cm^{-2}).

3.3. *pH Adjustment*

Previous studies have reported mixed results regarding pH effects on PFAS degradation during electrochemical oxidation [16]. Niu et al. [25] reported that pH affects •OH generation, oxygen overpotential and life span of BDD anodes. Further, lower pH can inhibit oxygen evolution, which can improve PFAS degradation. In most cases, however, pH effects have been reported to be slight to modest. Reasons for the lack of strong pH effects may be attributed, in part, to the lack of true pH-stat conditions. This is due to the surface acidity of the anode, which decreases the solution pH over time. While treating PFOA, Lin et al. [26] recorded bulk pH decreases of 0.5 to 1.5 units during 90 min of electrolysis.

Zhuo et al. [15] tested PFOA degradation by BDD anodes at different pH values (3, 9, 12). They found degradation rates were slightly higher at pH 3 versus 12. Lin et al. [26] similarly treated PFOA by electrochemical oxidation at pH 3, 5, 7, 9, and 11 and found pH 5 produced the highest degradation rate. Nienhauser et al. [27] treated a 4-carbon perfluorobutanesulfonic acid (PFBS), a 6-carbon perfluorohexanesulfonic acid (PFHxS) and an 8-carbon PFOA with EC-BDD at pH 3, 7 and 12; they found little pH effect for PFOA but slower kinetics at pH 12 for the 4- and 6-carbon PFAS. Given these results, we adjusted pH to determine if reaction rates could be increased. Our results of lowering pH provided mixed results in influencing degradation kinetics.

Using two BDD electrodes, the difference between a neutral pH and acidic (pH = 2.5) was negligible (Figure 3A). Moreover, using BDD/Pt/Ti electrodes, we compared degradation rates with and without pH adjustment. Results from two freshly made PFOA solutions showed that pH adjustment did not affect degradation kinetics (Figure 3B). Although pH did not influence degradation, it is noteworthy that the BDD/Pt/Ti electrodes (Figure 3B) produced the fastest degradation kinetics with a first-order rate of 1.478 h^{-1} and was greater than the degradation kinetics observed with two BDD (Figure 3A). While this combination was effective, the lifespan of the BDD anode was shortened due to mineral deposits.

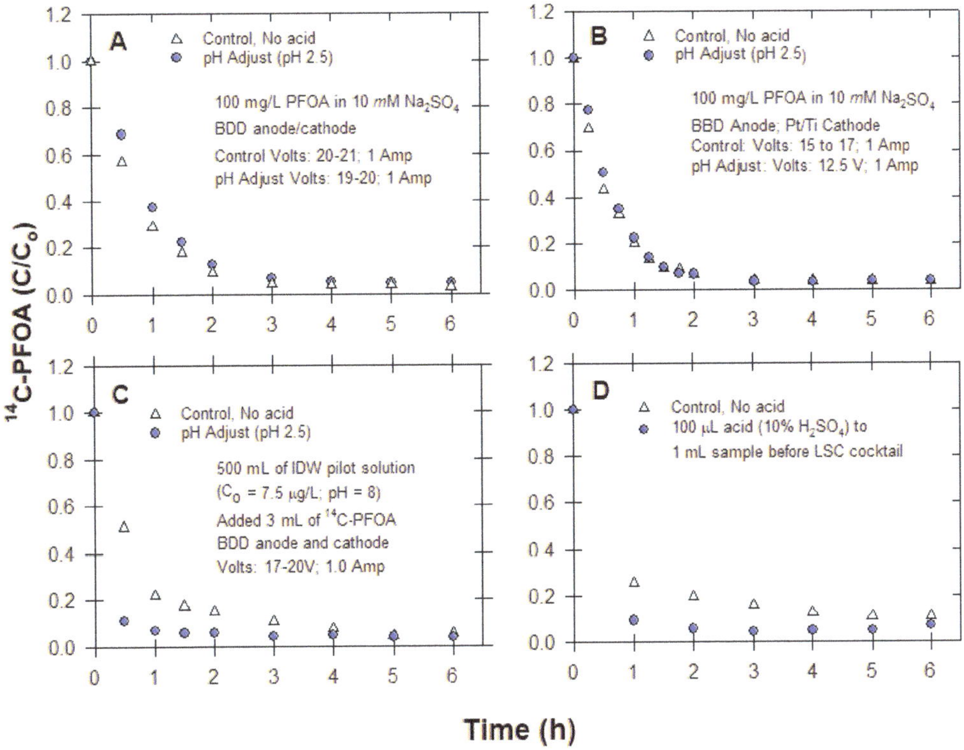

Figure 3. (**A**) Effect of pH adjustment on PFOA kinetics; (**B**) degradation of ^{14}C-PFOA with BDD anode and Pt/Ti cathode, with and without pH adjustment; (**C**) degradation of ^{14}C-PFOA in alkaline PFOA solution (pH 8) with BDD anode and cathode, with and without pH adjustment (pH adjust = 2.5); (**D**) effects of acidifying samples before mixing with liquid scintillation counting (LSC) cocktail.

An example where pH appeared to influence degradation rates was when we treated a solution from our pilot-scale IDW barrel (i.e., pilot-scale) with both BDD anodes and cathodes, we observed faster kinetics by lowering the pH (Figure 3C). Given that the IDW barrel solution was alkaline (pH ~8), and not neutral as observed in the previous acidifying experiments (i.e., experiments Figure 3A,B), we determined if the pH effect on PFOA degradation kinetics was real, as reported by Zhuo et al. [15], or an artifact of using temporal changes in ^{14}C-activity as an indicator of degradation. In other words, electrochemical oxidation of ^{14}C-PFOA produces $^{14}CO_2$ through direct anodic oxidation (Figure 1A) but the produced CO_2 is more soluble in an alkaline solution than a neutral or acid solution and thus a higher pH may cause some $^{14}CO_2$ to stay in solution rather than escape as a gas. If some $^{14}CO_2$ stays in solution, then monitoring temporal changes in ^{14}C-activity includes both ^{14}C species (^{14}C-PFOA + $^{14}CO_2$) and would yield a degradation rate that is slower than the actual degradation rate. By comparing ^{14}C-activity of samples acidified versus not acidified before mixing with scintillation cocktail, we showed that an alkaline solution indeed retained some $^{14}CO_2$, if not first acidified (Figure 3D). Thus, when the pH of the ^{14}C-PFOA solution being treated is not lowered prior to EC treatment, individual samples need to be acidified to drive out any dissolved $^{14}CO_2$ in solution.

3.4. Electrolyte

Supporting electrolytes are used to provide an electro-conductive medium and minimize the voltage drop and resistance of the electrochemical reactor [15,26]. High electrical conductivity leads to faster electron transport and a better degradation rate for organic pollutants. Zhuo et al. [28] found that the electrochemical oxidation of 6:2 FTS decreased in the trend of $NaClO_4$ > NaCl > Na_2SO_4. To avoid the potential for organochloride byproducts, chlorate, or perchlorate from forming, we compared Na_2SO_4 versus KH_2PO_4 and found no differences in rate constants (Figure 4). It is important to note that by increasing the salt concentration, lower voltages produced similar amperage. Thus, if amperage or current density was similar (8 mA cm^{-2}, Figure 4), kinetic rates were similar.

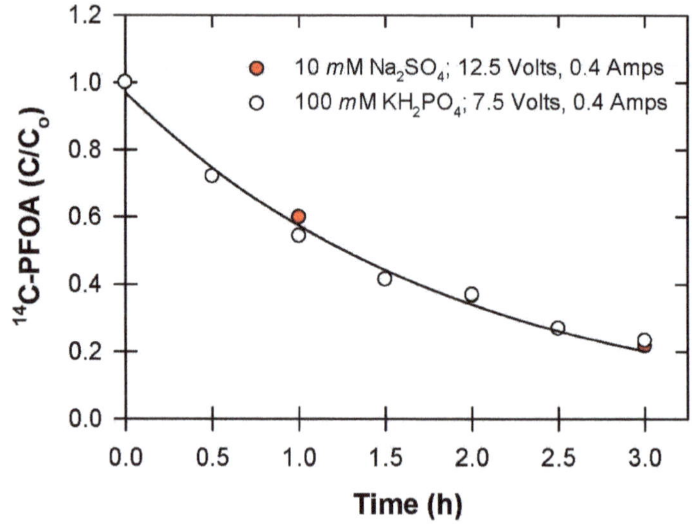

Figure 4. Effect of ionic salt and concentration on PFOA degradation by BDD treatment.

3.5. Reseeding PFAS to BDD Electrodes

To gain insight into the robustness of the EC-BDD system to continually degrade PFAS, we reseeded PFOA back into the batch reactor. By monitoring disintegrations per minute (dpms) following reseeding of ^{14}C-PFOA into the BDD-treated solution, the initial drop in dpms increased after the second and third reseedings (52% → 65% → 72%, Figure 5A). We also noticed a similar behavior when we electrochemically treated a 40 mg L^{-1} fluoride solution before adding ^{14}C-PFOA (Figure 5B). Results showed that the addition of fluoride increased the initial decrease in ^{14}C-PFOA (Figure 5B) but caused a plateau; similarly, in a similar but separate experiment using a higher PFOA concentration, the defluorination rate was higher during the first 2 h with BDD electrodes that had been previously exposed to fluoride; but with time, defluorination plateaued after 4 h (Figure 5C). These observations indicate that at least initially, preconditioning the electrodes with fluoride may have altered the electrode surface.

Previous researchers have documented that adsorption of fluoride to the BDD electrodes can occur. Guan et al. [29] observed in their electrochemical system that PFOA decomposed on the BDD anode to generate radicals containing fluorine, which reacted and bonded with the BDD surface. They confirmed this surface adsorption with XPS measurements. Guan et al. [29] also showed that the fluorinated surface showed water repulsion (i.e., more hydrophobic) and suggested that enrichment of the BDD with fluoride may be altering PFAS reaction rates over time. While the fluoride pretreatment may have initially increased defluorination (Figure 5C), it did not last, as observed by the plateau (Figure 5C) and may have promoted additional anodic wear. Anodic wear of BDD can lead to the formation of oxygenated sites and observable wear of BDD electrodes [30–33]. This changing surface chemistry can affect the reproducibility of electrode performance in water treatment [32].

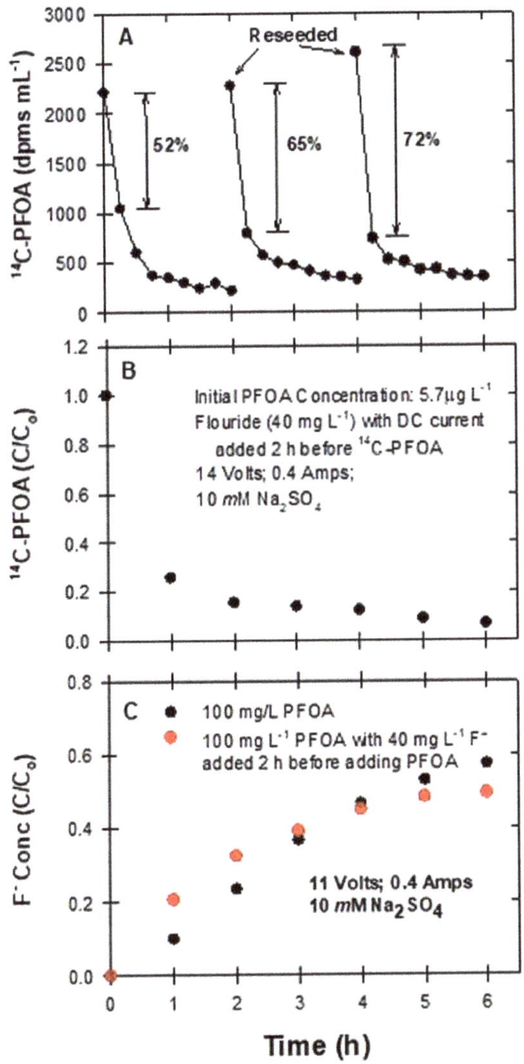

Figure 5. (**A**) Temporal changes in ^{14}C-PFOA following reseeding; (**B**) effect of adding fluoride (40 mg L^{-1}) and treating with BDD electrodes for 2h before adding ^{14}C-PFOA; (**C**) fluoride generation from PFOA, with and without preconditioning of BDD electrodes with fluoride (40 mg L^{-1}).

3.6. Defluorination and Fluoride Mass Balance

By using ^{14}C-labeled PFOA and measuring temporal loss of ^{14}C-activity and fluoride generation, we were able to quantify the transformation rate of the parent structure (i.e., ^{14}C-PFOA) as well as the transformation of the subsequently formed shorter chain degradation products (Figure 6). Results showed nearly complete transformation of the PFOA molecule with approximately 60% defluorination (Figure 6). Both PFOA removal and F- generation plateaued after approximately 3 h with defluorination occurring at a slightly lower rate than PFOA degradation (Figure 6).

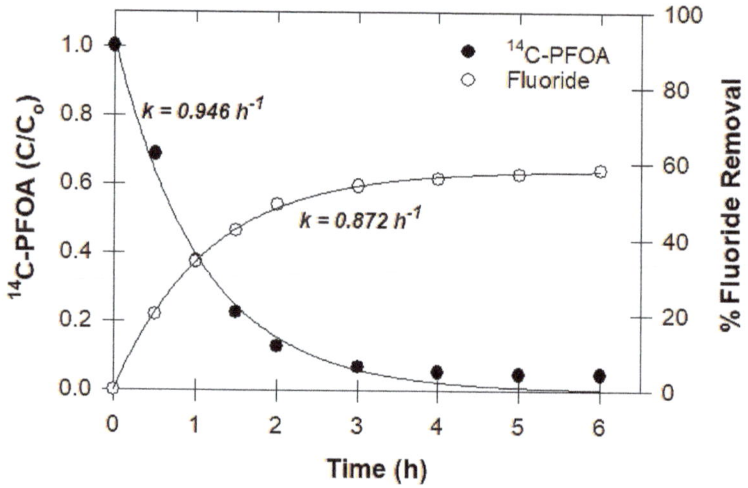

Figure 6. Temporal changes in ^{14}C-PFOA and fluoride concentrations. Experimental conditions: BDD anode and cathode; 100 mg L^{-1} PFOA, acidified, 20 V, 1 A.

Uwayezu et al. [34] used similar direct current EC-BDD conditions in treating PFOA, and reported 99.5% PFOA degradation and 50% fluoride recovery. They also confirmed the detection of shorter chain (4, 5, 6, and 7 carbon) fluorinated carboxylates. Zhou et al. [15] similarly documented the formation of shorter chain perfluorinated carboxylates following BDD treatment of both PFOA and PFOS. Given that shorter chain perfluorinated compounds are being formed by the EC-BDD treatment (Figure 1) and we observed defluorination rates slowed down after 3 h of treatment (Figure 6), we investigated defluorination rates of various chain length perfluorinated carbonates by using them as starting substrates (Table 1). In these experiments, we used the same starting molar concentration (0.24 *m*M). Our goal was to determine if any of the shorter chain perfluorinated compounds were resistant to electrochemical oxidation via the EC-BDD treatment. These results confirmed that the shorter chain degradation products formed during the destruction of the parent PFOA or PFOS (Figure 1) were also mineralized by the EC-BDD treatment method.

Results showed that fluoride generated from the perfluorinated compounds were generally linear during the first 3 to 5 h (Figure 7A) with the longer chain length compounds producing a higher percentage of defluorination. Although defluorination rates were slightly slower for the shorter chain compounds, our results confirm that these compounds were not resistant to EC-BDD oxidation. Radjenovic et al. [16] reported that longer chain PFAS are more hydrophobic and have a higher affinity to adsorb to electrode surfaces than shorter chain PFAS, and that this would enhance electrooxidation and defluorination. Barisci and Suri [35] specifically treated short and long chain pefluorocarboxylic acids with Si/BDD electrodes and found overall defluorination was greater with longer carbon chain lengths. Short chain PFAS are characterized as having lower molecular polarizabilities [36], which is the ability of a compound to form a dipole in an electrical field, and is the one molecular descriptor that corresponds with observed reaction rates of PFAS treated by electrochemical oxidation [16,36].

Given that we have observed both zero-order and first-order removal rates of PFOA under varying conditions (Figure 2), we determined initial concentration effects on defluorination rates. Results showed that higher fluoride recovery rates were observed at higher concentrations (Figure 7B), but all concentrations plateaued after approximately 3 h.

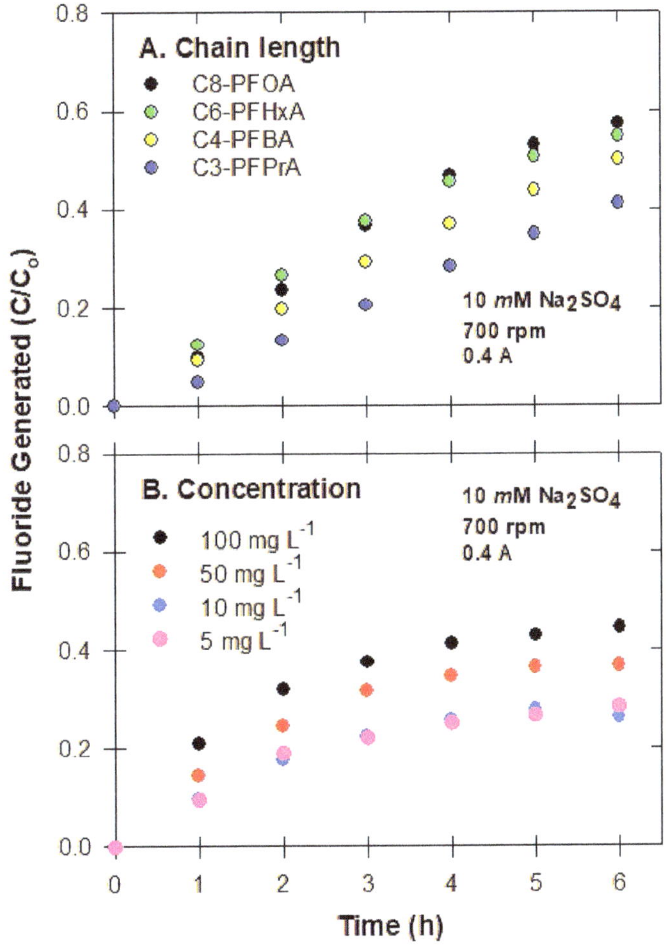

Figure 7. (**A**) Effects of initial concentration on defluorination rates; (**B**) defluorination rates of various carbon chain length perfluorinated carbonates.

3.7. Pilot-Scale Experiments

The custom printed 3D reactor was designed to be compatible with 2 or 4 BDD electrodes and suitable for treating a 208-L container (Figure 8). Before pilot-testing of the 3D reactor with 189 L of PFAS water, the 3D reactor was tested by treating ^{14}C-PFOA in the laboratory (2 L). By using only ^{14}C-PFOA (no unlabeled PFOA), the initial concentration was in the 8 to 10 µg L^{-1} (ppb) range, which was calculated from the specific activity (µCi $mmol^{-1}$) of the PFOA.

Laboratory results using ^{14}C-PFOA and the 3D reactor with different power sources (pilot-scale vs. laboratory-scale) showed that reaction rates could be influenced by adjusting the number of electrodes, power sources and polarity (Figure 9). By using 4 BDD electrodes over 2 BDD, reaction rates increased from $k = 0.738\ h^{-1}$ to $1.256\ h^{-1}$ (Figure 9). Similarly, using two power sources was superior to a single power source ($k = 0.855$ vs. $1.577\ h^{-1}$) and by switching the polarity of the electrodes every 30 s, degradation rates were greater than rates obtained using a constant polarity ($k = 0.854$ vs. $1.256\ h^{-1}$) (Figure 9).

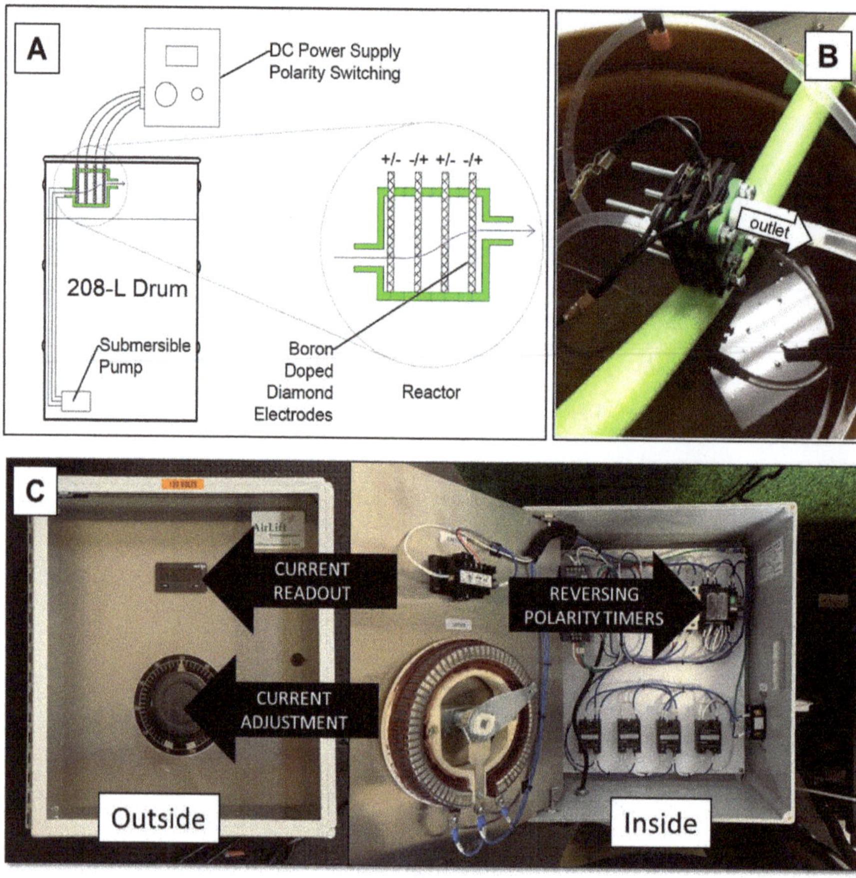

Figure 8. (**A**) Schematic of 208 L barrel experiment with components consisting of barrel, direct-current power supply, and 3D printed, four-BDD electrode chamber; (**B**) photograph of 3D printed electrode chamber; (**C**) photograph of direct current power supply.

Results from our 208 L pilot-scale experiment using LC/MS/MS analysis indicated that micrograms per liter (ppb) PFAS concentrations can easily be degraded by electrochemical oxidation using BDD anodes. The pilot-scale reactor and pump moved liquid at 2.33 L min^{-1}, meaning the 189 L of simulated IDW was cycled through the 3D reactor every 81 min. Results showed that PFOS was transformed relatively quickly and was below detection in less than 200 h. By contrast, PFOA was transformed approximately three times slower and was reduced to 0.114 μg L^{-1} at the end of the experiment. In a preliminary batch experiment, we also observed faster kinetics with PFOS over PFOA (Figure S1). Using a BDD flow-through reactor, Maldonando et al. [37] also observed faster kinetics with PFOS than PFOA. Nienhauser et al. [27] also found that 4, 6, and 8-carbon PFASs with a sulfonated head group degraded 2–3 times faster than their carboxylic counterparts. Radjenovic et al. [16] noted however, how reactions rates of mixed PFAS could be influenced by starting concentrations. For instance, Liang et al. [38] showed that at PFOA and PFOS concentrations of 10 mg L^{-1}, PFOS degradation was slower in the presence of PFOA, whereas Wang et al. [36] working with microgram per liter concentrations observed similar rate constants of perfluoroalkyl acids when treated individually or in mixtures. Fenti et al. [39], who used a Magnéli-phase Ti_nO_{2n-1} mesh anode in a flow-through electrochemical reactor reported faster kinetics for PFOA (0.033 h^{-1}) than PFOS (0.022 h^{-1}); these rate constants reported by Fenti et al. [39] are in the range similar to those observed in our pilot-scale experiment (Figure 10). Given EPA's new health advisory levels for PFOA and PFOS (0.004 ng L^{-1} for PFOA and 0.02 ng L^{-1} for PFOS), additional treatment time would be needed for PFOA destruction under the current treatment.

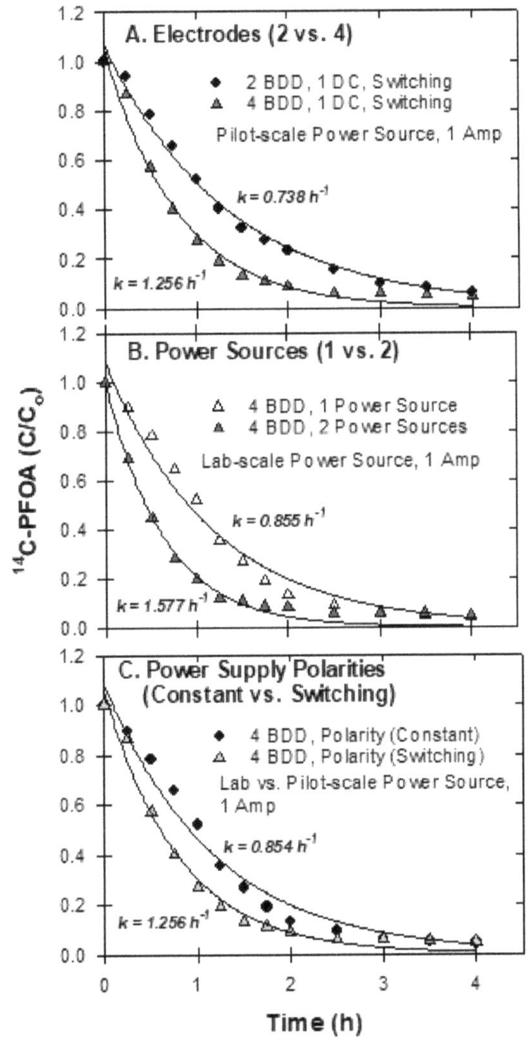

Figure 9. Temporal changes in ^{14}C-PFOA when treated with 3D printed reactor using: (**A**) 2 versus 4 BDD electrodes; (**B**) 1 versus 2 power sources; and (**C**) constant versus switching polarities.

At the end of the pilot-scale experiment, the BDD reactor was disassembled and photographed (Figure 11). Although polarity was switched every thirty seconds, the inflow of the IDW solution through the 3D reactor remained constant and, therefore, all IDW solution entered the same port. This resulted in a buildup of salt on the first BDD electrode (Figure 11). Adding an additional valve to periodically reverse flow direction during the treatment of the IDW may help to alleviate salt buildup in future tests.

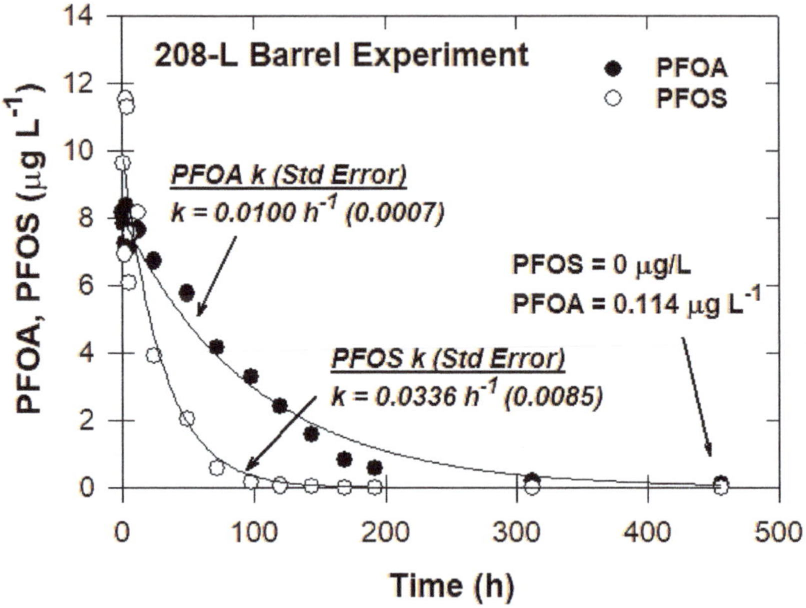

Figure 10. Temporal changes in PFOA and PFOS concentrations in 208 L barrel pilot-scale experiment.

Given only synthetic solutions of IDW were tested in this study, we acknowledge treating actual IDW is a logical progression and will be more challenging. Other researchers have been successful in treating real world solutions, such as landfill leachates. Maldonado et al. [36] found that PFOS and PFOA degradation was six times slower in a landfill leachate matrix than in a water matrix using similar conditions (current density, flow through cell); they attributed the slower kinetics to the presence of co-contaminants in the landfill leachate. While treating an ion exchange regenerant, Maldonado et al. [40] found PFAS reaction kinetics were three-fold slower than a synthetic solution and attributed the difference to the presence of co-contaminants.

While the costs of BDD electrodes remain high and a potential roadblock to full scale use, alternative electrodes, such as Magnéli-phase Ti_4O_7 anodes offer a lower cost alternative [36]. The use of the electrodes in batch, flow-by and flow-through modes is also a consideration, especially when treating larger quantities of IDW. While more elaborate BDD electrode reactor assemblies have been successfully used in the past [40,41], with some being commercially available (ElectroCell, Amherst, NY, USA), we showed herein, that a low-cost, 3D-printed reactor was functional and capable of treating larger volumes (189 L) of PFAS-contaminated water at a modest cost. Material costs for the printed 3D reactor were less than USD 100. The cost of the additional main components included the four commercial BDD electrodes (USD 1180, NeoCoat®, La Chaux-de Fonds, Switzerland), and the custom designed direct current system (USD 2668, 3E, Windsor Heights, IA, USA).

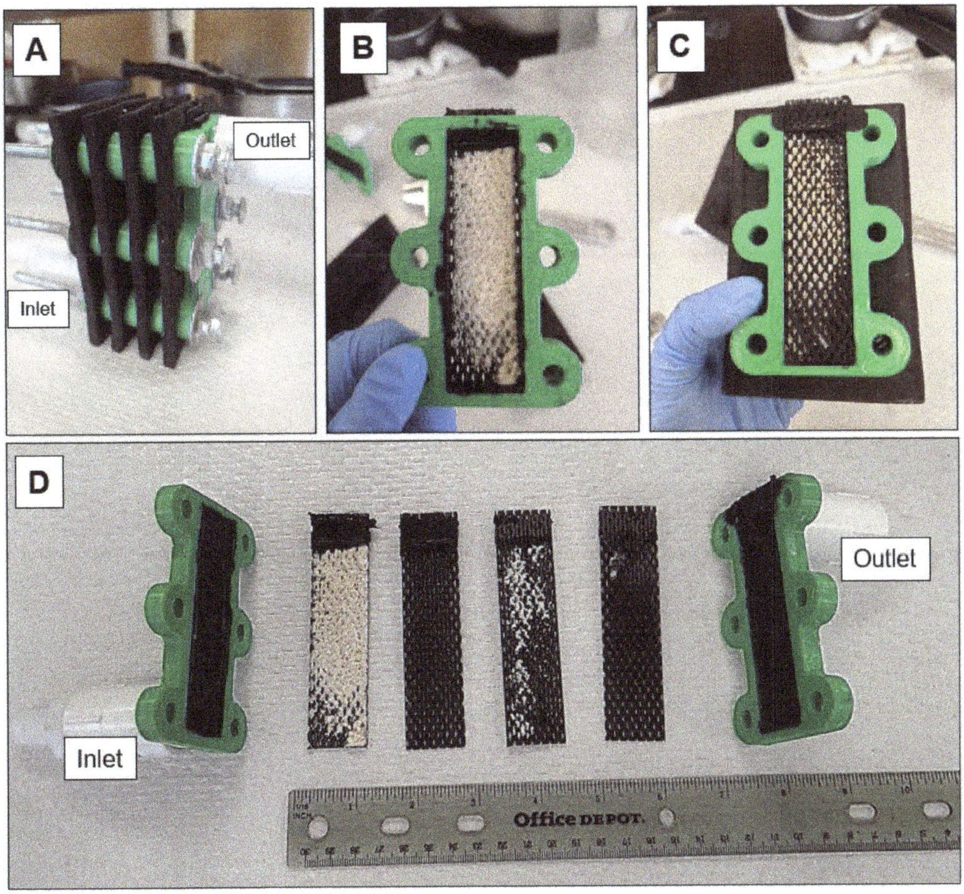

Figure 11. Photographs of pilot-scale BDD reactor after treatment: (**A**) before disassembly. (**B**) first BDD electrode (front side facing incoming solution) and (**C**) back side of first electrode. (**D**) Front side of all 4 BDD electrodes.

4. Conclusions

A series of laboratory and pilot scale experiments were performed to quantify the efficacy of electrochemical oxidation of PFAS using boron-doped diamond electrodes. By using ^{14}C-labeled PFOA, we showed that low concentrations of PFOA (μg L^{-1}) could easily be measured with a liquid scintillation counter and reaction kinetics quantified without the concern of cross contamination. By tracking temporal changes in both ^{14}C-activity and fluoride concentrations, we showed oxidation of the carboxylic head (-$^{14}COOH$ to $^{14}CO_2$) was possible, and up to 60% of the bonded fluorine was released into the solution. We showed that the shorter chain degradation products formed during the destruction of the parent contaminant were also degraded and defluorinated by the EC-BDD treatment. LC/MS/MS analysis indicated micrograms per liter (ppb) PFAS concentrations were easily degraded by electrochemical oxidation using BDD electrodes. By increasing current density (amperage), observed kinetics changed from current-controlled (zero-order) to mass-transfer controlled (first-order). Reversing polarity improved the lifespan of the BDD electrodes. A low-cost, 3D printed, four-electrode BDD reactor was successful in treating 189 L of simulated IDW in a 208-L container. Periodically alternating the flow direction of the IDW into the 3D reactor may reduce salt buildup and improve performance.

Supplementary Materials: The following supporting information can be downloaded at: https://www.mdpi.com/article/10.3390/w14172708/s1, Figure S1: Temporal changes in PFOS and PFOA concentrations following treatment with EC-BDD.

Author Contributions: Conceptualization, A.Y., E.W., C.E.H., M.C. and S.C.; methodology, A.Y., E.W. and M.C.; validation, A.Y. and E.W.; formal analysis, S.C.; resources, M.C. and S.C.; writing—original draft preparation, A.Y., E.W. and S.C.; writing—review and editing, A.Y., E.W., C.E.H., M.C. and S.C.; funding acquisition, M.C. and S.C. All authors have read and agreed to the published version of the manuscript.

Funding: This research was supported by the University of Nebraska–Lincoln School of Natural Resources and Water Science Laboratory with funding from the U.S. Environmental Protection Agency Small Business Innovation Research Program (EPA SBIR, Contract: 68HERC20C0033) and the Nebraska Department of Economic Development's matching SBIR grant program (Contract: 20-01-074).

Institutional Review Board Statement: Not Applicable.

Informed Consent Statement: Not Applicable.

Data Availability Statement: Not applicable.

Acknowledgments: We thank James Reece (AirLift Environmental) for assistance with research and data collection.

Conflicts of Interest: The authors declare no conflict of interest.

References

1. Copp, T. DoD: At least 126 bases report water contaminants linked to cancer, birth defects. *Military Times*, 26 April 2018.
2. Moody, C.A.; Field, J.A. Determination of perfluorocarboxylates in groundwater impacted by fire-fighting activity. *Environ. Sci. Technol.* **1999**, *33*, 2800–2806. [CrossRef]
3. Moody, C.A.; Hebert, G.N.; Strauss, S.H.; Field, J.A. Occurrence and persistence of perfluorooctanesulfonate and other perfluorinated surfactants in groundwater at a fire-training area at Wurtsmith Air Force Base, Michigan, USA. *J. Environ. Monit.* **2003**, *5*, 341–345. [CrossRef] [PubMed]
4. Schultz, M.M.; Barofsky, D.F.; Field, J.A. Quantitative determination of fluorotelomer sulfonates in groundwater by LC MS/MS. *Environ. Sci. Technol.* **2004**, *38*, 1828–1835. [CrossRef] [PubMed]
5. Suthersan, S.S.; Horst, J.; Ross, I.; Kalve, E.; Quinnan, J.; Houtz, E.; Burdick, J. Responding to emerging contaminant impacts: Situational management. *Groundw. Monit. Remediat.* **2016**, *36*, 22–32. [CrossRef]
6. Key, B.L.; Howell, R.D.; Criddle, C.S. Fluorinated organics in the biosphere. *Environ. Sci. Technol.* **1997**, *31*, 2445–2454. [CrossRef]
7. O'Hagan, D. Understanding organofluorine chemistry. An introduction to the C-F bond. *Chem Soc. Rev.* **2008**, *37*, 308–318. [CrossRef]
8. Conder, J.M.; Hoke, R.A.; Wolf, W.D.; Russell, M.H.; Buck, R.C. Are PFCAs Bioaccumulative? A critical review and comparison with regulatory criteria and persistent lipophilic compounds. *Environ. Sci. Technol.* **2008**, *42*, 995–1003. [CrossRef]
9. Lau, C.; Anitole, K.; Hodes, C.; Lai, D.; Pfahles-Hutchens, A.; Seed, J. Perfluoroalkyl acids: A review of monitoring and toxicological findings. *Toxicol. Sci.* **2007**, *99*, 366–394. [CrossRef]
10. Johansson, N.; Eriksson, P.; Viberg, H. Neonatal exposure to PFOS and PFOA in mice results in changes in proteins which are important for neuronal growth and synaptogenesis in the developing brain. *Toxicol. Sci.* **2009**, *108*, 412–418. [CrossRef]
11. Liu, W.; Xu, X.L.; Li, X.; Jin, Y.H.; Sasaki, K.; Saito, N.; Sato, I.; Tsuda, S. Human nails analysis as biomarker of exposure to perfluoroalkyl compounds. *Environ. Sci. Technol.* **2011**, *45*, 8144–8150. [CrossRef]
12. United States Environmental Protection Agency. EPA Announces New Drinking Water Health Advisories for PFAS Chemicals, $1 Billion in Bipartisan Infrastructure Law Funding to Strengthen Health Protections | US EPA. 2022. Available online: https://www.epa.gov/newsreleases/epa-announces-new-drinking-water-health-advisories-pfas-chemicals-1-billion-bipartisan (accessed on 30 July 2022).
13. Crunden, E.A. Defense department hits the brakes on PFAS incineration. E&E PM News. 2022. Available online: https://www.eenews.net/articles/defense-department-hits-the-brakes-on-pfas-incineration/ (accessed on 30 July 2022).
14. United States Environmental Protection Agency. PFAS Innovative Treatment Team (PITT). 2020. Available online: https://www.epa.gov/chemical-research/pfas-innovative-treatment-team-pitt (accessed on 15 March 2022).
15. Zhuo, Q.; Deng, S.; Yang, B.; Huang, J.; Wang, B.; Zhang, T.; Yu, G. Degradation of perfluorinated compounds on a boron-doped diamond electrode. *Electrochim. Acta* **2012**, *77*, 17–22. [CrossRef]
16. Radjenovic, J.; Duinslaeger, N.; Avvl, S.S.; Chaplin, B.P. Facing the challenge of poly- and perfluoroalkyl substances in water: Is electrochemical oxidation the answer? *Environ. Sci. Technol.* **2020**, *54*, 14815–14829. [CrossRef] [PubMed]

17. United States Environmental Protection Agency. Method 533: Determination of Per- and Polyfluoroalkyl Substances in Drinking Water by Isotope Dilution Anion Exchange Solid Phase Extraction and Liquid Chromatography/Tandem Mass Spectrometry | US EPA. 2019. Available online: https://www.epa.gov/dwanalyticalmethods/method-533-determination-and-polyfluoroalkyl-substances-drinking-water-isotope (accessed on 30 July 2022).
18. Carter, K.E.; Farrel, J. Oxidative destruction of perfluorooctane sulfate using boron-doped diamond film electrodes. *Environ. Sci. Technol.* **2009**, *42*, 6111–6115. [CrossRef] [PubMed]
19. Liao, Z.; Farrell, J. Electrochemical oxidation of perfluorobutane sulfonate using boron-doped diamond film electrodes. *J. Appl. Electrochem.* **2009**, *39*, 1993–1999. [CrossRef]
20. Matzek, L.W.; Tipton, M.J.; Farmer, A.T.; Steen, A.D.; Carter, K.E. Understanding electrochemically activated persulfate and its application to Ciprofloxacin abatement. *Environ. Sci. Technol.* **2018**, *52*, 5875–5883. [CrossRef]
21. Schaefer, C.E.; Andaya, C.; Urtiaga, A.; McKenzie, E.R.; Higgins, C.P. Electrochemical treatment of perfluorooctanoic acid (PFOA) and perfluorooctane sulfonic acid (PFOS) in groundwater impacted by aqueous film forming foams (AFFFs). *J. Hazard. Mater.* **2015**, *295*, 170–175. [CrossRef]
22. Schaefer, C.E.; Andaya, C.; Burant, A.; Condee, C.W.; Urtiagad, A.; Strathmann, T.J.; Higgins, C.P. Electrochemical treatment of perfluorooctanoic acid and perfluorooctane sulfonate: Insights into mechanisms and application to groundwater treatment. *Chem. Eng. J.* **2017**, *317*, 424–432. [CrossRef]
23. Kapalka, A.; Fóti, G.; Comninellis, C. Kinetic modelling of electrochemical mineralization of organic pollutants for wastewater treatment. *J. Appl. Electrochem.* **2008**, *38*, 7–16. [CrossRef]
24. Ochiai, T.; Iizuka, Y.; Nakata, K.; Murakami, T.; Tryk, D.A.; Fujishima, A.; Koide, Y.; Morito, Y. Efficient electrochemical decomposition of perfluorocarboxylic acids by the use of a boron-doped diamond electrode. *Diam. Relat. Mater.* **2011**, *20*, 64–67. [CrossRef]
25. Niu, J.; Li, Y.; Shang, E.; Xu, Z.; Liu, J. Electrochemical oxidation of perfluorinated compounds in water. *Chemosphere* **2016**, *146*, 526–538. [CrossRef]
26. Lin, H.; Niu, J.; Ding, S.; Zhang, L. Electrochemical degradation of per-fluorooctanoic acid (PFOA) by Ti/SnO_2-Sb, Ti/SnO2-Sb/PbO_2 and Ti/SnO_2-Sb/MnO_2 anodes. *Water Res.* **2012**, *46*, 2281–2289. [CrossRef] [PubMed]
27. Nienhauser, A.B.; Ersan, M.S.; Lin, Z.; Perreault, F.; Westerhoff, P. Boron-doped diamond electrodes degrade short- and long-chain per- and polyfluorinated alkyl substances in real industrial wastewaters. *J. Environ. Chem. Eng.* **2022**, *10*, 107192. [CrossRef]
28. Zhuo, Q.; Li, X.; Yan, F.; Yang, B.; Deng, S.; Huang, J.; Yu, G. Electrochemical oxidation of 1H, 1H, 2H, 2H-perfluorooctane sulfonic acid (6:2 FTS) on DSA electrode: Operating parameters and mechanism. *J. Environ. Sci.* **2014**, *26*, 1733–1739. [CrossRef] [PubMed]
29. Guan, B.; Zhi, J.; Zhang, X.; Murakami, T.; Fujishima, A. Electrochemical route for fluorinated modification of boron-doped diamond surface with perfluorooctanoic acid. *Electrochem. Commun.* **2007**, *9*, 2817–2821. [CrossRef]
30. Carrillo-Abad, J.; Perez-Herranz, V.; Urtiaga, A. Electrochemical oxidation of 6:2 fluorotelomer sulfonic acid (6:2 FTSA) on BDD: Electrode characterization and mechanistic investigation. *J. Appl. Electrochem.* **2018**, *48*, 589–596. [CrossRef]
31. Chaplin, B.P.; Wylie, I.; Zheng, H.; Carlisle, J.A.; Farrell, J. Characterization of the performance and failure mechanisms of boron-doped diamond ultrananocrystalline diamond electrodes. *J. Appl. Electrochem.* **2011**, *41*, 1329–1340. [CrossRef]
32. Chaplin, B.P.; Hubler, D.K.; Farrell, J. Understanding anodic wear at boron doped diamond film electrodes. *Electrochim. Acta* **2013**, *89*, 122–131. [CrossRef]
33. Duo, I.; Levy-Clement, C.; Fujishima, A.; Comninellis, C. Electron transfer kinetics on boron-doped diamond Part I: Influence of anodic treatment. *J. Appl. Electrochem.* **2004**, *34*, 935–943. [CrossRef]
34. Uwayezu, J.N.; Carabante, I.; Lejon, T.; van Hees, P.; Karlsson, P.; Hollman, P.; Kumpiene, J. Electrochemical degradation of per- and poly-fluoroalkyl substances using boron-doped diamond electrodes. *J. Environ. Manag.* **2021**, *290*, 112573. [CrossRef]
35. Barisci, S.; Suri, R. Electrooxidation of short and long chain perfluorocarboxylic acids using boron doped diamond electrodes. *Chemosphere* **2020**, *243*, 125349. [CrossRef]
36. Wang, Y.; Pierce, R.D.; Shi, H.; Li, C.; Huang, Q. Electrochemical degradation of perfluoroalkyl acids by titanium suboxides anodes. *Environ. Sci. Water Res. Technol.* **2020**, *6*, 144–152. [CrossRef]
37. Maldonado, Y.; Landi, G.M.; Ensch, M.; Becker, M.F.; Witt, S.E.; Rusinek, C.A. A flow-through cell for the electrochemical oxidation of perfluoroalkyl substances in landfill leachates. *J. Water Process Eng.* **2021**, *43*, 102210. [CrossRef]
38. Liang, S.; Pierce, R.D., Jr.; Lin, H.; Chiang, S.-Y.D.; Huang, Q.J. Electrochemical oxidation of PFOA and PFOS in concentrated waste streams. *Remediation* **2018**, *28*, 127–134. [CrossRef]
39. Fenti, A.; Jin, Y.; Rhoades, A.J.H.; Dooley, G.P.; Iovino, P.; Salvestrini, S.; Musmarra, D.; Mahendra, S.; Peaslee, G.F.; Blotevogel, J. Performance testing of mesh anodes for in situ electrochemical oxidation of PFAS. *Chem. Eng. J. Adv.* **2022**, *9*, 100205. [CrossRef]
40. Maldonado, Y.; Becker, M.F.; Nickelsen, M.G.; Witt, S.E. Laboratory and semi-pilot scale study on the electrochemical treatment of perfluoroalkyl acids from ion exchange still bottoms. *Water* **2021**, *13*, 2873. [CrossRef]
41. Witt, S.; Rancis, N.; Ensch, M.; Maldonado, V. Electrochemical destruction of "Forever Chemicals": The right solution at the right time. *Electrochem. Soc. Interface* **2020**, *29*, 73–76. [CrossRef]

water

MDPI

Article

Investigation on Energetic Efficiency of Reactor Systems for Oxidation of Micro-Pollutants by Immobilized Active Titanium Dioxide Photocatalysis

Simon Mehling [1,*], Tobias Schnabel [2] and Jörg Londong [1]

[1] Professorship of Urban Water Management, Department of Civil Engineering, Bauhaus-Universität Weimar, 99423 Weimar, Germany

[2] Professorship of Environmental Engineering, Engineering Department, University of Applied Science Hof, 95028 Hof, Germany

* Correspondence: simon.theodor.mehling@uni-weimar.de

Abstract: In this work, the degradation performance for the photocatalytic oxidation of eight micro-pollutants (amisulpride, benzotriazole, candesartan, carbamazepine, diclofenac, gabapentin, methly-benzotriazole, and metoprolol) within real secondary effluent was investigated using three different reactor designs. For all reactor types, the influence of irradiation power on its reaction rate and energetic efficiency was investigated. Flat cell and batch reactor showed almost similar substance specific degradation behavior. Within the immersion rotary body reactor, benzotriazole and methyl-benzotriazole showed a significantly lower degradation affinity. The flat cell reactor achieved the highest mean degradation rate, with half time values ranging from 5 to 64 min with a mean of 18 min, due to its high catalysts surface to hydraulic volume ratio. The EE/O values were calculated for all micro-pollutants as well as the mean degradation rate constant of each experimental step. The lowest substance specific energy per order (EE/O) values of 5 kWh/m^3 were measured for benzotriazole within the batch reactor. The batch reactor also reached the lowest mean values (11.8–15.9 kWh/m^3) followed by the flat cell reactor (21.0–37.0 kWh/m^3) and immersion rotary body reactor (23.9–41.0 kWh/m^3). Catalyst arrangement and irradiation power were identified as major influences on the energetic performance of the reactors. Low radiation intensities as well as the use of submerged catalyst arrangement allowed a reduction in energy demand by a factor of 3–4. A treatment according to existing treatment goals of wastewater treatment plants (80% total degradation) was achieved using the batch reactor with a calculated energy demand of 7000 Wh/m^3.

Keywords: photocatalysis; micro-pollutant treatment; titanium dioxide; reactor design; energy per order; immobilized catalyst

Citation: Mehling, S.; Schnabel, T.; Londong, J. Investigation on Energetic Efficiency of Reactor Systems for Oxidation of Micro-Pollutants by Immobilized Active Titanium Dioxide Photocatalysis. *Water* **2022**, *14*, 2681. https://doi.org/10.3390/w14172681

Academic Editors: Huijiao Wang, Dionysios (Dion) Demetriou Dionysiou and Yujue Wang

Received: 12 July 2022
Accepted: 25 August 2022
Published: 29 August 2022

1. Introduction

Within our industrialized society, thousands of chemical compounds are used daily for industrial and household purposes. Within the European Union, for example, over 100 thousand unique substances are registered within the European Community (EC) inventory [1]. Some of these substances, such as pharmaceuticals, industrial chemicals, and pesticides, enter the wastewater stream and can be discharged into downstream aquatic systems due to inadequate treatment performance of conventional wastewater treatment technologies [2,3]. Despite their low concentrations with a range of ng/L to µg/L, these micro-pollutants represent a challenge for future water and drinking water protection [4]. An additional treatment stage for municipal wastewater is necessary as an important barrier to the emission of these pollutants. This (4th) treatment stage is usually arranged after the biological treatment stage of wastewater treatment plants. The state of the art of treatment mostly consists of adsorptive processes, which use granulated or powdered activated carbon, and oxidative processes, which use ozone [5–7]. The cumulative energy demand of a large-scale ozonation can be assumed to be 50–100 Wh/m^3 [8]. For this application, a biological post treatment, in most case an aerated sand filtration, is necessary due to transformation products, which requires a cumulative energy consumption of 20–50 Wh/m^3 [8]. Powdered activated carbon (PAC) applications also require the use of sand filtration, which can be associated with an energy demand of 120 Wh/m^3. Further relevant energy consumption occurs during the production of activated carbon and can be quantified as 350 Wh/m^3 (primary energy/treated wastewater) [8]. These values can be seen as a first point of comparison towards the state of the art.

Advanced oxidation processes (AOP), which use cold oxidation by hydroxyl radicals as a core mechanism, are not yet used on a broad industrial scale within municipal wastewater treatment [9]. The main advantage of these applications is the higher oxidation potential of the hydroxyl radicals, which in principle allows a complete mineralization of treated pollutants to CO_2, water, and mineral salts [9,10]. During photocatalysis on titanium dioxide, irradiation of the catalyst with UV-A light (320–400 nm) leads to charge separation and, thus, to the formation of electron holes and free electrons [11]. The use of this process is possible in several variations. Existing developments can be differentiated in terms of the light source (wavelength, natural/artificial, type), the catalyst (TiO_2-based/other material, immobilized/suspended, size), and the reactor design and operation [12]. The emphasis of this work is the use of immobilized TiO_2 and UVA-LEDs as light source for the treatment of micro-pollutants in wastewater. A comparison of reactor systems is possible by a number of performance indicators [12,13]. In the context of economic competitiveness, the energy efficiency is an important parameter if artificial irradiation is used Wang et al. 2021 [14]. The electrical energy required to degrade a pollutant by a factor of 10 in relation to the volume of water treated is called electrical energy per order (EE/O) and is usually given as a characteristic value [14]. Within the literature, values from $5 * 10^4$ up to 10^{-1} kWh/m^3 are reported for different pollutants and reactor systems [14]. Most of these systems described in [12] are in the range of 10–50 kWh/m^3. In addition to the pollutant treated and the type of reactor, degradation rates and, thus, energy efficiencies get influenced/decreased by a variety of other parameters, especially matrix influences [15,16]. A realistic assessment of the suitability of photocatalytic processes for wastewater treatment consequently requires an investigation of a wide range of micro-pollutants in real matrix. [17] achieved EE/O values of 0.21 kWh/m^3 for diclofenac and 2.1 kWh/m^3 for carbamazepine using suspended $\boldsymbol{TiO_2}$ and UVC-lamps when treating river water with a photocatalytic membrane reactor module. In contrast, [18] stated, that no strictly photocatalytic experiment measured EE/O values lower than 10, except the experiments of [19], which was conducted using ultrapure water at pH 3.8. [20] measured EE/O values of 15–57.3 kWh/m^3 for the treatment of azithromycin, trimethopirm, oflaxacin, and sulfamethoxazole within secondary effluent via suspended $\boldsymbol{TiO_2}$ photocatalysis.

The content of this work is the comparison of three reactor types for the photocatalytic degradation of eight micro-pollutants based on experimental measurement data within secondary effluent. The catalyst used was immobilized $\boldsymbol{TiO_2}$, which is suitable for wastewater treatment due to its low resource cost and toxicity, if nanoparticulate emission can be avoided. The focus lays on the energetic efficiency of these applications, since this parameter has a significant influence on the economic competitiveness compared to conventional processes of the 4th treatment stage. The energetic comparison is carried out in this work using the EE/O methodology. It must be noted that, at least for the German context, a degradation of 90% is substantially beyond currently discussed cleaning targets. Existing requirements for the degradation performance of these applications require an 80% elimination of a selection of indicator substances relative to the raw wastewater [21,22]. A detailed comparison with conventional treatment technologies requires a detailed configuration of a photocatalytic treatment stage considering variable degradation performances of upstream biological stages and, thus, exceeds the scope of this work. As a rough estimate, it can be assumed that a 20% average micro-pollutant degradation is achieved within the biological stage of a wastewater treatment plant [2,8]. Assuming 75% elimination within a (photocatalytic) 4th treatment stage (relative to secondary effluent), this results in a total degradation of 80%. Assuming pseudo 1st order reaction kinetics, which applies for the photocatalytic treatment of most micropollutants, a non-linear reduction of the energy demand of about 40% for a reduced elimination rate (90% to 75%) is to be expected.

2. Methods

2.1. Chemicals and Materials

A more detailed description of the catalyst material used is given in previous works [23,24]. It consists of a suspension of anatase-modified titanium dioxide nanoparticles of 14 nm diameter and an anatase content of the titanium dioxide of 90%, which is coated onto a stainless-steel mesh. The V4A steel mesh had a mesh size of 200 µm and a wire thickness of 100 µm. A 10 µg/L terbutryn (Sigma Aldrich, St. Louis, MO, USA) solution in methanol was used within analysis as an internal standard. Further analytical purity chemicals were used for calibration: amisulpride (Gentham Life Science, Corsham, UK), benzotriazole (Gentham Life Science), candesartan (BLD Pharma, Hyderabad, India), carbamazepine (Sigma Aldrich), diclofenac (Cayman Chemical Company, Ann Arbor, MI, USA), gabapentin (Gentham Life Science), 1-methylbenzotriazole (Chempure, Bengaluru, India), and metoprolol (Sigma Aldrich).

2.2. Experimental Set-Up and Procedure

The investigations were carried out in different time series in 2019, 2020, 2021, and 2022. The investigated wastewater was always taken during dry weather conditions from the effluent of the wastewater treatment plant Weimar-Tiefurt and were filtered (50 µm) before use. The treatment plant has a capacity of 100,000 inhabitants. It consists of a mechanical treatment stage and a double channel cascade biological treatment system with horizontal secondary sedimentation. In total, three different reactor systems were used:

- Laboratory scale batch reactor
- Laboratory scale flat cell reactor
- Semi-technical photocatalytic immersion rotary body reactor

Data regarding the individual cleaning performance are already published for the flat cell [24] and the immersion rotary body reactor [23].

As shown in Figure 1, the laboratory batch reactor consists of an aluminum cylinder with a diameter of 14 cm and a hydraulic volume of 500 mL. The irradiated catalyst area is 0.0154 m^2. The reactor is stirred continuously at 300 rpm. Six LEDs with a total electrical power of 12 W are installed and the power can be manually controlled. The catalyst is placed horizontally, submerged in the reactor 5 mm below the water surface. The reactor was investigated in batch mode. Before starting the experiment, the reactor was flushed with wastewater. Afterwards, a 30-min concentration equilibration was performed whereby the reactor was mixed and not irradiated. Subsequently, a photocatalytic treatment was carried out for 90 min, with 5 mL of sample material being taken every 15 min.

The flat cell reactor studied, as shown in Figure 2, consists of a pleated steel mesh coated with TiO_2 and eight UV-A LEDs (365 nm). The catalyst is placed inside a stainless-steel enclosure with glass windows on both sides. The edge length is 11.5 cm, and the hydraulic volume is 200 mL. The pleated design of the catalyst allowed a surface area of 0.0181 m^2 to be arranged on a base area of 0.0128 m^2. On each side of the reactor, 4 UV-A LEDs (1.8 W each) were evenly spaced at 8 cm distance. The reactor was constantly fed via peristaltic pump. Within the experiments, a hydraulic residence time of 12.17 min was set. After a run time of 30 min, which ensured a stationary reactor condition, the inlet and outlet of the reactor were sampled.

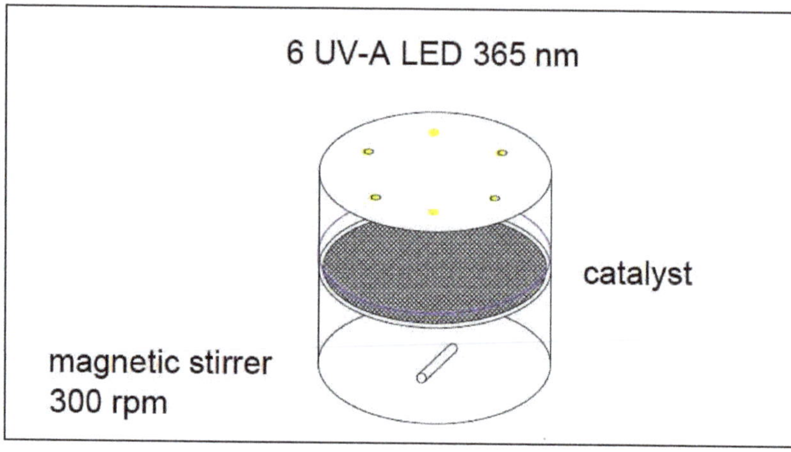

Figure 1. Schematic 3D-Model of the photocatalytic batch reactor.

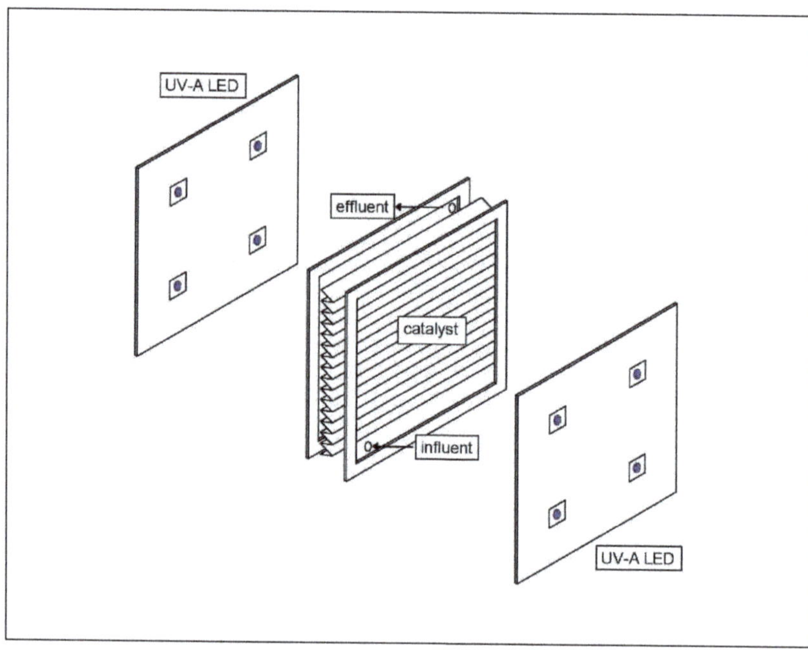

Figure 2. Schematic 3D-Model of the photocatalytic flat cell reactor.

The photocatalytic immersion body reactor in Figure 3 consists of 12 rotating discs of 36 cm diameter on which the catalyst material is placed on both sides. Total catalyst area is 2.44 m^2, of which about 50% is irradiated via UV-A LED arrays. The hydraulic part of the reactor is divided into 12 cascades, which are flown through in series with a total volume of 25 L. Investigation and sampling were carried out in the following steps: replace reactor volume with fresh wastewater, start reactor feed and operation and let run for 90 min, stop the reactor feed and operation, and sample. For this reactor, one sample is taken from every second of the twelve cascades. In contrast to previous laboratory applications, continuous feeding continues to take place directly from the effluent of the Weimar-Tiefurt wastewater treatment plant, whereas in the laboratory scale sample, the material was fed to the reactors via a tank, which was taken as a quick sample (about 20 L).

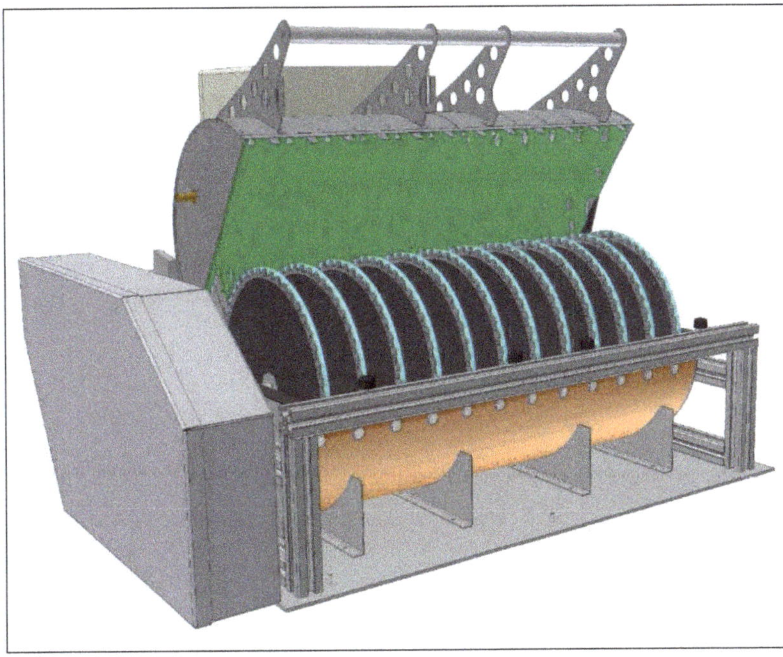

Figure 3. Schematic 3D-Model of the photocatalytic immersion body reactor (© Lynatox GmbH, 2020).

The experimental program is divided according to the reactor type. For the batch reactor, the degradation of micro-pollutants was investigated under the influence of the electrical irradiation power at constant treatment time. For the other continuous flow reactors, a variation of the irradiation power at a constant feed rate was also carried out. An overview of the investigated parameters can be seen in Table 1. The investigated radiation power per reactor volume was varied for different reactor systems. There are overlapping ranges for all reactor systems with a total magnitude of about 3.08–57.60 kW/m^3. The LED power relative to the irradiated catalyst area was also similar for the batch and immersion rotary body reactor (100–300 W/m^2). In the experiment with the flat-cell reactor, higher LED powers per catalyst area of up to 636 W/m^2 were additionally investigated.

The values of the catalyst area-related irradiation power must be understood as a coefficient for reactor configuration and do not represent the actual average irradiation intensity on the respective catalysts. Due to the emission angle of the LEDs being 120° and the LED gap being 8 cm for the flat-cell reactor, considerable lateral losses are to be expected here. Furthermore, adsorption and reflection dynamics or other geometric characteristics of irradiation sources are not considered within this value.

Table 1. Experimental reactor operation and geometric parameters.

Batch Reactor				
Number	LED-power per irradiated catalyst area [W/m^2]	Irradiated catalyst area per hydraulic volume [m^2/m^3]	LED-power per hydraulic volume [kW/m^3]	Treatment time [min]
1.1	99.94	30.79	3.08	90
1.2	199.88		6.15	90
1.3	299.82		9.23	90
Immersion Rotary Body Reactor				
Number	LED-power per irradiated catalyst area [W/m^2]	Irradiated catalyst area per hydraulic volume [m^2/m^3]	LED-power per hydraulic volume [kW/m^3]	Hydraulic retention time [min]
2.1	87.83	48.86	4.29	59.76
2.2	197.89		9.67	64.10
2.3	303.72		14.84	63.83
Flat-Cell Reactor				
Number	LED-power per irradiated catalyst area [W/m^2]	Irradiated catalyst area per hydraulic volume [m^2/m^3]	LED-power per hydraulic volume [kW/m^3]	Hydraulic retention time [min]
3.1	159.12	90.50	14.40	12.17
3.2	318.23		28.80	12.17
3.3	477.35		43.20	12.17
3.4	636.46		57.60	12.17

2.3. Analytical Methods

The micro-pollutant analysis for the flat-cell and immersion rotary body reactor experiments was conducted using LC-MS/MS analogous to previous publications [24]. The HPLC was a Dionex R3000 system (ThermoFischer, Waltham, MA, USA) with gradient pump and autosampler. The HPLC eluent was ultrapure water with 1 mmol/l ammonium acetate and acetonitrile with 0.1% acetic acid. A Synergy 2.5 μm hydro PP column (Phenomenex, Aschaffenburg, Germany) measuring 100 2 mm was used as analytical separation column. The flow was 0.25 mL/min, and the gradient was run from 4% acetonitrile to 96% acetonitrile in 28 min. The injection volume was 100 μL sample, which was mixed with 10 μL internal standard. An API 4000 triple quadrupole mass spectrometer (Sciex, Framingham, MA, USA) with electro spray ionization was used for detection. For each substance, two multiple reaction monitoring mass transitions were evaluated. The micro-pollutant measurements of the batch reactor experiments were carried out by the commercial laboratory "GWA Umweltanalytik Luisenthal" following [25]. The following substances were analyzed: amisulpride, benzotriazole, candesartan, carbamazepine, diclofenac, gabapentin, 1-methylbenzotriazole, and metoprolol. All analytical measurements were carried out at least in duplicate.

2.4. Kinetic and Modelling

The kinetics of heterogeneous photocatalytic degradation can be described by the Langmuir–Hinshelwood (L–H) model [26,27], where the reaction rate of the parent substances can be calculated as follows:

$$r = -\frac{dC}{dt} = k_{LH}\frac{K_L C_p}{1 + K_L C_p + \sum_{i=1}^{n} K_i C_i (i = 1, n)} \quad (1)$$

The reaction rate r is determined by the specific reaction constant k_{LH}, the Langmuir constant K_L, and the equilibrium concentration C_p. The constant k_{LH} is significantly influenced by the formation and recombination rate of radicals. These are again significantly influenced by the activity of the catalyst, the radiation power, and the concentration of electron acceptors in the liquid matrix. The constant K_L describes the substance-specific adsorption capacity of the reactant on the catalyst material. K_i and C_i are the Langmuir constant and the concentration of every substance within the solution. However, the application of this model in practice is very difficult, because, on the one hand, in a wastewater matrix in the rarest cases, all ingredients can be determined and the adsorption of the investigated micro-pollutants on the catalyst within the present concentration ranges cannot be measured. However, according to the observations of many authors, the practically observed reaction follows the first order [26]. Thus, the following simplification is assumed:

$$r = -\frac{dC}{dt} = k_{LH}K_L C_p = kC_p \quad (2)$$

For easier readability of measured reaction rates, they have been presented within this work as half-lives according to Formula (3):

$$T_{1/2} = \frac{\ln 2}{k} \quad (3)$$

In order to compare the efficiency of different reactor systems, it is necessary to balance the amount of energy used, in this case electrical irradiation power, per unit volume of treated wastewater. The volume-specific energy requirement is calculated according to the following formula:

$$E_{spez} = \frac{P_A * A_V * V_R}{Q} \quad (4)$$

Here, P_A is the electrical radiation power per catalyst surface, A_V is the catalyst area used per volume, V_R is the reactor volume, and Q is the volumetric flow rate. Assuming a 90% degradation rate, the feed rate can be calculated from reactor volume and reaction rate on a substance-specific basis. Under this assumption, Formula (5) can be represented as follows:

$$EE/O_i = P_A * A_V * T_i \quad (5)$$

where, T_i is the treatment or hydraulic retention time to achieve a given degradation ratio of a micropollutant within a specific experimental set-up. By means of Formula (6), an energy balance for a batch reactor is also possible. For flow reactors this balancing is only valid if a plug-flow behavior can be assumed. This proof as well as the corresponding 1st order reaction rate constants have been provided experimentally for the flat cell reactors within [24]. The 1st order reaction rate constants for immersion rotary body were determined in [25]. The time T_i was calculated using Formula (6):

$$T_i = \frac{\ln(1-\eta_i)}{k_i} \quad (6)$$

Here η_i is the degradation ration (0.9 for calculation of the EE/O-values) and k_i reaction rate constant. The first order reaction rate constants for the batch reactor were calculated using the following formula:

$$y = A1 * \exp\left(-\frac{(x - x0)}{t1}\right) \quad (7)$$

where x is equal to the treatment time within the batch reactor. The correlation was conducted using the Origin Pro software.

In Figure 4, the concentration curves and parameter estimations for the batch reactor are shown exemplary. For carbamazepine, inconsistent concentration curves were measured. As shown in Figure 4A, in two experiments, an increase of concentration occurred during the photocatalytic treatment. These increases may be explained, by the presence of oxcarbazepine, among other predecessor substances, within these wastewater samples. Consequently, carbamazepine was not included in the evaluation of the batch tests. For all reactors, 1st order reaction kinetic was applied and experimental proven. All experimental parameters (Table S1), measured concentrations (Table S2) as well as the corresponding half-live constants and fitting parameters (Table S3) are shown in the Supplementary Materials.

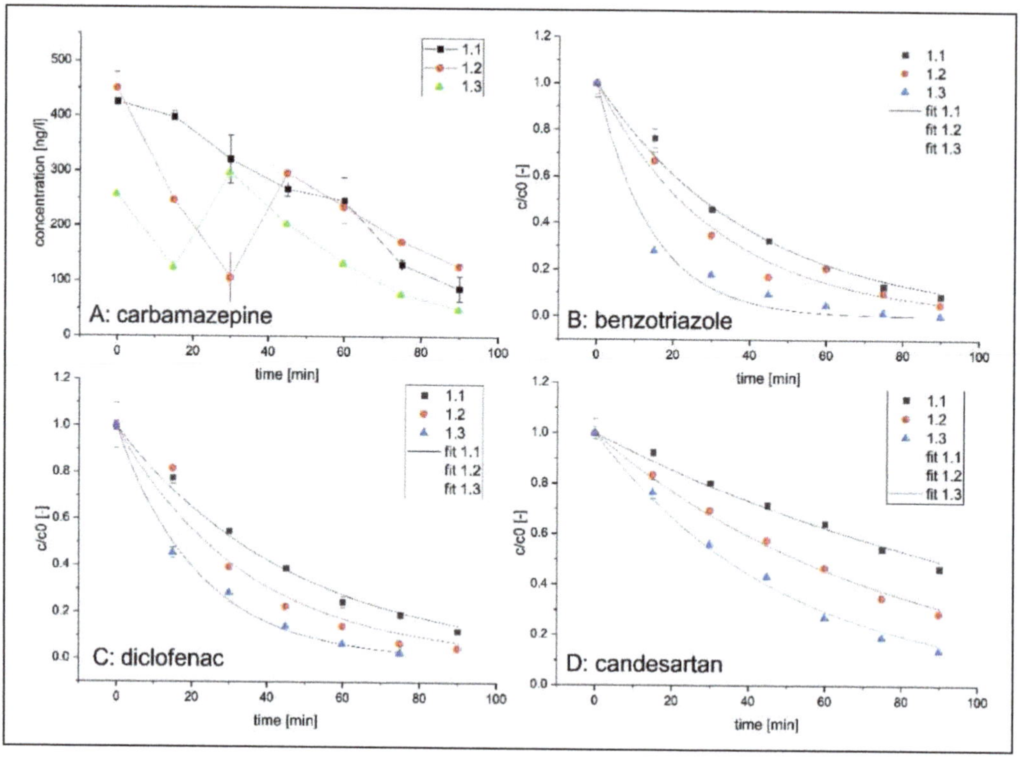

Figure 4. Concentration curves of carbamazepine and concentration curves and first-order fitting curves for benzotriazole, diclofenac, and candesartan for batch reactor experiments.

3. Results

In Figure 5, the measured micro-pollutant concentrations in the effluent of the wastewater treatment plant Weimar-Tiefurt are shown. There were high loads of benzotraizole, diclofenac, and gabapentin. The initial concentrations of the flat cell reactor experiments were elevated for amisulpride, benzotriazole, candesartan, and carbamazepine compared to the other samples. Especially the parameters benzotriazole and gabapentin showed high fluctuations. Compared to typical effluent concentrations reported in [2], similar values were measured for all the micropollutants except diclofenac and carbamazepine, which were detected in higher concentrations. Overall, representative micro-pollutant concentrations of the secondary effluent were observed.

The photocatalytic reaction constants for all three reactor systems operated with similar electrical irradiation power per hydraulic volume are presented as halftime values in Figure 6. A higher reaction rate of many substances is evident for the flat cell reactor. For benzotriazole and diclofenac, similar degradation rates were measured for flat cell and batch reactor. Similar half time values were achieved for carbamazepine, amisulpride, and gabapentin within flat cell and immersion rotary body reactor. Interestingly, benzotriazole and methlybenzotriazole showed significantly lower degradation affinity within the immersion rotary body reactor. Overall, a reactor- and substance specific reaction behaviour was found.

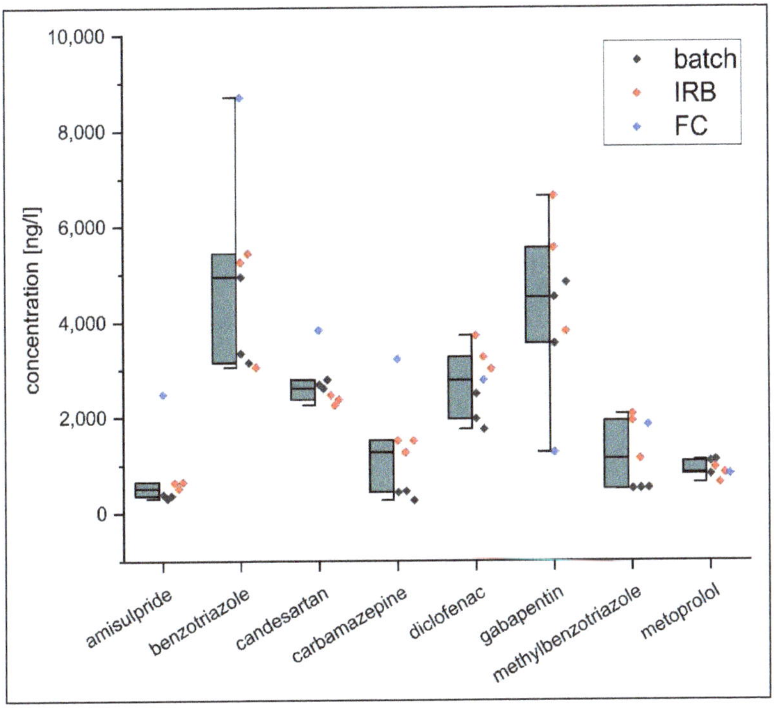

Figure 5. Initial concentrations of micro-pollutants shown as boxplot and individual data point for all experimental steps measured in samples of the secondary effluent of the Weimar-Tiefurt treatment plant (IRB: immersion rotary body reactor; FC: flat cell reactor).

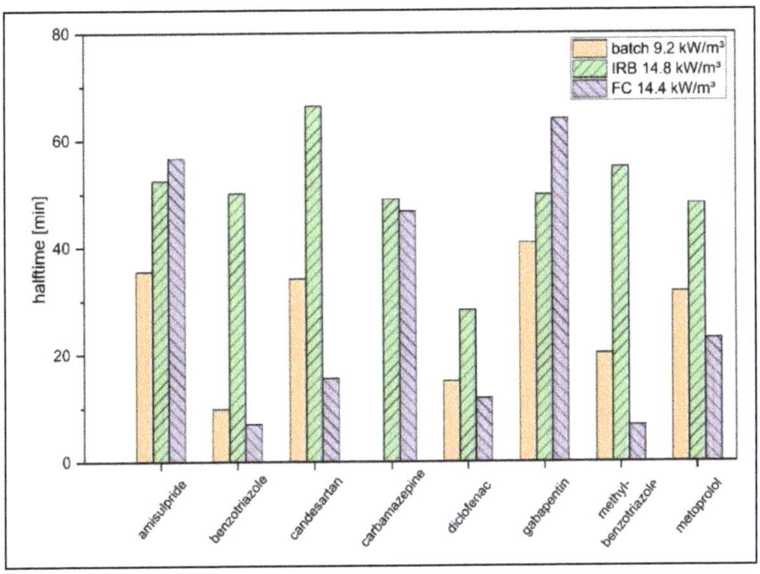

Figure 6. Halftime constants for the degradation of micro-pollutants within batch, flat cell (FC), and immersion rotary body reactor (IRB) with similar electrical irradiation powers of 9.2 to 1483 kW/m^3.

In Figure 7A the mean 1st order reaction constants for all experimental steps are shown in relation to each electrical irradiation power. Similar to Figure 6, overall higher values were achieved for the flat cell reactor. The progression of curves indicates an overlapping behaviour at similar radiation powers for batch and flat cell reactor. The immersion rotary body reactor showed overall lower mean reaction rates. For all reactors, an increase in radiation power was associated with higher degradation rates. For the flat cell, as well as the immersion rotary body reactor, efficiency losses at higher radiation powers can be observed. This effect can be caused by the quantum efficiency of the catalysts, diffusion processes, or substances specific adsorption affinities. In Figure 7B, the influence of varying irradiated catalyst area to hydraulic volume (A/V) ratios was corrected assuming a linear relation to the mean 1st order constants. The A/V values of the batch (30.8 m^2/m^3), flat cell (90.5 m^2/m^3), and immersion rotary body (48.9 m^2/m^3) were scaled to an intermediate value of 50 m^2/m^3. It is evident that the comparatively high reaction performance of the flat cell reactor is mainly caused by its higher A/V ratio. The reaction rates of flat cell and batch reactor show similar magnitudes; however, for the batch reactor, these are achieved with significantly lower irradiation powers. The immersion rotary body reactor shows overall about a 2–3 times lower reaction rate even at significantly higher irradiation powers than the batch reactor.

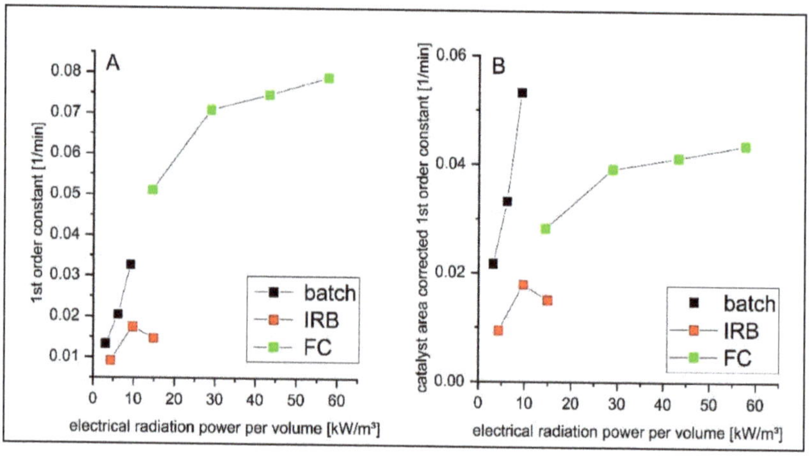

Figure 7. Mean 1st order reaction constant for the degradation of micro-pollutants with the batch, flat cell (FC), and immersion rotary body reactor (IRB) for varying electrical irradiation powers (**A**); mean 1st order reaction constant with a linear correction to a surface area to volume ratio of 50 m^2/m^3 (**B**) (all mean values without carbamazepine).

The EE/O values were calculated for all experimental steps using the mean 1st order reaction constant of every micro-pollutant. The result is shown in Figure 8 for all micro-pollutants and the three reactor types at similar radiation power per hydraulic volume. The lowest EE/O values of 5–10 kWh/m^3 were achieved for benzotriazole and diclofenac within the batch and flat cell reactor. Within the batch reactor, the other investigated substances were degraded with EE/O values of 10–20 kWh/m^3. For the immersion rotary body reactor, overall higher EE/O values were measured ranging from 30 to 60 kWh/m^3. The flat cell showed the biggest substance specific range of energy demand with values from 5 to 50 kWh/m^3.

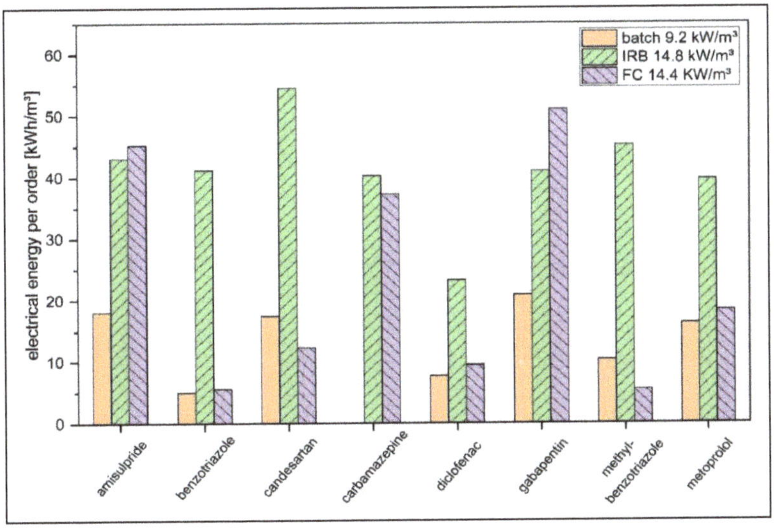

Figure 8. Calculated EE/O value for the photocatalytic oxidation of all micro-pollutants within all three reactor systems at a radiation power of 9.2–14.8 kW/m^3.

In Figure 9, the mean EE/O values for every experiment are shown in relation to the LED power depicted as the electrical irradiation power per hydraulic volume. For all three reactors, lower EE/O values were archived when using lower radiation lowers and consequentially higher treatment durations. This effect resulted in a 25–45% reduction in energy consumption. Significant differences are evident in the overall energy efficiency of the investigated reactor systems. The batch reactor achieved the lowest EE/O values of 11.8–15.9 kWh/m^3. The EE/O values of the flat cell and immersion rotary body reactor were 2–3 times higher. Analogous to Figure 7A similar values at overlapping radiation powers are indicated by the curve progression of batch and flat cell reactor.

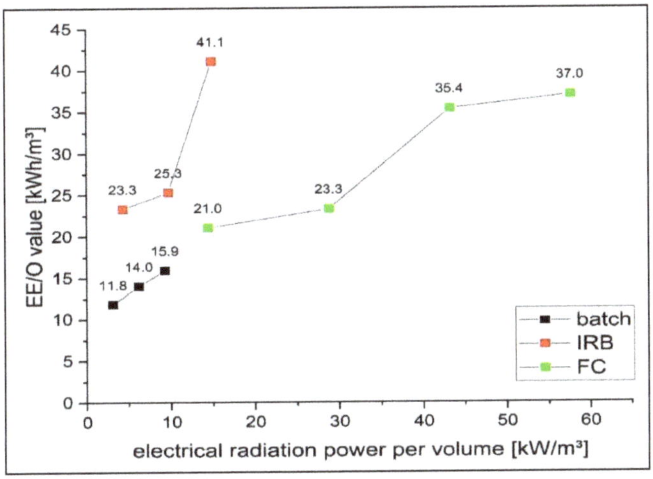

Figure 9. Mean EE/O values for the degradation of all eight micro-pollutants within the batch, flat cell, and immersion rotary body reactor as a function of the electrical irradiation per surface area (all mean values without carbamazepine).

Given the use of similar catalyst material, wastewater samples, and reactor parameters, this result must be significantly influenced by the geometrical arrangement of the catalyst and associated hydraulic impacts. Within the immersion rotary body reactor, the catalyst is only periodically submerged within the medium. The catalyst within the batch reactor is permanently submerged. The hydraulic volume below the catalyst is in a turbulent state due to the magnetic mixer. In the flat cell reactor, the fluid is forced through the catalysts grid via the pressure gradient of the inlet pump. Based on these results, a submerged arrangement of catalysts proves beneficial in terms of energetic efficiency, whereby no difference could be found between external sources of turbulence and flow path through the catalysts mesh itself.

The energy demand for variable mean degradation rates was calculated based on the experimental results of all reactor types (Figure 10). For all reactors, the 1st order constant, which resulted in the lowest EE/O value, was chosen for calculation. Typical treatment target values, as described within the introduction, are indicated graphically. Using the batch reactor, a 75% degradation rate can be achieved with an electrical energy input of 7000 Wh/m^3. The flat cell and immersion rotary body reactor had an energy demand of 12,000–14,000 Wh/m^3 to achieve a similar degradation. However, following the trend described in Figure 9, values similar to the batch reactor are to be expected when using the flat cell reactor with lower radiation intensities.

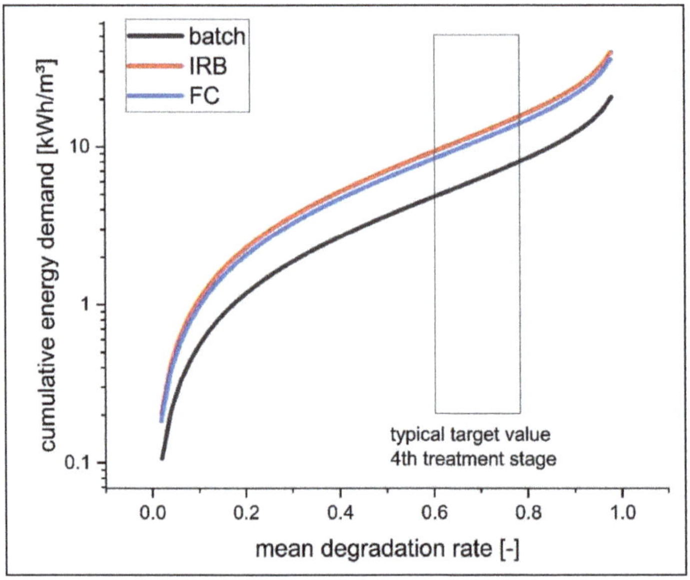

Figure 10. Energy demand per volume treated of the batch, flat cell (FC), and immersion rotary body reactor (IRB) in relation to the mean degradation rate of all measured micro-pollutants.

Compared to the state of the art, with typical values of 100–300 Wh/m^3, further improvement is still needed to achieve a beneficial use of photocatalytic technics. However, compared to the immersion body reactor, significant improvement in energetic efficiency could be achieved at a laboratory scale. When comparing the results to previous measurement of EE/O values within wastewater, similar EE/O values were achieved within the immersion rotary body and flat cell reactor system [20]. The results of the batch reactor test showed an overall improvement in energetic efficiency with EE/O values ranging from 5 to 20 kWh/m^3. When comparing the EE/O values aggregated in [12], significantly lower energy demands (up to 240 Wh/m^3) were measured for the degradation of various indicator substances (e.g., methylene blue, methylene orange, malachite green). However, a comparability of results is not given due the use of ultra-pure water and the increased degradation affinity of these indicator substances. In [28], an energy demand of 38–47 kWh/m^3 was determined for the degradation of phenol within secondary effluent using suspended TiO_2 catalysts.

4. Conclusions

In this work, the degradation performance for the photocatalytic oxidation of eight micro-pollutants with real secondary effluent was investigated using three different reactor designs. Flat cell and batch reactor showed similar substance specific degradation behaviors. Within the immersion rotary body reactor, benzotriazole and methyl-benzotriazole showed a significantly lower degradation affinity. The flat cell reactor achieved the highest mean degradation rate due to its high catalysts surfaces to hydraulic volume ratio. The EE/O values were calculated for all micro-pollutants as well as the mean degradation rate constant of each experimental step. The lowest substance specific EE/O values of 5 kWh/m^3 were measured for benzotriazole within the batch reactor. The batch reactor also reached the lowest mean values (11.8–15.9 kWh/m^3) followed by the flat cell reactor (21.0–37.0 kWh/m^3) and immersion rotary body reactor (23.9–41.0 kWh/m^3). Catalyst arrangement and irradiation power were identified as major influences on the energetic performance of the reactors. A high catalyst surface to hydraulic volume ratio, which led to high degradation rates within the flat cell reactor showed no beneficial impact on energetic efficiency. A treatment according to existing treatment goals of wastewater treatment plants can be achieved using the batch reactor with a calculated energy demand of 7000 Wh/m^3. Compared to existing literature values, an improved energetic efficiency was measured for the batch reactor configuration. This highlights the potential of this reactor type within photocatalytic wastewater treatment. However, a major difference compared to the energetic efficiency of ozonation and activated carbon still persists.

As mentioned in the introduction, the batch and flat cell reactor types were only developed on a laboratory scale. Especially in terms of radiation, transport optimizations, such as equalizing radiation fields, minimizing absorption of the liquid and reflections, are possible. Therefore, further research work regarding these reactor types is necessary. Following the results of [29], further increases in efficiency seem possible, when integration controlled periodic illumination (CPI) within the reactor systems. When comparing the energy demand with the state of the art (especially ozone oxidation), it also needs to be mentioned that a direct comparability is not given due to deviating oxidation, transformation, and mineralization characteristics of each reaction cascade. The beneficial effect of AOP on the formation of transformation products as described in [30] needs to be seriously considered when comparing treatment options.

Supplementary Materials: The following are available online at https://www.mdpi.com/article/10.3390/w14172681/s1, Table S1: Experimental program; Table S2: Concentrations; Table S3: Kinetics.

Author Contributions: Conceptualization, S.M.; methodology, S.M. and T.S.; investigation, S.M.; data curation, S.M.; writing—original draft preparation, S.M. and T.S.; writing—review and editing, S.M.; supervision, J.L. All authors have read and agreed to the published version of the manuscript.

Funding: This research was funded by Federal Ministry of Education and Research grant number [02WQ1492A].

Institutional Review Board Statement: Not applicable.

Informed Consent Statement: Not applicable.

Data Availability Statement: Supporting data can be found in Supplementary Materials.

Conflicts of Interest: The authors declare no conflict of interest.

References

1. ECHA. EC Inventory. Edited by European Chemicals Agency, 2022. Available online: https://echa.europa.eu/information-on-chemicals/ec-inventory (accessed on 7 June 2022).
2. Margot, J.; Rossi, L.; Barry, D.A.; Holliger, C. A review of the fate of micropollutants in wastewater treatment plants. *WIREs Water* **2015**, *2*, 457–487. [CrossRef]
3. Kanaujiya, D.K.; Paul, T.; Sinharoy, A.; Pakshirajan, K. Biological Treatment Processes for the Removal of Organic Micropollutants from Wastewater: A Review. *Curr. Pollut. Rep.* **2019**, *5*, 112–128. [CrossRef]
4. Gallé, T.; Pittois, D.; Bayerle, M.; Braun, C. An immission perspective of emerging micropollutant pressure in Luxembourgish surface waters: A simple evaluation scheme for wastewater impact assessment. *Environ. Pollut.* **2019**, *253*, 992–999. [CrossRef] [PubMed]
5. Guillossou, R.; Le Roux, J.; Brosillon, S.; Mailler, R.; Vulliet, E.; Morlay, C.; Nauleau, F.; Rocher, V.; Gaspéri, J. Benefits of ozonation before activated carbon adsorption for the removal of organic micropollutants from wastewater effluents. *Chemosphere* **2020**, *245*, 125530. [CrossRef] [PubMed]
6. Guillossou, R.; Le Roux, J.; Mailler, R.; Vulliet, E.; Morlay, C.; Nauleau, F.; Gasperi, J.; Rocher, V. Organic micropollutants in a large wastewater treatment plant: What are the benefits of an advanced treatment by activated carbon adsorption in comparison to conventional treatment? *Chemosphere* **2019**, *218*, 1050–1060. [CrossRef]
7. Bourgin, M.; Beck, B.; Boehler, M.; Borowska, E.; Fleiner, J.; Salhi, E.; Teichler, R.; von Gunten, U.; Siegrist, H.; McArdell, C.S. Evaluation of a full-scale wastewater treatment plant upgraded with ozonation and biological post-treatments: Abatement of micropollutants, formation of transformation products and oxidation by-products. *Water Res.* **2018**, *129*, 486–498. [CrossRef]
8. Abegglen, C.; Siegrist, H. *Mikroverunreinigungen aus Kommunalem Abwasser Verfahren zur Weitergehenden Elimination auf Kläranlagen (Micropollutants from Municipal Wastewater Process for further Elimination in Sewage Treatment Plants)*; Federal Office for the Environment BAFU: Ittigen, Switzerland, 2012.
9. Kanakaraju, D.; Glass, B.D.; Oelgemöller, M. Advanced oxidation process-mediated removal of pharmaceuticals from water: A review. *J. Environ. Manag.* **2018**, *219*, 189–207. [CrossRef]
10. Kisch, H. *Semiconductor Photokatalysis—Principles and Applications*; Wiley: Hoboken, NJ, USA, 2015.
11. Pitre, S.P.; Yoon, T.P.; Scaiano, J.C. Titanium dioxide visible light photocatalysis: Surface association enables photocatalysis with visible light irradiation. *Chem. Commun.* **2017**, *53*, 4335–4338. [CrossRef]
12. Sundar, K.P.; Kanmani, S. Progression of Photocatalytic reactors and it's comparison: A Review. *Chem. Eng. Res. Des.* **2019**, *154*, 135–150. [CrossRef]
13. Sacco, O.; Vaiano, V.; Sannino, D. Main parameters influencing the design of photocatalytic reactors for wastewater treatment: A mini review. *J. Chem. Technol. Biotechnol.* **2020**, *95*, 2608–2618. [CrossRef]
14. Wang, D.; Mueses, M.A.; Márquez, J.A.C.; Machuca-Martínez, F.; Grčić, I.; Moreira, R.P.M.; Puma, G.L. Engineering and modeling perspectives on photocatalytic reactors for water treatment. *Water Res.* **2021**, *202*, 117421. [CrossRef]
15. Mecha, A.C.; Chollom, M.N. Photocatalytic ozonation of wastewater: A review. *Environ. Chem. Lett.* **2020**, *18*, 1491–1507. [CrossRef]
16. Mehrjouei, M.; Müller, S.; Möller, D. A review on photocatalytic ozonation used for the treatment of water and wastewater. *Chem. Eng. J.* **2015**, *263*, 209–219.
17. Benotti, M.J.; Stanford, B.D.; Wert, E.C.; Snyder, S.A. Evaluation of a photocatalytic reactor membrane pilot system for the removal of pharmaceuticals and endocrine disrupting compounds from water. *Water Res.* **2009**, *43*, 1513–1522. [CrossRef]
18. Silva, D.B.; Buttiglieri, G.; Babić, S. State-of-the-art and current challenges for $\boldsymbol{TiO_2}$/UV-LED photocatalytic degradation of emerging organic micropollutants. *Environ. Sci. Pollut. Res.* **2021**, *28*, 103–120. [CrossRef]
19. Kelly, J.; Morrison, G.; Skillen, N.; Manesiotis, P.; Robertson, P. An investigation of the role of pH in the rapid photocatalytic degradation of MCPA and its primary intermediate by low-power UV LED irradiation. *Chem. Eng. J.* **2019**, *359*, 112–118. [CrossRef]
20. Biancullo, F.; Moreira, N.F.; Ribeiro, A.R.; Manaia, C.M.; Faria, J.L.; Nunes, O.C.; Castro-Silva, S.M.; Silva, A.M. Heterogeneous photocatalysis using UVA-LEDs for the removal of antibiotics and antibiotic resistant bacteria from urban wastewater treatment plant effluents. *Chem. Eng. J.* **2019**, *367*, 304–313. [CrossRef]
21. GSchV (1/1/2020). Gewässerschutzverordnung (Water Protection Ordinance) Switzerland. Available online: https://fedlex.data.admin.ch/filestore/fedlex.data.admin.ch/eli/cc/1998/2863_2863_2863/20210101/en/pdf-a/fedlex-data-admin-ch-eli-cc-1998-2863_2863_2863-20210101-en-pdf-a-1.pdf (accessed on 1 April 2022).
22. Competence Center for Trace Substances Baden-Württemberg. *Handlungsempfehlungen für die Vergleichskontrolle und den Betrieb von Verfahrenstechniken zur gezielten Spurenstoffelimination (Recommendations for Action for Comparative Control and the Operation of Process Technologies for Targeted Trace Substance Elimination)*; Kompetenzzentrum Spurenstoffe: Stuttgart, Germany, 2018.
23. Mehling, S.; Schnabel, T.; Londong, J. Photocatalytic ozonation in an immersion rotary body reactor for the removal of micropollutants from the effluent of wastewater treatment plants. *Water Sci. Technol.* **2021**, *85*, 535–548. [CrossRef]
24. Schnabel, T.; Mehling, S.; Londong, J.; Springer, C. Hydrogen peroxide-assisted photocatalytic water treatment for the removal of anthropogenic trace substances from the effluent of wastewater treatment plants. *Water Sci. Technol.* **2020**, *82*, 2019–2028. [CrossRef]

25. *DIN 38407-47:2017-07*; German Standard Methods for the Examination of Water, Waste Water and Sludge. International Organization for Standardization: Geneva, Switzerland, 2017.
26. Lin, Y.; Ferronato, C.; Deng, N.; Wu, F.; Chovelon, J.-M. Photocatalytic deg-radation of methylparaben by TiO_2: Multivariable experimental design and mechanism. *Appl. Catal. B Environ.* **2009**, *88*, 32–41. [CrossRef]
27. Asenjo, N.G.; Santamaría, R.; Blanco, C.; Granda, M.; Álvarez, P.; Menéndez, R. Correct use of the Langmuir–Hinshelwood equation for proving the absence of a synergy effect in the photocatalytic degradation of phenol on a suspended mixture of titania and activated carbon. *Carbon* **2013**, *55*, 62–69. [CrossRef]
28. Mecha, A.C.; Onyango, M.S.; Ochieng, A.; Momba, M.N. Ultraviolet and solar photocatalytic ozonation of municipal wastewater: Catalyst reuse, energy requirements and toxicity assessment. *Chemosphere* **2017**, *186*, 669–676. [CrossRef] [PubMed]
29. Liang, R.; Van Leuwen, J.C.; Bragg, L.M.; Arlos, M.J.; Fong, L.C.L.C.; Schneider, O.M.; Jaciw-Zurakowsky, I.; Fattahi, A.; Rathod, S.; Peng, P.; et al. Utilizing UV-LED pulse width modulation on TiO_2 advanced oxidation processes to enhance the decomposition efficiency of pharmaceutical micropollutants. *Chem. Eng. J.* **2019**, *361*, 439–449. [CrossRef]
30. Rueda-Marquez, J.J.; Levchuk, I.; Ibañez, P.F.; Sillanpää, M. A critical review on application of photocatalysis for toxicity reduction of real wastewaters. *J. Clean. Prod.* **2020**, *258*, 120694. [CrossRef]

water

MDPI

Article

Enhanced Degradation of Rhodamine B through Peroxymonosulfate Activated by a Metal Oxide/Carbon Nitride Composite

Yuanmin Mo [1,2], Wei Xu [1,2], Xiaoping Zhang [1,2,3,4,*] and Shaoqi Zhou [1,2,3]

1 School of Environment & Energy, South China University of Technology, Guangzhou Higher Education Mega Centre, Guangzhou 510006, China; yuanminmo@163.com (Y.M.); xuwei06042022@163.com (W.X.); fesqzhou@yeah.net (S.Z.)
2 The Key Laboratory of Pollution Control and Ecosystem Restoration in Industry Clusters of Ministry of Education, Guangzhou 510006, China
3 Guangdong Provincial Key Laboratory of Solid Wastes Pollution Control and Recycling, Guangzhou 510006, China
4 Guangdong Provincial Engineering and Technology Research Center for Environmental Risk Prevention and Emergency Disposal, Guangzhou 510006, China
* Correspondence: xpzhang@scut.edu.cn; Tel.: +86-13678920429

Abstract: The development of high catalytic performance heterogeneous catalysts such as peroxymonosulfate (PMS) activators is important for the practical remediation of organic pollution caused by Rhodamine B (RhB). An economical and facile synthesized composite of copper–magnesium oxide and carbon nitride (CM/g-C_3N_4) was prepared by the sol-gel/high-temperature pyrolysis method to activate PMS for RhB degradation. CM/g-C_3N_4 exhibited a splendid structure for PMS activation, and the aggregation of copper–magnesium oxide was decreased when it was combined with carbon nitride. The introduction of magnesium oxide and carbon nitride increased the specific surface area and pore volume of CM/g-C_3N_4, providing more reaction sites. The low usage of CM/g-C_3N_4 (0.3 g/L) and PMS (1.0 mM) could rapidly degrade 99.88% of 10 mg/L RhB, and the RhB removal efficiency maintained 99.30% after five cycles, showing the superior catalytic performance and reusability of CM/g-C_3N_4. The synergistic effect of copper and g-C_3N_4 improved the PMS activation. According to the analyses of EPR and quenching experiments, $SO_4^{\bullet-}$, $^{\bullet}OH$ and $O_2^{\bullet-}$ radicals and $^1O_2$ were generated in the activation of PMS, of which $SO_4^{\bullet-}$ and $^1O_2$ were important for RhB removal. The toxicity of RhB was alleviated after being degraded by the CM/g-C_3N_4/PMS system. This study provides an efficient and promising strategy for removing dyes in water due to the hybrid reaction pathways in the CM/g-C_3N_4/PMS system.

Keywords: RhB degradation; heterogeneous catalyst; copper–magnesium oxide/carbon nitride composite; PMS activation

Citation: Mo, Y.; Xu, W.; Zhang, X.; Zhou, S. Enhanced Degradation of Rhodamine B through Peroxymonosulfate Activated by a Metal Oxide/Carbon Nitride Composite. *Water* **2022**, *14*, 2054. https://doi.org/10.3390/w14132054

Academic Editors: Dionysios (Dion) Demetriou Dionysiou, Yujue Wang and Huijiao Wang

Received: 17 June 2022
Accepted: 25 June 2022
Published: 27 June 2022

1. Introduction

Rhodamine B (RhB) poses a potential threat to the aquatic environment and human health because of its high salt content and poor biodegradability [1]. RhB has been extensively applied in the fields of colored glass, as well as the textile and plastic industries. It can cause symptoms such as the red staining of human skin, mild congestion of blood vessels in the brain and breakage of myocardial fibers. According to the chemical carcinogenic risk assessment of the World Health Organization International Agency for Research on Cancer IARC), ingestion or skin contact with RhB will cause acute and chronic poisoning. Therefore, exploring efficient treatment technology is important for removing the highly toxic RhB.

Many methods have emerged to remove RhB, including adsorption [2], ion exchange [3], advanced oxidation processes (AOPs) [4,5], membrane filtration [6] and biodegradation [7].

However, owing to its special chemical structural stability, light resistance, corrosion resistance and bacteriostasis, RhB cannot be fully degraded by traditional physical methods and biological technologies [4]. Among them, AOPs have garnered extensive attention for their high efficiency and mineralization in removing pollutants [8–10]. Traditional AOPs use hydroxyl radicals (•OH) to remove pollutants [11]. However, the problems of weak redox potential (1.8–2.7 V), short lifetime (<1 μs) and a narrow range of optimum pH restrict the application of •OH [12]. Recently, AOPs based on sulfate radicals ($SO_4^{\bullet-}$) have received more and more attention for their high redox potential (2.5–3.1 V), long lifetime (30–40 ms) and a wide pH range [13]. Peroxymonosulfate (PMS) is one of the widely used oxidants activated by transition metal ions [14,15], heat [16], alkali materials [17], ultrasound [18], UV light sources (254 nm) [19] and heterogeneous catalysts [20] to produce $SO_4^{\bullet-}$ for pollutant degradation. Among various methods, PMS activated by heterogeneous catalysts is more promising for practical applications because of its mild reaction conditions, low energy consumption and easy recycling [20]. Previous research has employed heterogeneous catalysts including CuO [21], $CuCo_2O_4$ [22], CuO@FeO_x@Fe-0 [23] and CuO-Co_3O_4@CeO_2 [24] to activate PMS for pollutant degradation. All these studies have demonstrated that PMS activated by metal oxide for the generation of active substance is a prospective advanced oxidation process for the decomposition of toxic pollutants.

Copper is one of the transition metals for PMS activation. Copper oxide, as an efficient, economical and low toxic heterogeneous catalyst, has a high catalytic performance [25]. However, the problems with the aggregation and the dissolution of copper ion decrease the activity and limit the practical application of copper oxide. To overcome such drawbacks, Du et al. prepared CuO/rGO for 2,4,6-trichlorophenol removal, which exhibited much better catalytic performance than CuO [26]. Kiain et al. found that CuO@AC could be reused in three cycles, and the leaching of copper ions was negligible [27]. Li et al. synthesized BC-CuO for methylene blue, acid orange 7, atrazine and ciprofloxacin removal in a highly saline system, which achieved a high catalytic performance [28]. All these results indicated that loading the copper oxide on the supporter is an efficient way to improve the catalytic performance of copper oxide. However, the applications of these catalysts were restricted by the complex synthesis, high cost of precursor materials and insufficient reaction sites. Therefore, it is necessary to develop a method of low-cost and facile synthesis and increase the reaction site to enhance the catalytic performance of copper oxide.

Carbon nitride (g-C_3N_4), as a low-cost and stable material, is commonly applied in many fields [29–31], which could be easily prepared by being calcined in the air, without any chemical or physical pretreatment. The g-C_3N_4 can not only be used as a supporter to disperse the metal oxide, with more reaction sites exposed, but also activate PMS to some extent. Therefore, the introduction of g-C_3N_4 may decrease the aggregation and improve the catalytic performance of copper oxide. It is reported that the introduction of magnesium oxide can enhance the specific surface area and electron transfer capacity of the catalyst [32–34], which is beneficial for PMS activation. Furthermore, the large quantities of surface basic sites on magnesium oxide can facilitate the formation of an M–OH complex [34], thus increasing the reaction sites between the catalyst and PMS, which is the vital step for PMS activation. Hence, the combination of copper oxide, magnesium oxide and g-C_3N_4 is an effective way to improve the catalytic performance of copper oxide. Moreover, it is an economical and eco-friendly catalyst worth exploring.

In order to develop a high catalytic performance catalyst with a low cost and abundant reaction sites for RhB degradation, copper oxide was combined with magnesium oxide and g-C_3N_4 to form a composite (CM/g-C_3N_4). The composite was initially employed as a heterogeneous catalyst for PMS activation, which showed splendid catalytic performance and stability. When compared with previous reports, low-cost magnesium oxide and g-C_3N_4 were introduced into the catalyst at the same time for increasing the reaction sites and decreasing the aggregation of copper oxide. Furthermore, the hybrid reaction pathways (radicals and non-radical) were produced in the CM/g-C_3N_4/PMS system.

In this study, the effects of preparation procedures on the catalytic performance of composite were investigated, and the effects of catalyst dosage, PMS concentration, RhB concentration and initial pH value on the RhB removal were also explored. The degradation intermediates of RhB were analyzed by LC-MS, and the toxicity was predicted by the Toxicity Estimation Software Tool. The activation mechanism of PMS was proposed according to the results of XPS, the EPR test and quenching experiments. The reusability of the catalyst was conducted through five cycle experiments.

2. Material and Methods

2.1. Chemicals and Reagents

The information relating to the chemicals and reagents is described in supplementary material—Text S1.

2.2. Preparation of Catalyst

A preparation of pure g-C_3N_4: 5 g melamine was placed in a 30 mL quartz crucible and calcined at 550 °C (5 °C/min) for 4 h in an air atmosphere. After cooling down, it was fully ground with an agate mortar to obtain yellow powder.

Preparation of the copper–magnesium oxide/g-C_3N_4 composite (CM/g-C_3N_4): First, 0.250 M $Cu(NO_3)_2{\cdot}3H_2O$ and 0.250 M $Mg(NO_3)_2{\cdot}6H_2O$ were dissolved in 40 mL of deionized water. Then, 0.50 M citric acid was added to the above solution and stirred for 10 min. Thereafter, the mixture was heated at 80 °C in the air until the water was completely evaporated to obtain a blue colloid. The obtained colloid was heated to 700 °C at a rate of 5 °C/min under a nitrogen atmosphere for 4 h to obtain black powder. Finally, the black powder and melamine were ground and mixed at a mass ratio of 1:2, and calcined in the air at 550 °C for 4 h (Scheme S1). In order to investigate the effect of the preparation procedures on the catalytic performance of CM/g-C_3N_4, the materials with different g-C_3N_4 precursors (urea and pure g-C_3N_4), mass ratios of copper–magnesium oxide to melamine (1:1, 1:3, 1:4, 1:5), pyrolysis temperatures (500 °C, 600 °C, 800 °C, 900 °C), pyrolysis times (2 h, 3 h, 5 h, 6 h) and molar ratios of copper to magnesium precursor (1:0.25, 1:0.5, 1:1.5, 1:2) were also synthesized according to the preparation processes of CM/g-C_3N_4. The composites prepared with urea or pure g-C_3N_4 as carbon nitride precursors were denoted as CM/g-C_3N_4 (urea) and CM/g-C_3N_4 (2), respectively. The preparations of copper oxide, copper–magnesium oxide, copper oxide/g-C_3N_4 composite and magnesium oxide/g-C_3N_4 composite were the same as the above processes, unless no $Mg(NO_3)_2{\cdot}6H_2O$ and melamine, melamine $Mg(NO_3)_2{\cdot}6H_2O$ and $Cu(NO_3)_2{\cdot}3H_2O$ were added. They were denoted as CuO, CM, C/g-C_3N_4 and M/g-C_3N_4, respectively.

2.3. Degradation Experiment

The information relating to the degradation experiment was described in supplementary material—Text S2. Three groups of parallel samples were carried out in each experiment. All results are displayed as an average value and standard error.

2.4. Characterizations

The information relating to the characterizations is described in supplementary material—Text S3.

2.5. Analytical Methods

RhB concentration was analyzed by a UV-visible spectrophotometer at 554 nm. The total organic carbon (TOC) concentration was calculated by a TOC analyzer (Muti N/C 2100). Active substances were observed by using an electron paramagnetic resonance spectrometer (EPR, Bruker E 500-10/12). The intermediates of RhB degradation were identified using a liquid chromatography–mass spectrometry machine (LC-MS, Agilent 1290II). The ionization mode was chosen as ESI (positive ion mode), and the source temperature was 350 °C. The mixture of 0.10% formic acid and CH_3CN (60:40, v/v) was the mobile phase, and the flow rate was 0.20 mL/min. The leaching concentrations of copper and magnesium ions were analyzed by an inductively coupled plasma optical emission spectrometer (ICP-OES, Agilent 720ES). As to the calculation and optimization of RhB molecule, the Fukui Function based on DFT calculation was adopted in Materials Studio software (MS). The Fukui function analysis was analyzed by LDA/PWC in Dmol3 of MS, without taking into account the influence of spin in the calculation [32]. The Fukui function representing free radical (f^0) attacks was used to evaluate the reaction sites of active substances on the molecules. The toxicity of RhB and its intermediates was evaluated by the Toxicity Estimation Software Tool (TEST) version 5.1.1 based on quantitative structure–activity relationships (QSAR) methodologies [35].

The RhB degradation kinetics were fit by the pseudo first order model and the apparent rate constant (k) was calculated according to Equation (1).

$$\mathrm{Ln}\ (C_t/C_0) = -\ kt \quad (1)$$

where C_t is the RhB concentration at a certain reaction time (t) and C_0 is the initial RhB concentration; t is the reaction time; k is the apparent rate constant. The crystallite size of CM/g-C_3N_4 was calculated from XRD data using the Scherrer Equation (2) [36].

$$D = \frac{K\lambda}{B \cos\theta} \quad (2)$$

where D is the crystallites' size (nm), K is the Scherrer constant (0.9), λ is the wavelength of X-ray sources, B is the FWHM (radians), and θ is the peak position (radians).

3. Result and Discussion

3.1. Degradation of RhB in Different Systems

As shown in Figure 1a, the removal of RhB by CM/g-C_3N_4 and PMS was 9.47 and 13.42% respectively, indicating that CM/g-C_3N_4 and PMS show poor RhB removal efficiency. Pure g-C_3N_4 could remove 28.01% of RhB in 5 min by activating PMS, demonstrating that g-C_3N_4 is not only a supporter but also can activate PMS to some extent. CM showed a slightly better catalytic performance than CuO, and the catalytic performance of CM was enormously improved when it was combined with g-C_3N_4 to form the CM/g-C_3N_4 composite, with the removal of RhB increasing from 19.03 to 99.53%. Based on the characterizations of SEM and BET, the introduction of g-C_3N_4 could increase the specific surface area and pore volume of CM, and decrease the aggregation of CM, with more reaction sites being exposed. Furthermore, more hydroxyl groups, which were important for PMS activation, were produced on the surface of CM/g-C_3N_4 than that of CM (Sections 3.3.3 and 3.5). Therefore, CM/g-C_3N_4 showed a better catalytic performance than CM. As shown in Figure 1b, the catalytic performances of C/g-C_3N_4 and M/g-C_3N_4 were worse than that of CM/g-C_3N_4. The reason is that the structure of CM/g-C_3N_4 is more conductive for PMS activation than those of C/g-C_3N_4 and M/g-C_3N_4 (Section 3.3). These results indicate that the interaction among copper, magnesium and g-C_3N_4 promotes the catalytic performance of CM/g-C_3N_4. Thus, the CM/g-C_3N_4 is an efficient PMS activator.

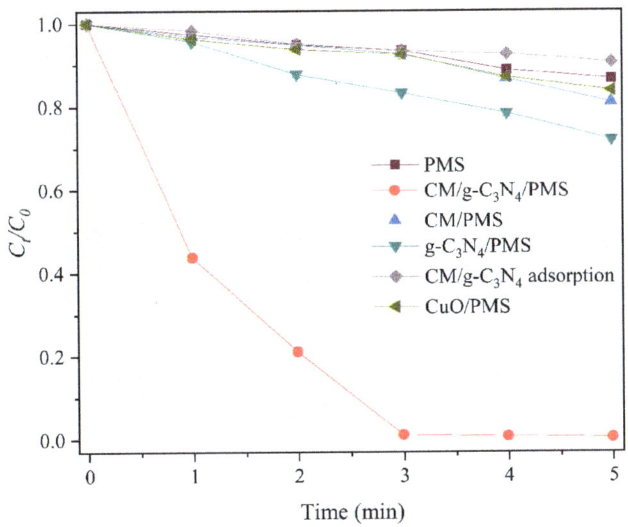

(a)

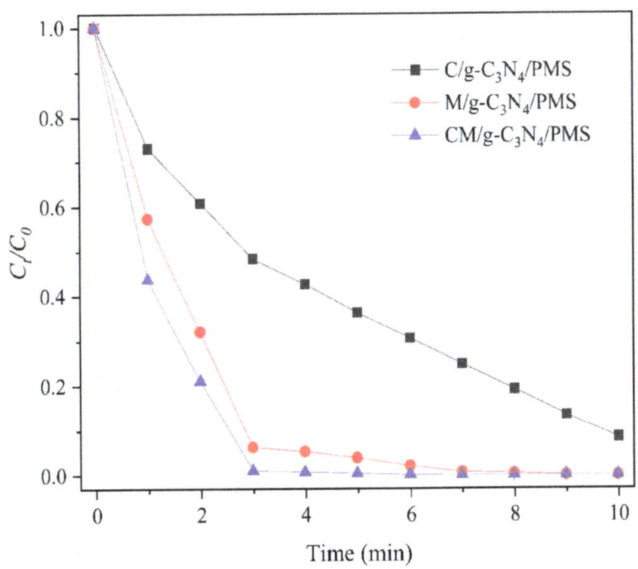

(b)

Figure 1. Degradation of RhB under different systems (**a**,**b**). Reaction conditions: [PMS] = 2.0 mM, [catalyst] = 0.50 g/L, [RhB] = 10.0 mg/L, 25 °C.

3.2. Effect of Preparation Procedures on the Catalytic Performance of CM/g-C_3N_4

The effects of preparation conditions, such as g-C_3N_4 formed by different precursors, mass ratios of copper–magnesium oxide to melamine, pyrolysis temperatures and times, and molar ratios of copper to magnesium precursor, on the catalytic performance of CM/g-C_3N_4 were investigated. CM/g-C_3N_4 presented better catalytic performance than CM/g-

C_3N_4 (urea) and showed a similar catalytic performance to CM/g-C_3N_4 (2) (Figure S1a). This is because the decomposition temperature of melamine is higher than that of urea. CM/g-C_3N_4 will form more carbon nitride than CM/g-C_3N_4 (urea) in the preparation process, which is beneficial for PMS activation. Guan et al. (2020) also found that g-C_3N_4 prepared by melamine showed a better catalytic performance than that prepared by urea [37]. To simplify the synthesis process, melamine rather than pure g-C_3N_4 was utilized as the g-C_3N_4 precursor.

As the mass ratio of copper–magnesium oxide to melamine increased from 1:1 to 1:2 the degradation of RhB by CM/g-C_3N_4/PMS increased in 4 min, while RhB degradation decreased in an increase of the mass ratio from 1:2 to 1:5 (Figure S1b). In a low addition amount, g-C_3N_4 will completely collapse the basic structure [38], and the dispersion of copper–magnesium oxide on the g-C_3N_4 would be inadequate. Furthermore, for a high addition amount of g-C_3N_4, the dispersion of copper–magnesium oxide was excessive causing the unsaturated contents of Cu, O and Mg in the CM/g-C_3N_4. All these phenomena would decrease the catalytic performance of CM/g-C_3N_4.

The pyrolysis temperature and time have an influence on the catalytic performance of catalysts. When the pyrolysis temperature of CM/g-C_3N_4 was less than 700 °C, the removal of RhB increased with an increase in temperature. However, the removal of RhB decreased as the pyrolysis temperature of CM/g-C_3N_4 reached higher than 700 °C (Figure S1c). It was possible that the structure of the catalyst would be destroyed or collapsed at a much higher pyrolysis temperature [39], thereby reducing reaction sites and decreasing the catalytic performance. The removal of RhB increased in 3 min with CM/g-C_3N_4 heated from 2 to 4 h, while it decreased with CM/g-C_3N_4 heated over 4 h (Figure S1d). This was possible because the structure of the catalyst would be sintered in the long pyrolysis time.

The best molar ratio of copper to magnesium precursors of CM/g-C_3N_4 was explored (Figure S1e). Within 3 min, the removal efficiency of RhB was 48.30, 71.80, 98.93, 73.80 and 66.70% for the molar ratios of copper to magnesium precursor of CM/g-C_3N_4 of 1:0.25, 1:0.5, 1:1, 1:1.5 and 1:2, respectively. The sample was prone to agglomeration and produced other impurities as the molar ratio of copper to magnesium precursor increased or decreased [40]. It could be inferred that the crystal structure of CM/g-C_3N_4 was well constructed in the 1:1 molar ratio of copper to magnesium precursor.

Based on the above results, the catalyst synthesized with a 1:2 mass ratio of copper–magnesium oxide to melamine and a 1:1 molar ratio of copper to magnesium precursor at 700 °C for 4 h, was the best PMS activator.

3.3. *Characterizations of the Catalysts*

3.3.1. SEM and TEM

For investigating the effect of the microstructure of catalysts on their catalytic performance, SEM characterization was performed, and the distribution of various elements of CM/g-C_3N_4 was also tested by EDS mapping characterization. As shown in Figure 2a,b, CM was composed of particles with aggregation and an irregular shape, and the pure g-C_3N_4 was a sheet packing structure. The aggregation of CM decreased when it was combined with g-C_3N_4 (Figure 2c). The above results showed that CM was well dispersed on the g-C_3N_4, thus exposing more reaction sites and improving the catalytic performance. The C/g-C_3N_4 were accumulated in a cluster, and no obvious particles had been seen, indicating that the CuO was covered and piled up with the g-C_3N_4; thus, fewer reaction sites were exposed (Figure 2e). The structures of CuO and g-C_3N_4 in C/g-C_3N_4 had been changed. On the contrary, M/g-C_3N_4 was accumulated in the form of particles, and no obvious flaky structure had been seen (Figure 2f). The research by Hoai Ta et al. showed a similar phenomenon [41]. Compared with C/g-C_3N_4 and M/g-C_3N_4, g-C_3N_4 kept its structure in the CM/g-C_3N_4. These results indicated that the interaction among copper, magnesium and g-C_3N_4 would not make the structure of g-C_3N_4 change. The decreasing aggregation and the structure of CM/g-C_3N_4 was more beneficial for exposing the reaction site to activate PMS than that of C/g-C_3N_4 and M/g-C_3N_4. Therefore, the catalytic performance of CM/g-C_3N_4 was the best among them. As shown in Figure 2d, the structure of CM/g-C_3N_4 did not change significantly after the reaction, but small cracks appeared in some places of CM/g-C_3N_4. It may be caused by the corrosion of CM/g-C_3N_4 during the PMS activation. The results of SEM-EDS mapping showed that C, N, O, Cu and Mg elements were uniformly distributed on the CM/g-C_3N_4, rather than a random mixture of all substances (Figure 2g).

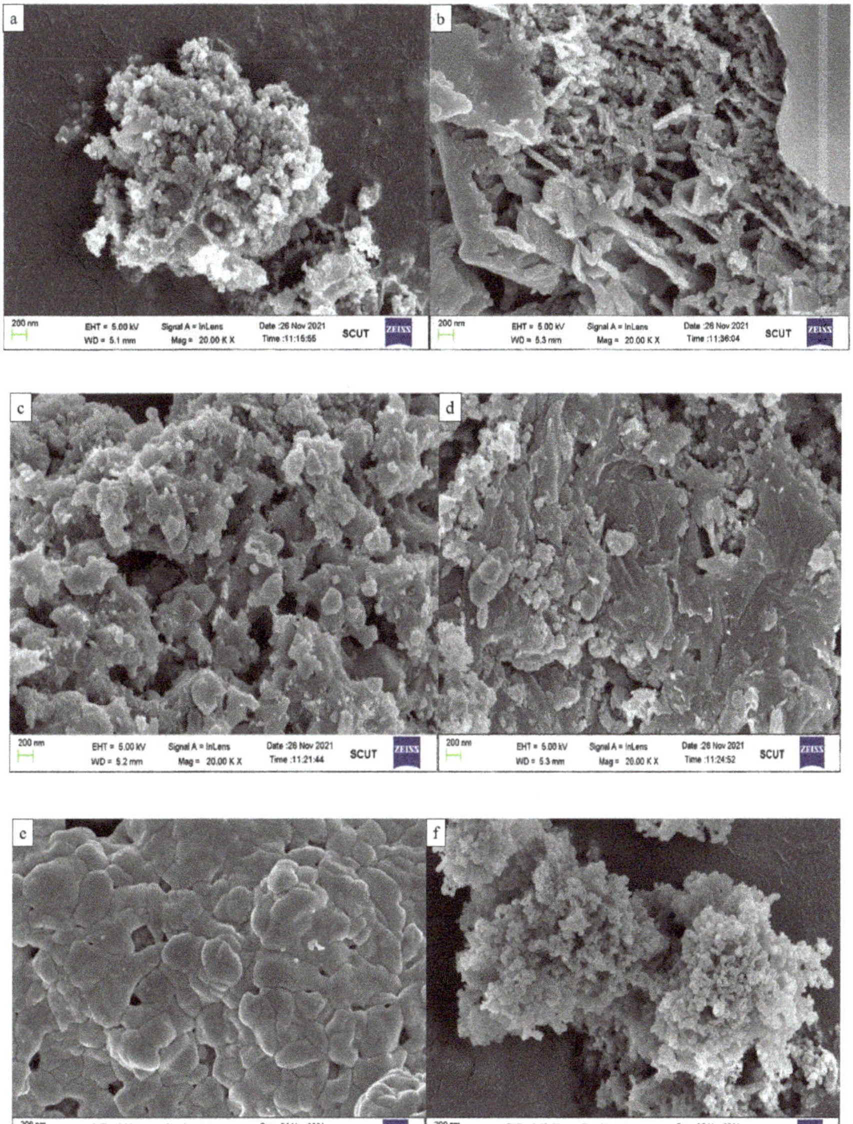

Figure 2. *Cont.*

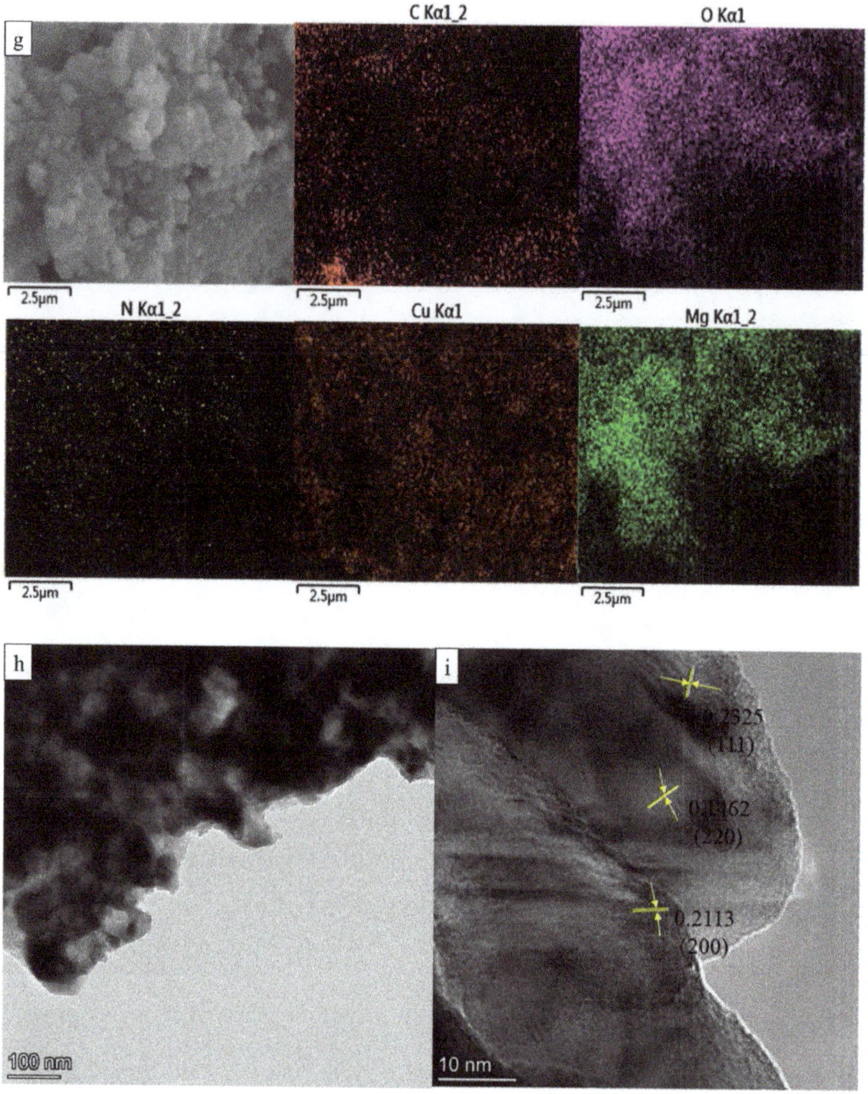

Figure 2. SEM image of CM (**a**), g-C_3N_4 (**b**), CM/g-C_3N_4 before (**c**) and after the reaction (**d**), C/g-C_3N_4 (**e**), M/g-C_3N_4 (**f**), EDS mapping of CM/g-C_3N_4 (**g**), TEM image (**h**) and HRTEM image (**i**) of CM/g-C_3N_4.

Figure 2h clearly shows that the CM was uniformly dispersed on the g-C_3N_4. The HRTEM image of CM/g-C_3N_4 with clear lattice fringes is exhibited in Figure 2i. The lattice spacing was about 0.2325, 0.2113 and 0.1462 nm, corresponding to CuO (111) and MgO (200), (220) crystal planes, respectively. It was consistent with the XRD results (Section 3.3.2).

3.3.2. XRD

The XRD spectra of different catalysts are presented in Figure 3. The peaks at 13.2° and 27.3° in pure g-C_3N_4 were respectively attributed to the in-plane repeating units of the continuous heptazine framework ((100) crystal plane) and the interlayer stacking of the periodic conjugated aromatic structure ((002) crystal plane) [42–44]. The diffraction peaks of M/g-C_3N_4 at 42.74° and 62.18° corresponded to the (200) and (220) crystal planes of MgO, respectively. The (111) crystal plane attributed to CuO could be observed in the C/g-C_3N_4 at 38.74°. SEM characterization showed that the morphology of g-C_3N_4 in the C/g-C_3N_4 and M/g-C_3N_4 had changed, therefore the g-C_3N_4 diffraction peaks were not found in these two materials. The corresponding peaks of CuO and MgO appeared in the CM, and the aggregation of CM led to the low intensity. The crystal plane diffraction peaks of g-C_3N_4 (002), CuO (111) and MgO (200, 220) were detected at 25.18°, 38.7°, 42.84° and 62.52° of CM/g-C_3N_4, respectively. It was in agreement with the HRTEM result. Due to the irregular arrangement of tri-s-triazine units [45,46], CM/g-C_3N_4 showed a weak (100) crystal plane diffraction peak. The diffraction peak of the (002) crystal plane in CM/g-C_3N_4 was weakened and shifted when compared with pure g-C_3N_4. This phenomenon indicated that the layered g-C_3N_4 was peeled off, and copper–magnesium oxide was successfully introduced into the structure of g-C_3N_4. When compared with pure CuO (JCPDS: 80-1916) and MgO (JCPDS: 79-0612), the corresponding crystal planes in the CM/g-C_3N_4 were slightly shifted, showing that there was an interaction between CuO and MgO in the CM/g-C_3N_4. Furthermore, the XRD peak of CM/g-C_3N_4 was more broadening than that of CM. It could be ascribed to the interaction among copper oxide, magnesium oxide and g-C_3N_4. According to the calculation of the Scherrer equation, the size distribution of CM/g-C_3N_4 was 22.01–63.54 nm, and the average size was 36.94 nm. The XRD spectra of CM/g-C_3N_4 before and after the reaction were similar, but the peak of (002) crystal plane shifted from 25.18° to 26.2° as the intensity decreased. Furthermore, the diffraction peaks of (200) and (220) crystal planes at 42.74° and 62.18° of CM/g-C_3N_4 disappeared after the reaction. The results show that the g-C_3N_4 and MgO in CM/g-C_3N_4 have important effects on the PMS activation.

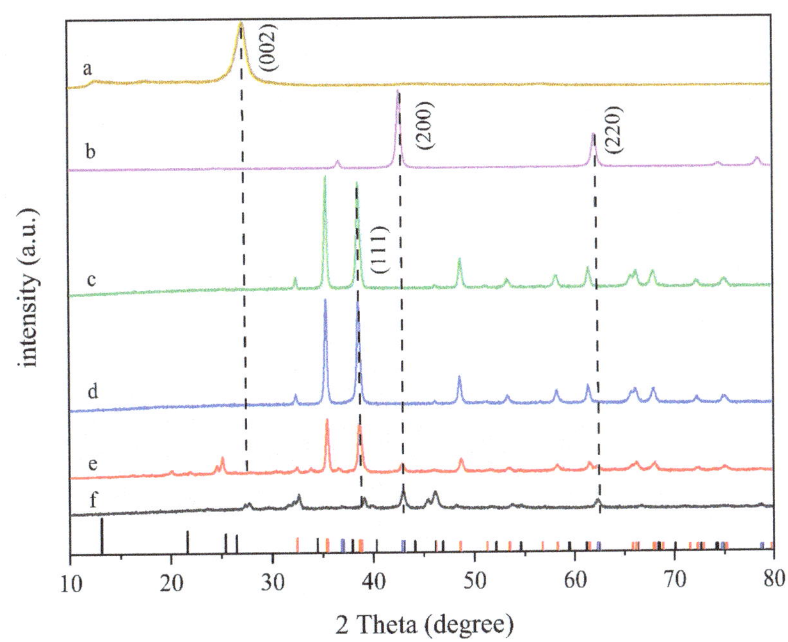

Figure 3. XRD spectra of different catalysts, g-C_3N_4, M/g-C_3N_4, C/g-C_3N_4, used CM/g-C_3N_4, fresh CM/g-C_3N_4 and CM. Black: g-C_3N_4 (JCPDS: 87-1526); Red: CuO (JCPDS: 80-1916); Blue: MgO (JCPDS: 79-0612).

3.3.3. FTIR

The FTIR spectrum of each sample is shown in Figure 4. The peak at 3375 cm^{-1} could be assigned to the stretching vibration of a hydroxyl group (−OH) [47]. This peak was observed in the M/g-C_3N_4. It could be ascribed to the MgO in M/g-C_3N_4, which was in favor of -OH being absorbed on the surface. The peak of −OH was also observed in the CM/g-C_3N_4, with a higher intensity than that seen in the M/g-C_3N_4, but there were no peaks in the CM and C/g-C_3N_4. The result indicated that the large number of −OH on the CM/g-C_3N_4 surface could be attributed to the interaction between MgO and g-C_3N_4. It may be because the aggregation of CM was not beneficial for −OH produced on the surface. However, the aggregation was decreased when CM was combined with g-C_3N_4, and in the action of MgO, −OH could be produced on the surface of CM/g-C_3N_4. −OH has been certified as an important PMS activation site [48], thus the excellent catalytic performance of CM/g-C_3N_4 can be attributed to the numerous −OH (demonstrated in Section 3.5). The peak at 1635 cm^{-1} of CM/g-C_3N_4 could be ascribed to the −OH bending vibration in adsorbed water [49], which is also important for PMS activation. However, this peak was not observed on the other catalysts. The peaks at 1224–1647 and 3175 cm^{-1} of pure g-C_3N_4 were attributed to the stretching vibration of the heterocyclic ring (C−N=C) and N−H [50–55], respectively. This peak of the heterocyclic ring shifted and became weaker in the CM/g-C_3N_4 (1083–1176 cm^{-1}), possibly because of the influence of Cu, Mg and O elements on the structure of the C−N=C. The peak at 810 cm^{-1} was associated with the vibration of the triazine unit, but it was not observed in the CM/g-C_3N_4 due to its irregular arrangement. These phenomena were in agreement with the XRD result. Due to the shape of the g-C_3N_4 changing in the C/g-C_3N_4 and M/g-C_3N_4, their FTIR spectrums showed no characteristic peak about g-C_3N_4, which was consistent with the results of SEM and XRD. The peaks below 700 cm^{-1} of C/g-C_3N_4, M/g-C_3N_4 and CM/g-C_3N_4 corresponded to the stretching and bending vibrations of M−O (M: Cu and Mg) [39], respectively.

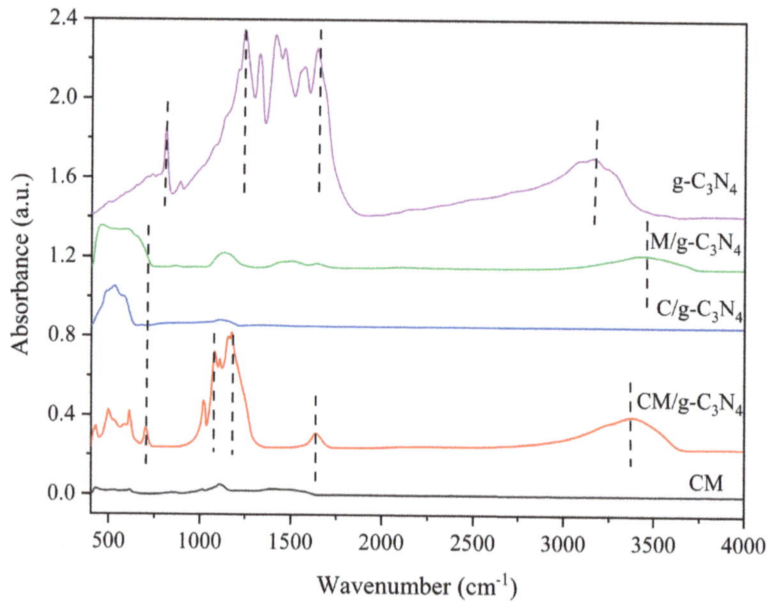

Figure 4. FTIR spectra of different catalysts (g-C_3N_4; M/g-C_3N_4; C/g-C_3N_4; CM/g-C_3N_4; CM).

3.3.4. BET

A Brunauer-Emmett-Teller (BET) gas adsorption analysis was used to investigate the porous structure and specific surface area (SSA) of the catalyst. All samples showed type

IV isotherms and H_3 type hysteresis loops, indicating that the mesoporous structure was dominant in these samples (Figure 5). When compared with CM and C/g-C_3N_4, the SSA and pore volume of CM/g-C_3N_4 increased, and the average pore diameter slightly reduced (Table 1). These results demonstrated that the introduction of g-C_3N_4 and MgO increased the SSA and pore volume of CM/g-C_3N_4, thus more reaction sites were exposed and the structure of CM/g-C_3N_4 was more conducive to activating PMS. According to the SEM characterizations of CM, C/g-C_3N_4 and CM/g-C_3N_4, the metal oxide in CM/g-C_3N_4 was mostly dispersed. Thus, it could be inferred that the aggregation of the catalyst had a relationship with the SSA. However, the M/g-C_3N_4 with the largest SSA (58.47 m^2/g) was accumulated in the form of particles, illustrating that the relationship between SSA and the aggregation of catalyst was not linear. When compared with other materials, such as CuMg-MMO [40] and CuO-CN [45], the SSA, pore volume and average pore diameter of the materials prepared in this article were all the best.

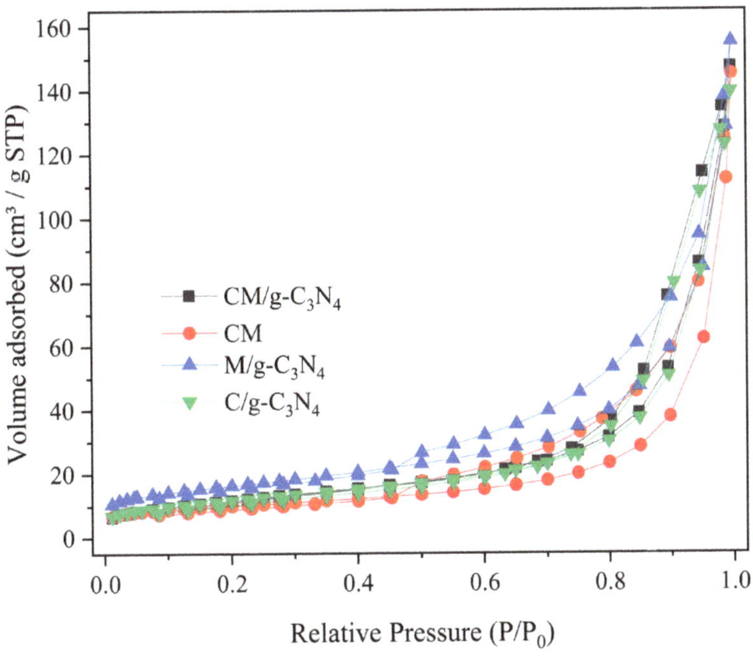

Figure 5. N_2 adsorption–desorption isotherms of different catalysts (g-C_3N_4; M/g-C_3N_4; C/g-C_3N_4; CM/g-C_3N_4; CM).

Table 1. Specific surface area and pore characteristics of different catalysts (g-C_3N_4; M/g-C_3N_4; C/g-C_3N_4; CM/g-C_3N_4; CM).

	Specific Surface Area (m^2/g)	Pore Volume (cm^3/g)	Pore Size (nm)
CM	35.45	0.1803	20.34
CM/g-C_3N_4	42.72	0.2092	19.59
C/g-C_3N_4	42.18	0.1992	18.89
M/g-C_3N_4	58.47	0.2041	13.96

3.4. Factors Impacting RhB Degradation by CM/g-C_3N_4/PMS

3.4.1. Catalyst Dosage, PMS Concentration and RhB Concentration

As shown in Figure S2a, the removal efficiency of RhB increased from 36.88 to 99.90% at 5 min, and the k value raised from 0.0867 to 1.2916 min^{-1} with increasing catalyst dosage (Table S1). More catalysts can provide more reaction sites for PMS activation, thus more active substances can degrade RhB in the reaction system. As shown in Figure S2b the removal of RhB raised from 91.83 to 99.55% at 5 min for the PMS concentration of 0.50–1.0 mM, with the k value increased from 0.4939 to 1.1703 min^{-1} (Table S1). As the PMS concentration increased from 1.0 to 3.0 mM, the k value decreased from 1.1703 to 0.8455 min^{-1}. In fact, as the concentration of PMS increases, the activation reaction can be promoted rapidly. However, the self-quenching reaction of PMS will reduce the number of active substances and limit the expression of catalysts due to the overdose of PMS molecules in the reaction system [56]. Figure S2c showed that the degradation efficiency of RhB decreased from 99.90 to 65.69% in the RhB initial concentration range from 5.0 to 20.0 mg/L at 5 min, and the k value dropped from 1.4365 to 0.2187 min^{-1}. This may be because the number of active substances were constant in an assured concentration of catalyst and PMS. The higher the content of RhB, the stronger the competition is among RhB molecules for active substances, which slows down the reaction rate of the CM/g-C_3N_4/PMS system. Considering environmental protection and the concentration of RhB in real life, 0.30 mg/L CM/g-C_3N_4 and 1.0 mM PMS were suitable for 10.0 mg/L RhB degradation.

The k value of the system (0.30 mg/L CM/g-C_3N_4, 1.0 mM PMS, 10.0 mg/L RhB, k = 1.1703 min^{-1}) was much higher than those of other reaction systems under the optimal conditions, such as the Fe_3O_4/Co_3S_4/PMS/RhB system (0.302 min^{-1}), 5%-T/LDOs/PMS/RhB system (0.4537 min^{-1}), Co-G/PMS/RhB system (0.9438 min^{-1}) and $Co_{1+x}Fe_{2-x}O_4$ /PMS/RhB system (0.260 min^{-1}) (Table S2). It indicated that CM/g-C_3N_4 can efficiently and rapidly activate PMS to degrade RhB. The mineralization efficiency of RhB in the CM/g-C_3N_4/PMS system reached 43.43% within 5 min and slightly increased (48.93%) in 30 min (Figure S3). It is probable that RhB was decomposed into small molecule intermediates that can undergo sufficient degradation on the surface of CM/g-C_3N_4, then stabilizing the TOC value of the system.

3.4.2. pH

The effects of initial pH (3.0–11.0) on the RhB degradation were investigated under the optimal conditions of 0.30 g/L CM/g-C_3N_4, 1.0 mM PMS and 10.0 mg/L RhB. As shown in Figure S4, the removal of RhB was inhibited under the strong acid (pH = 3.0), weak acid (pH = 5.0) and strong alkaline (pH = 11.0) conditions, with the removal efficiency falling below 40.00% within 6 min. The reason was possible in that PMS was mainly presented in the form of H_2SO_5 (SO_5^{2-}) under the acid (strong alkaline) condition [57]. These two forms were not activated by the catalyst to produce active substances for pollutant degradation. Under the neutral (pH = 7.0) and weak alkaline (pH = 9.0) conditions, over 99.00% of RhB was removed in 6 min. This means that the CM/g-C_3N_4/PMS system has good potential for RhB degradation in practical applications because of the neutral or weakly alkaline pH of most actual wastewater.

3.5. Possible Activation Mechanisms of PMS

Quenching experiments were utilized to investigate the reactive oxygen species (ROS) produced in the CM/g-C_3N_4/PMS system. Figure 6a,b show that the introduction of MeOH (a scavenger for $SO_4^{\bullet-}$ and $^{\bullet}OH$) and IPA (a scavenger for $^{\bullet}OH$) inhibited the removal of RhB. The removal of RhB dropped from 99.88 to 60.50% (MeOH, 300.0 mM) and 82.08% (IPA, 300.0 mM), respectively. Furthermore, the removal efficiency of RhB respectively decreased from 99.88 to 85.78 and 18.14% by adding p-BQ (10.0 mM, $O_2^{\bullet-}$ scavenger) and L-histidine (10.0 mM, $^1O_2$ scavenger) into the system (Figure 6c,d). EPR was carried out to verify ROS in the oxidation reactions. As shown in Figure 7a,b, typical signals of DMPO-OH and DMPO-O_2^- were respectively observed in the CM/g-C_3N_4/PMS/DMPO and

$CM/g\text{-}C_3N_4/PMS/DMPO/MeOH$ systems [40]. However, the DMPO-SO_4^- signal could not be detected in the $CM/g\text{-}C_3N_4/PMS/DMPO$ system. It may be because of the low sensitivity and short life of DMPO-SO_4^- signal [39,40,58], but the result of the quenching experiment had shown that $SO_4^{\bullet-}$ existed in the reaction system. A typical TEMP-$^1O_2$ triplet signal peak with strong intensity was detected in the $CM/g\text{-}C_3N_4/PMS/TEMP$ system (Figure 7c), showing the existence of $^1O_2$ [59]. All these results indicated that $SO_4^{\bullet-}$, $^{\bullet}OH$, $O_2^{\bullet-}$ and $^1O_2$ were generated in the $CM/g\text{-}C_3N_4/PMS$ system, and both $SO_4^{\bullet-}$ and $^1O_2$ were the predominant ROS responsible for the removal of RhB.

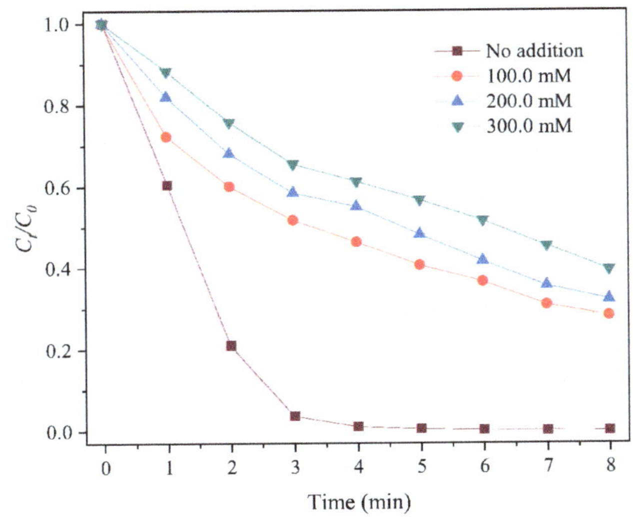

(**a**)

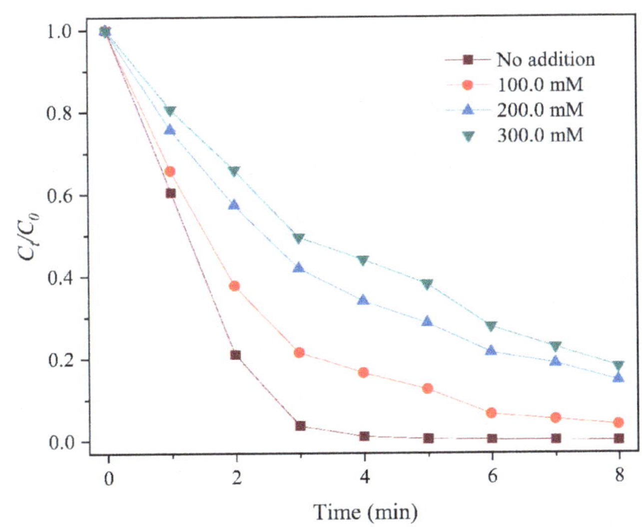

(**b**)

Figure 6. *Cont.*

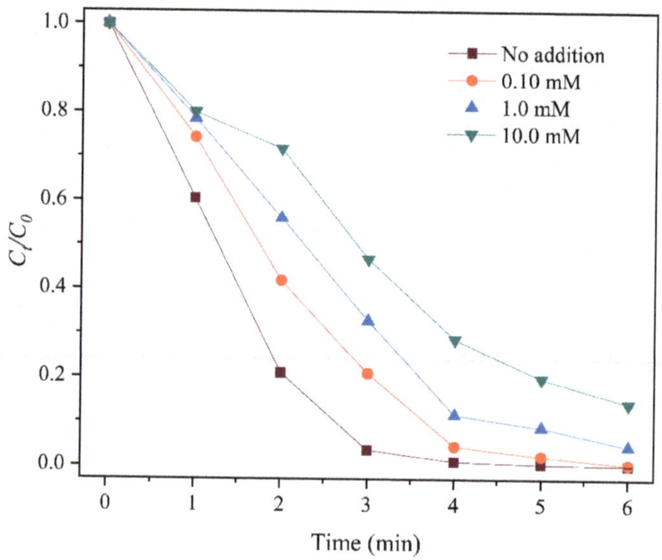

(**c**)

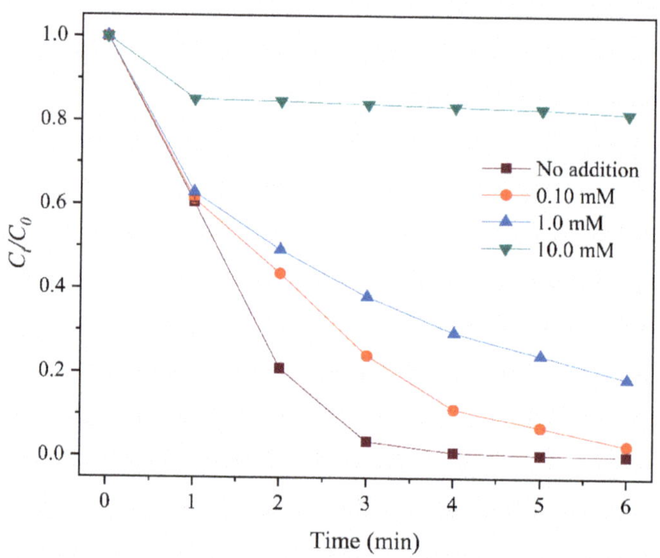

(**d**)

Figure 6. Effects of different scavengers on RhB removal. MeOH (**a**), IPA (**b**), p-BQ (**c**) and L-Histidine (**d**). Reaction conditions: [PMS] = 1.0 mM, [CM/g-C_3N_4] = 0.30 g/L, [RhB] = 10.0 mg/L, 25 °C.

The nitrogen was introduced into the CM/g-C_3N_4/PMS system to verify the contribution of dissolved oxygen in the $O_2^{\bullet-}$ generation. As shown in Figure S5, the removal of RhB decreased within 4 min, indicating that dissolved oxygen had contributed to $O_2^{\bullet-}$ generation to some extent.

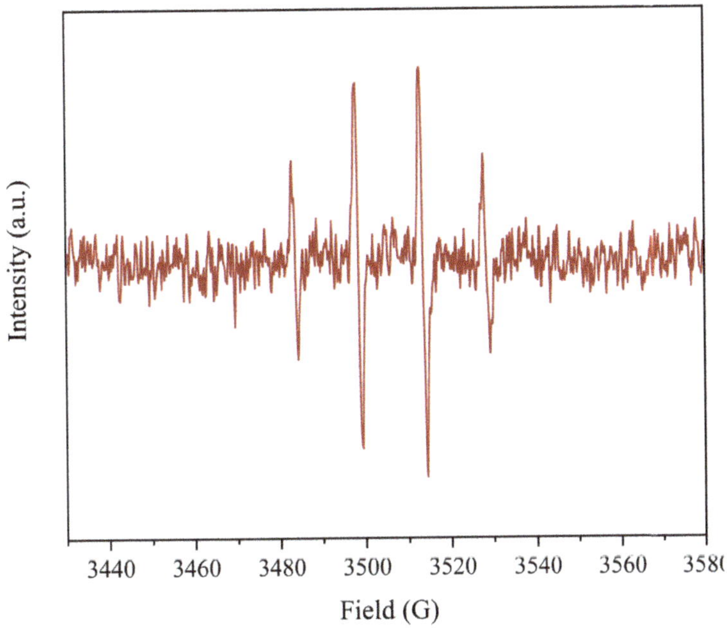

(a)

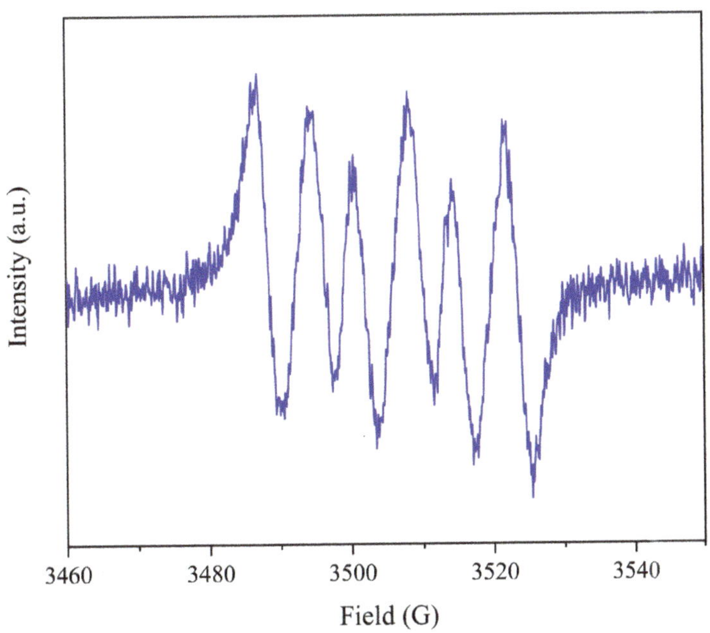

(b)

Figure 7. *Cont.*

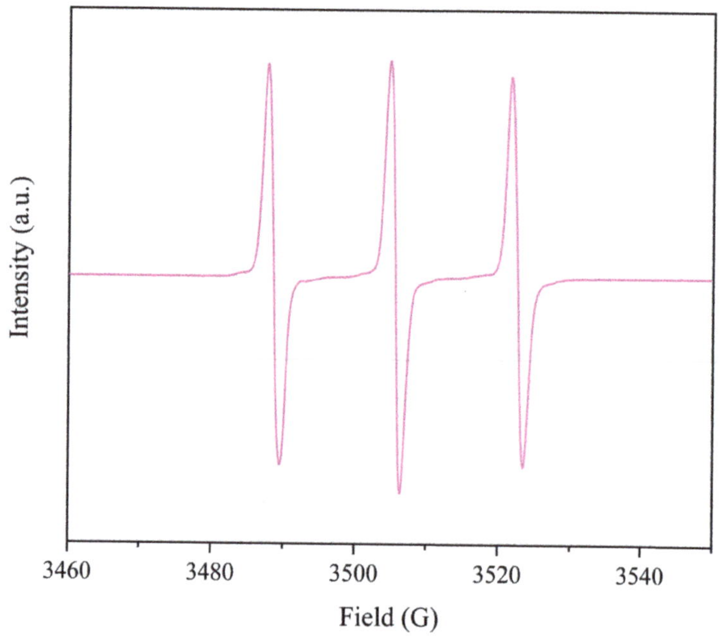

(c)

Figure 7. EPR spectra of $SO_4^{\bullet-}$ and $^{\bullet}OH$ (**a**), $O_2^{\bullet-}$ (**b**) and $^1O_2$ (**c**) in the CM/g-C_3N_4/PMS system. Reaction conditions: [PMS] = 1.0 mM, [CM/g-C_3N_4] = 0.30 g/L, [RhB] = 10.0 mg/L, DMPO/TEMP = 0.10 M, 25 °C.

In order to clarify the surface components of CM/g-C_3N_4 before and after the reaction, XPS analysis was performed, and the atomic area ratios of different components were shown in Table S3. The full spectrum shows that CM/g-C_3N_4 contained C, O, N, Cu and Mg elements, and there was little difference in its spectrum before and after the reaction (Figure 8a), showing the stability of CM/g-C_3N_4. The C1s spectrum showed three peaks at 284.8 (C−C), 286.47 (C−O) and 288.72 ev (C=O) (Figure S6a,b). The atomic area ratios of C−C, C−O and C=O before (after) the reaction were 79.00 (77.42%), 11.68 (10.4%) and 9.32% (12.18%), respectively (Table S3). The result indicates that C−O may participate in the activation process of PMS. For O1s' spectrum, three binding energy peaks at 529.91, 531.93 and 532.84 ev could be classified as lattice oxygen (O^{2-}), hydroxyl groups (−OH) and adsorbed oxygen [60,61], with atomic area ratios of 9.58, 43.27 and 41.99%, respectively (Figure 8b, Table S3). After the reaction, the atomic area ratios of O^{2-} and −OH increased to 25.14 and 45.79%, with adsorbed oxygen being reduced to 13.78% (Figure 8c, Table S3). The result demonstrates that adsorbed oxygen was important for PMS activation. Adsorbed oxygen can promote the formation of −OH on the surface of the catalyst and facilitate the activation of PMS [62,63]. Previous studies have shown that −OH is an important reaction site for PMS activation, and $H_2PO_4^-$ has a strong chelating influence on the –OH of catalyst surfaces [64–66], thus impacting the PMS activation. Therefore, $H_2PO_4^-$ was introduced into the CM/g-C_3N_4/PMS system to investigate the role of −OH. As shown in Figure S7, the RhB removal was inhibited at the $H_2PO_4^-$ concentration of 1.0–8.0 mM at 4 min. This indicated that –OH on the CM/g-C_3N_4 surface was an important reaction site for PMS activation. Hence, the adsorbed oxygen on the surface of CM/g-C_3N_4 can provide more reaction sites for activating PMS. The higher atomic area ratio of –OH after the reaction demonstrated that the rate of consuming –OH was slower than that of producing –OH. This is mainly because of the constant formation of -OH by adsorbed oxygen. For N1s' spectrum, two peaks with the binding energy of 398.7 and 400.6 eV were related to sp^2 hybridized N atoms (C=N−C) and $N(C)^3$ groups (C−N), respectively (Figure 8d,e) [45]. The atomic area ratio of C=N−C groups decreased from 86.45 to 40.03%, and that of C−N groups increased from 13.55 to 59.97% (Table S1). This shows that the g-C_3N_4 of CM/g-C_3N_4 has an important contribution to PMS activation, and the conversion of copper species on the surface of g-C_3N_4 may occur [63]. For the Cu2p spectrum, the binding peaks located at 933.12 and 934.41 ev could be respectively ascribed to Cu^+ and Cu^{2+} [40,62], with atomic area ratios of 42.59 and 57.41% (Figure 8f, Table S1). After the reaction, the atomic area ratios of Cu^+ and Cu^{2+} were 51.14 and 48.86% respectively (Figure 8g, Table S1), indicating that the conversion between Cu^+ and Cu^{2+} were a major factor in activating PMS. For Mg1s' spectrum, the binding energy peak before and after the reaction was at 1304.59 ± 0.5 eV (Figure S6c,d). It is a typical MgO peak.

Based on the above analyses, the possible activation mechanism of PMS by CM/g-C_3N_4 could be sufficiently expressed by Equations (3)–(17) (Scheme 1). They show the redox cycle of Cu^{2+}/Cu^+-activated PMS to produce $SO_4^{\bullet-}$, OH^- and $SO_5^{\bullet-}$ (Equations (3) and (4)). The results of FTIR and XPS analyses had shown that the interaction between MgO and g-C_3N_4 and the adsorbed oxygen on the surface of CM/g-C_3N_4 would enhance the production of -OH, which promoted the formation of Cu−OH complexes. PMS could react with these substances to form $SO_4^{\bullet-}$ and $SO_5^{\bullet-}$ (Equations (5) and (6)) [67]. The results of RhB removed by the g-C_3N_4/PMS system and XPS analysis had indicated that g-C_3N_4 was not only a supporter but also participated in the PMS activation. The electron-rich structure of g-C_3N_4 could provide electrons to promote the reduction of PMS. Moreover, PMS could act as an electron donor due to the presence of O−O bonds and the C atoms adjacent to the N atoms in g-C_3N_4 could act as the electron acceptors (Equations (7) and (8)) [68]. The generated $SO_4^{\bullet-}$ reacted with water molecules or OH^- to produce $^{\bullet}OH$ (Equation (9)). In the oxidation system, PMS would be hydrolyzed to produce $O_2^{\bullet-}$ [40]. Furthermore, as mentioned earlier, dissolved oxygen also contributed to $O_2^{\bullet-}$ generation (Equations (10) and (12)). The generated $O_2^{\bullet-}$ could consume PMS to accelerate the production of $SO_4^{\bullet-}$ for improving the oxidation performance of system (Equation (13)). The self-reaction of $SO_5^{\bullet-}$ generated in the PMS activation and the self-decomposition of PMS would produce $^1O_2$ (Equations (14) and (15) [69]. Moreover, the reaction between $O_2^{\bullet-}$ ($SO_5^{\bullet-}$) and water molecules would also generate $^1O_2$ (Equations (16) and (17)) [70–72]. RhB was transformed into small molecule intermediates and finally mineralized into H_2O and CO_2 under the action of $SO_4^{\bullet-}$, $^{\bullet}OH$, $O_2^{\bullet-}$ and $^1O_2$.

$$\equiv Cu^+ + HSO_{5^-} \rightarrow \equiv Cu^{2+} + SO_4^{\bullet-} + OH^- \quad (3)$$

$$\equiv Cu^+ + HSO_{5^-} \rightarrow \equiv Cu^{2+} + SO_4^{\bullet-} + OH^- \quad (4)$$

$$\equiv Cu^+ - OH + HSO_{5^-} \rightarrow \equiv Cu^{2+} - OH + SO_4^{\bullet-} + OH^- \quad (5)$$

$$\equiv Cu^{2+} - OH + HSO_{5^-} \rightarrow \equiv Cu^+ - OH + SO_5^{\bullet-} + H^+ \quad (6)$$

$$HSO_{5^-} + g - C_3N_4 + e^- \rightarrow SO_4^{\bullet-} + OH^- \quad (7)$$

$$HSO_{5^-} + g - C_3N_4 \rightarrow SO_5^{\bullet-} + H^+ + e^- \quad (8)$$

$$SO_4^{\bullet-} + H_2O/OH^- \rightarrow HSO_{4^-}/SO_4^{2-} + ^{\bullet}OH \quad (9)$$

$$HSO_{5^-} \rightarrow SO_5^{2-} + H^+ \quad (10)$$

$$SO_5^{2-} + H_2O \rightarrow O_2^{\bullet-} + SO_4^{2-} + H^+ \quad (11)$$

$$O_2 + e^- \rightarrow O_2^{\bullet-} \quad (12)$$

$$HSO_{5^-} + O_2^{\bullet-} \rightarrow SO_4^{\bullet-} + O_2 + OH^- \quad (13)$$

$$SO_5^{\bullet-} + SO_5^{\bullet-} \rightarrow 2SO_4^{2-} + ^1O_2 \quad (14)$$

$$HSO_{5^-} + SO_5^{2-} \rightarrow HSO_{4^-} + SO_4^{2-} + ^1O_2 \quad (15)$$

$$2O_2^{\bullet-} + 2H_2O \rightarrow {}^1O_2 + H_2O_2 + 2OH^- \quad (16)$$

$$2SO_5^{\bullet-} + H_2O \rightarrow 2HSO_{4^-} + 1.5{}^1O_2 \quad (17)$$

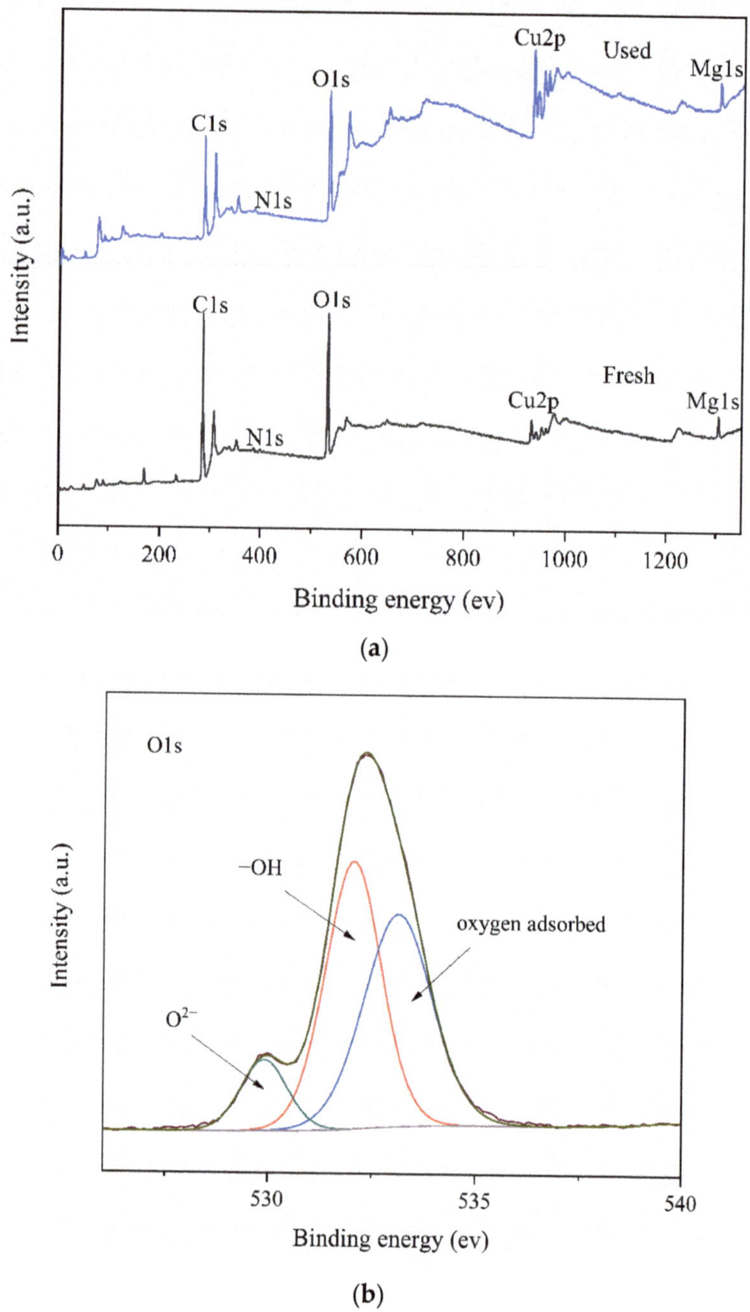

Figure 8. *Cont.*

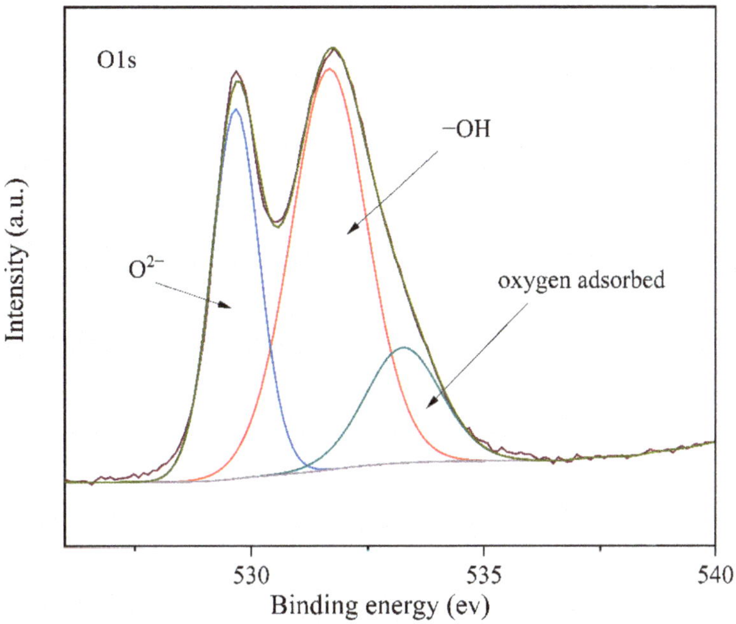

(**c**)

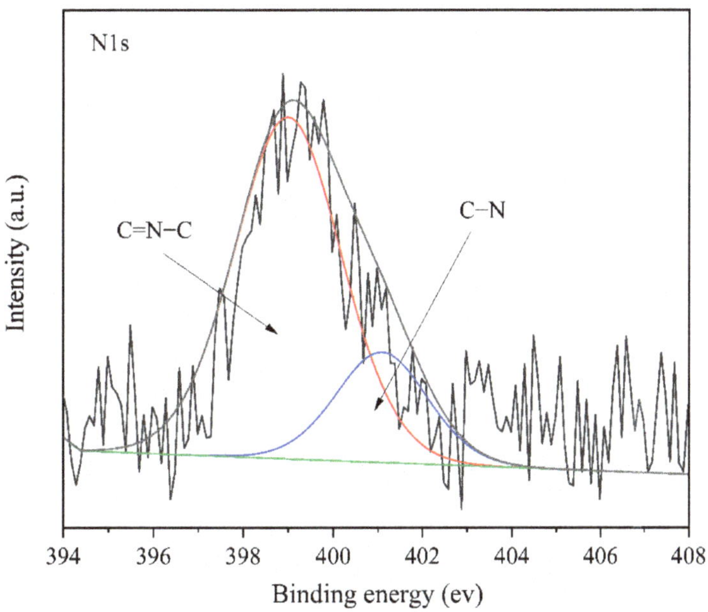

(**d**)

Figure 8. *Cont.*

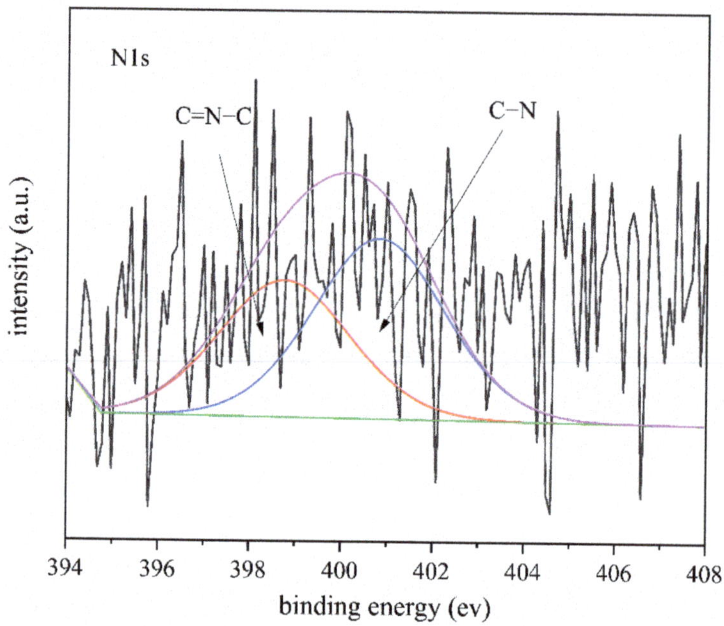

(e)

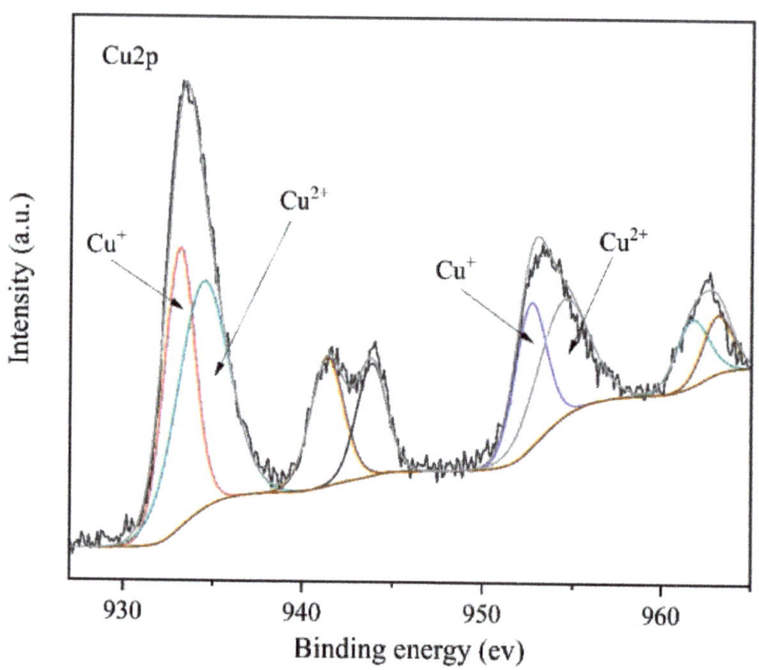

(f)

Figure 8. *Cont.*

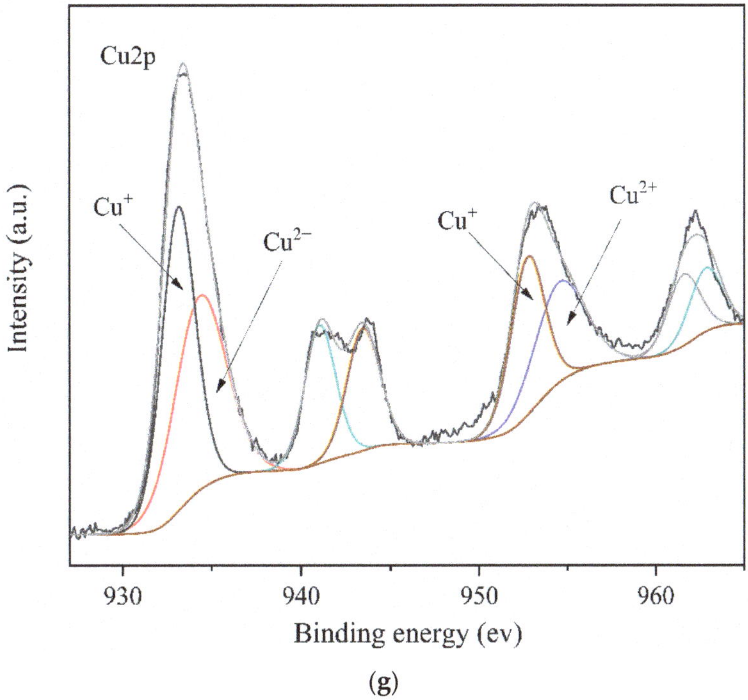

Figure 8. X-ray photoelectron spectroscopy spectra of full survey (**a**), O1s (**b**,**c**), N 1s (**d**,**e**), Cu 2p (**f**,**g**) in the CM/g-C_3N_4 before (**b**,**d**,**f**) and after (**c**,**e**,**g**) the reaction.

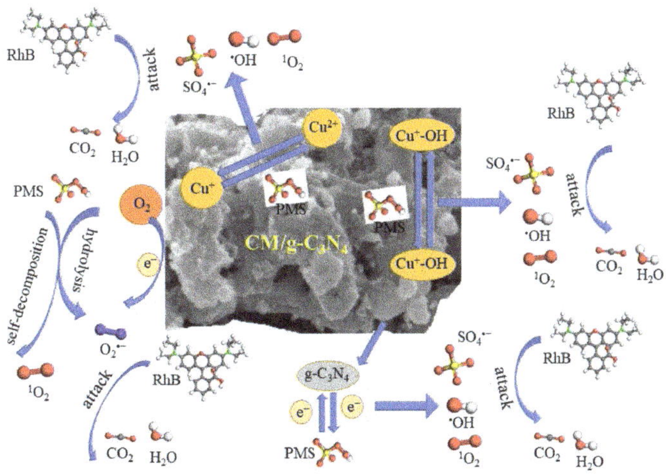

Scheme 1. The possible activation mechanism of PMS.

3.6. Degradation Pathway of RhB and Toxicity Estimation

The oxidation intermediates of RhB were detected by LC-MS analysis. Generally, the degradation of RhB occurs through a series of reactions, including N-demethylation, deamination, dealkylation, decarboxylation, chromophore cleavage, ring-opening and mineralization [73–75]. The DFT calculation was used to analyze the molecular structure of RhB (Figure S8). The result indicated that C10 was vulnerable to being attacked by active substances, and the atoms shown in Table S4 were also under attack. Depending on the previous reports, the LC-MS analysis and the DFT calculation, nineteen intermediates including isomers were identified during the reaction [75–84]. As shown in Figure 9, the degradation pathway of RhB in the CM/g-C_3N_4/PMS system was proposed. First, under the intense attack of various active species, RhB (m/z = 443) was transformed into a series of intermediates (P1–P6) by N-demethylation and deamination. Then, P7 and P8 intermediates were produced by the decarboxylation and further oxidation of P6. Moreover, the chromophores of intermediates (P1–P8) were cleaved to form the P9–P13 intermediates. By a ring-opening reaction, the P9–P13 intermediates further formed small molecule products (P14–P19). Finally, RhB and its intermediates were mineralized into CO_2 and H_2O. Many small intermediates of RhB were produced in the CM/g-C_3N_4/PMS system, showing the good degradation capacity of this system.

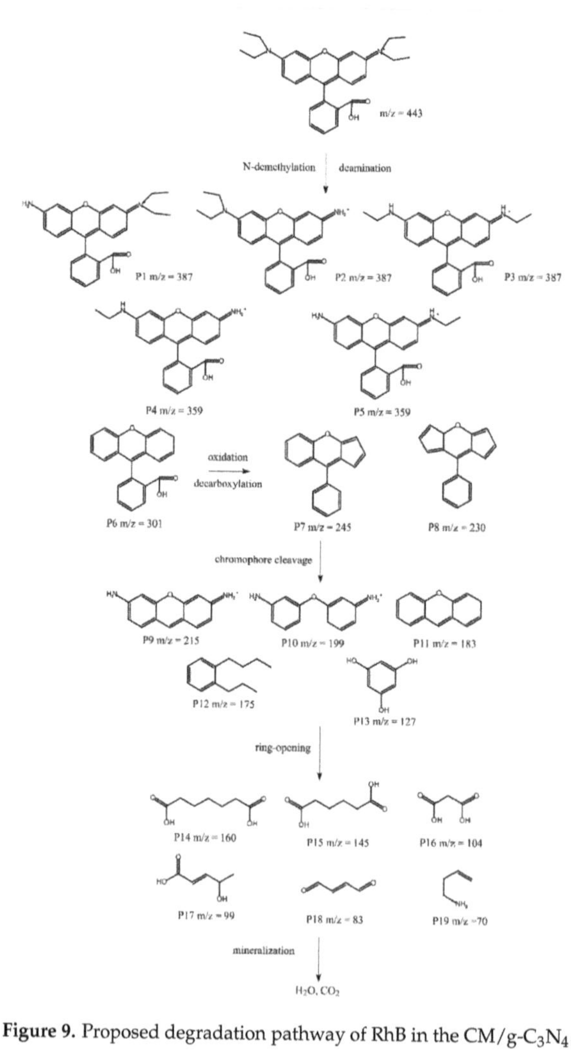

Figure 9. Proposed degradation pathway of RhB in the CM/g-C_3N_4/PMS system.

The toxicity estimation is important for predicting the damage of pollutants to the environment and humans. Therefore, the toxicities of RhB and its intermediates were examined by assessing oral rat LD_{50} and bioaccumulation factors on the TEST. The results are shown in the Figure 10. The predicted oral rat LD_{50} of RhB was 924.86 mg/kg. Except for P5, P7 and P19, the oral rat LD_{50} of other intermediates decreased after the degradation of RhB, indicating that the toxicities of intermediates were reduced. The predicted bioaccumulation factor was 15.48 in the RhB, and decreased significantly in the intermediates, indicating the lower accumulation of intermediates in the organisms. In terms of oral rat LD_{50} and bioaccumulation factors, the CM/g-C_3N_4/PMS system was able to sufficiently reduce RhB toxicity by transforming it into a series of intermediates with lower toxicity.

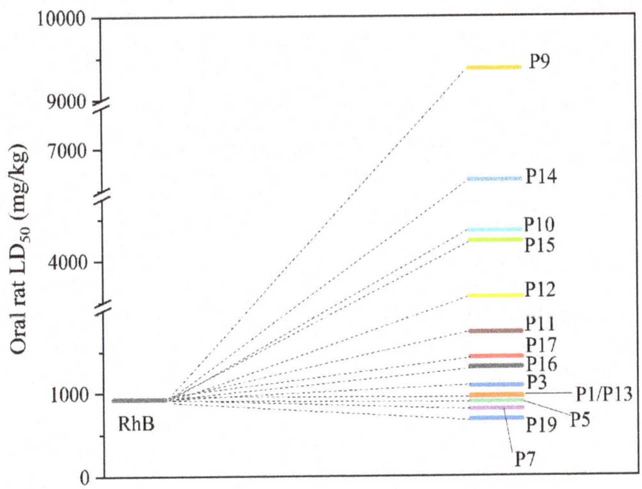

(a)

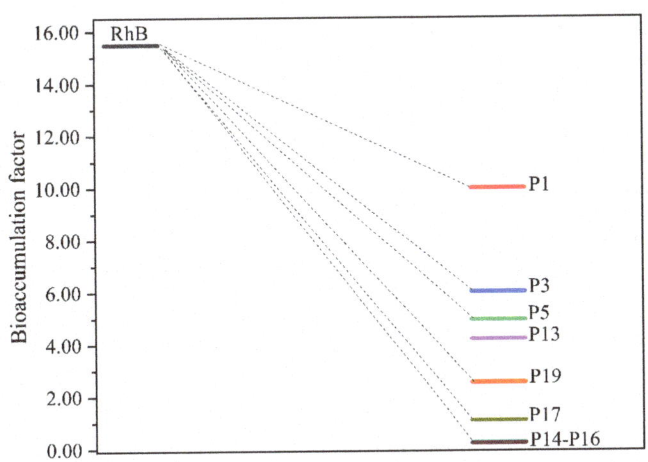

(b)

Figure 10. Oral rat LD_{50} (**a**) and bioaccumulation factors (**b**) of RhB and its intermediates.

3.7. Stability of the CM/g-C_3N_4

The stability and repeatability of a material is an important parameter to measure its practical applications. Therefore, the CM/g-C_3N_4 was reused five times under the same conditions (0.30 g/L CM/g-C_3N_4, 1.0 mM PMS, 10.0 mg/L RhB). As shown in Figure S9, the removal of RhB decreased from 99.88 to 99.30% after five cycles, showing the good repeatability of CM/g-C_3N_4. This indicates that the CM/g-C_3N_4 was stable for long-term running. The leaching concentrations of copper and magnesium ions were monitored by ICP-OES. The results showed that the loss of copper and magnesium ions were 5.16 and 31.97 mg/L after the reaction, respectively. As a non-toxic metal, magnesium ions are widely present in the water matrix with a concentration of about 0.0–130.0 mg/L. Unlike transition metal ions, the leaching of magnesium ions would not cause secondary pollution. In addition, the copper ion leaching concentration of CM/g-C_3N_4 was lower than those of other copper-based heterogeneous catalysts (Cu_2FeSnS_4 (16.70 mg/L), such as Cu_2S (131.60 mg/L) [85]). All the above results show that the eco-friendly and stable CM/g-C_3N_4 system has great application potential in the field of environmental remediation.

3.8. Practical Application of the CM/g-C_3N_4/PMS System

The CM/g-C_3N_4/PMS system could degrade other pollutants, such as CR, MB and TC (Figure 11a). The removal of CR, MB and TC were all over 73.00%, indicating that the CM/g-C_3N_4/PMS system can oxidize various contaminants. Figure 11b reveals that RhB could be degraded in the tap water and the wastewater in 10 min, indicating that the CM/g-C_3N_4/PMS system can efficiently remove RhB in the actual environment. All these results indicate that the CM/g-C_3N_4 system has the practical application potential for pollutant degradation.

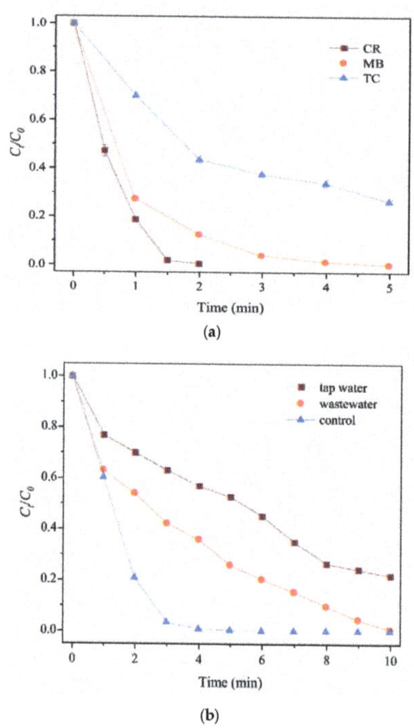

Figure 11. Degradation of different pollutants in the CM/g-C3N4/PMS system (**a**) and RhB removal in different water matrixes (**b**). Reaction conditions: [PMS] = 1.0 mM, [CM/g-C3N4] = 0.30 g/L, [RhB/CR/MB/TC] = 10.0 mg/L, 25 °C.

4. Conclusions

CM/g-C_3N_4 was synthesized and initially used as a heterogeneous catalyst to activate PMS for the degradation of RhB. Owing to the splendid structure and abundant reaction sites, CM/g-C_3N_4 exhibited much a better catalytic performance than CuO, CM, g-C_3N_4, M/g-C_3N_4 and C/g-C_3N_4. Under the low usage reaction conditions, RhB could be rapidly degraded at a natural pH in 5 min. PMS was activated by the copper and g-C_3N_4 in CM/g-C_3N_4, with $SO_4^{\bullet -}$ and $^1O_2$ as the main active substances for the removal of RhB. CM/g-C_3N_4 showed good stability and reusability. This work enriches the family of copper-based heterogeneous catalysts used in the PMS activation, and provides a guiding role of enhancing the catalytic performance of copper oxide. The copper-based catalyst with a higher specific surface area, used for industrial applications, should be considered in the future work.

Supplementary Materials: The following supporting information can be downloaded at: https://www.mdpi.com/article/10.3390/w14132054/s1. Scheme S1: Preparation of CM/g-C_3N_4. Figure S1: Effects of different g-C_3N_4 precursors (a), mass ratios of copper-magnesium oxide to melamine (b), pyrolysis temperatures (c), pyrolysis times (d) and molar ratios of copper to magnesium precursor (e) on the catalytic performance of material. Figure S2: Effects of catalyst dosage (a), PMS concentration (b) and RhB concentration (c) for RhB degradation. Figure S3: Mineralization efficiency of CM/g-C_3N_4/PMS system. Figure S4: Effect of initial pH for RhB degradation in the CM/g-C_3N_4/PMS system. Figure S5: Effect of nitrogen on the degradation of RhB in the CM/g-C_3N_4/PMS system. Figure S6: X-ray photoelectron spectroscopy spectra of C1s (a, b) and Mg 1s (c, d) in the CM/g-C_3N_4 before (a, c) and after (b, d) the reaction. Figure S7: Effect of $H_2PO_4^-$ on the degradation of RhB in the CM/g-C_3N_4/PMS system. Figure S8: The molecular structure of the optimized RhB (a) (oxygen, red; carbon, gray; nitrogen, green; hydrogen, white). Figure S9: Degradation of RhB with the recycled CM/g-C_3N_4. Figure S10: Degradation of different pollutants in the CM/g-C_3N_4/PMS system (a) and RhB removal in different water matrixes (b). Table S1: k value of CM/g-C_3N_4/PMS system in different conditions. Table S2: RhB degradation in different systems. Table S3: Atomic area ratios of different compositions in the CM/g-C_3N_4 before and after the reaction. Table S4: Partial Fukui indexes of RhB molecule.

Author Contributions: Conceptualization, Y.M.; Formal analysis, Y.M.; Funding acquisition, X.Z.; Methodology, W.X.; Project administration, X.Z.; Resources, X.Z.; Software, W.X.; Supervision, X.Z. and S.Z.; Writing—original draft, Y.M.; Writing—review & editing, X.Z. All authors have read and agreed to the published version of the manuscript.

Funding: This research was funded by the National Key Research and Development Program of China, (Grant No. 2016YFC0400702-2), the National Natural Science Foundation of China (Grant No. 21377041) and the Guangdong Science and Technology Program (Grant No.2020B121201003).

Institutional Review Board Statement: Not applicable.

Informed Consent Statement: Not applicable.

Data Availability Statement: Not applicable.

Conflicts of Interest: The authors declare that they have no known competing financial interests or personal relationships that could have appeared to influence the work reported in this paper.

References

1. Solis, M.; Solis, A.; Ines Perez, H.; Manjarrez, N.; Flores, M. Microbial decolouration of azo dyes: A review. *Process Biochem.* **2012**, *47*, 1723–1748. [CrossRef]
2. Ren, Q.; Nie, M.; Yang, L.; Wei, F.; Ding, B.; Chen, H.; Liu, Z.; Liang, Z. Synthesis of MOFs for RhB Adsorption from Wastewater. *Inorganics* **2022**, *10*, 27. [CrossRef]
3. Sivarajasekar, N.; Baskar, R. Agriculture waste biomass valorisation for cationic dyes sequestration: A concise review. *J. Chem. Pharm. Res.* **2015**, *7*, 737–748.
4. Shi, X.; Hong, P.; Huang, H.; Yang, D.; Zhang, K.; He, J.; Li, Y.; Wu, Z.; Xie, C.; Liu, J.; et al. Enhanced peroxymonosulfate activation by hierarchical porous Fe_3O_4/Co_3S_4 nanosheets for efficient elimination of rhodamine B: Mechanisms, degradation pathways and toxicological analysis. *J. Colloid Interface Sci.* **2022**, *610*, 751–765. [CrossRef]
5. Yang, S.; Zhang, S.; Li, X.; Du, Y.; Xing, Y.; Xu, Q.; Wang, Z.; Li, L.; Zhu, X. One-step pyrolysis for the preparation of sulfur-doped biochar loaded with iron nanoparticles as an effective peroxymonosulfate activator for RhB degradation. *New J. Chem.* **2022**, *46*, 5678–5689. [CrossRef]

6. Foo, K.Y.; Hameed, B.H. Decontamination of textile wastewater via TiO_2/activated carbon composite materials. *Adv. Colloid Interface Sci.* **2010**, *159*, 130–143. [CrossRef]
7. Noman, E.; Al-Gheethi, A.A.; Talip, B.; Mohamed, R.; Kassim, A.H. Mycoremediation of Remazol Brilliant Blue R in greywater by a novel local strain of Aspergillus iizukae 605EAN: Optimisation and mechanism study. *Int. J. Environ. Anal. Chem.* **2020** *100*, 1650–1668. [CrossRef]
8. Yousefi, S.; Ghanbari, D.; Salavati-Niasari1, M.; Hassanpour, M. Photo-degradation of organic dyes: Simple chemical synthesis of $Ni(OH)_2$ nanoparticles, Ni/$Ni(OH)_2$ and Ni/NiO magnetic nanocomposites. *J. Mater. Sci.-Mater. Electron.* **2016**, *27*, 1244–1253 [CrossRef]
9. Yang, Q.; Zhang, Y.; Liang, J.; Luo, Y.; Liu, Q.; Yang, Y.; Sun, X. Facile hydrothermal synthesis of co-glycerate as an efficient peroxymonosulfate activator for rhodamine B degradation. *Colloids Surf. A-Physicochem. Eng. Asp.* **2022**, *648*, 129239. [CrossRef]
10. Yousefi, S.; Alshamsi, H.; Amiri, O.; Salavati-Niasari, M. Synthesis, characterization and application of Co/Co_3O_4 nanocomposites as an effective photocatalyst for discoloration of organic dye contaminants in wastewater and antibacterial properties. *J. Mol. Liq.* **2021**, *337*, 116405. [CrossRef]
11. Syuhei, Y.; Kohei, M.; Hidenori, Y. Catalytic oxidation of benzene to phenol with hydrogen peroxide over Fe-terpyridine complexes supported on a cation exchange resin. *Catal. Commun.* **2018**, *116*, S1566736718303224.
12. Amanollahi, H.; Moussavi, G.; Giannakis, S. Enhanced vacuum UV-based process (VUV/H_2O_2/PMS) for the effective removal of ammonia from water: Engineering configuration and mechanistic considerations. *J. Hazard. Mater.* **2021**, *402*, 123789. [CrossRef]
13. Duan, X.; Sun, H.; Shao, Z.; Wang, S. Nonradical reactions in environmental remediation processes: Uncertainty and challenges. *Appl. Catal. B Environ.* **2018**, *224*, 973–982. [CrossRef]
14. Ling, S.K.; Wang, S.; Peng, Y. Oxidative degradation of dyes in water using Co^{2+}/H_2O_2 and Co^{2+}/peroxymonosulfate. *J. Hazard. Mater.* **2010**, *178*, 385–389. [CrossRef]
15. Rodriguez-Narvaez, O.M.; Pacheco-Alvarez, M.O.A.; Wrobel, K.; Paramo-Vargas, J.; Bandala, E.R.; Brillas, E.; Peralta-Hernandez, J.M. Development of a Co^{2+}/PMS process involving target contaminant degradation and PMS decomposition. *Int. J. Environ. Sci. Technol.* **2020**, *17*, 17–26. [CrossRef]
16. Ulucan-Altuntas, K.; Guvenc, S.Y.; Can-Guven, E.; Ilhan, F.; Varank, G. Degradation of oxytetracycline in aqueous solution by heat-activated peroxydisulfate and peroxymonosulfate oxidation. *Environ. Sci. Pollut. Res.* **2022**, *29*, 9110–9123. [CrossRef]
17. Huang, S.; Guo, X.; Duan, W.; Cheng, X.; Zhang, X.; Li, Z. Degradation of high molecular weight polyacrylamide by alkali-activated persulfate: Reactivity and potential application in filter cake removal before cementing. *J. Pet. Sci. Eng.* **2019**, *174*, 70–79. [CrossRef]
18. Yin, R.; Guo, W.; Wang, H.; Du, J.; Zhou, X.; Wu, Q.; Zheng, H.; Chang, J.; Ren, N. Enhanced peroxymonosulfate activation for sulfamethazine degradation by ultrasound irradiation: Performances and mechanisms. *Chem. Eng. J.* **2018**, *335*, 145–153. [CrossRef]
19. Cui, C.; Jin, L.; Jiang, L.; Han, Q.; Lin, K.; Lu, S.; Zhang, D.; Cao, G. Removal of trace level amounts of twelve sulfonamides from drinking water by UV-activated peroxymonosulfate. *Sci. Total Environ.* **2016**, *572*, 244–251. [CrossRef]
20. Li, T.; Du, X.; Deng, J.; Qi, K.; Zhang, J.; Gao, L.; Yue, X. Efficient degradation of Rhodamine B by magnetically recoverable Fe3O4-modified ternary CoFeCu-layered double hydroxides via activating peroxymonosulfate. *J. Environ. Sci.* **2021**, *108*, 188–200. [CrossRef]
21. Wan, Q.; Chen, Z.; Cao, R.; Wang, J.; Huang, T.; Wen, G.; Ma, J. Oxidation of organic compounds by PMS/CuO system: The significant discrepancy in borate and phosphate buffer. *J. Clean. Prod.* **2022**, *339*, 130773. [CrossRef]
22. Deng, J.; Ya, C.; Ge, Y.; Cheng, Y.; Chen, Y.; Xu, M.; Wang, H. Activation of peroxymonosulfate by metal (Fe, Mn, Cu and Ni) doping ordered mesoporous Co_3O_4 for the degradation of enrofloxacin. *Rsc Adv.* **2018**, *8*, 2338–2349. [CrossRef]
23. He, Y.; Zhang, J.; Zhou, H.; Yao, G.; Lai, B. Synergistic multiple active species for the degradation of sulfamethoxazole by peroxymonosulfate in the presence of CuO@FeOx@Fe-0. *Chem. Eng. J.* **2020**, *380*, 122568. [CrossRef]
24. Li, W.; Li, Y.; Zhang, D.; Lan, Y.; Guo, J. CuO-Co_3O_4@CeO_2 as a heterogeneous catalyst for efficient degradation of 2,4-dichlorophenoxyacetic acid by peroxymonosulfate. *J. Hazard. Mater.* **2020**, *381*, 122568. [CrossRef]
25. Ji, F.; Li, C.; Liu, Y.; Liu, P. Heterogeneous activation of peroxymonosulfate by Cu/ZSM5 for decolorization of Rhodamine B. *Sep. Purif. Technol.* **2014**, *135*, 1–6. [CrossRef]
26. Du, X.; Zhang, Y.; Si, F.; Yao, C.; Du, M.; Hussain, I.; Kim, H.; Huang, S.; Lin, Z.; Hayat, W. Persulfate non-radical activation by nano-CuO for efficient removal of chlorinated organic compounds: Reduced graphene oxide-assisted and CuO (001) facet-dependent. *Chem. Eng. J.* **2019**, *356*, 178–189. [CrossRef]
27. Kiani, R.; Mirzaei, F.; Ghanbari, F.; Feizi, R.; Mehdipour, F. Real textile wastewater treatment by a sulfate radicals-Advanced Oxidation Process: Peroxydisulfate decomposition using copper oxide (CuO) supported onto activated carbon. *J. Water Process Eng.* **2020**, *38*, 101623. [CrossRef]
28. Li, Z.; Liu, D.; Huang, W.; Wei, X.; Huang, W. Biochar supported CuO composites used as an efficient peroxymonosulfate activator for highly saline organic wastewater treatment. *Sci. Total Environ.* **2020**, *721*, 137764. [CrossRef]
29. Yin, Z.; Han, M.; Hu, Z.; Feng, L.; Liu, Y.; Du, Z.; Zhang, L. Peroxymonosulfate enhancing visible light photocatalytic degradation of bezafibrate by Pd/g-C_3N_4 catalysts: The role of sulfate radicals and hydroxyl radicals. *Chem. Eng. J.* **2020**, *390*, 124532. [CrossRef]

30. Xu, M.; Han, L.; Dong, S. Facile Fabrication of Highly Efficient g-C_3N_4/Ag_2O Heterostructured Photocatalysts with Enhanced Visible-Light Photocatalytic Activity. *ACS Appl. Mater. Interfaces* **2013**, *5*, 12533–12540. [CrossRef]
31. Kim, M.; Hwang, S.; Yu, J.S. Novel ordered nanoporous graphitic C_3N_4 as a support for Pt-Ru anode catalyst in direct methanol fuel cell. *J. Mater. Chem.* **2007**, *17*, 1656–1659. [CrossRef]
32. Liu, S.; Wei, X.; Lin, S.; Guo, M. Preparation of aerogel $Mg(OH)_{(2)}$ nanosheets by a combined sol-gel-hydrothermal process and its calcined MgO towards enhanced degradation of paraoxon pollutants. *J. Sol-Gel Sci. Technol.* **2021**, *99*, 122–131. [CrossRef]
33. Ali, J.; Jiang, W.; Shahzad, A.; Ifthikar, J.; Yang, X.; Wu, B.; Oyekunle, D.T.; Jia, W.; Chen, Z.; Zheng, L.; et al. Isolated copper ions and surface hydroxyl groups as a function of non-redox metals to modulate the reactivity and persulfate activation mechanism of spinel oxides. *Chem. Eng. J.* **2021**, *425*, 130679. [CrossRef]
34. Chen, C.; Liu, L.; Li, Y.; Li, W.; Zhou, L.; Lan, Y.; LI, Y. Insight into heterogeneous catalytic degradation of sulfamethazine by peroxymonosulfate activated with $CuCo_2O_4$ derived from bimetallic oxalate. *Chem. Eng. J.* **2020**, *384*, 123257. [CrossRef]
35. Guo, H.; Wang, Y.; Yao, X.; Zhang, Y.; Wang, Y. A comprehensive insight into plasma-catalytic removal of antibiotic oxytetracycline based on graphene-TiO_2-Fe_3O_4 nanocomposites. *Chem. Eng. J.* **2021**, *425*, 130614. [CrossRef]
36. Yousefi, A.; Alireza, N. Photodegradation pathways of phenazopyridine by the CdS-WO_3 hybrid system and its capability for the hydrogen generation. *Mater. Res. Bull.* **2022**, *148*, 111669. [CrossRef]
37. Guan, C.; Jiang, J.; Pang, S.; Chen, X.; Webster, R.D.; Lim, T.-T. Facile synthesis of pure g-C_3N_4 materials for peroxymonosulfate activation to degrade bisphenol A: Effects of precursors and annealing ambience on catalytic oxidation. *Chem. Eng. J.* **2020**, *387*, 123726. [CrossRef]
38. Song, H.; Liu, Z.; Guan, Z.; Yang, F.; Xia, D.; Li, D. Efficient persulfate non-radical activation of electron-rich copper active sites induced by oxygen on graphitic carbon nitride. *Sci. Total Environ.* **2021**, *762*, 143127. [CrossRef]
39. Zhang, S.; Gao, H.; Xu, X.; Cao, R.; Yang, H.; Xu, X.; Li, J. MOF-derived CoN/N-C@SiO_2 yolk-shell nanoreactor with dual active sites for highly efficient catalytic advanced oxidation processes. *Chem. Eng. J.* **2020**, *381*, 122670. [CrossRef]
40. Dan, J.; Rao, P.; Wang, Q.; Dong, L.; Chu, W.; Zhang, M.; He, Z.; Gao, N.; Deng, J.; Chen, J. MgO-supported CuO with encapsulated structure for enhanced peroxymonosulfate activation to remove thiamphenicol. *Sep. Purif. Technol.* **2022**, *280*, 119782. [CrossRef]
41. Qui Thanh Hoai, T.; Namgung, G.; Noh, J.-S. Facile synthesis of porous metal-doped ZnO/g-C_3N_4 composites for highly efficient photocatalysts. *J. Photochem. Photobiol. A Chem.* **2019**, *368*, 110–119.
42. Kang, Y.; Yang, Y.; Yin, L.-C.; Kang, X.; Liu, G.; Cheng, H.M. An Amorphous Carbon Nitride Photocatalyst with Greatly Extended Visible-Light-Responsive Range for Photocatalytic Hydrogen Generation. *Adv. Mater.* **2015**, *27*, 4572–4577. [CrossRef]
43. Liang, Q.; Li, Z.; Huang, Z.-H.; Kang, F.; Yang, Q.H. Holey Graphitic Carbon Nitride Nanosheets with Carbon Vacancies for Highly Improved Photocatalytic Hydrogen Production. *Adv. Funct. Mater.* **2015**, *25*, 6885–6892. [CrossRef]
44. Lee, S.J.; Begildayeva, T.; Jung, H.J.; Koutavarapu, R.; Yu, Y.; Choi, M.; Choi, M.Y. Plasmonic ZnO/Au/g-C_3N_4 nanocomposites as solar light active photocatalysts for degradation of organic contaminants in wastewater. *Chemosphere* **2021**, *263*, 128262. [CrossRef]
45. Song, H.; Guan, Z.; Xia, D.; Xu, H.; Yang, F.; Li, D.; Li, X. Copper-oxygen synergistic electronic reconstruction on g-C_3N_4 for efficient non-radical catalysis for peroxydisulfate and peroxymonosulfate. *Sep. Purif. Technol.* **2021**, *257*, 117957. [CrossRef]
46. Gao, Y.; Zhu, Y.; Lyu, L.; Zeng, Q.; Xing, X.; Hu, C. Electronic Structure Modulation of Graphitic Carbon Nitride by Oxygen Doping for Enhanced Catalytic Degradation of Organic Pollutants through Peroxymonosulfate Activation. *Environ. Sci. Technol.* **2018**, *52*, 14371–14380. [CrossRef]
47. Tsoncheva, T.; Ivanova, L.; Rosenholm, J.; Linden, M. Cobalt oxide species supported on SBA-15, KIT-5 and KIT-6 mesoporous silicas for ethyl acetate total oxidation. *Appl. Catal. B Environ.* **2009**, *89*, 365–374. [CrossRef]
48. Wu, M.; Li, L.; Xue, Y.; Xu, G.; Tang, L.; Liu, N.; Huang, W.-Y. Fabrication of ternary GO/g-C_3N_4/MoS_2 flower-like heterojunctions with enhanced photocatalytic activity for water remediation. *Appl. Catal. B Environ.* **2018**, *228*, 103–112. [CrossRef]
49. Harish, S.; Archana, J.; Sabarinathan, M.; Navaneethan, M.; Nisha, K.D.; Ponnusamy, S.; Muthamizhchelvan, C.; Ikeda, H.; Aswal, D.K.; Hayakawa, Y. Controlled structural and compositional characteristic of visible light active ZnO/CuO photocatalyst for the degradation of organic pollutant. *Appl. Surf. Sci.* **2017**, *418*, 103–112. [CrossRef]
50. Li, D.; Zan, J.; Wu, L.; Zuo, S.; Xia, D. Heterojunction tuning and catalytic efficiency of g-C_3N_4-Cu_2O with glutamate. *Ind. Eng. Chem. Res.* **2019**, *58*, 4000–4009. [CrossRef]
51. Zhang, J.; Zhang, M.; Zhang, G.; Wang, X. Synthesis of Carbon Nitride Semiconductors in Sulfur Flux for Water Photoredox Catalysis. *ACS Catal.* **2012**, *2*, 940–948. [CrossRef]
52. Parvari, R.; Ghorbani-Shahna, F.; Bahrami, A.; Azizian, S.; Assari, M.J.; Farhadian, M. A novel core-shell structured alpha-Fe_2O_3/Cu/g-C_3N_4 nanocomposite for continuous photocatalytic removal of air ethylbenzene under visible light irradiation. *J. Photochem. Photobiol. A Chem.* **2020**, *399*, 112643. [CrossRef]
53. Tian, Y.; Li, Q.; Zhang, M.; Nie, Y.; Tian, X.; Yang, C.; Li, Y. pH-dependent oxidation mechanisms over FeCu doped g-C_3N_4 for ofloxacin degradation via the efficient peroxymonosulfate activation. *J. Clean. Prod.* **2021**, *315*, 128207. [CrossRef]
54. Zhang, W.; Zhou, L.; Deng, H. Ag modified g-C_3N_4 composites with enhanced visible-light photocatalytic activity for diclofenac degradation. *J. Mol. Catal. A Chem.* **2016**, *423*, 270–276. [CrossRef]
55. Zhu, J.N.; Zhu, X.Q.; Cheng, F.F.; Li, P.; Xiong, W.W. Preparing copper doped carbon nitride from melamine templated crystalline copper chloride for Fenton-like catalysis. *Appl. Catal. B Environ.* **2019**, *256*, 117830. [CrossRef]
56. Li, H.; Guo, J.; Yang, L.; Lan, Y. Degradation of methyl orange by sodium persulfate activated with zero-valent zinc. *Sep. Purif. Technol.* **2014**, *132*, 168–173. [CrossRef]

57. Lu, H.; Sui, M.; Yuan, B.; Wang, J.; Lv, Y. Efficient degradation of nitrobenzene by Cu-Co-Fe-LDH catalyzed peroxymonosulfate to produce hydroxyl radicals. *Chem. Eng. J.* **2019**, *357*, 140–149. [CrossRef]
58. Timmins, G.S.; Liu, K.J.; Bechara, E.J.H.; Kotake, Y.; Swartz, H.M. Trapping of free radicals with direct in vivo EPR detection: A comparison of 5,5-dimethyl-1-pyrroline-N-oxide and 5-diethoxyphosphoryl-5-methyl-1-pyrroline-N-oxide as spin traps for HO and SO_4. *Free Radic. Biol. Med.* **1999**, *27*, 329–333. [CrossRef]
59. Zhang, J.; Zhao, W.; Wu, S.; Yin, R.; Zhu, M. Surface dual redox cycles of Mn(III)/Mn(IV) and Cu(I)/Cu(II) for heterogeneous peroxymonosulfate activation to degrade diclofenac: Performance, mechanism and toxicity assessment. *J. Hazard. Mater.* **2021**, *410*, 124623. [CrossRef]
60. Wang, Y.; Ji, H.; Liu, W.; Xue, T.; Liu, C.; Zhang, Y.; Liu, L.; Wang, Q.; Qi, F.; Xu, B.; et al. Novel $CuCo_2O_4$ Composite Spinel with a Meso-Macroporous Nanosheet Structure for Sulfate Radical Formation and Benzophenone-4 Degradation: Interface Reaction, Degradation Pathway, and DFT Calculation. *ACS Appl. Mater. Interfaces* **2020**, *12*, 20522–20535. [CrossRef]
61. Wang, R.; An, H.; Zhang, H.; Zhang, X.; Feng, J.; Wei, T.; Ren, Y. High active radicals induced from peroxymonosulfate by mixed crystal types of $CuFeO_2$ as catalysts in the water. *Appl. Surf. Sci.* **2019**, *484*, 1118–1127. [CrossRef]
62. Chen, C.; Liu, L.; Guo, J.; Zhou, L.; Lan, Y. Sulfur-doped copper-cobalt bimetallic oxides with abundant Cu(I): A novel peroxymonosulfate activator for chloramphenicol degradation. *Chem. Eng. J.* **2019**, *361*, 1304–1316. [CrossRef]
63. Li, Y.; Li, J.; Pan, Y.; Xiong, Z.; Yao, G.; Xie, R.; Lai, B. Peroxymonosulfate activation on $FeCo_2S_4$ modified g-C_3N_4 ($FeCo_2S_4$-CN): Mechanism of singlet oxygen evolution for nonradical efficient degradation of sulfamethoxazole. *Chem. Eng. J.* **2020**, *384*, 123361. [CrossRef]
64. Wu, S.; Liang, G.; Guan, X.; Qian, G.; He, Z. Precise control of iron activating persulfate by current generation in an electrochemical membrane reactor. *Environ. Int.* **2019**, *131*, 105024. [CrossRef]
65. Li, W.; Wang, Z.; Liao, H.; Liu, X.; Zhou, L.; Lan, Y.; Zhang, J. Enhanced degradation of 2,4,6-trichlorophenol by activated peroxymonosulfate with sulfur doped copper manganese bimetallic oxides. *Chem. Eng. J.* **2021**, *417*, 128121. [CrossRef]
66. Chen, K.; Zhang, X.-M.; Yang, X.-F.; Jiao, M.-G.; Zhou, Z.; Zhang, M.-H.; Wang, D.-H.; Bu, X.-H. Electronic structure of heterojunction MoO_2/g-C_3N_4 catalyst for oxidative desulfurization. *Appl. Catal. B Environ.* **2018**, *238*, 263–273. [CrossRef]
67. Li, H.; Yang, Z.; Lu, S.; Su, L.; Wang, C.; Huang, J.; Zhou, J.; Tang, J.; Huang, M. Nano-porous bimetallic CuCo-MOF-74 with coordinatively unsaturated metal sites for peroxymonosulfate activation to eliminate organic pollutants: Performance and mechanism. *Chemosphere* **2021**, *273*, 129643. [CrossRef]
68. Wagner, G.W.; Yang, Y.C. Rapid nucleophilic/oxidative decontamination of chemical warfare agents. *Ind. Eng. Chem. Res.* **2002**, *41*, 1925–1928. [CrossRef]
69. Ding, D.; Yang, S.; Chen, L.; Cai, T. Degradation of norfloxacin by CoFe alloy nanoparticles encapsulated in nitrogen doped graphitic carbon (CoFe@N-GC) activated peroxymonosulfate. *Chem. Eng. J.* **2020**, *392*, 123725. [CrossRef]
70. Oh, W.-D.; Chang, V.W.C.; Hu, Z.-T.; Goei, R.; Lim, T.-T. Enhancing the catalytic activity of g-C_3N_4 through Me doping (Me=Cu, Co and Fe) for selective sulfathiazole degradation via redox-based advanced oxidation process. *Chem. Eng. J.* **2017**, *323*, 260–269. [CrossRef]
71. Anipsitakis, G.P.; Dionysiou, D.D. Degradation of organic contaminants in water with sulfate radicals generated by the conjunction of peroxymonosulfate with cobalt. *Environ. Sci. Technol.* **2003**, *37*, 4790–4797. [CrossRef] [PubMed]
72. Shao, P.; Tian, J.; Yang, F.; Duan, X.; Gao, S.; Shi, W.; Luo, X.; Cui, F.; Luo, S.; Wang, S. Identification and Regulation of Active Sites on Nanodiamonds: Establishing a Highly Efficient Catalytic System for Oxidation of Organic Contaminants. *Adv. Funct. Mater.* **2018**, *28*, 1705295. [CrossRef]
73. Rasalingam, S.; Peng, R.; Koodali, R.T. An insight into the adsorption and photocatalytic degradation of rhodamine B in periodic mesoporous materials. *Appl. Catal. B Environ.* **2015**, *174*, 49–59. [CrossRef]
74. Natarajan, T.S.; Thomas, M.; Natarajan, K.; Bajaj, H.C.; Tayade, R.J. Study on UV-LED/TiO_2 process for degradation of Rhodamine B dye. *Chem. Eng. J.* **2011**, *169*, 126–134. [CrossRef]
75. Zhou, P.; Li, W.; Zhang, J.; Zhang, G.; Cheng, X.; Liu, Y.; Huo, X.; Zhang, Y. Removal of Rhodamine B during the corrosion of zero valent tungsten via a tungsten species-catalyzed Fenton-like system. *J. Taiwan Inst. Chem. Eng.* **2019**, *100*, 202–209. [CrossRef]
76. Vigneshwaran, S.; Karthikeyan, P.; Park, C.M.; Meenakshi, S. Boosted insights of novel accordion-like (2D/2D) hybrid photocatalyst for the removal of cationic dyes: Mechanistic and degradation pathways. *J. Environ. Manag.* **2020**, *273*, 111125. [CrossRef]
77. Chen, X.; Zhou, J.; Yang, H.; Wang, H.; Li, H.; Wu, S.; Yang, W. PMS activation by magnetic cobalt-N-doped carbon composite for ultra-efficient degradation of refractory organic pollutant: Mechanisms and identification of intermediates. *Chemosphere* **2022**, *287*, 132074. [CrossRef]
78. Pang, Y.; Kong, L.; Chen, D.; Yuvaraja, G.; Mehmood, S. Facilely synthesized cobalt doped hydroxyapatite as hydroxyl promoted peroxymonosulfate activator for degradation of Rhodamine B. *J. Hazard. Mater.* **2020**, *384*, 121447. [CrossRef]
79. Govarthanan, M.; Mythili, R.; Kim, W.; Alfarraj, S.; Alharbi, S.A. Facile fabrication of (2D/2D) MoS_2@MIL-88(Fe) interface-driven catalyst for efficient degradation of organic pollutants under visible light irradiation. *J. Hazard. Mater.* **2021**, *414*, 125522. [CrossRef]
80. Li, W.; Li, Y.; Zhang, D.; Lan, Y.; Guo, J. Enhanced kinetic performance of peroxymonosulfate/ZVI system with the addition of copper ions: Reactivity, mechanism, and degradation pathways. *J. Hazard. Mater.* **2020**, *393*, 121209. [CrossRef]

31. Chen, S.; Ma, L.; Du, Y.; Zhan, W.; Zhang, T.C.; Du, D. Highly efficient degradation of rhodamine B by carbon nanotubes-activated persulfate. *Sep. Purif. Technol.* **2021**, *256*, 117788. [CrossRef]
32. Xu, Y.; Hu, E.; Xu, D.; Guo, Q. Activation of peroxymonosulfate by bimetallic CoMn oxides loaded on coal fly ash-derived SBA-15 for efficient degradation of Rhodamine B. *Sep. Purif. Technol.* **2021**, *274*, 119081. [CrossRef]
33. Hu, L.; Deng, G.; Lu, W.; Lu, Y.; Zhang, Y. Peroxymonosulfate activation by Mn_3O_4/metal-organic framework for degradation of refractory aqueous organic pollutant rhodamine B. *Chin. J. Catal.* **2017**, *38*, 1360–1372. [CrossRef]
34. Liang, L.; Cheng, L.; Zhang, Y.; Wang, Q.; Meng, X. Efficiency and mechanisms of rhodamine B degradation in Fenton-like systems based on zero-valent iron. *RSC Adv.* **2020**, *10*, 28509–28515. [CrossRef]
35. Kong, L.; Fang, G.; Chen, Y.; Xie, M.; Zhu, F.; Ma, L.; Zhou, D.; Zhan, J. Efficient activation of persulfate decomposition by Cu_2FeSnS_4 nanomaterial for bisphenol A degradation: Kinetics, performance and mechanism studies. *Appl. Catal. B Environ.* **2019**, *253*, 278–285. [CrossRef]

water

MDPI

Article

Doping of TiO_2 Using Metal Waste (Door Key) to Improve Its Photocatalytic Efficiency in the Mineralization of an Emerging Contaminant in an Aqueous Environment

Dany Edgar Juárez-Cortazar [1], José Gilberto Torres-Torres [1], Aracely Hernandez-Ramirez [2], Juan Carlos Arévalo-Pérez [1], Adrián Cervantes-Uribe [1], Srinivas Godavarthi [1], Alejandra Elvira Espinosa de los Monteros [1], Adib Abiu Silahua-Pavón [1] and Adrián Cordero-Garcia [1,*]

1 Laboratory of Catalytic Nanomaterials Applied to the Development of Energy Sources and Environmental Remediation, Tabasco Applied Science and Technology Research Center (CICTAT), Universidad Juárez Autónoma de Tabasco, DACB, Km.1 Carretera Cunduacán-Jalpa de Méndez, Cunduacán 86690, Tabasco, Mexico; internacional_celeste@hotmail.com (D.E.J.-C.); gilberto.torres@ujat.mx (J.G.T.-T.); carlos.arevalo@ujat.mx (J.C.A.-P.); adrian.cervantes@ujat.mx (A.C.-U.); srinivas.godavarthi@gmail.com (S.G.); ale2962@gmail.com (A.E.E.d.l.M.); adibab45@gmail.com (A.A.S.-P.)

2 Facultad de Ciencias Químicas, UANL, Universidad Autónoma de Nuevo León, Ciudad Universitaria, San Nicolás de los Garza 66451, Nuevo León, Mexico; aracely.hernandezrm@uanl.edu.mx

* Correspondence: adrian.cordero@ujat.mx; Tel.: +52-993-591-0296

Citation: Juárez-Cortazar, D.E.; Torres-Torres, J.G.; Hernandez-Ramirez, A.; Arévalo-Pérez, J.C.; Cervantes-Uribe, A.; Godavarthi, S.; de los Monteros, A.E.E.; Silahua-Pavón, A.A.; Cordero-Garcia, A. Doping of TiO_2 Using Metal Waste (Door Key) to Improve Its Photocatalytic Efficiency in the Mineralization of an Emerging Contaminant in an Aqueous Environment. *Water* **2022**, *14*, 1389. https://doi.org/10.3390/w14091389

Academic Editors: Huijiao Wang, Dionysios (Dion) Demetriou Dionysiou and Yujue Wang

Received: 30 March 2022
Accepted: 21 April 2022
Published: 26 April 2022

Abstract: Photocatalysis is an effective advanced oxidation process to mineralize recalcitrant contaminants in aqueous media. TiO_2 is the most used photocatalyst in this type of process. To improve the deficiencies of this material, one of the most used strategies has been to dope TiO_2 with metallic ions. Chemical reagents are often used as dopant precursors. However, due to the depletion of natural resources, in this work it was proposed to substitute chemical reagents and instead use a metallic residue (door key) as a doping precursor. The materials were synthesized using the sol–gel method and calcined at 400 °C to obtain the crystal structure of anatase. The characterization of the materials was carried out using X-ray diffraction (XRD), transmission electron microscopy (TEM), diffuse reflectance spectroscopy (DRS), scanning electron microscopy–energy-dispersive X-ray analysis (SEM-EDX) methods X-ray photoelectron spectroscopy (XPS), and inductively coupled plasma optical emission spectroscopy (ICP-OES). The results obtained indicate that Cu^{+}/Cu^{2+} and Zn^{2+} ions coexist in the support, which modifies the physicochemical properties of TiO_2 and improves its photocatalytic efficiency. The synergistic effect of the dopants in TiO_2 allowed the mineralization of diclofenac in an aqueous medium when T-DK (1.0) was used as photocatalyst and simulated solar radiation as an activation source.

Keywords: emerging contaminant; photocatalysis; TiO_2 doped; discarded metal waste; sol–gel

1. Introduction

Heterogeneous photocatalysis is an advanced oxidation process that has proven to be effective for the mineralization of recalcitrant contaminants in an aqueous medium. The most efficient photocatalyst for the photocatalytic process is TiO_2 since it has optical, structural, and organoleptic properties, making it efficient in the mineralization of contaminants in an aqueous medium. However, this semiconductor has the disadvantage of having an Eg 3.2 eV, which means that it cannot be activated with UV light, so it absorbs only 5% of solar radiation [1–3]. This restricts the use of solar radiation as an activation source.

Various strategies have been developed to improve the spectral response in the visible region of photocatalytic oxides. One of the most widely used strategies has been to alter the physicochemical properties of oxide by incorporating metallic ions into its crystalline structure [4,5]. It has been shown that metal doping of TiO_2 causes the formation of new energy levels below the conduction band and delays the rapid recombination of electron–hole pairs, which increases the photocatalytic activity of this semiconductor [6,7]

To synthesize TiO_2 doped materials, transition metal salts are used. However, based on analysis by the World Metal Reserve, production from a mine has a shorter timescale than mineral deposit formation, suggesting that known primary metal supplies will be depleted. Importantly, the risk of contamination by heavy metals is higher since they are not chemically or biologically degradable. Once disposed of, they can remain in the environment for hundreds of years and cause environmental damage. It has been shown that the discharged metals are easily transported through groundwater, causing contamination of soil and rivers [8], as well as the degradation and death of vegetation, animals, and even direct damage to humans [9,10]. Therefore, in our research group, we are convinced that the recycling of metals that we use in our daily lives helps conserve the natural riches of the environment and favors the reduction of environmental pollution. Among metal scrap, door keys are among the most common forms of consumer waste. The door key used as a doping precursor was taken from a metal scrap yard; hence, it had advanced metal wear. Due to this, its origin of manufacture is unknown. However, it is known that most door keys are alloys of steel or brass.

Among the metals that compose these alloys, nickel, copper, and zinc are chemical elements widely used to dope TiO_2 nanoparticles [11–13]. Such is the case of a study carried out by Raguram, T. et al. [11], who synthesized TiO_2-Ni nanoparticles using the sol–gel method. The results reported by the researchers indicate that the doping of Ni^{2+} in TiO_2 benefited the absorption of visible light, reaching a maximum of 61.04% in the degradation of the methylene blue dye in an aqueous medium. On the other hand, Hemraj, Y. et al. [12] synthesized TiO_2 nanoparticles doped with different Cu^{2+} contents (0 to 3.0% mol) using the sol–gel method for semiconductor synthesis. The results show that the TiO_2-Cu nanoparticle (3.0%) has a photocatalytic efficiency superior to pure TiO_2. They mentioned that the improvement in the photocatalytic performance for the photodegradation of methyl orange was due to the redshift of the bandgap energy (Eg) and the decrease in the recombination rate of the electron–hole pair of TiO_2-Cu nanoparticles. In general, it has been mentioned that noble metals such as Ni and Cu are good candidates to act as traps for photogenerated electrons because the Fermi level of these ions is lower than that of TiO_2. Regarding Zn use in TiO_2 doping, Shao, M. et al. [13] concluded that adding Zn to TiO_2 reduces Eg and increases optical absorption in the visible region, reaching a maximum rate of 88.14% in degradation of tetracycline. In addition to the above, many researchers have focused on multi-doping metals (bimetallic or trimetallic) over the last few years. They mention that the second metal can alter the electronic properties and the formation of active surface structures, increasing the photocatalytic activity of TiO_2 concerning monometallic doping [14–17]. The previous work provides evidence of the feasibility of using transition metals as monometallic or bimetallic dopants in TiO_2 nanoparticles using chemical reagents as doping precursors. However, we did not find any scientific report investigating the use of door-key waste for TiO_2 multidoping. As an additional motivation, it is important to mention that innovative strategies are currently being developed to increase the efficiency of contaminant removal and transformation [18–22]. In this context, an innovative synthesis route was designed to obtain TiO_2 doped with metallic ions, where the use of chemical reagents is avoided, and it was proposed to use a metallic residue (door key) as a doping precursor to improving the photocatalytic efficiency of TiO_2 in the mineralization of an emerging contaminant.

2. Materials and Methods

2.1. Chemicals and Reagents

To synthesize TiO_2, reagent grade chemicals were used without further purification. Titanium (IV) butoxide (97%), 1-butanol anhydrous (99.8%), and nitric acid ($\geq$65%) were purchased from Sigma-Aldrich. Deionized water was used for the hydrolysis step during the sol–gel synthesis process. A door key recovered from a metal scrapyard was used as a doping precursor.

2.2. Synthesis of Photocatalysts

To analyze the effect of metal ions that make up the door key on the photocatalytic activity of TiO_2, 3 samples were prepared with different percentages of metallic waste (0, 0.5, and 1.0 wt.%). The materials nomenclature is defined as T, T-DK (0.5), and T-DL (1.0) as well as the abbreviation T (TiO_2) + the metal residue percentage number (DK).

The synthesis of the photocatalysts was carried out using a non-traditional sol–gel synthesis route. The waste door key (0.5 or 1.0 wt.%) for the TiO_2 doping was dissolved at room temperature in 1.5 mL of HNO_3. *Solution A*. To synthesize TiO_2, doped titanium (IV) butoxide and anhydrous 1-butanol were mixed. *Solution B*. After stirring for 30 min at room temperature, *solution A* was added drop by drop to *solution B*, continuing to stir for 30 min. To begin the hydrolysis reaction in the *A–B* mixture, 10 mL of deionized H_2O was added per drop. The gel obtained was aged at ambient temperature for 24 h. Xerogel was subsequently obtained by filtering the gel with a vacuum pump to remove the solvents and then drying for 12 h at 60 °C. Finally, the xerogel was heat-treated at 400 °C for 4 h using a 4 °C min^{-1} heating ramp with a static air atmosphere. Pure TiO_2 was synthesized using a similar methodology as described earlier for comparison purposes. The only difference is that for pure TiO_2 synthesis, *solution A* had no metallic ions.

2.3. Characterization of Photocatalysts

The surface area of the photocatalysts was determined using the BET method. Their N_2 adsorption–desorption isotherms were obtained at −196 °C using Asap 2020 equipment after degassing samples at 250 °C for 12 h in a high vacuum.

The crystalline phase for each sample was determined using a Siemens D500 X-ray diffractometer (XRD) with Cu K_α radiation (λ = 0.15418 nm). The diffraction patterns were obtained within the range of 10 to 80° at an acquisition rate of 0.02 s^{-1} and 1 s per point. The average crystal size was calculated using the Scherrer equation.

$$\Phi = \frac{K\lambda}{\beta \cos\theta} \quad (1)$$

where Φ is the size of the crystal, K (0.91) is the form factor, λ is the wavelength of the X-rays, β is the width at half height (FWHM) of the main intensity peak, and θ is the Bragg angle.

The optical properties of the materials were obtained with diffuse reflection spectroscopy (DRS) using a Perkin-Elmer UV-Vis spectrophotometer with an integrating sphere. Absorption spectra were recorded in the range from 300 to 800 nm. A barium sulfate plate ($BaSO_4$) was used as a reference. Based on the absorption spectra, the bandgap energies of the photocatalysts were determined using the Kubelka–Munk theory and Tauc plot [23].

The morphological characterization of composites was conducted with transmission electron microscopy (HRTEM) using a JEOEL electronic microscope model JEM2100.

The semi-quantitative elementary composition of the materials was determined with energy-dispersive X-ray spectroscopy (EDX) using an electron scanning microscope with Rontec Xflash detector model Hitachi S-4700 Type IIc. In addition, the chemical composition of the catalysts was determined using inductively coupled plasma optical emission spectroscopy (ICP-OES) Varian 725-ES, after acidic digestion (Nitric Acid) of the materials.

The XPS spectra were obtained using X-ray photoelectron spectroscopy (XPS) with Thermo Scientific apparatus equipped with a non-monochromated Mg anode as an X-ray source operated at 12 kV under vacuum (2×10^{-7} mbar). The binding energies were calibrated at 284.6 eV with respect to the C1s peak of the carbon samples. The XPS spectrums were deconvoluted using the PeakFit software. The peaks were fitted to a linear background and a combination of Gaussian/Lorentzian functions.

2.4. Evaluation of Photocatalytic Activity

Diclofenac was used as a contaminant to evaluate the photocatalytic activity of photocatalysts. The photocatalytic degradation of 10 ppm of diclofenac (Sigma-Aldrich, Burlington, MA, USA, ≥98.5%) was carried out in a glass reactor with a volume of 0.5 L and photocatalyst load of 1.0 gL^{-1}. The reactor was placed in a SUNTEST XLS+ solar simulator equipped with a xenon arc lamp model XL-S-750 and a 320 nm cut-off filter. During the experiments, the temperature of the reaction system was maintained in the range of 25 to 35 °C. The suspension was stirred in the dark for 30 min while the air was bubbled into the system. The light was then turned on, and simultaneously the system continued to be supplied with air. To monitor the mineralization of diclofenac, samples were collected from the reactor every 30 min to up 180 min. Photolysis experiments were also carried out for comparative purposes. The progression of the mineralization of the contaminant was followed by the measurement of total organic carbon using a Shimadzu TOC-VSCH analyzer. The following equation determined the mineralization of diclofenac:

$$\%TOC = \frac{TOC_0 - TOC_f}{TOC_0} \times 100 \quad (2)$$

where TOC_0 is the mineralization at time zero and TOC_f is the final mineralization at each instant.

3. Results

3.1. Sample Characterization

3.1.1. XRD

The XRD patterns for pure and doped samples are illustrated in Figure 1. In all diffractograms, peaks are observed at angles 25.33, 37.8, 48.06, 53.9, and 55.08, characteristic of TiO_2-anatase (JCPDS card no. 21-1272). The crystalline phase obtained corresponds to that reported by other investigators. Other studies mention that the crystalline phase of anatase occurs when heat treatment varies between 200 and 500 °C. It was said that the amount and physicochemical properties (high ionic radii and electronic status) of the doping metal can influence the rate of transformation of anatase to the rutile phase [24,25]. Such is the case of a study conducted by Hampel et al. [25]; they observed that TiO_2 doped with 5 wt.% Cu ions, calcined at 450 °C, contained an anatase phase and a rutile phase fraction, which increased with increasing Cu concentration from 5 to 10 wt.%. In our case, the dopant concentration does not influence the anatase–rutile transition because the maximum theoretical concentration of doping ions in TiO_2 is less than 1 wt.%.

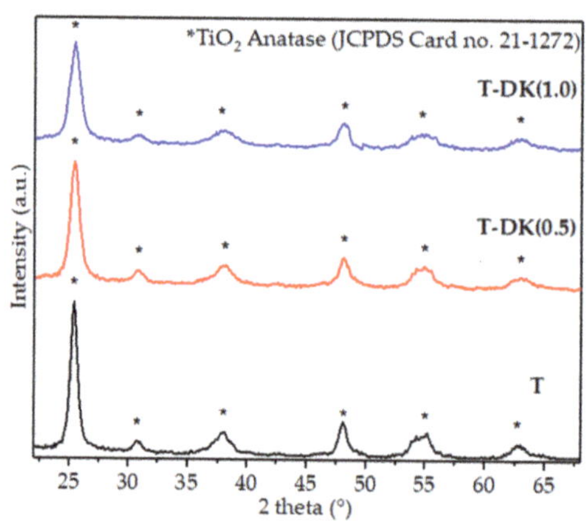

Figure 1. X-ray diffraction (XRD) patterns of TiO_2 and doped-TiO_2.

It is convenient to have TiO_2 nanoparticles with an anatase crystalline structure for this research. There is evidence that other polymorphic forms of this material are less effective in photocatalytic reactions. This is due to the higher adsorption capability of O_2 on the surface of the anatase phase than in the rutile phase. The adsorbed oxygen is reduced to superoxide radicals ($O_2^{\bullet-}$) by photoexciting electrons from BV to BC, thus functioning as an electron trap, inhibiting the recombination of e^-/h^+ pairs. This increases the half-life of the redox reactions that benefit the formation of hydroxyl radicals ($OH^{\bullet}$) with sufficient oxidation potential for the photocatalytic degradation of recalcitrant compounds. Along with the above, the redox potential of BV in anatase is more negative, making it more competitive than the rutile phase for oxidation reactions [26–28].

No characteristic peaks were associated with doping Cu, Zn, or Ni (metal oxide). It is known that X-ray diffractometers are not sensitive to levels of impurities lower than 5%. For this reason, the absence of characteristic peaks of the dopant ions can be attributed to the fact that the percentages of door lock used during the synthesis of the doped TiO_2 were less than 1 wt.%. Likewise, this also suggests that the doping ions could be highly dispersed in the titania support.

For doped TiO_2 nanoparticles, the maximum intensity of the anatase phase crystalline plane (101) decreases. This indicates that the crystallization of TiO_2 anatase is restricted due to the increase in the content of the dopant (0.5 to 1.0 wt.%). As shown in Table 1, the increase in the doping content leads to an average particle size smaller than the pure material. The decrease in crystalline nature has been reported by other authors who used Cu and Zn salts as doping precursors. They mentioned that a mechanism of fixation of doping ions at the grain boundary inhibits anatase nanoparticle growth [29–31].

Table 1. Material characterization results of TiO_2 and doped-TiO_2.

Catalyst		**Cu wt.%**			**Zn wt.%**		**Average Particle Size (nm)**		**Eg (eV)**
	Sol. A [1]	**EDX**	**ICP-OES**	**Sol. A** [1]	**EDX**	**ICP-OES**	**DRX**	**SEM**	
T	nd	nd	nd	nd	nd	nd	19.50	21	3.23
T-DK (0.5)	0.21 ±0.009	0.19 ±0.002	0.23 ±0.014	0.13 ±0.005	0.12 ±0.004	0.11 ±0.003	15.7	16	2.90
T-DK (1.0)	0.47 ±0.011	0.43 ±0.015	0.44 ±0.009	0.24 ±0.008	0.22 ±0.007	0.21 ±0.006	11.3	11	2.76

[1] Sol. A: Solution resulting from the digestion of the dopant precursor.

3.1.2. HRTEM and Elemental Analysis

In the HRTEM micrographs in Figure 2, pure and doped TiO_2 nanoparticles exhibit spherical morphology. After doping, it is also observed that the average crystal size decreases compared to pure TiO_2. This phenomenon has already been explained in the XRD diffractogram discussion. Based on the histograms in Figure 2, for T and T-DK (1.0), the average particle size was similar to those obtained with XRD.

Although it has been shown that the surface area is not decisive in the photocatalytic process, as a result of particle size decrease, the specific surface area increased gradually from 58.8 ± 0.8 m^2g^{-1} for T, 76 ± 0.4 m^2g^{-1} for T-DK (0.5), and 88.7 ± 0.5 m^2g^{-1} for T-DK (1.0). Increased dopants in TiO_2 inhibit particle growth; consequently, the narrow pore size distribution and the surface becomes larger [32].

From the HRTEM micrography in Figure 2, it is difficult to determine the presence of doping ions. Nevertheless, the elementary EDS and ICP-OES analysis results, summarized in Table 1, show that doped TiO_2 nanoparticles contain Cu and Zn doping ions. According to the EDS images in Figure 2, these dopants are homogeneously distributed over the surface of the TiO_2. The presence of Ni ions was not detected. As the dopant precursor was obtained from a junkyard, it had advanced metal wear, so there was no longer any nickel coating that this type of door key usually has. So, the absence of nickel in the elemental analysis performed by EDS and ICP-OES was an expected result. This means that only copper and zinc were considered doping ions.

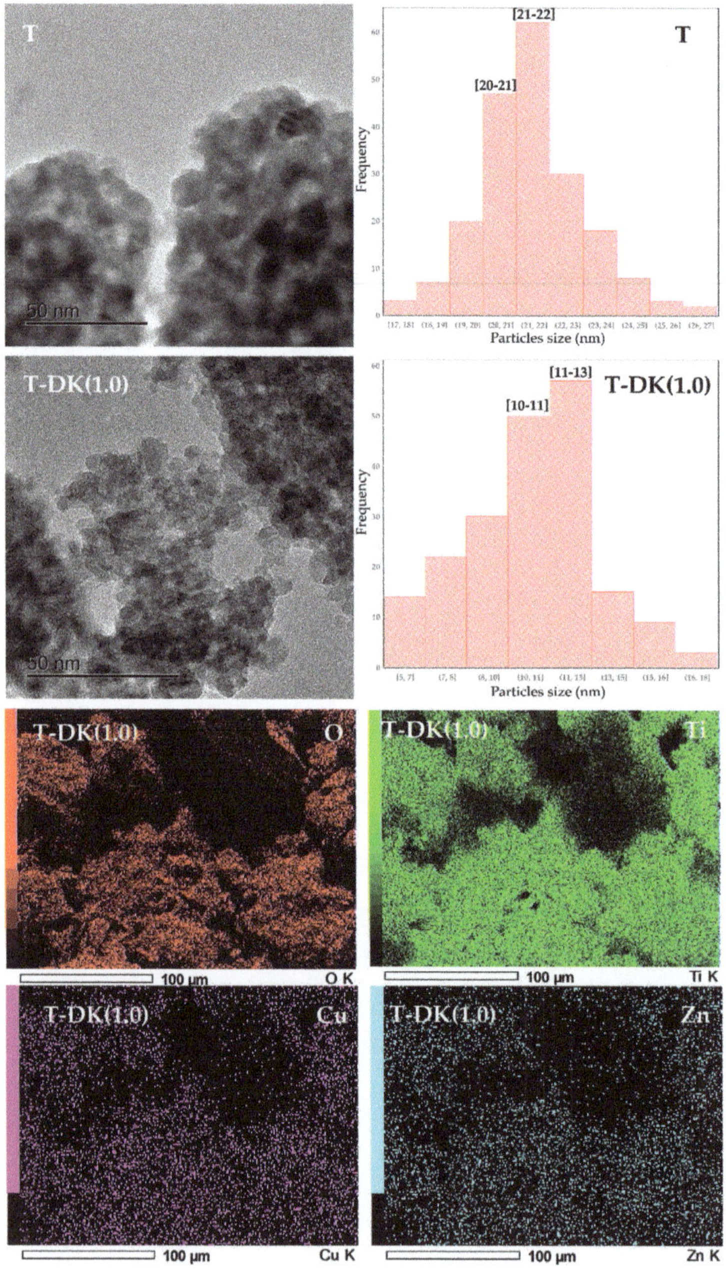

Figure 2. HRTEM micrographs, histograms, and EDS spectra of TiO_2 pure and T-DK (1.0).

Regarding the weight percentage of Cu and Zn in TiO_2, EDS and ICP-OES (Table 1) show differences between the theoretical weight percentage of Cu and Zn concerning the real percentage. To explain this difference, elemental analysis by ICP-OES was performed on solution A (solution resulting from the digestion of the metal residue). The results show that during the digestion carried out in an open system, part of the initial concentration of the doping ions is lost (Table 1). The efficacy of the digestion method, among other variables, will be investigated in future projects.

3.1.3. DRS

The optical properties of these materials were obtained using DRS. The absorption spectra are shown in Figure 3. Bandgap values (Eg) are shown in Table 1. Pure TiO_2 exhibits strong absorption at 396 nm (3.23 eV), assigned to the charge transfer of the metal-ligand in Ti^{4+}(3d)-O^{2-}(2p). For T-DK (0.5) and T-DK (1.0), the door key used as a dopant influenced the optical properties of TiO_2 since the optical absorption band decreased below 396 nm with the increase in the percentage by weight of dopant ions. The T-DK (0.5) material presented a bandgap of 2.90 eV, and the T-DK (1.0) material showed a bandgap of 2.76 eV. It is known that the value of the bandgap of ZnO (Eg $\approx$ 3.2 eV) is similar to that of TiO_2 (Eg $\approx$ 3.23 eV) [33], so zinc does not promote the decrease in Eg. The decrease is due to a redistribution of the electric charge of TiO_2 caused by copper ions in interstitial positions or to copper oxides (Eg $\approx$ 2.5 eV) formed [34]. The obtained Eg values indicate that in our materials, the valence electrons can be transferred to the conduction band using lower energy (visible light) than that required to activate pure TiO_2 (UV light).

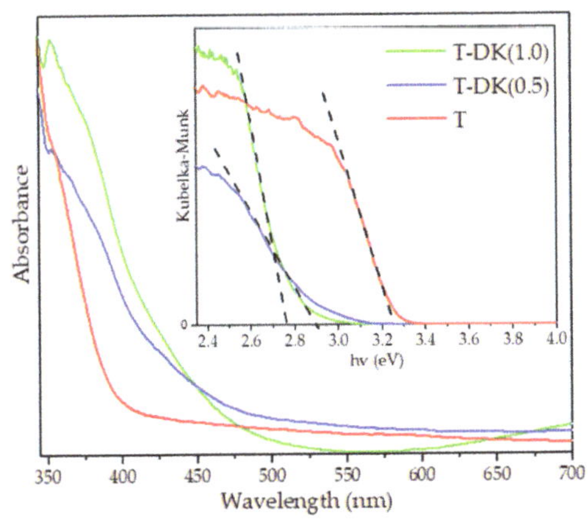

Figure 3. Absorption spectrum of T, T-DK (0.5), and T-DK (1.0). Insets: bandgap energy values: linear part of the plot extrapolated to the X-axis.

For T-DK (1.0), the absorption band at 600 nm suggests a $^2B_{1g}\rightarrow^2B_{2g}$ characteristic of octahedral coordination with tetragonal distortion around Cu^{2+} [34,35]. For low concentrations of copper, T-DK (0.5), no characteristic bands corresponding to metallic ion species were observed. Likewise, no interfacial charge transfer bands were observed from the valence band of TiO_2 to the valence band of zinc oxides. This is probably due to the low weight concentration of the dopant ions or their good dispersion on the titania surface.

3.1.4. X-ray Photoelectron Spectroscopy

The XPS spectra of T and T-DK (1.0) were obtained to determine the oxidation state of the material components. The oxidation states in the O 1s, Cu2p, and Zn 2p regions were obtained by deconvoluting their peaks. The PeakFit software version 4.1.2, AISN Software Inc, was used to this effect. The correlation coefficients (r^2) of the deconvoluted peaks were higher than 0.99. The spectrum in Figure 4a shows bands at 458.5 and 464.1 eV, characteristic of the binding energy of Ti $2p_{3/2}$ and Ti $2p_{1/2}$ in TiO_2 [36]. The observed band comprises two peaks in the O1s region of pure TiO_2. One of these, located at 530.02 eV, is assigned to ionic oxygen in the crystalline array (O-Ti^{4+}). The small peak at 531.82 eV is related to adsorbed OH groups, chemisorbed O species, or oxygen vacancies [37]. These same bands were observed in the Ti 2p (Figure 4c) and O 1s (Figure 4d) regions of the XPS spectra obtained for T-DK (1.0). However, a shift in the binding energies of the Ti 2p and O 1s bands is observed for this material. It is known that the binding energy depends on the oxidation state and the local chemical environment of titanium and oxygen [38,39]. Since the electronegativity of Cu (1.9) and Zn (1.65) is greater than that of Ti (1.54), the electron density around the oxygen ions decreases, which causes an increase in the binding energy.

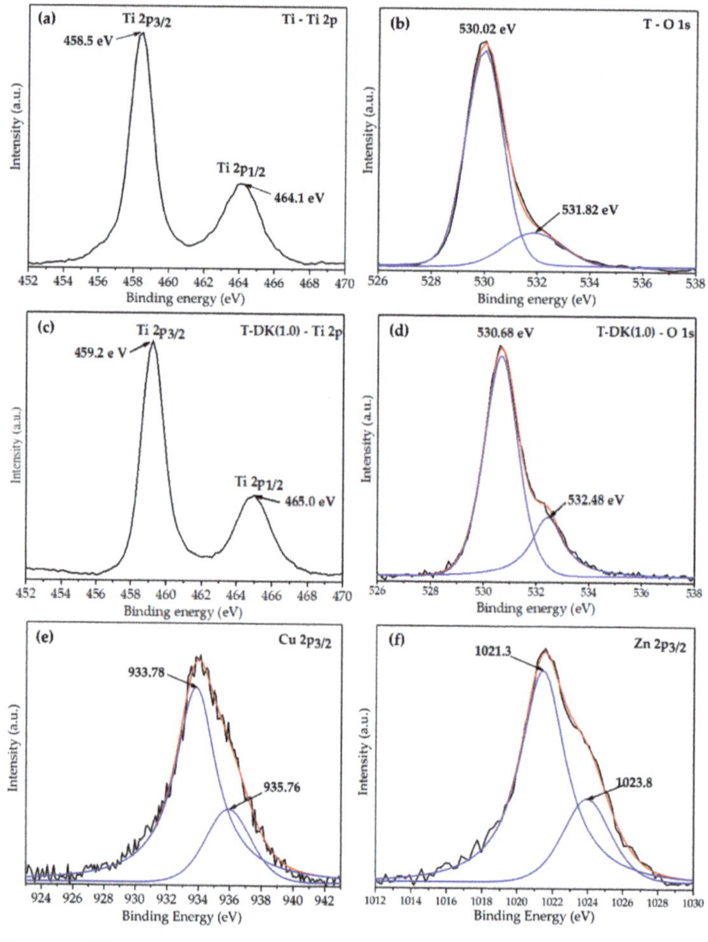

Figure 4. XPS spectra of TiO_2 and T-DK (1.0): (**a**,**b**) Ti 2p, (**c**,**d**) O 1s, (**e**) Cu 4d, and (**f**) Zn 4d.

To develop the above, the Cu 2p (Figure 4e) and Zn 2p (Figure 4f) XPS spectra from the T-DL (1.0) photocatalyst were analyzed. In Figure 4e, the binding energies at ~933.78 eV correspond to Cu^+ from Cu_2O, while the binding energies ~935.76 eV can be assigned to Cu^{2+} in the form of CuO or $Cu(OH)_2$ [25,39,40]. In the case of zinc ions, a typical peak of Zn $2p_{3/2}$ is observed in Figure 4f. The deconvolution of this peak produced two peaks at 1021.3 and 1023.8 eV. The strongest peak located at 1021.3 eV is associated with Zn^{2+} ions in the ZnO with an arrangement of wurtzite crystal. Peak at 1023.8 eV belongs to Zn^{2+} ions in $Zn(OH)_2$ [41–43].

The ionic radius of Ti^{4+}, Cu^{2+}/Cu^+, and Zn^{2+} is 0.061, 0.73/0.77, and 0.74 Å, respectively [44]. These data indicate a significant difference between the values of the ionic radii of the dopants with titanium. As a result, the substitution process of Ti ions by Cu^{2+}/Cu^{+} and Zn^{2+} is limited by the difference between their ionic radii. According to the principles of Hume-Rothery [45], lattice substitution between atoms can only happen if the difference between them is less than 20%. Theoretical studies for copper ions argue that, for the substitution of titanium ions in the crystalline structure, the maximum concentration of copper should be 0.3% [39]. As a result, we can report that some copper and zinc ions were aggregated as oxides on the surface of TiO_2 nanoparticles, creating a heterojunction between these semiconductor materials. Other doping ions held interstitial positions within the crystalline network of the TiO_2 anatase.

According to the results of the material characterization and the references used for their discussion, the interaction between the support and the copper and zinc ions is similar to that reported by other investigations in which the dopant precursor is a chemical reagent.

3.2. Photocatalytic Activity Measurement

To evaluate the photocatalytic activity of the synthesized materials, a solution of diclofenac at a concentration of 10 ppm was used. Total organic carbon (TOC) values are summarized in Table 2. All the materials showed photocatalytic activity in the degradation of diclofenac (Figure 5a,b). Nevertheless, for doped materials, the increased content of doping ions enhanced the photocatalytic activity of TiO_2. The best activity was achieved when T-DK (1.0) was used as a photocatalyst, obtaining up to 94% of mineralized diclofenac. The values for the rate constants (Table 2 and Figure 5b) are consistent with what is seen in Figure 5a. This means that with T-DK (1.0), the maximum mineralization is achieved, but this transformation is also carried out in less time than with T and T-DK (0.5).

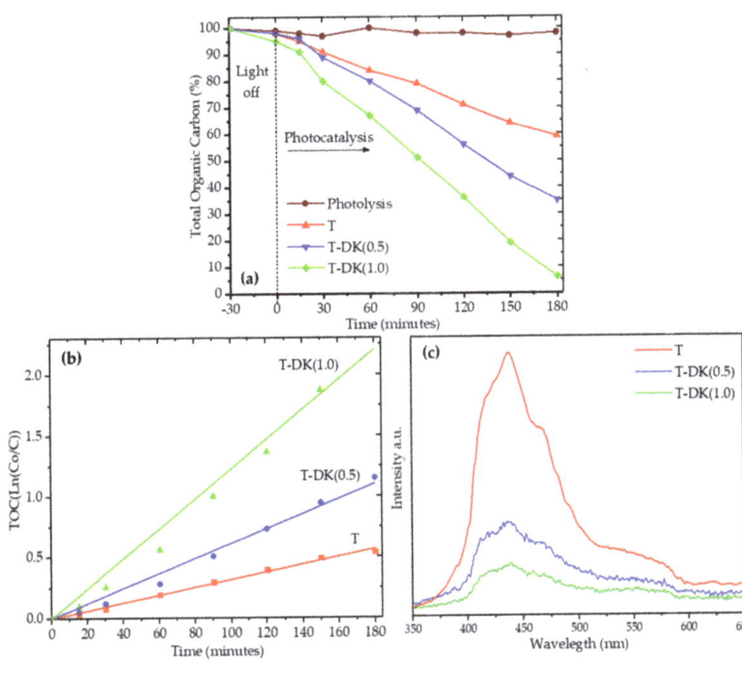

Figure 5. (**a**) Total organic carbon, (**b**) kinetic constants, (**c**) PL spectra of undoped and codoped TiO_2.

Table 2. Apparent first-order rate constants obtained for the diclofenac mineralization.

Catalyst	Residual TOC (%)	K_{app} (min^{-1})	$t_{1/2}$ (min)	R^2
T	59	3.2×10^{-3}	216.61	0.997
T-DK (0.5)	35	6.1×10^{-3}	113.63	0.993
T-DK (1.0)	6	1.23×10^{-2}	56.35	0.990

The correlation of material characterization results with photocatalytic test results suggests that the increase in photocatalytic activity is due to two factors:

i. First, the doping ions induced the reduction of Eg in TiO_2 (see Table 1). The decrease in the value of Eg for T-DK (0.5) and T-DK (1.0) indicates that these materials can be activated with visible light radiation. This means solar light absorption is more efficient in doped TiO_2 than pure TiO_2 [46]. This increases the efficiency of generating electron–hole pairs that initiate redox reactions that directly or indirectly produce the hydroxyl radicals that cause the pollutant to be mineralized. Due to the many possible reaction mechanisms during the diclofenac mineralization process, those considered the main ones in the photocatalytic mechanism are given below [24].

$$Catalyst + hv_{(Sunlight)} \rightarrow Catalyst\ \left(e_{CB}^{-} + h_{VB}^{+}\right)$$

$$h^{+} + H_2O \rightarrow HO^{\bullet} + H^{+}$$

$$h^{+} + OH^{-} \rightarrow HO^{\bullet}$$

Hydroxyl radical attack:

$$DCF + HO^{\bullet} \rightarrow Degradation\ products$$

Oxidation by the positive hole:

$$DCF + h^{+} \rightarrow Oxidation\ products$$

ii. Second, the synergistic effect of the dopant species inhibited the recombination of the e^-/h^+ pairs. This information was obtained by analyzing the charge carrier recombination of each synthesized material. In the emission spectra in Figure 5c, pure TiO_2 obtained the higher intensity emission spectra, meaning rapid recombination of the electron–hole pairs. Contrary to TiO_2, the intensity of the emission spectra was lower when the percentage by weight of the doping ions increased, which suggests a low recombination rate for the electron–hole pairs photogenerated in T-DK (0.5) and T-DK (1.0). According to the results obtained using XPS, in doped TiO_2 Cu^{2+} and Zn^{2+} ions in interstitial positions and oxides of the doping ions coexist. The low recombination of photogenerated charge carriers at T-DK (0.5) and T-DK (1.0) can be understood due to the positions of the band edges of the oxides in the heterojunction. The measured conduction band (CB) potential values of TiO_2 and CuO are –0.35 and +0.12 V (vs. SCE), respectively. The valence band of TiO_2 is lower than ZnO by about 0.36 V (vs. NHE), and this is superior to CuO by approximately 0.20 V (vs. SHE). The relative position difference of the energy band of CuO and ZnO charge transfer occurs between them and TiO_2. Thus, an electron photogenerated in TiO_2 is transferred from the conduction band of this semiconductor to ZnO and CuO, acting as an electron trap to inhibit their recombination. Concurrent with the above, hole transfer can arise from the valence band (VB) of TiO_2 to the VB of ZnO and CuO [43,44,47,48]. Therefore, in addition to the decrease in Eg, the coupled effect between energy bands of TiO_2, CuO, and ZnO was an essential factor to suppress the recombination of the electron–hole pairs, improving the photocatalytic activity of doped TiO_2.

An important characteristic of photocatalysts is their chemical stability after several recycling cycles. Therefore, the reuse of T-DK (1.0) was evaluated during five consecutive reuse cycles. The percentage results in Figure 6a show that during the five cycles of reuse, the mineralization efficiency of diclofenac was in the range of 92.5 to 95%, with a standard deviation of ±1.48. Consequently, no significant change in the mineralization efficiency of diclofenac was observed. The stability in the photocatalytic activity is because the concentration of the dopant ions (Figure 6a) and the crystal structure of the TiO_2-anatase (Figure 6b) does not change before or after the reuse cycles. A similar effect occurs for the photocatalyst morphology determined by SEM in Figure 6c,d. This means that the photocatalyst T-DK (1.0) has a high activity and stability during the five cycles of diclofenac photocatalytic mineralization.

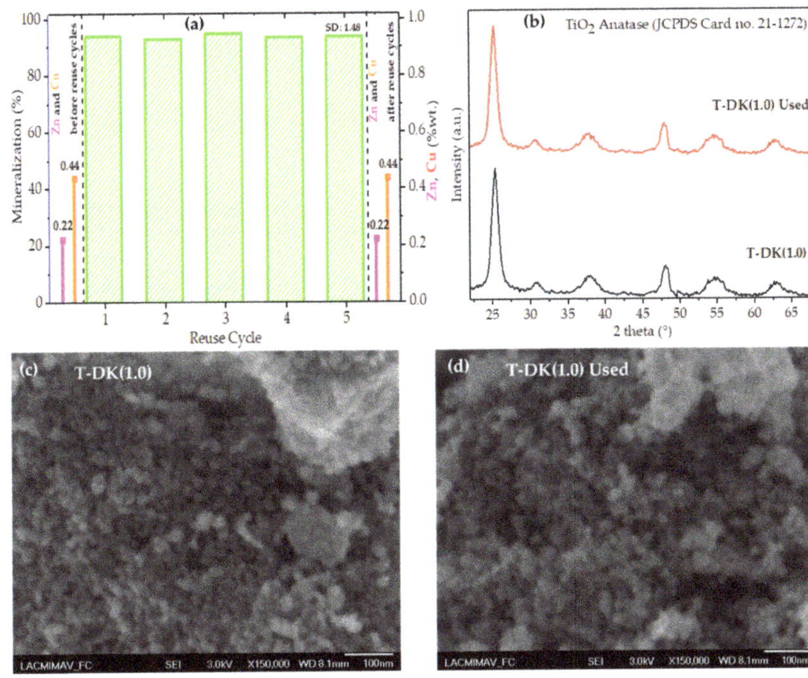

Figure 6. Reuse cycles using T-DK (1.0) and elemental analysis of Cu and Zn before and after the reuse cycle (**a**). DRX diffractograms (**b**) and SEM micrographs of T-DK (1.0) before and after five cycles of reuse (**c**,**d**).

4. Conclusions

This work shows that incorporating Cu and Zn ions into TiO_2 is possible when a door key is used as a doping precursor. Material characterization showed that these ions were integrated into the support as Cu^+/Cu^{2+} and Zn^{2+}, occupying interstitial positions or forming heterojunctions between copper and zinc oxides with titania. The modification with both ions had a dual effect dependent on the concentration of the doping metals. First, an inverse correlation was observed between the concentration of doping ions and the Eg. This resulted in greater light absorption within the visible range. Second, it was shown that the impurities in TiO_2 acted as an electron trap, so the recombination of the e^-/h^+ pairs was lower for T-DK (1.0) than for TiO_2. Therefore, the photocatalytic activity of the synthesized materials occurred in the following order: T < T-DK (0.5) < T-DK (1.0).

The results obtained in this work confirm previous state-of-the-art findings: Doping TiO_2 with Cu and Zn ions or other metal ions is an efficient strategy to improve the photocatalytic activity of titania. As an additional contribution, it is possible to use discarded door keys as a doping precursor to improving the photocatalytic activity of TiO_2. Finally, these results open the door for future research. Other variables can be considered: method of digestion of the metal residue, method of synthesis of materials, pH of synthesis, and calcination temperatures, among other variables. These variables were not considered in this first study, as the main objective was to investigate the feasibility of using door keys as doping precursors.

Author Contributions: Conceptualization, Visualization, Formal analysis, Investigation, Validation, Writing—original draft, Writing—review and editing, D.E.J.-C.; Writing—original draft, Writing—review and editing, Resources, J.G.T.-T.; Writing—original draft, Writing—review and editing, Resources, A.H.-R.; Writing—original draft, Writing—review and editing, J.C.A.-P.; Writing—original draft, Writing—review and editing, A.C.-U.; Writing—original draft, Writing—review and editing, A.E.E.d.l.M.; Writing—original draft, Writing—review and editing, S.G.; Writing—original draft, Writing—review and editing, A.A.S.-P.; Conceptualization, Formal analysis, Funding acquisition, Investigation, Methodology, Project administration, Resources, Supervision, Validation, Visualization, Writing—original draft, Writing—review and editing, A.C.-G. All authors have read and agreed to the published version of the manuscript.

Funding: This research received no external funding.

Institutional Review Board Statement: Not applicable.

Informed Consent Statement: Not applicable.

Data Availability Statement: Data are contained within the article.

Acknowledgments: This research was supported by the National Council of Science and Technology (CONACYT) through the scholarship granted to Dany Edgar Juárez Cortazar.

Conflicts of Interest: The authors declare no conflict of interest.

References

1. Chen, D.; Cheng, Y.; Zhou, N.; Chen, P.; Wang, Y.; Li, K.; Huo, S.; Cheng, P.; Peng, P.; Zhang, R.; et al. Photocatalytic degradation of organic pollutants using TiO_2-based photocatalysts: A review. *J. Clean. Prod.* **2020**, *268*, 121725. [CrossRef]
2. Zambrano, J.; Irusta-Mata, R.; Jiménez, J.J.; Bolado, S.; García-Encina, P.A. Chapter 24—Photocatalytic removal of emerging contaminants in water and wastewater treatments: A review. *Dev. Wastewater Treat. Res. Processes* **2022**, *1*, 543–572. [CrossRef]
3. Naciri, Y.; Hsini, A.; Bouziani, A.; Djellabi, R.; Ajmal, Z.; Laabd, M.; Navío, J.A.; Mills, A.; Bianchi, C.L.; Li, H.; et al. Photocatalytic oxidation of pollutants in gas-phase via Ag_3PO_4-based semiconductor photocatalysts: Recent progress, new trends, and future perspectives. *Crit. Rev. Environ. Sci. Technol.* **2021**, *52*, 1–44. [CrossRef]
4. Tanji, K.; Zouheir, M.; Naciri, Y.; Ahmoum, H.; Hsini, A.; Mertah, O.; El Gaidoumi, A.; Navio, J.A.; Hidalgo, M.C.; Kherbeche, A. Visible light photodegradation of blue basic 41 using cobalt doped ZnO: Box–Behnken optimization and DFT calculation. *J. Iran. Chem. Soc.* **2022**, *18*, 1–16. [CrossRef]
5. Tanji, K.; Zouheir, M.; Hachhach, M.; Ahmoum, H.; Jellal, I.; El Masaoudi, H.; Naciri, Y.; Huynh, T.-P.; Nouneh, K.; Benaissa, M.; et al. Design and simulation of a photocatalysis reactor for rhodamine B degradation using cobalt-doped ZnO film. *React. Kinet. Mech. Catal.* **2021**, *134*, 1017–1038. [CrossRef]
6. Zaleska, A. Doped-TiO_2: A Review. *Recent Patents Eng.* **2008**, *2*, 157–164. [CrossRef]
7. Sescu, A.M.; Favier, L.; Lutic, D.; Soto-Donoso, N.; Ciobanu, G.; Harja, M. TiO_2 Doped with Noble Metals as an Efficient Solution for the Photodegradation of Hazardous Organic Water Pollutants at Ambient Conditions. *Water* **2021**, *13*, 19. [CrossRef]
8. Hsini, A.; Naciri, Y.; Laabd, M.; Bouziani, A.; Navío, J.; Puga, F.; Boukherroub, R.; Lakhmiri, R.; Albourine, A. Development of a novel PANI@WO3 hybrid composite and its application as a promising adsorbent for Cr(VI) ions removal. *J. Environ. Chem. Eng.* **2021**, *9*, 105885. [CrossRef]
9. Briffa, J.; Sinagra, E.; Blundell, R. Heavy metal pollution in the environment and their toxicological effects on humans. *Heliyon* **2020**, *6*, e04691. [CrossRef]
10. Laabd, M.; Imgharn, A.; Hsini, A.; Naciri, Y.; Mobarak, M.; Szunerits, S.; Boukherroub, R.; Albourine, A. Efficient detoxification of Cr(VI)-containing effluents by sequential adsorption and reduction using a novel cysteine-doped PANi@faujasite composite: Experimental study supported by advanced statistical physics prediction. *J. Hazard. Mater.* **2022**, *422*, 126857. [CrossRef]
11. Raguram, T.; Rajni, K.S. Synthesis and analysing the structural, optical, morphological, photocatalytic and magnetic properties of TiO_2 and doped (Ni and Cu) TiO_2 nanoparticles by sol–gel technique. *Appl. Phys. A* **2019**, *125*, 288. [CrossRef]

12. Yadav, H.M.; Otari, S.; Koli, V.B.; Mali, S.S.; Hong, C.K.; Pawar, S.H.; Delekar, S.D. Preparation and characterization of copper-doped anatase TiO_2 nanoparticles with visible light photocatalytic antibacterial activity. *J. Photochem. Photobiol. A Chem.* **2014**, *280*, 32–38. [CrossRef]
13. Zhang, R.; Shao, M.; Xu, S.; Ning, F.; Zhou, L.; Wei, M. Photo-assisted synthesis of zinc-iron layered double hydroxides/TiO_2 nanoarrays toward highly-efficient photoelectrochemical water splitting. *Nano Energy* **2017**, *33*, 21–28. [CrossRef]
14. Bashiri, R.; Norani, M.M.; Kait, C.F.; Sufian, S. Study on Synthesis and Characterization of Cu-Ni Doped TiO_2 by Sol-Gel Hydrothermal. *Adv. Mater. Res.* **2014**, *925*, 248–252. [CrossRef]
15. Liu, B.; Chen, H.M.; Liu, C.; Andrews, S.C.; Hahn, C.; Yang, P. Large-Scale Synthesis of Transition-Metal-Doped TiO_2 Nanowires with Controllable Overpotential. *J. Am. Chem. Soc.* **2013**, *135*, 9995–9998. [CrossRef]
16. Shakil, R.; El-Sawy, A.M.; Tasnim, H.; Meguerdichian, A.G.; Jin, J.; Dubrosky, J.P.; Suib, S.L. Single-Doped and Multidoped Transition-Metal (Mn, Fe, Co, and Ni) ZnO and Their Electrocatalytic Activities for Oxygen Reduction Reaction. *Inorg. Chem.* **2018**, *57*, 9977–9987. [CrossRef]
17. Yadav, S.; Jaiswar, G. Review on Undoped/Doped TiO_2 Nanomaterial; Synthesis and Photocatalytic and Antimicrobial Activity. *J. Chin. Chem. Soc.* **2017**, *64*, 103–116. [CrossRef]
18. Naciri, Y.; Hsini, A.; Bouziani, A.; Tanji, K.; El Ibrahimi, B.; Ghazzal, M.; Bakiz, B.; Albourine, A.; Benlhachemi, A.; Navío, J.; et al. Z-scheme WO3/PANI heterojunctions with enhanced photocatalytic activity under visible light: A depth experimental and DFT studies. *Chemosphere* **2022**, *292*, 133468. [CrossRef]
19. Imgharn, A.; Anchoum, L.; Hsini, A.; Naciri, Y.; Laabd, M.; Mobarak, M.; Aarab, N.; Bouziani, A.; Szunerits, S.; Boukherroub, R.; et al. Effectiveness of a novel polyaniline@Fe-ZSM-5 hybrid composite for Orange G dye removal from aqueous media: Experimental study and advanced statistical physics insights. *Chemosphere* **2022**, *295*, 133786. [CrossRef]
20. Imgharn, A.; Ighnih, H.; Hsini, A.; Naciri, Y.; Laabd, M.; Kabli, H.; Elamine, M.; Lakhmiri, R.; Souhail, B.; Albourine, A. Synthesis and characterization of polyaniline-based biocomposites and their application for effective removal of Orange G dye using adsorption in dynamic regime. *Chem. Phys. Lett.* **2021**, *778*, 138811. [CrossRef]
21. Fahoul, Y.; Tanji, K.; Zouheir, M.; El Mrabet, I.; Naciri, Y.; Hsini, A.; Nahali, L.; Kherbeche, A. Novel River Sediment@ZnO Co nanocomposite for photocatalytic degradation and COD reduction of crystal violet under visible light. *J. Mol. Struct.* **2022**, *1253*, 132298. [CrossRef]
22. Mimouni, I.; Bouziani, A.; Naciri, Y.; Boujnah, M.; El Belghiti, M.A.; El Azzouzi, M. Effect of heat treatment on the photocatalytic activity of α-Fe_2O_3 nanoparticles: Towards diclofenac elimination. *Environ. Sci. Pollut. Res.* **2021**, *29*, 7984–7996. [CrossRef] [PubMed]
23. Aguilar, T.; Navas, J.; Alcántara, R.; Lorenzo, C.F.; Gallardo, J.; Blanco, G.; Martín-Calleja, J. A route for the synthesis of Cu-doped TiO_2 nanoparticles with a very low band gap. *Chem. Phys. Lett.* **2013**, *571*, 49–53. [CrossRef]
24. Mogal, S.I.; Mishra, M.; Gandhi, V.G.; Tayade, R.J. Metal Doped Titanium Dioxide: Synthesis and Effect of Metal Ions on Physico-Chemical and Photocatalytic Properties. *Mater. Sci. Forum* **2013**, *734*, 364–378. [CrossRef]
25. Hampel, B.; Pap, Z.; Sapi, A.; Szamosvolgyi, A.; Baia, L.; Hernadi, K. Application of TiO_2-Cu Composites in Photocatalytic Degradation Different Pollutants and Hydrogen Production. *Catalysts* **2020**, *10*, 85. [CrossRef]
26. Zhang, J.; Zhou, P.; Liu, J.; Yu, J. New understanding of the difference of photocatalytic activity among anatase, rutile and brookite TiO_2. *Phys. Chem. Chem. Phys.* **2014**, *16*, 20382–20386. [CrossRef]
27. Scanlon, D.O.; Dunnill, C.W.; Buckeridge, J.; Shevlin, S.A.; Logsdail, A.J.; Woodley, S.M.; Catlow, C.R.A.; Powell, M.J.; Palgrave, R.G.; Parkin, I.P.; et al. Band alignment of rutile and anatase TiO_2. *Nat. Mater.* **2013**, *12*, 798–801. [CrossRef]
28. Luttrell, T.; Halpegamage, S.; Tao, J.; Kramer, A.; Sutter, E.; Batzill, M. Why is anatase a better photocatalyst than rutile?—Model studies on epitaxial TiO_2 films. *Sci. Rep.* **2014**, *4*, 4043. [CrossRef]
29. Zhang, D.; Zeng, F. Photocatalytic oxidation of organic dyes with visible-light-driven codoped TiO_2 photocatalysts. *Russ. J. Phys. Chem. A* **2011**, *85*, 1077–1083. [CrossRef]
30. Tobaldi, D.; Piccirillo, C.; Rozman, N.; Pullar, R.; Seabra, M.; Škapin, A.S.; Castro, P.; Labrincha, J. Effects of Cu, Zn and Cu-Zn addition on the microstructure and antibacterial and photocatalytic functional properties of Cu-Zn modified TiO_2 nano-heterostructures. *J. Photochem. Photobiol. A Chem.* **2016**, *330*, 44–54. [CrossRef]
31. Khairy, M.; Zakaria, W. Effect of metal-doping of TiO_2 nanoparticles on their photocatalytic activities toward removal of organic dyes. *Egypt. J. Pet.* **2014**, *23*, 419–426. [CrossRef]
32. Niu, B.; Wang, X.; Wu, K.; He, X.; Zhang, R. Mesoporous Titanium Dioxide: Synthesis and Applications in Photocatalysis, Energy and Biology. *Materials* **2018**, *11*, 1910. [CrossRef]
33. Hernández, S.; Hidalgo, D.; Sacco, A.; Chiodoni, A.; Lamberti, A.; Cauda, V.; Tresso, E.; Saracco, G. Comparison of photocatalytic and transport properties of TiO_2 and ZnO nanostructures for solar-driven water splitting. *Phys. Chem. Chem. Phys.* **2015**, *17*, 7775–7786. [CrossRef] [PubMed]
34. Reddam, H.A.; Elmail, R.; Lloria, S.C.; Tomás, G.M.; Reddam, Z.A.; Coloma-Pascual, F. Synthesis of Fe, Mn and Cu modified TiO_2 photocatalysts for photodegradation of Orange II. *Boletín de la Sociedad Española de Cerámica y Vidrio* **2020**, *59*, 138–148. [CrossRef]
35. Mishra, A.P.; Sharma, N.; Jain, R.K. Microwave Synthesis, Spectral, Thermal and Antimicrobial Studies of Some Ni(II) and Cu(II) Schiff Base Complexes. *Open J. Synth. Theory Appl.* **2012**, *2*, 56–62. [CrossRef]

36. Arévalo-Pérez, J.C.; De La Cruz-Romero, D.; García, A.C.; Lobato-García, C.E.; Aguilar-Elguezabal, A.; Torres-Torres, J.G. Photodegradation of 17 α-methyltestosterone using TiO_2-Gd^{3+} and TiO_2-Sm^{3+} photocatalysts and simulated solar radiation as an activation source. *Chemosphere* **2020**, *249*, 126497. [CrossRef] [PubMed]
37. Cordero-García, A.; Palomino, G.T.; Hinojosa-Reyes, L.; Guzmán-Mar, J.L.; Maya-Teviño, L.; Hernández-Ramírez, A. Photocatalytic behaviour of WO_3/TiO_2-N for diclofenac degradation using simulated solar radiation as an activation source. *Environ. Sci. Pollut. Res.* **2016**, *24*, 4613–4624. [CrossRef]
38. Cordero-García, A.; Guzmán-Mar, J.; Hinojosa-Reyes, L.; Ruiz-Ruiz, E.; Hernández-Ramírez, A. Effect of carbon doping on WO_3/TiO_2 coupled oxide and its photocatalytic activity on diclofenac degradation. *Ceram. Int.* **2016**, *42*, 9796–9803. [CrossRef]
39. Mathew, S.; Ganguly, P.; Rhatigan, S.; Kumaravel, V.; Byrne, C.; Hinder, S.J.; Bartlett, J.; Nolan, M.; Pillai, S.C. Cu-Doped TiO_2: Visible Light Assisted Photocatalytic Antimicrobial Activity. *Appl. Sci.* **2018**, *8*, 2067. [CrossRef]
40. Obregón, S.; Muñoz-Batista, M.J.; Fernández-García, M.; Kubacka, A.; Colón, G. Cu–TiO_2 systems for the photocatalytic H_2 production: Influence of structural and surface support features. *Appl. Catal. B: Environ.* **2015**, *179*, 468–478. [CrossRef]
41. Xu, D.; Fan, D.; Shen, W. Catalyst-free direct vapor-phase growth of Zn1−xCuxO micro-cross structures and their optical properties. *Nanoscale Res. Lett.* **2013**, *8*, 46. [CrossRef] [PubMed]
42. Wang, M.; Jiang, L.; Kim, E.J.; Hahn, S.H. Electronic structure and optical properties of $Zn(OH)_2$: LDA+U calculations and intense yellow luminescence. *RSC Adv.* **2015**, *5*, 87496–87503. [CrossRef]
43. Thirupathi, B.; Smirniotis, P.G. Co-doping a metal (Cr, Fe, Co, Ni, Cu, Zn, Ce, and Zr) on Mn/TiO_2 catalyst and its effect on the selective reduction of NO with NH3 at low-temperatures. *Appl. Catal. B: Environ.* **2011**, *110*, 195–206. [CrossRef]
44. Adrover, E.; Boldrini, D.; Divins, N.J.; Casanovas, A.; Tonetto, G.; López, E.; Llorca, J. Study of Cu-Zn and Au/TiO2 catalysts on anodized aluminum monoliths for hydrogen generation and purification. *Int. J. Chem. React. Eng.* **2016**, *14*, 831–842. [CrossRef]
45. Mizutani, U. Hume-Rothery rules for structurally complex alloy phases. *MRS Bull.* **2012**, *37*, 169. [CrossRef]
46. Sulaiman, S.N.A.; Noh, M.Z.; Adnan, N.N.; Bidin, N.; Ab Razak, S.N. Ab Razak, Effects of photocatalytic activity of metal and non-metal doped TiO_2 for Hydrogen production enhancement—A Review. *J. Phys. Conf. Ser.* **2018**, *5*, 1027–1045. [CrossRef]
47. Rana, A.G.; Ahmad, W.; Al-Matar, A.; Shawabkeh, R.; Aslam, Z. Synthesis and characterization of Cu–Zn/TiO_2 for the photocatalytic conversion of CO_2 to methane. *Environ. Technol.* **2017**, *38*, 1085–1092. [CrossRef]
48. Paulino, P.; Salim, V.M.; Resende, N. Zn-Cu promoted TiO_2 photocatalyst for CO_2 reduction with H_2O under UV light. *Appl. Catal. B Environ.* **2016**, *185*, 362–370. [CrossRef]

water

MDPI

Review

Recent Advances of Emerging Organic Pollutants Degradation in Environment by Non-Thermal Plasma Technology: A Review

Yongjian He, Wenjiao Sang *, Wei Lu, Wenbin Zhang, Cheng Zhan and Danni Jia

School of Civil Engineering and Architecture, Wuhan University of Technology, Wuhan 430070, China; hyjwhut@126.com (Y.H.); wluemail@163.com (W.L.); zhangwenbin1114@126.com (W.Z.); wlzhancheng@126.com (C.Z.); jdnwz9998@163.com (D.J.)
* Correspondence: whlgdxswj@126.com

Abstract: Emerging organic pollutants (EOPs), including endocrine disrupting compounds (EDCs), pharmaceuticals and personal care products (PPCPs), and persistent organic pollutants (POPs), constitute a problem in the environmental field as they are difficult to completely degrade by conventional treatment methods. Non-thermal plasma technology is a novel advanced oxidation process, which combines the effects of free radical oxidation, ozone oxidation, ultraviolet radiation, shockwave, etc. This paper summarized and discussed the research progress of non-thermal plasma remediation of EOPs-contaminated water and soil. In addition, the reactive species in the process of non-thermal plasma degradation of EOPs were summarized, and the degradation pathways and degradation mechanisms of EOPs were evaluated of selected EOPs for different study cases. At the same time, the effect of non-thermal plasma in synergy with other techniques on the degradation of EOPs in the environment was evaluated. Finally, the bottleneck problems of non-thermal plasma technology are summarized, and some suggestions for the future development of non-thermal plasma technology in the environmental remediation were presented. This review contributes to our better understanding of non-thermal plasma technology for remediation of EOPs-contaminated water and soil, hoping to provide reference for relevant practitioners.

Keywords: advanced oxidation processes; discharge plasma; reactive species; environmental remediation; combination system; degradation mechanism

Citation: He, Y.; Sang, W.; Lu, W.; Zhang, W.; Zhan, C.; Jia, D. Recent Advances of Emerging Organic Pollutants Degradation in Environment by Non-Thermal Plasma Technology: A Review. *Water* **2022**, *14*, 1351. https://doi.org/10.3390/w14091351

Academic Editors: Dionysios (Dion) Demetriou Dionysiou, Yujue Wang and Huijiao Wang

Received: 29 March 2022
Accepted: 19 April 2022
Published: 21 April 2022

1. Introduction

Over the past few decades, increasing industrial, agricultural, and human activities have promoted the use of chemicals [1,2]. In addition, the development of human society and the wrong understanding of personal safety have increased the chemical load in water and soil environment. In many developing countries, untreated sewage is also used for agricultural purposes, resulting in substandard or untreated wastewater that adds many pollutants to the food chain [3,4]. Therefore, appropriate technologies are needed to eliminate pollutants in water and soil environment for the sake of protection for both human health and the environment.

Emerging organic pollutants (EOPs) are a kind of organic pollutants which have no environmental monitoring standards or emission standards and have negative effects on ecology and human health [5]. EOPs include endocrine disrupting compounds (EDCs), pharmaceutical and personal care products (PPCPs), and persistent organic pollutants (POPs) [6,7]. EOPs may be candidates for future regulation because of their potential risks to the environment and human health, the continuous entry into the environment and the fact that even the most modern wastewater treatment plants (WWTPs) cannot completely convert/remove these compounds [8]. In recent years, with the improvement of environmental analysis, these substances have been frequently detected in the environment, such as sewage, surface water, drinking water [9], and soil [10]. The continuous detection of EOPs brings new challenges to environmental pollution control and makes the treatment of EOPs become an international research hotspot.

Currently, there are many methods for organic pollutants degradation in water environment: bioremediation [11,12], advanced oxidation processes (AOPs) [13–15], adsorption process [16,17], membrane treatment [18,19], and combination process [20]. Compared with water remediation, organic pollution in soil is so subtle that it is difficult to detect. In addition, soil remediation tends to be costlier and takes longer to complete. Therefore, it is urgent to develop effective and convenient remediation technology for organic contaminated soil. At present, many methods have been developed for different organic contaminated soils, including physical remediation (e.g., thermal desorption [21], soil vapor extraction [22]), chemical remediation (e.g., soil washing [23], electrochemical remediation [24], chemical oxidation remediation [25]), and biological remediation (e.g., microbial remediation [26], phytoremediation [27]). With the exception of AOPs, most of these technologies either transfer contaminants from one phase to another rather than complete degradation and mineralization, or are not efficient when the concentration of organic pollutants is at low levels. However, bioremediation often requires long treatment time and it is difficult to reduce the pollution level below the standard.

Non-thermal plasma may be a viable alternative to more common AOPs due to its comparable energy requirements for contaminant degradation and its ability to operate without any additional chemicals [28]. Non-thermal plasma technology has been used as a method to degrade EOPs, including EDCs [29,30] (e.g., pesticides, industrial chemicals, steroids), PPCPs [31–34] (e.g., antibiotics, antidepressant, anti-inflammatories, antimicrobials, surfactants), and POPs [35,36] (e.g., polychlorinated biphenyls, polycyclic aromatic hydrocarbons, organochlorine pesticides). In the review of Magureanu et al. [37], research on the degradation of various pharmaceutical compounds by non-thermal plasma was discussed, and the removal efficiency of target compounds and the energy yield of plasma technology were compared and discussed. Russo et al. [38] summarized the research works on the removal and mineralization of organic pollutants in water by the combination of non-thermal plasma and catalyst. They concluded that the catalyst played an important role in improving the performance of the plasma system. Zhang et al. [39] introduced several typical non-thermal plasma sources for remediation of organic contaminated soil. The effects of different important parameters (such as applied voltage, reactor configuration, soil properties, type of feed air, and gas flow rate) on the remediation performance were discussed. Guo et al. [40] supplemented the research on the mechanism and process of repairing organic contaminated soil by discharge plasma on the basis of previous reviews. Figure 1 shows bibliography data of papers related to non-thermal plasma technology for the degradation of pollutants published in the last decade. As shown in Figure 1a, the number of published papers using non-thermal plasma as a means of pollutant degradation in the last decade has always maintained high and is gradually increasing. Especially, PPCPs and POPs have become hotspots in this field (as shown in Figure 1b).

During the past decade, a number of studies have reported the degradation of EOPs in the environment by non-thermal plasma. At present, the review of non-thermal plasma mainly focuses on water remediation [41–45], and a few reviews have also summarized its application in soil remediation [39,40], but there is a lack of comprehensive review on water and soil remediation. Therefore, this paper focuses on the degradation of different types of EOPs (i.e., EDCs, PPCPs, and POPs) in water and soil environment by the non-thermal plasma technology. Since the reactive species in the plasma system play an indispensable role in the degradation of pollutants, the generation of reactive species in non-thermal plasma is reviewed. In addition, this review also summarizes the research progress of degradation pathways of different kinds of EOPs by non-thermal plasma. An important part of this work is devoted to the combination of non-thermal plasma with various other technologies to compensate for the shortcomings of plasmas alone. In this review, the bottleneck problems of non-thermal plasma and the future prospects are also presented.

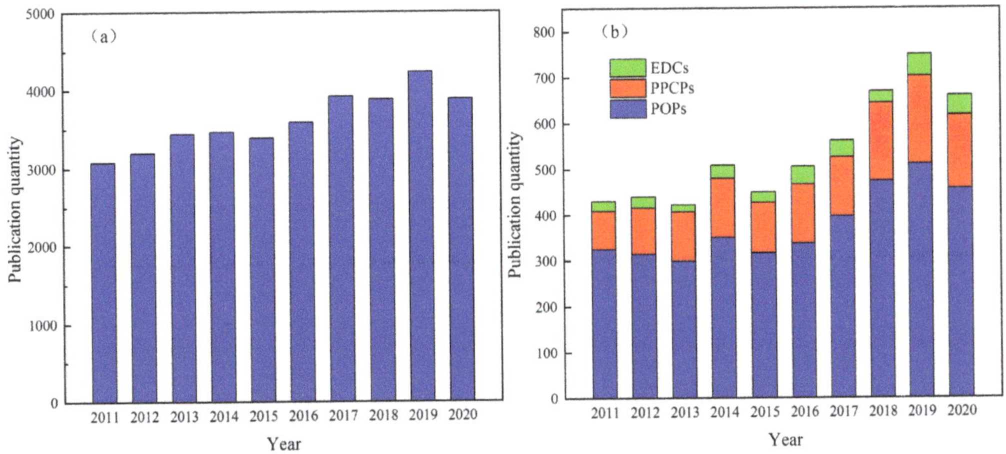

Figure 1. Statistics of papers about non-thermal plasma technology as a method for the degradation of (**a**) pollutants and (**b**) EOPs (EDCs, PPCPs, and POPs) published in the last ten years (data obtained from Web of Science).

2. Degradation of Emerging Organic Pollutants by Non-Thermal Plasma Technology

Different kinds of EOPs may cause harm to human health and ecological environment; so, it is necessary to find appropriate methods to deal with EOPs. EDCs, also known as environmental hormones, can bind to hormone receptors in organisms and disrupt normal metabolism in the endocrine system [46]. Many researchers have reported their toxic effects on human health and the environment [47]. PPCPs are the most widely used chemical reagents in animal husbandry, agriculture, and human daily life. However, they have the potential to cause serious ecotoxicological problems and pose a great threat to ecosystems or organisms [48–50]. It is worth mentioning that antibiotics with a certain concentration level in the environment for a long time may not only have toxic effects on some sensitive organisms, but also lead to the generation, maintenance, transfer and transmission of antibiotic-resistant bacteria (ARB), and antibiotic resistance genes (ARGs) under selective pressure [51]. Since the adoption of the "Stockholm Convention" by the United Nations Environment Program (UNEP) in 2001, awareness of the potential risks of POPs in the environment and the need to remove POPs from the environment have become more urgent. POPs are generally considered to have three basic physical and chemical properties: persistence, lipophilicity, and long-distance mobility. These properties enable them to perform bioamplification and bioaccumulation in animals and seriously harm the health of humans and the natural environment. Therefore, it is imperative to develop environmentally friendly removal methods. Faced with this environmental problem, many researchers have focused on AOPs to eliminate EOPs that are resistant to conventional treatment processes [52–54]. Non-thermal plasma may be a viable alternative to more common AOPs due to its comparable energy requirements for contaminant degradation and its ability to operate without any additional chemicals. The specific research results of non-thermal plasma in water and soil remediation are introduced below.

2.1. Water Remediation

Researchers have paid extensive attention to the degradation of EOPs in recent years and, in order to reduce the risk of EOPs in water environment, non-thermal plasma has been a widely studied and applied technology. Summaries of some representative studies are compiled in Table 1.

Table 1. Overview of work done in the degradation of EOPs in water by non-thermal plasma.

EDCs				
Target Compounds	**Non-Thermal Plasma System**	**Experimental Condition**	**Results**	**Ref.**
17α-ethinylestradiol	Corona discharge plasma	Discharge voltage: 16 kV; Frequency: 20 Hz; Solution volume: 100 mL; Initial concentration: 100 μg/L	Degradation efficiency: ~100% (0.25 min); Energy yield: 4.80 g/kWh	[55]
Bisphenol A	Pulsed discharge plasma	Discharge voltage: 20 kV; Frequency: 50 Hz; Air flow rate: 4.0 L/min; Initial concentration: 20 mg/L	Degradation efficiency: 75.1% (60 min); TOC removal efficiency: 29.8%	[56]
Bisphenol A	Dielectric barrier discharge plasma	Discharge voltage: 5 kV; Gas flow rate: 0.5 L/min; Solution volume: 500 mL; Initial concentration: 20 mg/L	Degradation efficiency: 67.8% (60 min); TOC removal efficiency: 34.2%	[57]
Bisphenol A	Dielectric barrier discharge plasma	Discharge voltage: 5 kV; Gas flow rate: 0.5 L/min; Solution volume: 500 mL; Initial concentration: 20 mg/L	Degradation efficiency: 69.5% (60 min); Energy yield: 0.117 g/kWh	[58]
Bisphenol A	Dielectric barrier discharge plasma	Discharge voltage: 16.8 kV; Discharge power: 412.5 W; Solution volume: 500 mL; Initial concentration: 50 mg/L; Initial pH = 6	Degradation efficiency: 100% (25 min); Rate constant: 0.152 min^{-1}	[59]
P-methylphenol	Dielectric barrier discharge plasma	Discharge voltage: 110 V; O_2 flow rate: 100 mL/min; Liquid flow rate: 100 mL/min; Solution volume: 400 mL; Initial concentration:100 mg/L; Initial pH: 6.1	Degradation efficiency: 49.13% (8 min)	[60]
Dichlorvos	Dielectric barrier discharge plasma	Discharge voltage: 80 kV; Solution volume: 20 mL; Frequency: 50 Hz; Initial concentration: 850 ppb	Degradation efficiency: 78.98% (8 min); Rate constant: 0.00303 s^{-1}	[61]
Malathion		Discharge voltage: 80 kV; Solution volume: 20 mL; Frequency: 50 Hz; Initial concentration: 1320 ppb	Degradation efficiency: 69.62% (8 min); Rate constant: 0.00239 s^{-1}	
Endosulfan		Discharge voltage: 80 kV; Solution volume: 20 mL; Frequency: 50 Hz; Initial concentration: 350 ppb	Degradation efficiency: 57.71% (8 min); Rate constant: 0.00192 s^{-1}	
Dimethyl phthalate	Dielectric barrier discharge plasma	Discharge voltage: 18 kV; Air flow rate: 2.5 L/min; Solution volume: 500 mL; Initial concentration: 30 mg/L	Degradation efficiency: 61.4% (30 min); Rate constant: 0.029 min^{-1}; Energy yield: 0.29 g/kWh	[62]
4-chlorophenol	Corona discharge plasma	Discharge voltage: 20 kV; Frequency: 100 Hz; Flow rate: 200 mL/min; Solution volume: 200 mL; Initial concentration: 100 mg/L	Degradation efficiency: 86.2% (40 min);	[63]

Table 1. *Cont.*

PPCPs				
Target Compounds	**Non-Thermal Plasma System**	**Experimental Condition**	**Results**	**Ref.**
Enrofloxacin	Pulsed discharge plasma	Discharge voltage: 20 kV; Frequency: 50 Hz; Air flow rate: 4 L/min; Solution volume: 150 mL; Initial concentration: 20 mg/L	Degradation efficiency: 76% (60 min); Energy yield: 18.7 mg/kJ	[64]
Tetracycline	Surface discharge plasma	Discharge voltage: 7 kV; Frequency: 6 kHz; Air flow rate: 1.0 L/min; Solution volume: 900 mL; Initial concentration: 40 mg/L	Degradation efficiency: 82.6% (15 min); Energy yield: 16.1 mg/kJ	[65]
Pefloxacin	Dielectric barrier discharge plasma	Discharge power: 0.7 W; Frequency: 50 Hz; O_2 flow rate: 125 mL/min; Solution volume: 100 mL; Initial concentration: 120 mg/L	Degradation efficiency: 96.1% (25 min); Energy yield: 39.6 g/kWh	[66]
Oxytetracycline	Dielectric barrier discharge plasma	Discharge voltage: 4.8 kV; Frequency: 10 kHz; Air flow rate: 1.0 L/min; Solution volume: 200 mL; Initial concentration: 50 mg/L	Degradation efficiency: 88.2% (20 min); TOC removal efficiency: 36.7%; COD removal efficiency: 21.2%; Energy yield: 0.27 mg/kJ	[67]
Amoxicillin	Dielectric barrier discharge plasma	Discharge voltage: 15 kV; Frequency: 50 Hz; Initial concentration: 16 mg/L; Initial pH: 4.5	Degradation efficiency: 75% (18 min)	[68]
Chloramphenicol	Pulsed discharge plasma	Discharge voltage: 19 kV; Air flow rate: 4 L/min; Solution volume: 150 mL; Initial concentration: 20 mg/L; Initial pH: 6.9	Degradation efficiency: ~45% (60 min); Energy yield: 10.77 mg/kWh	[69]
Tetracycline	Dielectric barrier discharge plasma	Discharge voltage: 7 kV; Discharge power: 38.8 W; Air flow rate: 2.0 L/min; Solution volume: 450 mL; Initial concentration: 50 mg/L; Initial pH: 7.2	Degradation efficiency: ~52% (5 min); Energy yield: 3.79 g/kWh	[70]
Ofloxacin	Pulsed discharge plasma	Discharge voltage: 18 kV; Solution volume: 150 mL; Initial concentration: 20 mg/L	Degradation efficiency: 65.0% (60 min); Rate constant: 0.017 min^{-1}	[71]
Oxytetracycline	Pulsed discharge plasma	Discharge voltage: 18 kV; Air flow rate: 4 L/min; Solution volume: 150 mL; Initial concentration: 40 mg/L; Initial pH: 3.2	Degradation efficiency: 59% (60 min); TOC removal efficiency: 23.8%	[72]
Acetaminophen	Dielectric barrier discharge plasma	Discharge voltage: 18 kV; Air flow rate: 200 L/h; Solution volume: 180 mL; Initial concentration: 20 ppm	Degradation efficiency: 50% (18 min)	[73]

Table 1. *Cont.*

PPCPs				
Target Compounds	**Non-Thermal Plasma System**	**Experimental Condition**	**Results**	**Ref.**
Diclofenac	Dielectric barrier discharge plasma	Discharge power: 150 W; Ar flow rate: 20 L/h; Solution volume: 500 mL; Initial concentration: 50 mg/L; Initial pH: 5.6;	Degradation efficiency: 100% (30 min)	[74]
Ibuprofen		Discharge power: 150 W; Ar flow rate: 20 L/h; Solution volume: 500 mL; Initial concentration: 50 mg/L; Initial pH: 5.7;	Degradation efficiency: 100% (20 min)	
Ibuprofen	Corona discharge plasma	Discharge voltage: 18 kV; Discharge power: 2.1 W; Frequency: 25 Hz; Solution volume: 330 mL; Initial concentration: 22.8 mg/L	Degradation efficiency: 100% (20 min); Energy yield: 20.2 g/kWh	[75]
Ibuprofen	Dielectric barrier discharge plasma	Discharge power: 70 W; Air flow rate: 0.50 m^3/h; Initial concentration: 18 mg/L; Initial pH: 6.72	Degradation efficiency: 95.6% (5 min)	[76]
Diclofenac	Corona discharge plasma	Discharge voltage: 80 kV; Frequency: 20 Hz; Solution volume: 300 mL; Initial concentration: 0.5 mg/L	Degradation efficiency: >80% (15 min)	[77]
Amoxicillin	Dielectric barrier discharge plasma	Discharge power: ~2 W; O_2 flow rate: 600 sccm; Solution volume: 200 mL; Initial concentration: 100 mg/L; Initial pH: 8	Degradation efficiency: ~100% (10 min)	[78]
Oxacillin			Degradation efficiency: ~100% (30 min)	
Ampicillin			Degradation efficiency: ~100% (30 min)	
Diclofenac	Pulsed corona discharge plasma	Discharge voltage: 18 kV; Frequency: 22 Hz; O_2 flow rate: 1 L/min; Solution volume: 55 mL; Initial concentration: 50 mg/L	Degradation efficiency: 100% (15 min); TOC removal efficiency: 50% (30 min)	[79]
Enalapril	Dielectric barrier discharge plasma	Discharge voltage: 18 kV; Discharge power: 2 W Frequency: 50 Hz; O_2 flow rate: 600 sccm; Solution volume: 300 mL; Initial concentration: 50 mg/L	Degradation efficiency: ~90% (15 min); TOC removal efficiency: >40% (120 min); Energy yield: 4.33 g/kWh	[80]
Paracetamol	Dielectric barrier discharge plasma	Discharge voltage: 5.9 kV; Frequency: 500 Hz; Air flow rate: 100 sccm; Solution volume: 40 mL; Initial concentration: 25 mg/L	Degradation efficiency: 90% (60 min); Energy yield: 2.9 g/kWh	[81]
Verapamil	Dielectric barrier discharge plasma	Discharge voltage: 18 kV; Frequency: 50 Hz; N_2:O_2 = 80:20; Gas flow rate: 30 mL/min; Solution volume: 70 mL; Initial concentration: 0.1 mM	Degradation efficiency: >99% (40 min)	[82]

Table 1. *Cont.*

PPCPs				
Target Compounds	**Non-Thermal Plasma System**	**Experimental Condition**	**Results**	**Ref.**
Dimethyl phthalate	Liquid phase plasma	Discharge voltage: 250 V; Frequency: 30 kHz; Solution volume: 600 mL; Initial concentration: 20 ppm	Degradation efficiency: 63% (180 min); Rate constant: 0.00433 min^{-1}	[83]
Caffeine	Dielectric barrier discharge plasma	Discharge power: 60 W; Frequency: 10 kHz; Solution volume: 20 mL; Initial concentration: 100 μg/mL	Degradation efficiency: 72.6% (4 min)	[84]
Caffeine	Dielectric barrier discharge plasma	Discharge power: 75 W; Solution volume: 100 mL; Initial concentration: 50 mg/L	Degradation efficiency: 41% (24 min); TOC removal efficiency: >10%	[85]
Perfluorooctanoic acid	Dielectric barrier discharge plasma	Discharge voltage: 22 kV; Solution volume: 250 mL; Initial concentration: 20 mg/L	Degradation efficiency: 73.5% (60 min); TOC removal efficiency: 28.9%; Energy yield: 46.39 mg/kWh	[86]
Diatrizoate	Corona discharge plasma	Discharge voltage: 11 kV; Frequency: 3 kHz; Solution volume: 7.5 mL; Initial concentration: 200 μg/L	Degradation efficiency: 90% (20 min); Energy yield: 0.140 g/kWh	[87]
Pentoxifylline	Dielectric barrier discharge plasma	Discharge voltage: 12 kV; Frequency: 120 Hz; O_2 flow rate: 600 sccm; Solution volume: 200 mL; Initial concentration: 100 mg/L; Initial pH: 7	Degradation efficiency: 92.5% (60 min); Energy yield: 16 g/kWh	[88]
Carbamazepine	Dielectric barrier discharge plasma	Discharge voltage: 8 kV; Frequency: 500 Hz; Initial concentration: 0.219 μg/L	Degradation efficiency: 90%	[28]
POPs				
Target Compounds	**Non-Thermal Plasma system**	**Experimental Condition**	**Results**	**Ref.**
3,3′,4,4′-tetrachlorobiphenyl	Dielectric barrier discharge plasma	Discharge voltage: 10 kV; Feed gas: He; Solution volume: 3 mL; Initial concentration: 0.2 mM	Degradation efficiency: 80% (2 min)	[89]
2,2′,4,4′-tetrachlorobiphenyl	Pulsed corona discharges plasma	Discharge voltage: 45 kV; Frequency: 60 Hz; Initial conductivity: 100 μs/cm; Initial concentration: 68 ppb	Degradation efficiency: 70% (60 min)	[90]

Non-thermal plasma technology has been proved to be an effective method to degrade EOPs in water environment. For example, Yang et al. [59] studied the degradation of bisphenol A (BPA) in water by dielectric barrier discharge (DBD) plasma. The results showed that BPA was completely degraded (100%) within 25 min when the discharge voltage reached 16.8 kV. Satisfactory BPA degradation performance was achieved in a relatively short treatment time, which proved the superiority of non-thermal plasma treatment. Aggelopoulos et al. [91] investigated the degradation of enrofloxacin (ENRO) in aqueous solution in a gas-liquid nanosecond-pulsed dielectric barrier discharge (NSP-DBD) plasma reactor. Under the optimal pulse voltage and pulse frequency, ENRO was degraded completely (100%) after 20 min, and the corresponding energy yield was 1.1 g/kWh.

3,3′,4,4′-tetrachlorobiphenyl (PCB77) was selected as the target pollutant for DBD treatment in order to verify the effectiveness of non-thermal plasma in the degradation of POPs in aqueous solution [89]. Their study showed that non-thermal plasma can effectively degrade PCB77 in aqueous solution. After DBD plasma treatment for 2 min, more than 75% of PCB77 was degraded. In addition, the biotoxicity of PCB77 degradation products was also evaluated, and it was found that DBD degradation products of PCB77 are almost non-toxic, which demonstrated that non-thermal plasma is an efficient, green, and environmentally friendly treatment technology to remove POPs from the environment.

The discharge voltage applied in non-thermal plasma significantly affects the degradation/mineralization efficiency of EOPs. On the one hand, with the increase of discharge voltage, the intensity of ultraviolet radiation increases, which leads to the improvement of pollutant oxidation. On the other hand, the number of reactive species, especially •OH, increases with the discharge voltage, which may lead to enhanced degradation of pollutants. However, as the discharge voltage increases excessively, the energy yield generally decreases. It is possible that the increased discharge voltage leads to increased energy waste, indicating that more electrical energy is being converted to heat [92,93]. Therefore, reasonable control of discharge voltage is the key of non-thermal plasma technology.

The pH value of aqueous solution is also one of the key factors for the degradation of EOPs by non-thermal plasma technology. It not only affects the properties of EOPs, but also affects the generation of reactive species in the non-thermal plasma system, which is probably the main reason why the results of different studies seem contradictory. The pH dependence of degradation efficiency has been extensively studied by many researchers. In general, the formation of •OH is more intense under neutral or alkaline conditions. Some previous reports also support that some reactive species (e.g., H_2O_2 and O_3) can decompose more quickly under alkaline conditions, forming •OH, leading to higher efficiency of pollutant degradation [94–96]. However, in some studies, it was found that better degradation efficiency was achieved under acidic conditions. In the study of Li et al. [97], as the pH value of the solution increased from 2.0 to 10.0, the reaction rate constants of tetracycline (TC), sulfadiazine (SD) and ciprofloxacin (CIP) decreased by 40.9%, 60.0%, and 65.0%, respectively. The lower degradation efficiency under alkaline conditions can be explained by the deprotonation of pollutant molecules, which may lead to the contraction of the bond length of pollutant molecules, thus increasing the stability of molecular structure and enhancement of hydrophilicity of pollutant molecules, thus reducing the interaction with reactive species. In addition, in the relatively high pH environment, the generated •OH will be quenched by OH^- [98], thus inhibiting the degradation of pollutants. Therefore, pH is not a simple parameter, and the optimal pH value for different studies may be different, and largely depends on the chemical structure and properties of pollutants.

The reaction of EOPs with different molecular structures to oxidative attacks may be different and largely depends on the substituents on the benzene ring. Generally speaking, EOPs with stable molecular structure have strong resistance to reactive species oxidation. Li et al. [97] investigated the degradation of three antibiotics with different substituents and chemical properties by non-thermal discharge plasma oxidation, namely TC, SD, and CIP. The results showed that the three antibiotics could degrade effectively, but the reaction kinetics were different. The authors speculated that the significant difference in degradation performance of the three antibiotics may be due to their different molecular structures. To verify this experimental conclusion, the authors further determined the relationship between the chemical structure of these antibiotics and their removal efficiency by using Gaussian calculations. The ionization potential (IP) of organic compounds was calculated using the following equation (Equation (1)). Compared with SD and CIP, TC had the lowest ionization potential and was therefore more easily oxidized by reactive oxygen species (ROS). Therefore, organic pollutants with different structures have different degradation effects, which was also found in the study of Kim et al. [31]. Unfortunately, they did not explain the specific reasons. Banaschik et al. [77] explained why some pharmaceutical compounds are more recalcitrant than others. Aromatic ring systems, unsaturated bonds, and electron donating functional groups (+I/+M) increased molecular reactivity towards plasma treatment and also towards other AOPs that are relying on the generation of •OH.

$$IP = 1.3124 \times (-\varepsilon HOMO) + 0.514\ eV \quad (1)$$

2.2. Soil Remediation

At present, there is more and more research on remediation of EOPs contaminated soil by non-thermal plasma technology, and the excellent treatment effect has gradually attracted the attention of researchers. Summaries of some representative studies are presented in Table 2.

Non-thermal plasma has been proved to be effective in repairing EOPs-contaminated soil. Aggelopoulos et al. [101] studied the degradation of atrazine (ATZ) in soil by DBD discharge plasma, and the results showed that ATZ with initial concentration of 100 and 10 mg/kg could be degraded in dry soil with a degradation efficiency of 86.9% and 98.1% after plasma treatment for 60 min, respectively. Lou et al. [105] studied the remediation of chloramphenicol (CAP)-contaminated soil by DBD plasma. The results showed that the degradation efficiency of CAP was close to 81% after 20 min of plasma treatment, which demonstrated the feasibility of non-thermal plasma in removing pharmaceutical compounds from soil. Li et al. [112] used pulsed DBD plasma system to rehabilitate phenanthrene (PHE)-contaminated soil. Under the condition of 0.6 L/min air flow and 110 V voltage, the removal efficiency can reach 87.3% within 20 min, and the energy yield is 0.01 mg/kJ.

The researchers found that water content in soil is one of the most important factors affecting the removal efficiency of non-thermal plasma. A certain amount of water molecules can promote the production of •OH [114]. Wang et al. [100] demonstrated that in dry soil (0% moisture), about 65.6% of glyphosate was degraded after 45 min of DBD plasma treatment, and when soil moisture increased to 10%, the proportion increased to 86.5% over the same treatment time. However, soil moisture increased further to 20%, while glyphosate degradation efficiency decreased to 76.5%. The researchers attributed this to the fact that as the water content increased, the pores of the soil became clogged with water molecules, resulting in reduced transport of reactive species through the soil, resulting in less effective degradation of pollutants. In addition, some researchers have suggested that the presence of water contributes to this by helping dissolve organic matter in soil particles, allowing the dissolved organic matter to compete with the target pollutant for reactive species [115].

Different types of power supplies used to drive various plasma reactors also have a great impact on the efficiency of soil remediation. Recently, more and more researchers have focused on the plasma driven by nanosecond pulse power supply. Nanosecond pulsed plasmas have the following advantages: (1) higher electron energy [116,117], since most of the energy in the discharge process is used to accelerate the electron energy rather than neutral gas; (2) stronger chemical activity [118], because the nanosecond pulse discharge can produce more high-energy electrons, the inelastic collision in the discharge process is more intense, and more reactive species can be produced, the chemical activity is stronger than other discharges such as alternating current (AC) and direct current (DC); (3) better uniformity and stability [119,120]. Aggelopoulos et al. [106] studied the remediation of CIP-contaminated soil by nanosecond-pulsed DBD plasma. Under the optimal conditions (pulse voltage 17.4 kV, pulse frequency 200 Hz), CIP was completely degraded in soil (~99%), and treatment time was only 3 min, with a corresponding energy efficiency of 4.6 mg/kJ, which is quite high for soil remediation. However, the plasma discharge in this study was carried out in the gas phase above the soil surface, which would lead to poor permeability of ultraviolet radiation and reactive species produced by plasma in the soil, thus affecting the treatment effect. Hatzisymeon et al. [107] designed a new type of discharge plasma reactor to alleviate this problem. In such reactors, the reactive species produced by the discharge and plasma were produced directly in the contaminated medium, rather than in the gas phase above the contaminated medium. Under optimized conditions, the energy efficiency of the system was 21.2 mg/kJ. In addition, this system could degrade CIP almost completely over a wide range of soil thicknesses (2.4 to 9.4 mm). At the actual concentration of CIP contamination in soil (20 mg/kg), the degradation process was very rapid and complete. At high initial contaminant concentrations (200 mg/kg), considerable degradation efficiency was also achieved. Water content of up to 10% did not appear to significantly affect process efficiency, which is important for implementation under practical conditions.

Table 2. Overview of work done in the degradation of EOPs in soil by non-thermal plasma.

EDCs				
Target Compounds	**Non-Thermal Plasma System**	**Experimental Condition**	**Results**	**Ref.**
P-nitrophenol	Pulsed discharge plasma	Discharge voltage: 20 kV; Frequency: 100 Hz; Soil mass: 2.0 g; Initial concentration: 800 mg/kg	Degradation efficiency: 78.1% (10 min)	[99]
Glyphosate	Dielectric barrier discharge plasma	Discharge voltage: 18.0 kV; Air flow rate: 1.5 L/min^{-1}; Soil mass: 5.0 g; Initial concentration: 200 mg/kg; Soil moisture: 10%	Degradation efficiency: 93.9% (45 min); Energy yield: 0.47 g/kWh; Rate constant: 0.062 min^{-1}	[100]
Atrazine	Dielectric barrier discharge plasma	Discharge voltage: 44.8 kV; Discharge power: 0.087 W; Air flow rate: 1.0 L/min; Frequency: 300 Hz; Soil mass: 5 g; Initial concentration: 100 mg/kg	Degradation efficiency: 86.9% (60 min)	[101]
Pentachlorophenol	Pulsed corona discharge plasma	Discharge voltage: 16.0 kV Frequency: 67 Hz; Air flow rate: 1.0 L/min; Soil mass: 5 g; Initial concentration: 200 mg/kg; Pollution time: 4 h	Degradation efficiency: 85% (60 min)	[102]
P-nitrophenol (PNP) and pentachlorophenol (PCP) mixtures	Pulsed corona discharge plasma	Discharge voltage: 18 kV; Frequency: 50 Hz; Specific energy density: 485 J/g_{soil}; Air flow rate: 0.5 L/min; Soil mass: 5 g; Initial concentration: 300 mg/kg (PNP) and 300 mg/kg (PCP)	Degradation efficiency: 86.0% (PNP) and 94.1% (PCP); Energy yield: 4.01 g/kWh; TOC removal efficiency: 39.2%	[103]
Trifluralin	Dielectric barrier discharge plasma	Discharge voltage: 26.8 kV; Frequency: 100 Hz; Air flow rate: 0.075 L/min; Soil mass: 10.1 g; Initial concentration: 200 ppm	Degradation efficiency: 97.3% (10 min); Energy yield: 4.1 mg/kJ	[104]
PPCPs				
Target Compounds	**Non-Thermal Plasma System**	**Experimental Condition**	**Results**	**Ref.**
Chloramphenicol	Dielectric barrier discharge plasma	Discharge voltage: 18.4 kV; O_2 flow rate: 0.5 L/min; Soil mass: 2.5 g; Initial concentration: 200 mg/kg; Soil moisture content: 10%	Degradation efficiency: 81.0% (20 min)	[105]

Table 2. *Cont.*

PPCPs				
Target Compounds	**Non-Thermal Plasma System**	**Experimental Condition**	**Results**	**Ref.**
Ciprofloxacin	Dielectric barrier discharge plasma	Discharge voltage: 17.4 kV; Frequency: 200 Hz; Air flow rate: 1.0 L/min; Soil mass: 5 g; Initial concentration: 200 mg/kg; Soil moisture content: 5%	Degradation efficiency: 99% (3 min); Energy yield: 4.6 mg/kJ	[106]
Ciprofloxacin	Dielectric barrier discharge plasma	Discharge voltage: 26.8 kV; Frequency: 100 Hz; Air flow rate: 0.075 L/min; Soil mass: 6.1 g; Initial concentration: 200 mg/kg; Soil moisture content: 0.1%	Degradation efficiency: 95.2% (5 min); Energy yield: 21.2 mg/kJ	[107]
POPs				
Target Compounds	**Non-Thermal Plasma System**	**Experimental Condition**	**Results**	**Ref.**
Pyrene	Dielectric barrier discharge plasma	Discharge voltage: 36.2 kV; Electrode gap: 14 mm; Initial concentration: 100 mg/kg	Degradation efficiency: 61.6% (60 min)	[108]
Fluorene	Corona discharge plasma	Discharge voltage: 30 kV; Frequency: 50 Hz; Electrode gap: 20 mm; Initial concentration: 200 mg/kg	Degradation efficiency: 78.7% (60 min)	[109]
Naphthalene	Dielectric barrier discharge plasma	Discharge voltage: 40 kV; Air flow rate: 0.85 L/min; Moisture contents: 4%; Soil depth: 1 mm; Initial concentration: 100 mg/kg	Degradation efficiency: 96.32% (30 min)	[110]
Phenanthrene			Degradation efficiency: 89.08% (30 min)	
Pyrene			Degradation efficiency: 88.59% (30 min)	
Pyrene	Dielectric barrier discharge plasma	Discharge voltage: 35.8 kV; Frequency: 9 kHz; Air flow rate: 0.85 L/min; Initial concentration: 10 mg/kg	Degradation efficiency: 85.09% (30 min); Energy yield: 0.8 μg/kJ	[111]
Phenanthrene	Dielectric barrier discharge plasma	Discharge voltage: 110 V; Frequency: 150 Hz; Air flow rate: 0.6 L/min; Soil mass: 5 g; Initial concentration: 200 mg/kg	Degradation efficiency: 87.3% (20 min); Energy yield: 0.01 mg/kJ	[112]
Polychlorinated biphenyls	Dielectric barrier discharge plasma	Discharge power: 21 W; Gas flow rate: 120 mL/min; Initial concentration: 1.78×10^4 μg/kg	Degradation efficiency: 84.6% (90 min)	[113]
Polycyclic aromatic hydrocarbons	Pulsed corona discharge plasma	Discharge voltage: 18 kV; Frequency: 70 Hz; Air flow rate: 0.8 L/min; Initial concentration: 100 mg/kg	Degradation efficiency: 70% (40 min)	[36]

2.3. Comparison with Other AOPs

Comparisons between non-thermal plasma and other AOPs are difficult because experimental conditions (such as molecular structure, initial concentration, treatment volume, etc.) vary greatly and these parameters significantly affect the degradation process. In order to find a more accurate evaluation of degradation efficiency and energy yield, Hama Aziz et al. [74,121] focused on the degradation of 2,4-dichlorophenoxyacetic acid (2,4-D), 2,4-dichlorophenol (2,4-DCP), diclofenac (DCF), and ibuprofen (IBP) by several AOPs: ozonation, photocatalysis, and non-thermal plasma. The common reactor design of all experiments can directly compare the degradation efficiency and energy yield obtained by different methods. The specific comparison results are shown in Table 3. Comparing these AOPs from the perspective of degradation efficiency and energy yield, it is found that there is an obvious gap in the degradation of pollutants with different molecular structures, which indicates that different AOPs will be affected by the molecular structure.

Table 3. Comparison of EOPs degradation by non-thermal plasma and other AOPs.

Target Pollutant	Technology	Initial Concentration (mg/L)	Solution Volume (mL)	Treatment Time (min)	Degradation Efficiency	Energy Yield G_{50} (g/kWh)	References
2,4-D	Ozonation	100	500	20	~100%	6.60	[121]
	Photocatalysis			90	67%	0.92	
	Non-thermal plasma			15	~100%	8.80	
2,4-DCP	Ozonation	100	500	10	~100%	18.09	
	Photocatalysis			90	73%	0.35	
	Non-thermal plasma			15	~100%	4.64	
DCF	Ozonation	50	500	4	~100%	28.00	[74]
	Photocatalysis			90	~80%	~0.50	
	Non-thermal plasma			20	~100%	5.10	
IBP	Ozonation	50	500	25	~100%	2.50	
	Photocatalysis			90	~35%	~0.25	
	Non-thermal plasma			15	~100%	2.15	

Compared with photocatalysis, ozonation, and non-thermal plasma have the advantages of fast degradation rate, high degradation efficiency, and high energy yield. It can also be seen from Table 3 that ozonation and non-thermal plasma show similar performance. Taking 2,4-D as an example, the energy yield of non-thermal plasma is 8.8 g/kWh, while the energy yield of ozonation for the removal of 2,4-D is 6.6 g/kWh. It is worth noting that the mineralization efficiency of pollutants by ozonation is relatively low, and very good mineralization can be obtained by non-thermal plasma. Li et al. [122] compared the degradation of IBP by different AOPs. Except that the energy yield of photo-Fenton is slightly higher than that of water film DBD plasma, the energy yield of other AOPs is relatively low. However, photo-Fenton takes a longer time and the removal efficiency is not very high.

3. Mechanism of Emerging Organic Pollutants Degradation by Non-Thermal Plasma

3.1. Reactive Species in Non-Thermal Plasma Discharges

The ability of non-thermal plasma to produce highly reactive species in situ is well known. Their formation is mainly triggered by the collision of high-energy electrons produced in the discharge with gas atoms or molecules. Once the discharge process has occurred, reactive species can also be produced by radical recombination reactions or de-excitation of metastable substances [96]. The most abundant primary and secondary species formed in liquid or gas-liquid environments are hydroxyl radical (•OH), ozone (O_3), and hydrogen peroxide (H_2O_2), which are associated with the degradation of target pollutants. However, many other ROS and RNS are produced in plasma, and can also contribute to the decomposition of pollutants, such as singlet oxygen ($^1O_2$), atomic oxygen (O), superoxide anion radical ($\bullet O_2^-$), peroxide hydroxyl radical ($HO_2\bullet$), nitrite (NO_2^-), nitrate (NO_3^-), peroxynitrite ($ONOO^-$), etc. The redox potentials of common oxidants are shown in Figure 2. These reactive species react with pollutants in water or soil, or high-energy electrons in an electric field react directly with pollutants, degrading them into small intermediates and further splitting them into carbon dioxide and water molecules. In addition to oxidizing species, reductive species in discharge plasma may also contribute to the degradation of pollutants in water, such as aqueous electron (E^0 = −2.77 V) and H• radicals (E^0 = −2.30 V). Furthermore, some physical effects, such as ultraviolet radiation, heat, and shock wave, are often accompanied in the plasma discharge process. In conclusion, the degradation of organic compounds by discharge plasma depends mainly on the reactive species, while other physical effects are beneficial to the degradation process.

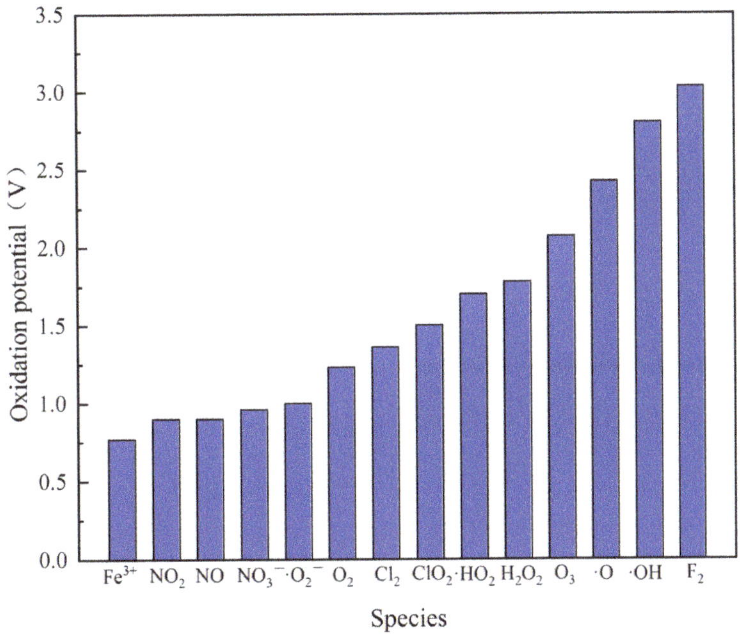

Figure 2. Comparison of oxidation potential of reactive species by non-thermal plasma.

•OH is the second most reactive substance after fluorine atom, and they attack most organic pollutant molecules with a rate constant of 10^6–10^9 M^{-1} s^{-1}, which is 10^6–10^{12} times faster than ozone [123,124]. The production of •OH in water or plasma in contact with water was discussed in detail [125]. Bruggeman et al. found that the production of •OH depends largely on plasma parameters, such as gas temperature (Tg), electron temperature (Te), ionization degree, electron and ion density, and gas composition. The results showed that •OH formed by electron dissociation (Equation (2)) and dissociation attachment (Equation (3)) of water molecules are dominant in plasma with Te higher than 2 eV. However, when Te is between 1 and 2 eV, electron-ion dissociation recombination (Equations (4) and (5)) and ion-ion dissociation recombination (Equations (6) and (7)) also play an important role in the mass production of •OH. Even at high enough ionization degree, it is also the main formation pathway of •OH.

$$e^- + H_2O \rightarrow HO\bullet + H + e^- \quad (2)$$

$$e^- + H_2O \rightarrow HO\bullet + H^- \quad (3)$$

$$e^- + H_2O^+ \rightarrow HO\bullet + H\bullet \quad (4)$$

$$e^- + H_3O^+ \rightarrow HO\bullet + H_2 + e^- \quad (5)$$

$$H^- + H_2O^+ \rightarrow HO\bullet + H_2 \quad (6)$$

$$H^- + H_3O^+ \rightarrow HO\bullet + H_2 + H\bullet \quad (7)$$

In the case of air or oxygen as the feed gas, O_3 is formed by non-thermal plasma, and the process has been well documented [126]. The electrons generated in the discharge excite and dissociate diatomic oxygen (Equation (8)), and the resulting atomic oxygen reacts with another oxygen molecule in the presence of a third object (M: M is a third collision partner O_2, O_3, O) to form ozone (Equation (9)).

$$O_2 + e^- \rightarrow 2O\bullet + e^- \tag{8}$$

$$O\bullet + O_2 + M \rightarrow O_3 + M \tag{9}$$

It is generally believed that H_2O_2 is mainly formed by the dimerization of •OH in discharge in contact with water (Equation (10)):

$$\bullet OH + \bullet OH \rightarrow H_2O_2 \tag{10}$$

H_2O_2 is a relatively stable oxidant and can accumulate in the liquid phase during plasma discharge. Locke and Shih [127] summarized the literature on the formation of H_2O_2 in water using various discharge techniques and experimental conditions. The results showed that the efficiency of H_2O_2 production depends largely on the experimental device, and the maximum energy yield is 80 g/kWh. In addition to the direct generation of •OH in the plasma, additional •OH may be generated by the interaction of dissolved O_3 with H_2O_2 (Equation (11)).

$$H_2O_2 + O_3 \rightarrow O_2 + HO\bullet + HO_2\bullet \tag{11}$$

Formation of more •OH is particularly beneficial for degradation because •OH is a powerful non-selective oxidant that reacts with most organic compounds, including short-chain carboxylic acids, and complex intermediates produced during degradation of organic molecules. Thus, increased concentrations of •OH can ensure further degradation/mineralization of organic pollutants.

In the study of plasma discharge technology using air/N_2 as raw gas, in addition to ROS, RNS are also formed, such as nitric oxide (NO), nitrogen dioxide (NO_2), nitrite (NO_2^-), nitrate (NO_3^-), and peroxynitrite ($ONOO^-$) [95]. Even though N_2 is a very stable molecule with high bonding energy, high-energy electrons generated during plasma discharge can dissociate the molecule from atomic nitrogen (Equation (12)). Nitrogen oxides are then rapidly formed by the interaction of atomic nitrogen with diatomic or triatomic oxygen (Equations (13) and (14)). The exchange between the two nitrogen oxides can be achieved by the interaction of nitric oxide with ozone and the photodissociation of nitrogen dioxide by ultraviolet radiation produced in the plasma (Equations (15) and (16)). Further dissolution and oxidation of nitrogen oxides in aqueous media lead to the formation of NO_2^-, NO_3^-, and $ONOO^-$.

$$N_2 + e^- \rightarrow 2N\bullet + e^- \tag{12}$$

$$N\bullet + O_2 \rightarrow NO\bullet + O\bullet \tag{13}$$

$$N\bullet + O_3 \rightarrow NO\bullet + O_2 \tag{14}$$

$$NO\bullet + O_3 \rightarrow NO_2 + O_2 \tag{15}$$

$$NO_2 + hv \rightarrow NO\bullet + O\bullet \tag{16}$$

3.2. *Degradation Pathways of Emerging Organic Pollutants*

Different degradation efficiencies and rates have been recorded using non-thermal plasma techniques for different classes of EOPs. This phenomenon is mainly due to the different chemical structure of EOPs. Different kinds of organic compounds require different amounts of energy to break bonds. Therefore, the complexity of degradation pathway is also a key factor in the final degradation results. During non-thermal plasma treatment, in addition to the residue of the target pollutant, some intermediates also exist in the water/soil, which will compete with the target pollutant for reactive species, resulting in poor degradation. However, little is known about the differences in degradation mechanisms and kinetics. Therefore, the degradation pathways of some different types of EOPs are summarized below.

3.2.1. Selected EDCs: Bisphenol A

Based on most relevant research, the results suggested that •OH, O_3 and •NO_2 play an important role in the degradation of BPA and most intermediates. Generally, there are two main pathways for the degradation of BPA in non-thermal plasma [56–59], as shown in Figure 3. On the one hand, •OH reacts with phenolic hydroxyl groups in BPA to form TP1 through hydrogen abstraction. The C atom on the ortho position of the hydroxyl group of TP1 undergoes hydroxylation and a recombination reaction under the attack of •OH and •NO_2 to generate TP2 and TP3. Then, TP4 and TP5 are generated by tautomerization of the keto and enol forms. On the other hand, the C atom between the two benzene rings is cleaved under the oxidation of •OH and O_3 to generate TP6 and TP7. In addition, the above intermediate products can be further oxidized and ring-opened to generate a series of small molecular organic compounds, and finally mineralized into CO_2 and H_2O.

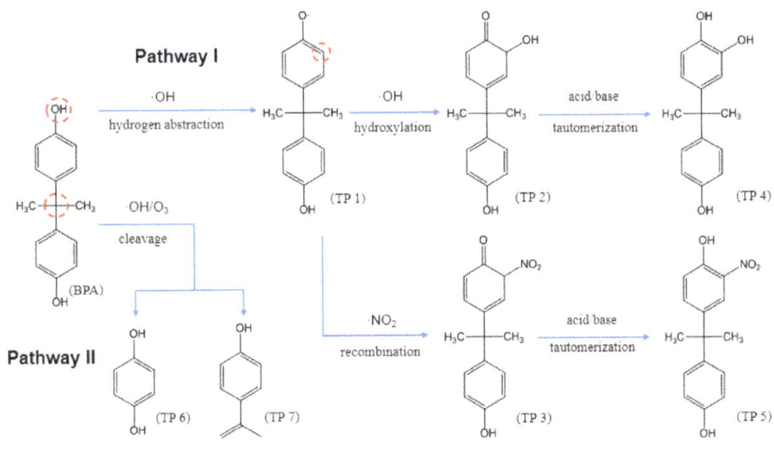

Figure 3. The proposed BPA degradation pathway.

3.2.2. Selected PPCPs: Ibuprofen

Based on density functional theory (DFT), the spatial configuration of organic molecules and electron cloud density distribution can be obtained by molecular orbital calculation, which is helpful to predict the degradation behavior of organic pollutants. Generally, there are three common pathways for IBP to undergo degradation by non-thermal plasma, as shown in Figure 4. First, •OH can cause IBP to lose its carboxyl structure to generate the product TP1. TP1 is further deprotonated to form an intermediate product TP2 under the oxidation of •OH, O_3 and other reactive species. Then, TP3 and TP4 are generated through hydroxylation and demethylation reactions. Secondly, the 10C in molecular structure of IBP is prone to hydroxylation to form TP5. Next, TP5 is deprotonated to form the product TP6, followed by demethylation and decarboxylation under the action of reactive species to generate TP7 and TP8 in turn. Li et al. [122] proposed the possible degradation pathway of IBP in DBD plasma based on DFT analysis. DFT was used to describe the molecular properties of IBP, which can reveal the potential of IBP in specific degradation reactions, as well as the optimal location of electrophilic or nucleophilic reactions in the molecule. The results showed that the 2 and 5 C positions of IBP are vulnerable to attacks by electrophiles and nucleophiles. Therefore, the third degradation pathway of IBP begins with the attack of the 2 or 5 C atom by the •OH. TP9 and P10 are generated by the substitution of •OH, and subsequent hydroxylation, demethylation, and deprotonation reactions may also occur to generate a series of intermediate products. Finally, the benzene ring undergoes a ring-opening process to form formic, acetic, and oxalic acids. These organic acid molecules continued to mineralize into CO_2 and H_2O.

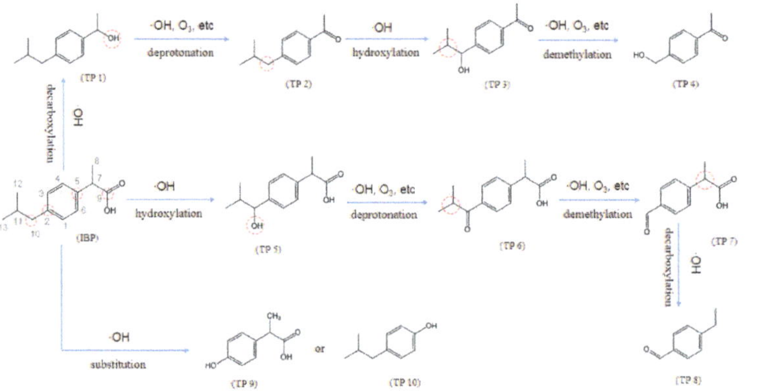

Figure 4. The proposed IBP degradation pathway.

3.2.3. Selected POPs: PCB77

The dechlorination of PCBs has always been the most important step in their degradation process. Huang et al. [89] applied DBD non-thermal plasma for the degradation of PCB77, and its degradation process and common reaction sites are shown in Figure 5. Firstly, •OH causes the dechlorination reaction of PCB77 and generates a series of the dechlorination product TP1. Due to the different reactivity of different sites on the benzene ring, the selectivity of the dechlorination reaction usually depends on the position of the chlorine atom, following the order of para > meta > ortho [128]. Secondly, •OH can lead to a benzene cycloaddition reaction, forming multiple C-OH when all C-Cl break. Since the O atom in the phenolic hydroxyl group can undergo p-π conjugation with the benzene ring, the p electron cloud is transferred to the benzene ring, which increases the electron cloud density on the benzene ring and is easy to attack by electrophiles [129,130]. Numerous sites on the benzene ring are attacked by various reactive species in the plasma, and then partially cleaved to form the intermediate product TP2. The aromatic ring structure in TP2 may be attacked to form unstable products through electrophilic addition. Finally, the above intermediate products are degraded into smaller organic molecules, CO_2 or inorganic salts.

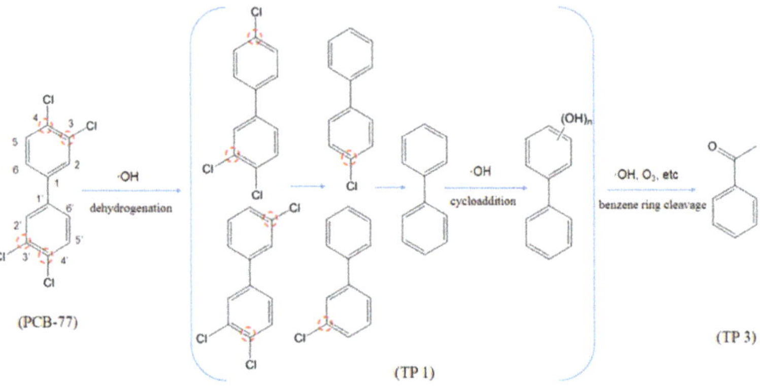

Figure 5. The proposed PCB77 degradation pathway.

4. Non-Thermal Plasma and Other Technologies Cooperate to Degrade Emerging Organic Pollutants in Environment

Because most single methods have some defects, it has been fully proved in various research fields to make up for the shortcomings of a single system by using it in combination with other methods [131,132]. Non-thermal plasma is a kind of AOPs which integrates many factors, such as reactive species, high-energy electrons, ultraviolet radiation, and so on, which shows good characteristics in the process of dealing with EOPs. However, non-thermal plasma also has some disadvantages, such as low utilization efficiency of active components and ultraviolet radiation, low energy utilization efficiency, and so on [72,133]. In view of these limiting factors, the synergistic plasma technology of oxidants and catalysts will become one of development trends in the future. Another barrier to plasma is the low mass transfer efficiency of the resulting reactive species from the plasma to the phase where pollutants are present. Therefore, when designing a plasma reactor, it is necessary to consider not only the generation of these reactive species, but also their efficient transfer to the target pollutants. Recently, microbubbles (MBs) have been considered as an effective method to improve the efficiency of plasma reactors. However, only few studies have been carried out, and its mechanism is still unclear. Therefore, it is necessary to study the effects of MBs on improving gas-liquid mass transfer in water treatment and the development of other means to improve the mass transfer efficiency of reactive species from plasma to medium is also the focus of plasma research in the future.

4.1. Oxidant

In recent years, the AOPs based on sulfate radical ($\bullet SO_4^-$) has attracted much attention because of its high oxidation activity and good adaptability to various EOPs. The redox potential of $\bullet SO_4^-$ (2.5–3.1 V) is higher than that of $\bullet OH$ (1.9–2.7 V), and the half-life of $\bullet SO_4^-$ (30–40 μs) is much longer than that of $\bullet OH$ (10^{-3} μs), which is beneficial to improve the degradation efficiency of pollutants [134]. Ultraviolet radiation, local high temperatures, and hydrated electrons produced by non-thermal plasma can activate persulfate (PS) to form$\bullet SO_4^-$, which helps improve the energy efficiency of the plasma system (Equations (17) and (18)) [135,136]. Tang et al. [65] studied the degradation of TC in water by gas surface discharge plasma-activated PS. With the increase of the PS dosage, the removal efficiency of TC in DBD plasma was also significantly improved. The calculated synergistic factor was 1.856, indicating that the addition of PS has an obvious synergistic effect. Wu et al. [133] found that the addition of peroxymonosulfate (PMS) into the DBD plasma can increase the degradation efficiency of benzotriazole (BTA) by 47%. In addition, energy production increased by 84%. The improvement of BTA degradation efficiency may be attributed to the activation of PMS by plasma, which increases the formation of reactive species and $\bullet SO_4^-$. In addition, some studies have shown that the introduction of PMS can realize the secondary utilization of O_3 (Equations (19)–(23)) [135].

$$HSO_5^- \overset{plasma}{\rightarrow} \bullet SO_4^- + \bullet OH \tag{17}$$

$$S_2O_8^{2-} \overset{plasma}{\rightarrow} 2\bullet SO_4^- \tag{18}$$

$$SO_5^{2-} + O_3 \rightarrow {}^-O_3SO_5^- \tag{19}$$

$$^-O_3SO_5^- \rightarrow \bullet SO_5^- + \bullet O_3^- \tag{20}$$

$$\bullet SO_5^- + O_3 \rightarrow \bullet SO_4^- + O_2 \tag{21}$$

$$\bullet O_3^- \rightarrow \bullet O^- + O_2 \tag{22}$$

$$\bullet O^- + H_2O \rightarrow \bullet OH + OH^- \tag{23}$$

In addition, oxidants, such as percarbonate (SPC), ferrate, and H_2O_2, were also used in non-thermal plasma. Tang et al. [137] explored the synergistic effect of SPC and plasma, and the results showed that the addition of SPC was beneficial to the production of H_2O_2

and the decomposition of O_3. When SPC was 52.0 mol/L and voltage was 4.8 kV, the removal efficiency of TC could reach 94.3% at 20 mg/L. Xu et al. [138] found that low concentration (0.1–1.0 mmol/L) of H_2O_2 could promote the degradation of norfloxacin (NOR) by plasma, while high concentration (1.0–2.0 mmol/L) of H_2O_2 could inhibit the degradation of NOR. Sang et al. [139] compared the effect of PMS, SPC, and ferrate on the degradation of Orange G (OG) by DBD plasma, and found that the degradation efficiency of OG by ferrate plasma was the best. The possible mechanism of degradation of EOPs by non-thermal plasma combined with different oxidants is shown in Figure 6.

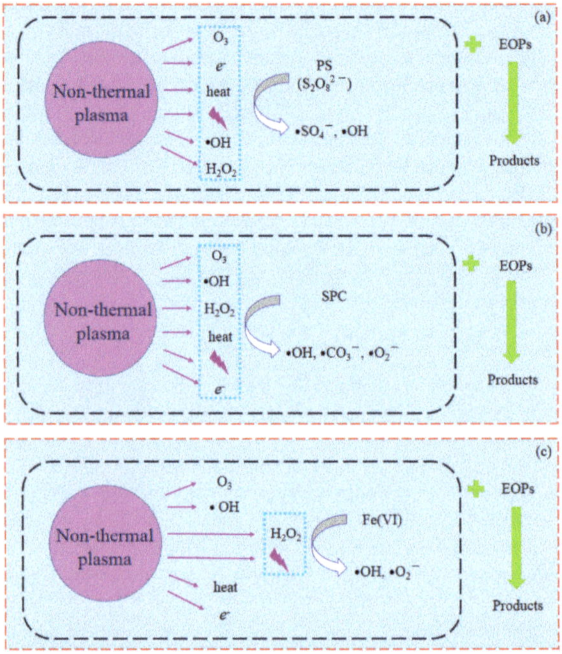

Figure 6. Possible degradation mechanisms of EOPs combined with different oxidants: (**a**) PS, (**b**) SPC, (**c**) Fe (VI).

4.2. Catalyst

4.2.1. Homogeneous Catalyst

Fe^{2+} is the most commonly used homogeneous catalyst in combination with non-thermal plasma. The addition of Fe^{2+} can form a Fenton system with the non-thermal plasma and produce more •OH to degrade EOPs (Equations (24) and (25)) [140,141]. Hao et al. [142] introduced iron ion (Fe^{2+}/Fe^{3+}) into the pulsed discharge plasma system. The experimental results showed that the addition of iron ion (Fe^{2+}/Fe^{3+}) could greatly improve the removal efficiency of 4-chlorophenol. At the same time, they found that the promoting effect of Fe^{2+} was greater than that of Fe^{3+}. Xu et al. [138] found that a low concentration of Fe^{2+} can promote the degradation of NOR by DBD plasma, while a high concentration can inhibit the degradation of NOR, which may be due to the reaction of excessive Fe^{2+} with •OH to form Fe^{3+} and OH^- (Equation (26)). At the same time, the Fe^{3+} can also react with H_2O_2 in aqueous solution (Equation (25)), thus reducing the concentration of H_2O_2.

$$H_2O_2 + Fe^{2+} \rightarrow \bullet OH + OH^- + Fe^{3+} \quad (24)$$

$$H_2O_2 + Fe^{3+} \rightarrow H^+ + HO_2\bullet + Fe^{2+} \quad (25)$$

$$Fe^{2+} + \bullet OH \rightarrow OH^- + Fe^{3+} \quad (26)$$

In addition, the oxidizability of O_3 produced by non-thermal plasma is lower than that of •OH, while the introduction of Fe^{2+} can react with O_3 to form •OH (Equations (27) and (28)), thus improving the •OH content and energy efficiency of the non-thermal plasma.

$$Fe^{2+} + O_3 \rightarrow O_2 + FeO^{2+} \quad (27)$$

$$H_2O + FeO^{2+} \rightarrow Fe^{3+} + \bullet OH + OH^- \quad (28)$$

4.2.2. Heterogeneous Catalyst

Non-thermal plasma co-heterogeneous catalytic oxidation technology refers to the addition of solid catalyst (generally used in powder form) in the discharge region of the plasma. Catalysts react with ultraviolet radiation or reactive species such as ozone and hydrogen peroxide generated in plasma discharge process, and then trigger a series of chain reactions to promote the generation of active free radicals and degradation of organic matter. Furthermore, the combined adsorption and catalytic action of catalysts with a large specific surface area can contribute to the removal of pollutants. Compared with other advanced oxidation technologies, plasma discharge has a variety of physical and chemical effects, such as high electric field, shock wave, etc., which can clean the surface of accelerators and contribute to the regeneration of catalysts [143]. At present, the solid catalysts used for plasma catalytic oxidation mainly include carbon catalysts represented by activated carbon, photocatalysts represented by TiO_2, other metal catalysts, metal or metal oxide catalysts supported on supports, etc. The possible mechanism of degradation of EOPs by non-thermal plasma combined with heterogeneous catalyst is shown in Figure 7.

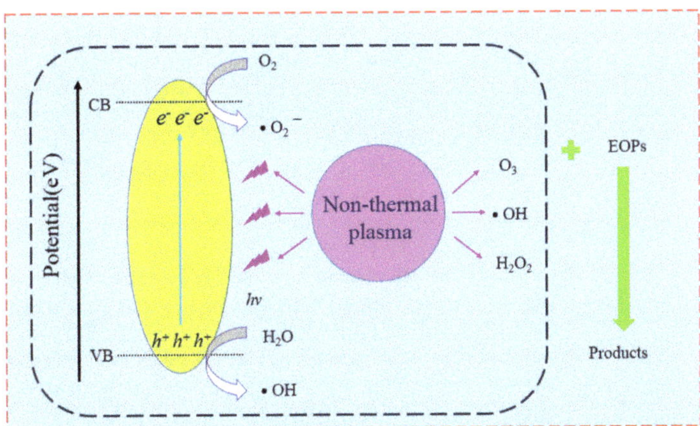

Figure 7. Possible degradation mechanisms of EOPs combined with heterogeneous catalyst.

Different from the homogeneous catalyst, the heterogeneous catalyst is easy to recover and separate; so, there are many studies on the coupling of the heterogeneous catalyst and plasma to degrade EOPs. Guo et al. [69] coupled pulsed discharge plasma (PDP) with Fe_3O_4 to promote the degradation of CAP. In PDP system, Fe_3O_4 not only catalyzed H_2O_2 to form a Fenton reaction, but also catalyzed O_3, which promotes the formation of •OH. When the addition amount of Fe_3O_4 is 0.26 g/L, under the conditions of higher peak voltage, lower initial solution concentration, and lower initial pH value, CAP is beneficial to decomposition and has the best catalytic performance. Cheng et al. [144] used α-MnO_2, β-MnO_2, and γ-MnO_2 to degrade CIP wastewater by DBD plasma-catalytic combined process. The results showed that the combination of DBD plasma and α-MnO_2 has the highest degradation efficiency of CIP, and the degradation efficiency could reach 93.1%, which was 10.8% and 18.1% higher than that of β-MnO_2 and γ-MnO_2 catalyst in the plasma catalytic system, respectively. The photocatalyst TiO_2 can enhance the degradation of EOPs in the plasma system by utilizing the ultraviolet radiation produced by the plasma (Equations (29) and (30)) [145]. Lee et al. [83] found that the addition of TiO_2 increased the decomposition rate of DMP, but the excess dosage of TiO_2 resulted in ultraviolet radiation blocking and decreased the decomposition rate of DMP. Jogi et al. [146] found that the addition of TiO_2 can increase the amount of O_3 produced by the DBD plasma, which may be one of the reasons for improving the efficiency of pollutant degradation.

$$TiO_2 + h\upsilon \rightarrow TiO_2(e_{cb}^- + h_{vb}^+) \quad (29)$$

$$H_2O + h_{vb}^+ \rightarrow \bullet OH + H^+ \quad (30)$$

A single metal oxide photocatalyst has a high recombination rate for electron-hole pairs generated by light energy and is only sensitive to ultraviolet light, while catalysts such as Fe_3O_4 can use H_2O_2 and O_3 but cannot use ultraviolet light [64,71]. Therefore, it is an ideal choice to prepare composite catalysts from metal oxides and other catalysts [147]. In order to make full use of the ultraviolet light generated by the plasma, Fe_3O_4 was supported on reduced graphene oxide (rGO) [71]. Compared with using Fe_3O_4 alone, rGO-Fe_3O_4 further improved the degradation efficiency and kinetic constant of ofloxacin in the discharge plasma system. After 60 min of treatment, the degradation efficiency and kinetic constant reached 99.9% and 0.108 min^{-1}, respectively.

In addition, composite catalysts composed of metal oxides and metal oxides, metal oxides, and metal elements are also used in non-thermal plasma. Ansari et al. [68] combined a ZnO/α-Fe_2O_3 composite catalyst with DBD plasma to degrade antibiotic amoxicillin (AMX). The results showed that the ZnO/α-Fe_2O_3 composite catalyst increased the degradation efficiency of AMX from 75.0% (sole DBD plasma) to 99.3% under optimal conditions. Wang et al. [148] found that compared with the plasma process, the combination of plasma and Mn/γ-Al_2O_3 catalyst significantly improved the degradation efficiency of tetracycline hydrochloride. Under the discharge power of 1.3 W, the degradation efficiency of tetracycline hydrochloride could reach 99.3%, while the degradation efficiency of the plasma treatment was only 69.7%.

At present, plasma-catalytic systems are mainly focused on water remediation, but there are few reports about soil remediation. However, it has been demonstrated that plasma-catalytic systems improve the efficiency of plasma systems in soil remediation. Wang et al. [99,149] investigated the degradation of p-nitrophenol (PNP) in soil using a pulsed discharge plasma-TiO_2 catalytic system. Compared with the single plasma system, the system showed higher degradation performance of PNP. Increased TiO_2 content promoted PNP degradation to a certain extent, while further increased TiO_2 content had negative effects. At higher TiO_2 content, particles aggregation may reduce the interface area between contaminants and catalyst surface sites, thereby reducing the number of active sites on the catalyst surface and resulting in reduced PNP degradation [150].

In conclusion, the combination of plasma discharge and catalyst reveals the catalytic promoting effect on pollutant removal and improves the efficiency of non-thermal plasma.

4.3. Microbubbles

MBs generally refer to bubbles with equivalent diameters less than 50 μm. Compared with millimeter-sized bubbles, MBs have some special properties that can enhance the discharge effect. Firstly, the gas-liquid mass transfer ability of MBs is strong. Unlike millimeter-sized bubbles, MBs will not rise rapidly from the water to the liquid surface and break. The rising speed of MBs in water is slower than that of millimeter-sized bubbles, and the residence time in water is longer, which greatly prolongs the gas-liquid contact time [151,152]. Figure 8 shows the characteristics of millimeter-sized bubbles and micron-sized bubbles in water. Secondly, MBs can stimulate the generation of free radicals in the process of fragmentation. Adiabatic compression occurs during bubble contraction, which generates local high temperature and high pressure around the bubble. Such conditions stimulate the decomposition of water molecules around the bubble into free radicals [153]. Takahashi et al. [152] demonstrated free-radical generation from the collapse of MBs in the absence of a harsh dynamic stimulus. Electron spin-resonance spectroscopy confirmed free-radical generation by the collapsing MBs. The increase of the surface charges (ζ potentials) of the MBs, which were measured during their collapse, supported the hypothesis that the significant increase in ion concentration around the shrinking gas-water interface provided the mechanism for radical generation.

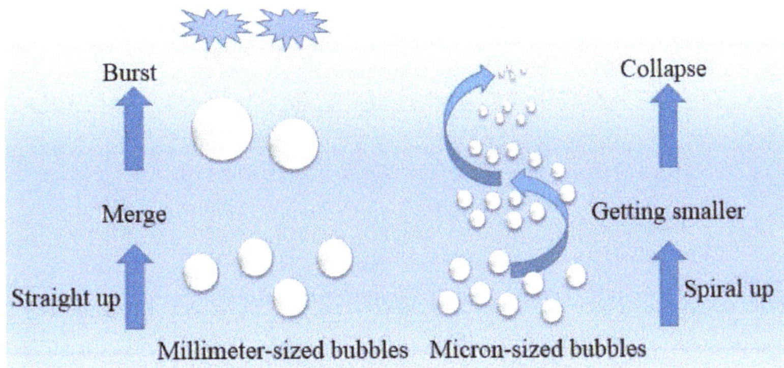

Figure 8. Comparison of millimeter-sized bubbles and micron-sized bubbles in water.

Studies showed that compared with plasma treatment alone, the presence of MBs can significantly improve the treatment efficiency, and preliminary studies speculate that the presence of MBs can improve gas-liquid mass transfer in plasma water treatment [154,155]. MBs were introduced with different carrier gases (air, N_2 and Ar) in the needle-plate pulsed discharge reactor to enhance the interface reaction [156]. Due to the surface ζ-potential, MBs can effectively enrich pollutants, and the large specific surface area also leads to a high gas-liquid interface area, which enhances the plasma reaction with pollutants. At the same time, these unique physical properties can also promote the mass transfer from gas to liquid in the system. Wang et al. [157] evaluated the degradation of ATZ in aqueous solution by DBD/MBs/PS system in order to develop a more efficient and environmentally friendly PS activation method. The observed ATZ removal efficiency (DBD/MBs/PS > DBD/PS > MBs/PS) confirmed the synergistic effect of DBD/MBs/PS. Based on the ATZ removal efficiencies of 64% and 56% in DBD/MBs/PS and DBD/PS systems, the mass transfer contribution rate was calculated as 13% in DBD/MBs/PS system. The synergies of DBD/MB/PMS systems are largely due to the interaction between DBD, MBs, and PMS. DBD plasma produces a large number of reactive species when gas molecules dissociate and form high-energy electrons. By combining DBD plasma and MBs, the plasma can be ignited inside the MBs, where an electron avalanche breakdown occurs and continues until the MBs crashes.

5. Practical Implications of This Study

Among a large number of research works in the past decade, non-thermal plasma technology has been proved to be a promising environmental remediation technology. However, there are still some challenges to be solved and breakthroughs to be made.

(1) One of the most serious challenges of non-thermal plasma technology is how to improve the energy yield of the treatment system and reduce the operation cost. According to the summary of existing research, the energy yields of different research results vary greatly, covering several orders of magnitude. In addition to the different molecular structures of pollutants, the design and experimental conditions of a non-thermal plasma reactor also have a great influence. Due to the particularity of in situ generation of reactive species by non-thermal plasma, the main improvement of

reactor design mainly lies in how to maximize the generation of reactive species and effectively transfer them to the medium where pollutants exist.

(2) In addition, non-thermal plasma technology is still in the laboratory research stage and the volume of the reactor is too small to be put into practical application. Therefore, how to expand the scale of the reactor is a major challenge for future research.

(3) Another key research challenge is that the current research mainly focuses on the degradation efficiency of pollutants and lacks the toxicity test of water and soil environment after non-thermal plasma treatment, which is very important for the practical application of non-thermal plasma technology in the future. Although it is generally shown that non-thermal plasma can reduce the concentration of pollutants, the toxicity of the degradation products and the ultimate impact of reactive species produced by non-thermal plasma that may persist are not clear. This is a problem that needs to be analyzed in more detail in the future.

Some suggestions for future development direction are also put forward.

(1) At present, most research focuses on artificially polluted water and soil, and research on actually polluted water and soil is limited. Therefore, more attention should be paid to the actually polluted water or soil, which may promote the practical application of non-thermal plasma in the future.

(2) It is well known that the traditional sewage treatment process of sewage treatment plants cannot completely remove these EOPs. In most cases, these EOPs are transferred to sludge due to their low solubility and high octanol/water distribution coefficient. Therefore, excess sludge is also an easy enrichment medium for these EOPs. In view of the excellent effect of non-thermal plasma in water and soil environment, it is necessary to extend this technology to the treatment of excess sludge in the future.

(3) The positive effects of catalysts on non-thermal plasma technology have been widely recognized in a large number of studies. However, at present, the recovery of solid catalysts in soil remediation seems to be less mentioned, and the influence of catalyst residues on soil properties and microbial behavior is still in the blank stage.

6. Conclusions

This review provides readers with a comprehensive overview of non-thermal plasma in environmental remediation. Based on the above discussion of this review, some important conclusions can be drawn.

Firstly, the research progress of non-thermal plasma on EOPs (i.e., EDCs, PPCPs, and POPs) in different environmental media is summarized. It is found that many parameters will affect the degradation efficiency of non-thermal plasma, and it is clear from previous studies that non-thermal plasma technology shows the potential for effective degradation and complete mineralization of EOPs, which cannot be successfully removed by conventional treatment methods. For soil remediation, direct ignition of plasma discharge in the interior of soil pores seems to promote the penetration of UV and short-lived reactive species in the soil matrix, thus further promoting effective remediation of the entire porous medium.

Secondly, the generation of reactive species and the degradation mechanism of EOPs in non-thermal plasma system are discussed. Similar to other AOPs, non-thermal plasma also emphasizes that •OH plays a central role in the degradation of pollutants. The degradation pathways of various EOPs are mainly the following reactions: •OH is directly generated in discharge or further decomposed by O_3 and H_2O_2. Then, according to the molecular structure of organic pollutants (e.g., electron density), the oxidative degradation pathway may also include the reaction with O_3.

Finally, some measures to improve the efficiency of non-thermal plasma are comprehensively summarized. In terms of water remediation, the formation of MBs under the water surface seems to have significant advantages, because it is a more effective means to provide reactive species to degrade EOPs in water and the existence of MBs enhances the dissolution and mass transfer of reactive species generated by plasma from gas phase to liquid phase. For soil remediation, the combination of non-thermal plasma and catalyst seems to be a promising technology to break through the bottleneck of single plasma.

In conclusion, non-thermal plasma technology has a good application prospect in environmental remediation.

Author Contributions: All authors contributed extensively to the work presented in this paper. Y.H.: investigation, visualization, writing—original draft; W.S.: writing—review and editing; W.L.: writing—review and editing; W.Z. and C.Z.: resources; D.J.: formal analysis. All authors have read and agreed to the published version of the manuscript.

Funding: This work was supported by the National Natural Science Foundation of China project 51108360.

Data Availability Statement: Not applicable.

Conflicts of Interest: The authors declare no conflict of interest.

References

1. Barbosa Ferreira, M.; Sales Solano, A.M.; Vieira Dos Santos, E.; Martínez-Huitle, C.A.; Ganiyu, S.O. Coupling of Anodic Oxidation and Soil Remediation Processes: A Review. *Materials* **2020**, *13*, 4309. [CrossRef] [PubMed]
2. Liu, J.; Ren, S.; Cao, J.; Tsang, D.C.W.; Beiyuan, J.; Peng, Y.; Fang, F.; She, J.; Yin, M.; Shen, N.; et al. Highly efficient removal of thallium in wastewater by $MnFe_2O_4$-biochar composite. *J. Hazard. Mater.* **2021**, *401*, 123311. [CrossRef] [PubMed]
3. Fu, Q.; Malchi, T.; Carter, L.J.; Li, H.; Gan, J.; Chefetz, B. Pharmaceutical and Personal Care Products: From Wastewater Treatment into Agro-Food Systems. *Environ. Sci. Technol.* **2019**, *53*, 14083–14090. [CrossRef] [PubMed]
4. Cheng, W.; Li, J.; Wu, Y.; Xu, L.; Su, C.; Qian, Y.; Zhu, Y.G.; Chen, H. Behavior of antibiotics and antibiotic resistance genes in eco-agricultural system: A case study. *J. Hazard. Mater.* **2016**, *304*, 18–25. [CrossRef]
5. Field, J.A.; Johnson, C.A.; Rose, J.B. What is "emerging"? *Environ. Sci. Technol.* **2006**, *40*, 7105. [CrossRef]
6. Liu, B.; Zhang, S.G.; Chang, C.C. Emerging pollutants—Part II: Treatment. *Water Environ. Res.* **2020**, *92*, 1603–1617. [CrossRef]
7. Ahmad, H.A.; Ahmad, S.; Cui, Q.; Wang, Z.; Wei, H.; Chen, X.; Ni, S.; Ismail, S.; Awad, H.M.; Tawfik, A. The environmental distribution and removal of emerging pollutants, highlighting the importance of using microbes as a potential degrader: A review. *Sci. Total Environ.* **2022**, *809*, 151926. [CrossRef]
8. Al-Maqdi, K.A.; Elmerhi, N.; Athamneh, K.; Bilal, M.; Alzamly, A.; Ashraf, S.S.; Shah, I. Challenges and Recent Advances in Enzyme-Mediated Wastewater Remediation—A Review. *Nanomaterials* **2021**, *11*, 3124. [CrossRef]
9. Ikehata, K.; Gamal El-Din, M.; Snyder, S.A. Ozonation and Advanced Oxidation Treatment of Emerging Organic Pollutants in Water and Wastewater. *Ozone Sci. Eng.* **2008**, *30*, 21–26. [CrossRef]
10. Cabrerizo, A.; Muir, D.C.G.; De Silva, A.O.; Wang, X.; Lamoureux, S.F.; Lafrenière, M.J. Legacy and Emerging Persistent Organic Pollutants (POPs) in Terrestrial Compartments in the High Arctic: Sorption and Secondary Sources. *Environ. Sci. Technol.* **2018**, *52*, 14187–14197. [CrossRef]
11. Ahmed, M.B.; Zhou, J.L.; Ngo, H.H.; Guo, W.; Thomaidis, N.S.; Xu, J. Progress in the biological and chemical treatment technologies for emerging contaminant removal from wastewater: A critical review. *J. Hazard. Mater.* **2016**, *323*, 274–298. [CrossRef] [PubMed]
12. Han, Y.; Yang, L.; Chen, X.; Cai, Y.; Zhang, X.; Qian, M.; Chen, X.; Zhao, H.; Sheng, M.; Cao, G.; et al. Removal of veterinary antibiotics from swine wastewater using anaerobic and aerobic biodegradation. *Sci. Total Environ.* **2020**, *709*, 136094. [CrossRef] [PubMed]
13. Zhou, L.; Liu, Z.; Guan, Z.; Tian, B.; Liu, Y. 0D/2D plasmonic Cu_2-xS/g-C_3N_4 nanosheets harnessing UV-vis-NIR broad spectrum for photocatalytic degradation of antibiotic pollutant. *Appl. Catal. B Environ.* **2019**, *263*, 118326. [CrossRef]
14. Sun, S.; Yao, H.; Li, X.; De Ng, S.; Zhang, W. Enhanced Degradation of Sulfamethoxazole (SMX) in Toilet Wastewater by Photo-Fenton Reactive Membrane Filtration. *Nanomaterials* **2020**, *10*, 180. [CrossRef] [PubMed]
15. Liu, X.; Huang, F.; He, Y.; Yu, Y.; Lv, Y.; Xu, Y.; Zhang, Y. Oxytetracycline degradation and toxicity evolution by catalytic oxidation process over sludge derived carbon. *J. Environ. Chem. Eng.* **2019**, *7*, 102889. [CrossRef]
16. Ma, J.; Jiang, Z.; Cao, J.; Yu, F. Enhanced adsorption for the removal of antibiotics by carbon nanotubes/graphene oxide/sodium alginate triple-network nanocomposite hydrogels in aqueous solutions. *Chemosphere* **2020**, *242*, 125188. [CrossRef]
17. Hernández-Abreu, A.B.; Álvarez-Torrellas, S.; Rocha, R.P.; Pereira, M.F.R.; Águeda, V.I.; Delgado, J.A.; Larriba, M.; García, J.; Figueiredo, J.L. Effective adsorption of the endocrine disruptor compound bisphenol a from water on surface-modified carbon materials. *Appl. Surf. Sci.* **2021**, *552*, 149513. [CrossRef]
18. Weng, X.; Ji, Y.; Ma, R.; Zhao, F.; An, Q.; Gao, C. Superhydrophilic and antibacterial zwitterionic polyamide nanofiltration membranes for antibiotics separation. *J. Membr. Sci.* **2016**, *510*, 122–130. [CrossRef]
19. Hao, W.; Niu, X.; Jia, Y.; Wang, C.; Lu, M. Retentions of bisphenol A and norfloxacin by three different ultrafiltration membranes in regard to drinking water treatment. *Chem. Eng. J.* **2016**, *294*, 410–416.

20. Zhang, J.; Giorno, L.; Drioli, E. Study of a hybrid process combining PACs and membrane operations for antibiotic wastewater treatment. *Desalination* **2006**, *194*, 101–107. [CrossRef]
21. Yuan, S.; Meng, T.; Lu, X. Microwave remediation of soil contaminated with hexachlorobenzene. *J. Hazard. Mater.* **2006**, *137*, 878–885. [CrossRef] [PubMed]
22. Qin, C.Y.; Zhao, Y.S.; Zheng, W.; Li, Y.S. Study on influencing factors on removal of chlorobenzene from unsaturated zone by soil vapor extraction. *J. Hazard. Mater.* **2010**, *176*, 294–299. [CrossRef] [PubMed]
23. Villa, R.D.; Trovó, A.G.; Nogueira, R.F.P. Soil remediation using a coupled process: Soil washing with surfactant followed by photo-Fenton oxidation. *J. Hazard. Mater.* **2010**, *174*, 770–775. [CrossRef] [PubMed]
24. Yeung, A.T.; Gu, Y. A review on techniques to enhance electrochemical remediation of contaminated soils. *J. Hazard. Mater.* **2011**, *195*, 11–29. [CrossRef]
25. Liao, X.; Zhao, D.; Yan, X.; Huling, S.G. Identification of persulfate oxidation products of polycyclic aromatic hydrocarbon during remediation of contaminated soil. *J. Hazard. Mater.* **2014**, *276*, 26–34. [CrossRef]
26. Kulkarni, M.; Chaudhari, A. Microbial remediation of nitro-aromatic compounds: An overview. *J. Environ. Manag.* **2007**, *85*, 496–512. [CrossRef]
27. Cao, X.; Cui, X.; Xie, M.; Zhao, R.; Xu, L.; Ni, S.; Cui, Z. Amendments and bioaugmentation enhanced phytoremediation and micro-ecology for PAHs and heavy metals co-contaminated soils. *J. Hazard. Mater.* **2022**, *426*, 128096. [CrossRef]
28. Gerrity, D.; Stanford, B.D.; Trenholm, R.A.; Snyder, S.A. An evaluation of a pilot-scale nonthermal plasma advanced oxidation process for trace organic compound degradation. *Water Res.* **2010**, *44*, 493–504. [CrossRef]
29. Bai, Y.; Chen, J.; Yang, Y.; Guo, L.; Zhang, C. Degradation of organophosphorus pesticide induced by oxygen plasma: Effects of operating parameters and reaction mechanisms. *Chemosphere* **2010**, *81*, 408–414. [CrossRef]
30. Sarangapani, C.; Danaher, M.; Tiwari, B.; Lu, P.; Bourke, P.; Cullen, P.J. Efficacy and mechanistic insights into endocrine disruptor degradation using atmospheric air plasma. *Chem. Eng. J.* **2017**, *326*, 700–714. [CrossRef]
31. Kim, K.S.; Kam, S.K.; Mok, Y.S. Elucidation of the degradation pathways of sulfonamide antibiotics in a dielectric barrier discharge plasma system. *Chem. Eng. J.* **2015**, *271*, 31–42. [CrossRef]
32. Sohrabi, A.; Haghighat, G.; Shaibani, P.M.; Van Neste, C.W.; Naicker, S.; Sadrzadeh, M.; Thundat, T. Elimination of pharmaceutical contaminants fluoxetine and propranolol by an advanced plasma water treatment. *Desalination Water Treat.* **2018**, *113*, 346–353. [CrossRef]
33. Xin, L.; Sun, Y.; Feng, J.; Wang, J.; He, D. Degradation of triclosan in aqueous solution by dielectric barrier discharge plasma combined with activated carbon fibers. *Chemosphere* **2016**, *144*, 855–863. [CrossRef] [PubMed]
34. Lei, W.; Jiang, X.; Liu, Y. Degradation of bisphenol A and formation of hydrogen peroxide induced by glow discharge plasma in aqueous solutions. *J. Hazard. Mater.* **2008**, *154*, 1106–1114.
35. Tichonovas, M.; Krugly, E.; Racys, V.; Hippler, R.; Kauneliene, V.; Stasiulaitiene, I.; Martuzevicius, D. Degradation of various textile dyes as wastewater pollutants under dielectric barrier discharge plasma treatment. *Chem. Eng. J.* **2013**, *229*, 9–19. [CrossRef]
36. Lu, N.; Wang, C.; Lou, C. Remediation of PAH-contaminated soil by pulsed corona discharge plasma. *J. Soil. Sediment.* **2017**, *17*, 97–105. [CrossRef]
37. Monica, M.; Nicolae, B.M.; Vasile, I.P. Degradation of pharmaceutical compounds in water by non-thermal plasma treatment. *Water Res.* **2015**, *81*, 124–136.
38. Russo, M.; Iervolino, G.; Vaiano, V.; Palma, V. Non-Thermal Plasma Coupled with Catalyst for the Degradation of Water Pollutants: A Review. *Catalysts* **2020**, *10*, 1438. [CrossRef]
39. Hao, Z.; Danyan, M.; Rongliang, Q.; Yetao, T.; Changming, D. Non-thermal plasma technology for organic contaminated soil remediation: A review. *Chem. Eng. J.* **2017**, *313*, 157–170.
40. Guo, H.; Wang, Y.; Liao, L.; Li, Z.; Pan, S.; Puyang, C.; Su, Y.; Zhang, Y.; Wang, T.; Ren, J.; et al. Review on remediation of organic-contaminated soil by discharge plasma: Plasma types, impact factors, plasma-assisted catalysis, and indexes for remediation. *Chem. Eng. J.* **2022**, *436*, 135239. [CrossRef]
41. Palma, D.; Richard, C.; Minella, M. State of the art and perspectives about non-thermal plasma applications for the removal of PFAS in water. *Chem. Eng. J. Adv.* **2022**, *10*, 100253.
42. Mohsen, A.; Mehdi, S.; Mohammad, H.E.; Amir, H.M.; Mohammad, H.S.; Hossein, F. Dielectric barrier discharge plasma with photocatalysts as a hybrid emerging technology for degradation of synthetic organic compounds in aqueous environments: A critical review. *Chemosphere* **2021**, *263*, 128065.
43. Hichem, Z.; Phuong, N.; Lotfi, K.; Abdeltif, A.; Aymen, A.A. Review on discharge Plasma for water treatment: Mechanism, reactor geometries, active species and combined processes. *J. Water Process Eng.* **2020**, *38*, 101664.
44. Mouele, E.S.M.; Tijani, J.O.; Badmus, K.O.; Pereao, O.; Babajide, O.; Fatoba, O.O.; Zhang, C.; Shao, T.; Sosnin, E.; Tarasenko, V.; et al. A critical review on ozone and co-species, generation and reaction mechanisms in plasma induced by dielectric barrier discharge technologies for wastewater remediation. *J. Environ. Chem. Eng.* **2021**, *9*, 105758. [CrossRef]
45. Pramila, M.; Evanjalin, M.V.; Moses, J.A.; Anandharamakrishnan, C. Water decontamination using non-thermal plasma: Concepts, applications, and prospects. *J. Environ. Chem. Eng.* **2020**, *8*, 104377.
46. Gmurek, M.; Olak-Kucharczyk, M.; Ledakowicz, S. Photochemical decomposition of endocrine disrupting compounds—A review. *Chem. Eng. J.* **2017**, *310*, 437–456. [CrossRef]

47. Pal, A.; He, Y.; Jekel, M.; Reinhard, M.; Gin, Y.H. Emerging contaminants of public health significance as water quality indicator compounds in the urban water cycle. *Environ. Int.* **2014**, *71*, 46–62. [PubMed]
48. Stülten, D.; Lamshöft, M.; Zühlke, S.; Spiteller, M. Isolation and characterization of a new human urinary metabolite of diclofenac applying LC–NMR–MS and high-resolution mass analyses. *J. Pharmaceut. Biomed.* **2008**, *47*, 371–376. [CrossRef]
49. Narayanan, M.; El-sheekh, M.; Ma, Y.; Pugazhendhi, A.; Natarajan, D.; Kandasamy, G.; Raja, R.; Saravana Kumar, R.M.; Kumarasamy, S.; Sathiyan, G.; et al. Current status of microbes involved in the degradation of pharmaceutical and personal care products (PPCPs) pollutants in the aquatic ecosystem. *Environ. Pollut.* **2022**, *300*, 118922.
50. Guo, M.; Feng, Y.; Li, X.; Yan, G.; Wang, X.; Li, X.; Zhang, S.; Yu, Y. Enhanced degradation of pharmaceuticals and personal care products (PPCPs) by three-dimensional electrocatalysis coupled biological aerated filter. *J. Environ. Chem. Eng.* **2021**, *9*, 106035. [CrossRef]
51. Bing, L.I.; Zhang, R. Biodegradation and adsorption of antibiotics in the activated sludge process. *Environ. Sci. Technol.* **2010**, *44*, 3468–3473.
52. Zhi, D.; Lin, Y.; Jiang, L.; Zhou, Y.; Huang, A.; Yang, J.; Luo, L. Remediation of persistent organic pollutants in aqueous systems by electrochemical activation of persulfates: A review. *J. Environ. Manag.* **2020**, *260*, 110125. [CrossRef] [PubMed]
53. Yu, J.; Zhu, Z.; Zhang, H.; Shen, X.; Qiu, Y.; Yin, D.; Wang, S. Persistent free radicals on N-doped hydrochar for degradation of endocrine disrupting compounds. *Chem. Eng. J.* **2020**, *398*, 125538. [CrossRef]
54. Maroga Mboula, V.; Héquet, V.; Andrès, Y.; Pastrana-Martínez, L.M.; Doña-Rodríguez, J.M.; Silva, A.M.T.; Falaras, P. Photocatalytic degradation of endocrine disruptor compounds under simulated solar light. *Water Res.* **2013**, *47*, 3997–4005. [CrossRef]
55. Choudhary, M.; Sarkar, P.; Kumar Sharma, S.; Kajla, A.; Neogi, S. Quantification of reactive species generated in pulsed electrical discharge plasma reactor and its application for 17α-ethinylestradiol degradation in different water matrices. *Sep. Purif. Technol.* **2021**, *275*, 119173. [CrossRef]
56. Guo, H.; Jiang, N.; Li, J.; Wu, Y. Synergistic degradation of bisphenol A by pulsed discharge plasma with granular activated carbon: Effect of operating parameters, synergistic mechanism and possible degradation pathway. *Vacuum* **2018**, *156*, 402–410. [CrossRef]
57. Mao, D.; Yan, X.; Wang, H.; Shen, Z.; Yi, C. Catalysis of rGO-WO3 nanocomposite for aqueous bisphenol A degradation in dielectric barrier discharge plasma oxidation process. *Chemosphere* **2020**, *262*, 128073. [CrossRef]
58. Wang, H.; Shen, Z.; Yan, X.; Guo, H.; Mao, D.; Yi, C. Dielectric barrier discharge plasma coupled with WO_3 for bisphenol A degradation. *Chemosphere* **2021**, *274*, 129722. [CrossRef]
59. Yang, J.; Zeng, D.; Hassan, M.; Ma, Z.; Dong, L.; Xie, Y.; He, Y. Efficient degradation of Bisphenol A by dielectric barrier discharge non-thermal plasma: Performance, degradation pathways and mechanistic consideration. *Chemosphere* **2022**, *286*, 131627. [CrossRef]
60. Wang, J.; Li, L.; Cao, H.; Yang, C.; Guo, Z.; Shi, Y.; Li, W.; Zhao, H.; Sun, J.; Xie, Y. Degradation of phenolic compounds by dielectric barrier plasma: Process optimization and influence of phenol substituents. *Chem. Eng. J.* **2020**, *385*, 123732. [CrossRef]
61. Sarangapani, C.; Misra, N.N.; Milosavljevic, V.; Bourke, P.; Cullen, P.J. Pesticide degradation in water using atmospheric air cold plasma. *J. Water Process Eng.* **2016**, *9*, 225–232. [CrossRef]
62. Wang, T.; Jia, H.; Guo, X.; Xia, T.; Qu, G.; Sun, Q.; Yin, X. Evaluation of the potential of dimethyl phthalate degradation in aqueous using sodium percarbonate activated by discharge plasma. *Chem. Eng. J.* **2018**, *346*, 65–76. [CrossRef]
63. Su, Y.; Liu, S.; Zhao, C.; Yang, X.; Liu, D. Needle electrode design of pulsed high voltage discharge reactor for performance enhancement of 4-chlorophenol degradation in highly conductive solution. *Chemosphere* **2021**, *266*, 129203. [PubMed]
64. Guo, H.; Jiang, N.; Wang, H.; Lu, N.; Shang, K.; Li, J.; Wu, Y. Pulsed discharge plasma assisted with graphene-WO_3 nanocomposites for synergistic degradation of antibiotic enrofloxacin in water. *Chem. Eng. J.* **2019**, *372*, 226–240.
65. Tang, S.; Yuan, D.; Rao, Y.; Li, N.; Qi, J.; Cheng, T.; Sun, Z.; Gu, J.; Huang, H. Persulfate activation in gas phase surface discharge plasma for synergetic removal of antibiotic in water. *Chem. Eng. J.* **2018**, *337*, 446–454. [CrossRef]
66. Hu, X.; Wang, B. Removal of pefloxacin from wastewater by dielectric barrier discharge plasma: Mechanism and degradation pathways. *J. Environ. Chem. Eng.* **2021**, *9*, 105720. [CrossRef]
67. Shoufeng, T.A.N.G.; Xue, L.I.; Zhang, C.; Yang, L.I.U.; Zhang, W.; Deling, Y.U.A.N. Strengthening decomposition of oxytetracycline in DBD plasma coupling with Fe-Mn oxide-loaded granular activated carbon. *Plasma Sci. Technol.* **2019**, *21*, 25504.
68. Ansari, M.; Hossein Mahvi, A.; Hossein Salmani, M.; Sharifian, M.; Fallahzadeh, H.; Hassan Ehrampoush, M. Dielectric barrier discharge plasma combined with nano catalyst for aqueous amoxicillin removal: Performance modeling, kinetics and optimization study, energy yield, degradation pathway, and toxicity. *Sep. Purif. Technol.* **2020**, *251*, 117270. [CrossRef]
69. Guo, H.; Li, Z.; Zhang, Y.; Jiang, N.; Wang, H.; Li, J. Degradation of chloramphenicol by pulsed discharge plasma with heterogeneous Fenton process using Fe_3O_4 nanocomposites. *Sep. Purif. Technol.* **2020**, *253*, 117540. [CrossRef]
70. Lou, J.; Wei, Y.; Zhang, M.; Meng, Q.; Jia, M. Removal of tetracycline hydrochloride in aqueous by coupling dielectric barrier discharge plasma with biochar. *Sep. Purif. Technol.* **2021**, *266*, 118515. [CrossRef]
71. Guo, H.; Li, Z.; Xie, Z.; Song, J.; Wang, H. Accelerated Fenton reaction for antibiotic ofloxacin degradation in discharge plasma system based on graphene-Fe_3O_4 nanocomposites. *Vacuum* **2020**, *185*, 110022. [CrossRef]
72. Guo, H.; Wang, Y.; Yao, X.; Zhang, Y.; Wang, H. A comprehensive insight into plasma-catalytic removal of antibiotic oxytetracycline based on graphene-TiO_2-Fe_3O_4 nanocomposites. *Chem. Eng. J.* **2021**, *425*, 130614. [CrossRef]

73. Zhang, G.; Sun, Y.; Zhang, C.; Yu, Z. Decomposition of acetaminophen in water by a gas phase dielectric barrier discharge plasma combined with TiO2-rGO nanocomposite: Mechanism and degradation pathway. *J. Hazard. Mater.* **2017**, *323*, 719–729. [CrossRef]
74. Hama Aziz, K.H.; Miessner, H.; Mueller, S.; Kalass, D.; Moeller, D.; Khorshid, I.; Rashid, M.A.M. Degradation of pharmaceutical diclofenac and ibuprofen in aqueous solution, a direct comparison of ozonation, photocatalysis, and non-thermal plasma. *Chem. Eng. J.* **2017**, *313*, 1033–1041. [CrossRef]
75. Magureanu, M.; Mandache, N.B.; Bradu, C.; Parvulescu, V.I. High efficiency plasma treatment of water contaminated with organic compounds. Study of the degradation of ibuprofen. *Plasma Process. Polym.* **2018**, *15*, 1700201. [CrossRef]
76. Zheng, B.; Li, C.; Zhang, J.; Zheng, Z. Dielectric barrier discharge induced the degradation of the emerging contaminant ibuprofen in aqueous solutions. *Desalin. Water Treat.* **2014**, *52*, 4469–4475. [CrossRef]
77. Banaschik, R.; Jablonowski, H.; Bednarski, P.J.; Kolb, J.F. Degradation and intermediates of diclofenac as instructive example for decomposition of recalcitrant pharmaceuticals by hydroxyl radicals generated with pulsed corona plasma in water. *J. Hazard. Mater.* **2018**, *342*, 651–660. [CrossRef]
78. Magureanu, M.; Piroi, D.; Mandache, N.B.; David, V.; Medvedovici, A.; Bradu, C.; Parvulescu, V.I. Degradation of antibiotics in water by non-thermal plasma treatment. *Water Res.* **2011**, *45*, 3407–3416. [CrossRef]
79. Dobrin, D.; Bradu, C.; Magureanu, M.; Mandache, N.B.; Parvulescu, V.I. Degradation of diclofenac in water using a pulsed corona discharge. *Chem. Eng. J.* **2013**, *234*, 389–396. [CrossRef]
80. Magureanu, M.; Dobrin, D.; Mandache, N.B.; Bradu, C.; Medvedovici, A.; Parvulescu, V.I. The Mechanism of Plasma Destruction of Enalapril and Related Metabolites in Water. *Plasma Process. Polym.* **2013**, *10*, 459–468. [CrossRef]
81. Korichi, N.; Aubry, O.; Rabat, H.; Cagnon, B.; Hong, D. Paracetamol Degradation by Catalyst Enhanced Non-Thermal Plasma Process for a Drastic Increase in the Mineralization Rate. *Catalysts* **2020**, *10*, 959. [CrossRef]
82. Krishna, S.; Ceriani, E.; Marotta, E.; Giardina, A.; Špatenka, P.; Paradisi, C. Products and mechanism of verapamil removal in water by air non-thermal plasma treatment. *Chem. Eng. J.* **2016**, *292*, 35–41. [CrossRef]
83. Lee, H.; Park, Y.; Kim, J.; Park, Y.; Jung, S. Degradation of dimethyl phthalate using a liquid phase plasma process with TiO_2 photocatalysts. *Environ. Res.* **2019**, *169*, 256–260. [CrossRef] [PubMed]
84. Bratty, M.A.; Al-Rajab, A.J.; Rehman, Z.U.; Sharma, M.; Alhazmi, H.A.; Najmi, A.; Muzafar, H.; Javed, S.A. Fast and efficient removal of caffeine from water using dielectric barrier discharge. *Appl. Water Sci.* **2021**, *11*, 1–10. [CrossRef]
85. Wang, J.; Sun, Y.; Jiang, H.; Feng, J. Removal of caffeine from water by combining dielectric barrier discharge (DBD) plasma with goethite. *J. Saudi Chem. Soc.* **2017**, *21*, 545–557. [CrossRef]
86. Wang, L.; Sun, L.; Yu, Z.; Hou, Y.; Peng, Z.; Yang, F.; Chen, Y.; Huang, J. Synergetic decomposition performance and mechanism of perfluorooctanoic acid in dielectric barrier discharge plasma system with Fe_3O_4@SiO_2-BiOBr magnetic photocatalyst. *Mol. Catal.* **2017**, *441*, 179–189. [CrossRef]
87. Corella Puertas, E.; Peyot, M.; Pineda, M.; Volk, K.; Coulombe, S.; Yargeau, V. Degradation of diatrizoate in a pin-to-liquid plasma reactor, its transformation products and their residual toxicity. *Sci. Total Environ.* **2021**, *782*, 146895. [CrossRef]
88. Magureanu, M.; Piroi, D.; Mandache, N.B.; David, V.; Medvedovici, A.; Parvulescu, V.I. Degradation of pharmaceutical compound pentoxifylline in water by non-thermal plasma treatment. *Water Res.* **2010**, *44*, 3445–3453. [CrossRef]
89. Huang, Q.; Fang, C. Degradation of 3,3′,4,4′-tetrachlorobiphenyl (PCB77) by dielectric barrier discharge (DBD) non-thermal plasma: Degradation mechanism and toxicity evaluation. *Sci. Total Environ.* **2020**, *739*, 139926. [CrossRef]
90. Sahni, M.; Finney, W.C.; Locke, B.R. Degradation of aqueous phase polychlorinated biphenyls (PCB) using pulsed corona discharges. *J. Adv. Oxid. Technol.* **2005**, *8*, 105–111. [CrossRef]
91. Aggelopoulos, C.A.; Meropoulis, S.; Hatzisymeon, M.; Lada, Z.G.; Rassias, G. Degradation of antibiotic enrofloxacin in water by gas-liquid nsp-DBD plasma: Parametric analysis, effect of H_2O_2 and CaO_2 additives and exploration of degradation mechanisms. *Chem. Eng. J.* **2020**, *398*, 125622. [CrossRef]
92. Ruan, J.J.; Li, W.; Shi, Y.; Nie, Y.; Wang, X.; Tan, T.E. Decomposition of simulated odors in municipal wastewater treatment plants by a wire-plate pulse corona reactor. *Chemosphere* **2005**, *59*, 327–333. [CrossRef] [PubMed]
93. Jiang, N.; Lu, N.; Shang, K.; Li, J.; Wu, Y. Effects of electrode geometry on the performance of dielectric barrier/packed-bed discharge plasmas in benzene degradation. *J. Hazard. Mater.* **2013**, *262*, 387–393. [CrossRef] [PubMed]
94. Joshi, R.P.; Thagard, S.M. Streamer-Like Electrical Discharges in Water: Part II. Environmental Applications. *Plasma Chem. Plasma Process.* **2013**, *33*, 17–49. [CrossRef]
95. Bing, S.; Sato, M.; Clements, J.S. Optical study of active species produced by a pulsed streamer corona discharge in water. *J. Electrostat.* **1997**, *39*, 189–202.
96. Lukes, P.; Dolezalova, E.; Sisrova, I.; Clupek, M. Aqueous-phase chemistry and bactericidal effects from an air discharge plasma in contact with water: Evidence for the formation of peroxynitrite through a pseudo-second-order post-discharge reaction of H_2O_2 and HNO_2. *Plasma Sources Sci. Technol.* **2014**, *23*, 15019. [CrossRef]
97. Li, H.; Li, T.; He, S.; Zhou, J.; Wang, T.; Zhu, L. Efficient degradation of antibiotics by non-thermal discharge plasma: Highlight the impacts of molecular structures and degradation pathways. *Chem. Eng. J.* **2020**, *395*, 125091. [CrossRef]
98. Singh, R.K.; Philip, L.; Ramanujam, S. Rapid degradation, mineralization and detoxification of pharmaceutically active compounds in aqueous solution during pulsed corona discharge treatment. *Water Res.* **2017**, *121*, 20–36. [CrossRef]
99. Wang, T.; Lu, N.; Li, J.; Wu, Y.; Su, Y. Enhanced degradation of p-nitrophenol in soil in a pulsed discharge plasma-catalytic system. *J. Hazard. Mater.* **2011**, *195*, 276–280. [CrossRef]

100. Wang, T.; Ren, J.; Qu, G.; Liang, D.; Hu, S. Glyphosate contaminated soil remediation by atmospheric pressure dielectric barrier discharge plasma and its residual toxicity evaluation. *J. Hazard. Mater.* **2016**, *320*, 539–546. [CrossRef]
101. Aggelopoulos, C.A.; Tataraki, D.; Rassias, G. Degradation of atrazine in soil by dielectric barrier discharge plasma—Potential singlet oxygen mediation. *Chem. Eng. J.* **2018**, *347*, 682–694. [CrossRef]
102. Wang, T.C.; Lu, N.; Li, J.; Wu, Y. Degradation of pentachlorophenol in soil by pulsed corona discharge plasma. *J. Hazard. Mater.* **2010**, *180*, 436–441. [CrossRef] [PubMed]
103. Wang, T.C.; Qu, G.; Li, J.; Liang, D. Remediation of p-nitrophenol and pentachlorophenol mixtures contaminated soil using pulsed corona discharge plasma. *Sep. Purif. Technol.* **2014**, *122*, 17–23. [CrossRef]
104. Hatzisymeon, M.; Tataraki, D.; Rassias, G.; Aggelopoulos, C.A. Novel combination of high voltage nanopulses and in-soil generated plasma micro-discharges applied for the highly efficient degradation of trifluralin. *J. Hazard. Mater.* **2021**, *415*, 125646. [CrossRef] [PubMed]
105. Lou, J.; Lu, N.; Li, J.; Wang, T.; Wu, Y. Remediation of chloramphenicol-contaminated soil by atmospheric pressure dielectric barrier discharge. *Chem. Eng. J.* **2012**, *180*, 99–105. [CrossRef]
106. Aggelopoulos, C.A.; Hatzisymeon, M.; Tataraki, D.; Rassias, G. Remediation of ciprofloxacin-contaminated soil by nanosecond pulsed dielectric barrier discharge plasma: Influencing factors and degradation mechanisms. *Chem. Eng. J.* **2020**, *393*, 124768. [CrossRef]
107. Hatzisymeon, M.; Tataraki, D.; Tsakiroglou, C.; Rassias, G.; Aggelopoulos, C.A. Highly energy-efficient degradation of antibiotics in soil: Extensive cold plasma discharges generation in soil pores driven by high voltage nanopulses. *Sci. Total Environ.* **2021**, *786*, 147420. [CrossRef]
108. Mu, R.; Liu, Y.; Li, R.; Xue, G.; Ognier, S. Remediation of pyrene-contaminated soil by active species generated from flat-plate dielectric barrier discharge. *Chem. Eng. J.* **2016**, *296*, 356–365. [CrossRef]
109. Zhan, J.; Liu, Y.; Cheng, W.; Zhang, A.; Li, R.; Li, X.; Ognier, S.; Cai, S.; Yang, C.; Liu, J. Remediation of soil contaminated by fluorene using needle-plate pulsed corona discharge plasma. *Chem. Eng. J.* **2018**, *334*, 2124–2133. [CrossRef]
110. Abbas, Y.; Lu, W.; Dai, H.; Fu, X.; Ye, R.; Wang, H. Remediation of polycyclic aromatic hydrocarbons (PAHs) contaminated soil with double dielectric barrier discharge plasma technology: Influencing parameters. *Chem. Eng. J.* **2020**, *394*, 124858. [CrossRef]
111. Abbas, Y.; Lu, W.; Wang, Q.; Dai, H.; Liu, Y.; Fu, X.; Pan, C.; Ghaedi, H.; Cheng, F.; Wang, H. Remediation of pyrene contaminated soil by double dielectric barrier discharge plasma technology: Performance optimization and evaluation. *Environ. Pollut.* **2020**, *260*, 113944. [CrossRef] [PubMed]
112. Li, R.; Liu, Y.; Cheng, W.; Zhang, W.; Xue, G.; Ognier, S. Study on remediation of phenanthrene contaminated soil by pulsed dielectric barrier discharge plasma: The role of active species. *Chem. Eng. J.* **2016**, *296*, 132–140. [CrossRef]
113. Li, X.; Zhang, H.; Luo, Y.; Teng, Y. Remediation of soil heavily polluted with polychlorinated biphenyls using a low-temperature plasma technique. *Front. Environ. Sci. Eng.* **2014**, *8*, 277–283. [CrossRef]
114. Shu, Z.; Lin, Z.; Yue, L.; Wang, W.; Wang, X. Modeling of the production of OH and O radicals in a positive pulsed corona discharge plasma. *Vacuum* **2008**, *83*, 238–243.
115. Hsu, I.Y.; Masten, S.J. The Kinetics of the Reaction of Ozone with Phenanthrene in Unsaturated Soils. *Environ. Eng. Sci.* **1997**, *14*, 207–218. [CrossRef]
116. Laroussi, M.; Lu, X.; Kolobov, V.; Arslanbekov, R. Power consideration in the pulsed dielectric barrier discharge at atmospheric pressure. *J. Appl. Phys.* **2004**, *96*, 3028–3030. [CrossRef]
117. Mildren, R.P.; Carman, R.J. Enhanced performance of a dielectric barrier discharge lamp using short-pulsed excitation. *J. Phys. D Appl. Phys.* **2001**, *34*, L1–L6. [CrossRef]
118. Walsh, J.L.; Kong, M.G. 10 ns pulsed atmospheric air plasma for uniform treatment of polymeric surfaces. *Appl. Phys. Lett.* **2007**, *91*, 1539. [CrossRef]
119. Jiang, H.; Shao, T.; Zhang, C.; Niu, Z.; Yu, Y.; Yan, P.; Zhou, Y. Comparison of AC and Nanosecond-Pulsed DBDs in Atmospheric Air. *IEEE Trans. Plasma Sci.* **2011**, *39*, 2076–2077. [CrossRef]
120. Williamson, J.M.; Trump, D.D.; Bletzinger, P.; Ganguly, B.N. Comparison of high-voltage ac and pulsed operation of a surface dielectric barrier discharge. *J. Phys. D Appl. Phys.* **2006**, *39*, 4400–4406. [CrossRef]
121. Hama Aziz, K.H.; Miessner, H.; Mueller, S.; Mahyar, A.; Kalass, D.; Moeller, D.; Khorshid, I.; Rashid, M.A.M. Comparative study on 2,4-dichlorophenoxyacetic acid and 2,4-dichlorophenol removal from aqueous solutions via ozonation, photocatalysis and non-thermal plasma using a planar falling film reactor. *J. Hazard. Mater.* **2018**, *343*, 107–115. [CrossRef] [PubMed]
122. Li, Z.; Wang, Y.; Guo, H.; Pan, S.; Puyang, C.; Su, Y.; Qiao, W.; Han, J. Insights into water film DBD plasma driven by pulse power for ibuprofen elimination in water: Performance, mechanism and degradation route. *Sep. Purif. Technol.* **2021**, *277*, 119415. [CrossRef]
123. Hoigné, J.; Bader, H. Rate constants of reactions of ozone with organic and inorganic compounds in water—II: Dissociating organic compounds. *Water Res.* **1983**, *17*, 185–194. [CrossRef]
124. Zhang, Y.; Han, C.; Zhang, G.; Dionysiou, D.D.; Nadagouda, M.N. PEG-assisted synthesis of crystal TiO_2 nanowires with high specific surface area for enhanced photocatalytic degradation of atrazine. *Chem. Eng. J.* **2015**, *268*, 170–179. [CrossRef]
125. Bruggeman, P.; Schram, D.C. On OH production in water containing atmospheric pressure plasmas. *Plasma Sources Sci. Technol.* **2010**, *19*, 45025–45033. [CrossRef]

126. Kogelschatz, U. Dielectric-Barrier Discharges: Their History, Discharge Physics, and Industrial Applications. *Plasma Chem. Plasma Process.* **2003**, *23*, 1–46. [CrossRef]
127. Locke, B.R.; Shih, K. Review of the methods to form hydrogen peroxide in electrical discharge plasma with liquid water. *Plasma Sources Sci. Technol.* **2011**, *20*, 34006. [CrossRef]
128. Zahran, E.M.; Bhattacharyya, D.; Bachas, L.G. Reactivity of Pd/Fe bimetallic nanotubes in dechlorination of coplanar polychlorinated biphenyls. *Chemosphere* **2013**, *91*, 165–171. [CrossRef]
129. Zhang, H.; Zhang, Q.; Miao, C.; Huang, Q. Degradation of 2, 4-dichlorophenol in aqueous solution by dielectric barrier discharge: Effects of plasma-working gases, degradation pathways and toxicity assessment. *Chemosphere* **2018**, *204*, 351–358. [CrossRef]
130. Zhang, Q.; Zhang, H.; Zhang, Q.; Huang, Q. Degradation of norfloxacin in aqueous solution by atmospheric-pressure non-thermal plasma: Mechanism and degradation pathways. *Chemosphere* **2018**, *210*, 433–439. [CrossRef]
131. Dudita, M.; Bogatu, C.; Enesca, A.; Duta, A. The influence of the additives composition and concentration on the properties of SnOx thin films used in photocatalysis. *Mater. Lett.* **2011**, *65*, 2185–2189. [CrossRef]
132. Rafieenia, R.; Sulonen, M.; Mahmoud, M.; El-Gohary, F.; Rossa, C.A. Integration of microbial electrochemical systems and photocatalysis for sustainable treatment of organic recalcitrant wastewaters: Main mechanisms, recent advances, and present prospects. *Sci. Total Environ.* **2022**, *824*, 153923. [CrossRef]
133. Wu, J.; Xiong, Q.; Liang, J.; He, Q.; Yang, D.; Deng, R.; Chen, Y. Degradation of benzotriazole by DBD plasma and peroxymonosulfate: Mechanism, degradation pathway and potential toxicity. *Chem. Eng. J.* **2020**, *384*, 123300. [CrossRef]
134. Shang, K.; Wang, X.; Li, J.; Wang, H.; Lu, N.; Jiang, N.; Wu, Y. Synergetic degradation of Acid Orange 7 (AO7) dye by DBD plasma and persulfate. *Chem. Eng. J.* **2017**, *311*, 378–384. [CrossRef]
135. Wang, X.; Wang, P.; Liu, X.; Hu, L.; Wang, Q.; Xu, P.; Zhang, G. Enhanced degradation of PFOA in water by dielectric barrier discharge plasma in a coaxial cylindrical structure with the assistance of peroxymonosulfate. *Chem. Eng. J.* **2020**, *389*, 124381. [CrossRef]
136. Liang, J.; Zhou, X.; Zhao, Z.; Yang, D.; Wang, W. Degradation of trimethoprim in aqueous by persulfate activated with nanosecond pulsed gas-liquid discharge plasma. *J. Environ. Manag.* **2021**, *278*, 111539. [CrossRef]
137. Tang, S.; Yuan, D.; Rao, Y.; Li, M.; Shi, G.; Gu, J.; Zhang, T. Percarbonate promoted antibiotic decomposition in dielectric barrier discharge plasma. *J. Hazard. Mater.* **2019**, *366*, 669–676. [CrossRef]
138. Xu, Z.; Xue, X.; Hu, S.; Li, Y.; Shen, J.; Lan, Y.; Zhou, R.; Yang, F.; Cheng, C. Degradation effect and mechanism of gas-liquid phase dielectric barrier discharge on norfloxacin combined with H_2O_2 or Fe^{2+}. *Sep. Purif. Technol.* **2020**, *230*, 115862. [CrossRef]
139. Sang, W.; Lu, W.; Mei, L.; Jia, D.; Cao, C.; Li, Q.; Wang, C.; Zhan, C.; Li, M. Research on different oxidants synergy with dielectric barrier discharge plasma in degradation of Orange G: Efficiency and mechanism. *Sep. Purif. Technol.* **2021**, *277*, 119473. [CrossRef]
140. Chen, J.; Feng, J.; Lu, S.; Shen, Z.; Du, Y.; Peng, L.; Nian, P.; Yuan, S.; Zhang, A. Non-thermal plasma and Fe^{2+} activated persulfate ignited degradation of aqueous crystal violet: Degradation mechanism and artificial neural network modeling. *Sep. Purif. Technol.* **2018**, *191*, 75–85. [CrossRef]
141. Koprivanac, N.; Kusi, H.; Vujevi, D.; Peternel, I.; Locke, B.R. Influence of iron on degradation of organic dyes in corona. *J. Hazard. Mater.* **2005**, *117*, 113–119. [CrossRef] [PubMed]
142. Hao, X.; Zhou, M.; Xin, Q.; Lei, L. Pulsed discharge plasma induced Fenton-like reactions for the enhancement of the degradation of 4-chlorophenol in water. *Chemosphere* **2007**, *66*, 2185–2192. [CrossRef] [PubMed]
143. Zhou, R.; Zhou, R.; Zhang, X.; Tu, S.; Yin, Y.; Yang, S.; Ye, L. An efficient bio-adsorbent for the removal of dye: Adsorption studies and cold atmospheric plasma regeneration. *J. Taiwan Inst. Chem. Eng.* **2016**, *68*, 372–378. [CrossRef]
144. Cheng, J.; Wang, D.; Wang, B.; Ning, H.; Zhang, Y.; Li, Y.; An, J.; Gao, P. Plasma-catalytic degradation of ciprofloxacin in aqueous solution over different MnO2 nanocrystals in a dielectric barrier discharge system. *Chemosphere* **2020**, *253*, 126595. [CrossRef] [PubMed]
145. Ghezzar, M.R.; Abdelmalek, F.; Belhadj, M.; Benderdouche, N.; Addou, A. Enhancement of the bleaching and degradation of textile wastewaters by Gliding arc discharge plasma in the presence of TiO_2 catalyst. *J. Hazard. Mater.* **2009**, *164*, 1266–1274. [CrossRef] [PubMed]
146. Jõgi, I.; Erme, K.; Raud, J.; Raud, S. The Effect of TiO_2 Catalyst on Ozone and Nitrous Oxide Production by Dielectric Barrier Discharge. *Catal. Lett.* **2020**, *150*, 992–997. [CrossRef]
147. Guo, H.; Jiang, N.; Wang, H.; Shang, K.; Lu, N.; Li, J.; Wu, Y. Enhanced catalytic performance of graphene-TiO_2 nanocomposites for synergetic degradation of fluoroquinolone antibiotic in pulsed discharge plasma system. *Appl. Catal. B Environ.* **2019**, *248*, 552–566. [CrossRef]
148. Wang, B.; Wang, C.; Yao, S.; Peng, Y.; Xu, Y. Plasma-catalytic degradation of tetracycline hydrochloride over $Mn/\gamma\text{-}Al_2O_3$ catalysts in a dielectric barrier discharge reactor. *Plasma Sci. Technol.* **2019**, *21*, 65503. [CrossRef]
149. Wang, T.C.; Lu, N.; Li, J.; Wu, Y. Plasma-TiO_2 catalytic method for high-efficiency remediation of p-nitrophenol contaminated soil in pulsed discharge. *Environ. Sci. Technol.* **2011**, *45*, 9301. [CrossRef]
150. Sohrabi, M.R.; Ghavami, M. Photocatalytic degradation of Direct Red 23 dye using UV/TiO2: Effect of operational parameters. *J. Hazard. Mater.* **2008**, *153*, 1235–1239. [CrossRef]
151. Takahashi, M. ζ Potential of Microbubbles in Aqueous Solutions: Electrical Properties of the Gas—Water Interface. *J. Phys. Chem. B* **2005**, *109*, 21858–21864. [CrossRef] [PubMed]

152. Takahashi, M.; Chiba, K.; Li, P. Free-radical generation from collapsing microbubbles in the absence of a dynamic stimulus. *J. Phys. Chem. B* **2007**, *111*, 1343–1347. [CrossRef] [PubMed]
153. Li, P.; Takahashi, M.; Chiba, K. Enhanced free-radical generation by shrinking microbubbles using a copper catalyst. *Chemosphere* **2009**, *77*, 1157–1160. [CrossRef] [PubMed]
154. Liu, Y.; Shen, X.; Sun, J.; Li, P.; Zhang, A. Treatment of aniline contaminated water by a self-designed dielectric barrier discharge reactor coupling with micro-bubbles: Optimization of the system and effects of water matrix. *J. Chem. Technol. Biotechnol.* **2019**, *94*, 494–504. [CrossRef]
155. Liu, Y.; Zhang, H.; Sun, J.; Liu, J.; Shen, X.; Zhan, J.; Zhang, A.; Ognier, S.; Cavadias, S.; Li, P. Degradation of aniline in aqueous solution using non-thermal plasma generated in microbubbles. *Chem. Eng. J.* **2018**, *345*, 679–687. [CrossRef]
156. Zhang, H.; Li, P.; Zhang, A.; Sun, Z.; Liu, J.; Héroux, P.; Liu, Y. Enhancing Interface Reactions by Introducing Microbubbles into a Plasma Treatment Process for Efficient Decomposition of PFOA. *Environ. Sci. Technol.* **2021**, *55*, 16067–16077. [CrossRef]
157. Wang, Q.; Zhang, A.; Li, P.; Héroux, P.; Zhang, H.; Yu, X.; Liu, Y. Degradation of aqueous atrazine using persulfate activated by electrochemical plasma coupling with microbubbles: Removal mechanisms and potential applications. *J. Hazard. Mater.* **2021**, *403*, 124087. [CrossRef]

MDPI
St. Alban-Anlage 66
4052 Basel
Switzerland
Tel. +41 61 683 77 34
Fax +41 61 302 89 18
www.mdpi.com

Water Editorial Office
E-mail: water@mdpi.com
www.mdpi.com/journal/water

www.ingramcontent.com/pod-product-compliance
Lightning Source LLC
Chambersburg PA
CBHW061813210926
51850CB00006B/692
* 9 7 8 3 0 3 6 5 6 6 1 1 5 *